Resonant Network Antennas for Radio-Frequency Plasma Sources

Theory, technology and applications

Online at: https://doi.org/10.1088/978-0-7503-5296-3

IOP Series in Plasma Physics

Series Editor

Richard Dendy

Culham Centre for Fusion Energy and the University of Warwick, UK

About the series

The IOP Plasma Physics ebook series aims at comprehensive coverage of the physics and applications of natural and laboratory plasmas, across all temperature regimes. Books in the series range from graduate and upper-level undergraduate textbooks, research monographs and reviews.

The conceptual areas of plasma physics addressed in the series include:

- Equilibrium, stability, and control.
- Waves: fundamental properties, emission, and absorption.
- Nonlinear phenomena and turbulence.
- Transport theory and phenomenology.
- Laser–plasma interactions.
- Non-thermal and suprathermal particle populations.
- Beams and non-neutral plasmas.
- High energy density physics.
- Plasma–solid interactions, dusty, complex, and non-ideal plasmas.
- Diagnostic measurements and techniques for data analysis.

The fields of application include:

- Nuclear fusion through magnetic and inertial confinement.
- Solar–terrestrial and astrophysical plasma environments and phenomena.
- Advanced radiation sources.
- Materials processing and functionalisation.
- Propulsion, combustion, and bulk materials management.
- Interaction of plasma with living matter and liquids.
- Biological, medical, and environmental systems.
- Low-temperature plasmas, glow discharges, and vacuum arcs.
- Plasma chemistry and reaction mechanisms.
- Plasma production by novel means.

A full list of titles published in this series can be found here: https://iopscience.iop.org/bookListInfo/iop-plasma-physics-series.

Resonant Network Antennas for Radio-Frequency Plasma Sources

Theory, technology and applications

Philippe Guittienne
Alan Howling
Ivo Furno
Ecole Polytechnique Fédérale de Lausanne (EPFL), Swiss Plasma Center (SPC), CH-1015 Lausanne, Switzerland

IOP Publishing, Bristol, UK

ISBN 978-0-7503-5296-3 (ebook)
ISBN 978-0-7503-5294-9 (print)
ISBN 978-0-7503-5297-0 (myPrint)
ISBN 978-0-7503-5295-6 (mobi)

DOI 10.1088/978-0-7503-5296-3

Version: 20240201

IOP ebooks

British Library Cataloguing-in-Publication Data: A catalogue record for this book is available from the British Library.

Published by IOP Publishing, wholly owned by The Institute of Physics, London

IOP Publishing, No.2 The Distillery, Glassfields, Avon Street, Bristol, BS2 0GR, UK

US Office: IOP Publishing, Inc., 190 North Independence Mall West, Suite 601, Philadelphia, PA 19106, USA

Contents

Part III Resonant network antennas in magnetized plasma

Part IV Technology, future developments, and appendices

Preface

Radio frequency plasmas provide the underlying technology for many of today's critical semiconductor industries. The demand for larger and more uniform plasma sources is reaching the limits of conventional capacitive and inductive RF plasma reactors, due to standing wave effects and the asymptotic impedance of large-area reactors. The inherent properties of resonant network antennas can overcome these limitations because of their spatially distributed internal resonances and real input impedance.

This book aims to show that resonant network antennas are versatile, alternative sources for inductively coupled and wave-driven plasma. The theory has developed alongside the technology (see also https://www.helyssen.com) to the extent that it is timely to document the progress in an accessible way, to aid antenna design for future RF plasma applications.

To maximize the usefulness of this book for the physicist, engineer, and student, we have taken care to provide all the necessary details for the reader. In particular:

- Equations are derived in full with all intermediate steps.
- Unfamiliar techniques, such as partial inductance calculations, the complex image method, and partial image theory are developed step-by-step from elementary principles by means of explanatory figures.
- The most useful and recurring antenna calculations are provided using a link to programs which reproduce the tensor solutions and many of the book's figures in appendix K.
- Basic concepts in plasma physics are explained, occasionally using a novel approach.

We assume an undergraduate science level familiar with complex numbers, complex impedance, and Maxwell's equations. The chapters and appendixes are cross-referenced throughout the book, but for the most part, the chapters can be read independently of each other. MKS SI units are used throughout. We use 'antennas' for the plural of 'antenna', with apologies to Latin scholars.

The readers will be grateful to Alex Howling, our illustrator, for bringing a touch of comic relief to each chapter.

Dr Philippe Guittienne
Dr Alan Howling
Professor Ivo Furno
Lausanne, Switzerland,
26 October 2023

Acknowledgements

We are indebted to Rémy Jacquier, our master engineer, for his contribution to the construction of, and experiments on, the various plasma reactors. We are grateful to the former group leader and co-founder (with Alan Howling and Pierre-Jean Paris in 1989) of the Industrial Plasma Applications Group, Dr Christoph Hollenstein (now retired), for his foresight and generosity in supporting the fledgling Helyssen plasma sources at the CRPP. Our thanks also go to: Helyssen Sàrl director Alain Dannaoui, Gennady Plyushchev, Riccardo Agnello, Fabio Avino, Basile Pouradier Duteil, Cédric Vivien, Claudio Marini, Pierre Demolon, Renat Karimov, Simon Vincent, Marcelo Baquero, Christine Stollberg, Lyes Kadi, Christian Schlatter, Edith Grueter, Yves Martin, Stefano Alberti, Duccio Testa, Pierre Fayet, Sylvain Lecoultre, Antonio d'Angola, Alain Simonin, Stephane Béchu, Hamza Drissikamili, Zacharie Jehan, Arthur Wuhrmann, Benjamin Forster, Andrii Shytikov, the technical services and workshops, and the Swiss Plasma Center directorate, especially Professor Ambrogio Fasoli, for their continual support.

The authors and the Basic Plasma Physics and Applications Group of the Swiss Plasma Center, Ecole Polytechnique Fédérale de Lausanne, wish to thank the Innosuisse Swiss Innovation Agency (formerly the CTI Commission for Technology and Innovation) for their support. Part of this work has been carried out within the framework of the EUROfusion Consortium, via the Euratom Research and Training Programme (Grant Agreement No 101 052 200—EUROfusion) and funded by the Swiss State Secretariat for Education, Research and Innovation (SERI). Views and opinions expressed are however those of the author(s) only and do not necessarily reflect those of the European Union, the European Commission, or SERI. Neither the European Union nor the European Commission nor SERI can be held responsible for them.

Individual projects are acknowledged in full in the publications referenced throughout the book.

Author biographies

Philippe Guittienne

Dr Philippe Guittienne is currently a physicist at the Swiss Plasma Center (SPC) in the Basic Plasma Physics and Applications group under Professor Ivo Furno, and founder of the Helyssen Sàrl company in 2003. Following an engineering degree in physics (1997) and a doctorate (2002) in condensed matter physics at the EPFL on magnetization reversal in ferromagnetic nanostructures, he completely changed his field of interest to birdcages for helicon sources, inspired by the PhD on MRI of his future wife, Jacqueline. On the basis of this intuition, he founded the Helyssen Sàrl start-up in 2003, and started a collaboration with Dr Christoph Hollenstein's group at SPC for the development of resonant antennas as plasma sources. This topic turned out to be a fruitful field of research, and during the years it became a central part of the Industrial Plasma Group (now included in Basic Plasma Physics and Applications), and is, indeed, the subject of this book.

Alan Arthur Howling

Dr Alan Arthur Howling is an Adjoint Scientifique/Senior Scientific Collaborator, co-founder of the group for industrial plasmas in 1989 with Dr Christoph Hollenstein. He is currently a researcher and lecturer in the Basic Plasma Physics and Applications group under Professor Ivo Furno at the Swiss Plasma Center, EPFL, Lausanne, Switzerland. A Gordon Warter Open Scholarship in 1978 to Pembroke College, Oxford University, led to a physics (Natural Sciences) degree in 1981, followed by an MSc and Gordon Francis prize (1982) in the Science and Applications of Electric Plasmas at Wolfson College, Oxford University, and then a doctorate titled 'Fluctuations in the edge plasma of the TOSCA tokamak' (1985) at both Oxford and UKAEA Culham Laboratory, as it was then called. A postdoc on the TCA tokamak in the Centre de Recherches en Physique des Plasmas (CRPP), EPFL Lausanne, Switzerland from 1986 to 1989 was the springboard to RF industrial plasmas in 1989. Applied research topics included negative ion polymerization (silanions) and particle formation in silane RF plasmas; design of large-area RF capacitive plasma reactors for solar cells and flat panel displays, including showerhead uniformity, discharge equilibration, plasmoids, and electromagnetic analysis of symmetric and anti-symmetric modes; RF plasma deposition of amorphous and micro-crystalline silicon; RF plasma diagnostics and glass substrate charging; design of resonant ladder networks for RF plasmas, introducing partial inductance and complex image methods; high voltage design of satellite slip-rings; bio-plasma with Alexandra Waskow and the effect of humidity on dielectric barrier discharges; finally concentrating on writing this swan-song book before retirement on 26 October 2023.

Ivo Furno

Professor Ivo Furno is currently Adjunct Professor at the EPFL and leader of the Basic Plasma Physics and Applications (BPPA) group of the Swiss Plasma Center. He graduated in Nuclear Engineering from the Politecnico di Torino, Italy, in 1995 and then received his PhD from the EPFL with a thesis on 'Fast transient transport phenomena measured by soft x-ray emission in TCV tokamak plasmas'. He continued with a postdoc at the Los Alamos National Laboratory, where he studied magnetic reconnection on the Reconnection Scaling Experiment (RSX), before re-joining the EPFL in 2006. His research is marked by the use of human-scale, dedicated plasma devices to investigate the fundamental physics of plasmas under conditions ranging from fusion plasmas to plasmas of relevance for solar physics and to non-equilibrium cold plasmas for industrial and biological applications. For his work on turbulence in magnetized plasmas, he was awarded the Fellowship of the American Physics Society. On the TCV tokamak, he contributed to the development of the first experiments on the so-called snowflake divertor, and, recently, he led the TCV team that obtained the first experimental demonstration of electron cyclotron microwave beam broadening by plasma turbulence. Since he took over the responsibility of the BPPA group, he has obtained numerous grants to develop industrial applications in collaboration with national academic institutions as well as with industrial partners. He developed the Resonant Antenna Ion Device (RAID) and launched its scientific program to study the physics of helicon waves and negative ion volume generation in helicon plasmas. As part of the new SPC activities beyond fusion, he started collaborating with CERN in the field of wakefield acceleration for the next-generation particle accelerator. The SPC is today an active member of the AWAKE Consortium. A new laboratory for societal, e.g. biological, applications of plasmas, such as plasma agriculture and plasma sterilization, was launched by Furno to expand the SPC infrastructure into atmospheric pressure plasmas for fundamental life science projects. It was Furno who originally proposed writing this book, titled *Resonant Network Antennas for Radio-Frequency Plasma Sources*.

The authors in front of the RAID device: Philippe Guittienne, Alan Howling, and Ivo Furno.

Glossary

Greek terms

α	attenuation constant per section
α_m	a Fourier coefficient
β	phase change per section (rad)
β_m	a Fourier coefficient
$\beta_{1,2}$	wavenumber for H, TG modes (m^{-1})
γ	propagation constant
Δf	FWHM bandwidth (Hz)
δ	Dirac delta function
$\delta\omega$	a small difference in angular frequency from the resonance frequency (rad s^{-1})
$\Delta\omega$	the FWHM half-power bandwidth in angular frequency (rad s^{-1})
ε_0	permittivity of free space (F m^{-1})
ε_p	relative permittivity of unmagnetized plasma (–)
$\bar{\varepsilon}_p$	tensor relative permittivity of magnetized plasma (–)
ζ_m	a Fourier coefficient
η	power transfer efficiency
θ	angle of propagation with respect to the magnetic field
λ	wavelength (m)
μ_0	permeability of free space (H m^{-1})
ν	effective collision frequency (s^{-1})
ν_m	electron–neutral collision frequency (s^{-1})
ξ_m	a Fourier coefficient
ρ	free charge density (C m^{-3})
ρ_{dc}	DC electrical resistivity (Ω m)
ρ_p	plasma complex electrical resistivity (Ω m)
σ_{dc}	DC electrical conductivity (S m^{-1})
σ_{en}	electron–neutral collision cross-section (m^2)
σ_p	complex electrical conductivity of unmagnetized plasma (S m^{-1})
$\bar{\sigma}_p$	tensor electrical conductivity of magnetized plasma (S m^{-1})
τ	RF period $=2\pi/\omega$ (s)
Φ	total magnetic flux linkage (Wb)
ϕ	azimuthal angle in cylindrical coordinates (r, ϕ, z)
ω	angular (RF) frequency (rad s^{-1})
$\hat{\omega}$	$(\omega + j\nu)$
ω_0	resonance angular frequency for an ideal (lossless) circuit (rad s^{-1})
ω_0'	resonance angular frequency for a real (lossy) circuit (rad s^{-1})
ω_m	angular frequency of the mth normal mode (rad s^{-1})
ω_{ci}	ion cyclotron angular frequency (rad s^{-1})
ω_{ce}	electron cyclotron angular frequency (rad s^{-1})

Symbols, abbreviations, and subscripts

$\bar{1}$	$N \times N$ identity matrix
1D, 2D	one-dimensional, two-dimensional
A	vector magnetic potential (Wb m^{-1})
A	vector column of upper-node voltages (V)
acw	anti-clockwise

ALD	atomic layer deposition
A_n	voltage phasor at the *n*th node of the upper stringer (V)
B	(wavefield) magnetic flux density vector (T)
B	vector column of lower-node voltages (V)
B	magnetic flux density magnitude (T)
$\mathbf{B}_0$, B_0	externally imposed constant, uniform magnetic flux density (T)
B_n	voltage phasor at the *n*th node of the lower stringer (V)
C	capacitance (F)
$\hat{C}$	capacitance matrix per unit length (F m^{-1})
CCD	charge-coupled device (CCD camera)
CCP	capacitively coupled plasma
c	contour
c	speed of light in vacuum
c	(subscript) collisional
cw	clockwise
CW	continuous wave; steady state
d	reactor height between a top-plate and a baseplate
d_s	source height above an interface
d**s**	elemental length vector along a contour
D	connection configuration term
DC	direct current, time-constant value
DLC	diamond-like carbon
e	the base of natural logarithms, Euler's number (2.718…)
e	electron (subscript)
E	(wavefield) electric field intensity vector (V m^{-1})
E-mode	coupling via the *E* field; see CCP
EM	electromagnetic
EM-mode	coupling via *EM* fields; see EMCP
EMC	electromagnetic compatibility
EMCP	electromagnetically coupled plasma
eq	equivalent circuit value
FTIR	Fourier transform infrared
FWHM	full width at half maximum
f	current feed point (subscript)
f	frequency (Hz)
f_{RF}	RF frequency (Hz)
f_{pe}	electron plasma frequency (Hz)
f_{pi}	ion plasma frequency (Hz)
f_p	plasma frequency (Hz)
$\hat{G}$	conductance per unit length (S m^{-1})
g	ground point (subscript)
h	height of an antenna above a baseplate
H	magnetic field strength vector (A m^{-1})
H	helicon
H-mode	coupling via the *H* field; see ICP
I	leg current column vector (A)
I	current (A)
ICP	inductively coupled plasma
Im	imaginary part
ISM	International Scientific and Medical standard

I_n	current phasor in the nth leg (A)
i_{line}	current phasor of current along the line (A)
i_{rf}	current phasor of injected RF current (A)
i_0	amplitude of the first harmonic of shell current density (A m^{-1})
i_s	shell current density (A m^{-1})
J_m	first kind of Bessel function, of order m (see Y_m)
J_n	current in the nth upper stringer section (A)
j	$\sqrt{-1}$
K_n	current in the nth lower stringer section (A)
$\mathbf{k}$	wavenumber vector (m^{-1})
k	magnitude of the wavenumber vector $\mathbf{k}$ (m^{-1})
k_0	wavenumber in vacuum; $k_0 = \omega/c$ (m^{-1})
k_d	wavenumber in a dielectric (m^{-1})
k_z	axial wavenumber, along z (m^{-1})
$k_\perp$	perpendicular wavenumber, perpendicular to z (m^{-1})
k_{B}	Boltzmann's constant ($1.38 \cdot 10^{-23}$ J K^{-1})
LHS	left-hand side
L	self partial inductance (H)
$\hat{L}$	self partial inductance per unit length (H m^{-1})
L_{leg}	self partial inductance of a leg (H)
L_{str}	self partial inductance of a stringer segment (H)
L^{loop}	loop self inductance of a wire loop, coil, or solenoid (H)
$\hat{L}^{\text{loop}}$	loop self inductance per unit length of a transmission line (H)
L_i^{loop}	contribution to loop self inductance of the ith wire segment (H)
l	length of a wire, or an antenna leg
leg	leg (or rung) across the ladder width; or a birdcage leg
MRI	magnetic resonance imaging
$\hat{M}$	mutual partial inductance matrix per unit length (H m^{-1})
M_{ij}	mutual partial inductance between wire filaments i and j (H)
m	mode number
m	azimuthal periodicity
m_e	electron mass (kg)
mes	measured, or measurement
NMR	nuclear magnetic resonance
N	total number of legs
N	refractive index
nc	(subscript) non-collisional, or collisionless
n_{e0}	time-constant electron number density (m^{-3})
OTR	oxygen transmission rate (c.c. m^{-2} atm $^{-1}$ day^{-1})
PEC	perfect electric conductor
PECVD	plasma-enhanced chemical vapour deposition
P_{rf}	RF input power (W)
$\bar{P}$	Maxwell's potential coefficient matrix (V)
p.u.l.	per unit length (signified by a hat above the symbol)
q	magnitude of the electron charge; $q = +1.602 \cdot 10^{-19}$ C
Q	quality factor
RAID	Resonant Antenna Ion Device
Re	real part
res	resonance (subscript or superscript)
RF, rf	radio frequency

RHS	right-hand side
R	resistance, real impedance (Ω)
$\hat{R}$	resistance per unit length (Ω/m)
R, r	radius (m)
r	resistance (Ω)
$\mathbf{S}$	Poynting's vector $\mathbf{E} \times \mathbf{H}$ (W m^{-2})
S	area (m^2)
S_n^J	source current at the nth node of the upper stringer (A)
S_n^K	source current at the nth node of the lower stringer (A)
sccm	a gas mass flow rate in standard cubic centimetres per second
Scrn	algebraic terms depending on currents induced in a PEC screen
SiO_x	silicon oxide with intermediate stoichiometry; x = 1–2
SLM	a gas mass flow rate in standard litres per second
str	stringer, along the length of a ladder
T	absolute temperature (K)
T_e	electron temperature
T	Trivelpiece–Gould mode (subscript)
T–G	Trivelpiece–Gould mode
TM	transverse magnetic
$\bar{U}_L$	lower shift matrix
V_{pp}	peak-to-peak voltage (V)
V_{rf}	phasor of the applied RF voltage (V)
v	velocity (m s^{-1})
υ	phase velocity (m s^{-1})
W-mode	coupling via helicon wave fields
$\bar{X}$	$N \times N$ matrix for parameters X_{nm}
$\mathbf{X}$	column vector array for parameters X_n
X	reactive (imaginary) impedance (Ω)
$\hat{\mathbf{x}}$	unit vector along the x-axis
Y	complex admittance (S)
Y_{eq}	complex equivalent circuit admittance at resonance (S)
$\hat{\mathbf{y}}$	unit vector along the y-axis
Y_m	second kind of Bessel function, of order m (see J_m)
Z	complex impedance (Ω)
Z_{str}	stringer complex impedance (Ω)
Z_{leg}	leg complex impedance (Ω)
Z_c	characteristic impedance of a transmission line (Ω)
Z_{in}^{eq}	equivalent circuit complex input impedance (Ω)
Z_{in}^{res}	complex input impedance at resonance (Ω)
$\hat{\mathbf{z}}$	unit vector along the z-axis

IOP Publishing

Resonant Network Antennas for Radio-Frequency Plasma Sources
Theory, technology and applications
Philippe Guittienne, Alan Howling and Ivo Furno

Chapter 1

Introduction

By way of a concise introduction, the title and contents of this book can be summarized in two sentences. First, a resonant network antenna is a mesh formed by the repetition of parallel assemblies of inductive and capacitive elements, whose overall circuit exhibits a set of resonant modes. Second, the strong oscillating currents at each resonance frequency are well suited for inductive coupling to generate radio-frequency (RF) plasma.

One particular concept of an inductive–capacitive regular mesh is shown in figure 1.1. Infinite combinations of antenna networks can be imagined, but in the main parts of this book we will concentrate on just two principal types, the birdcage antenna and the ladder antenna. Even with this restriction, there are many design options suitable for volumetric plasma sources (inductive and helicon) and large area plasmas, with linear dimensions from centimetres to more than 1 m. Alternative networks are briefly considered in the final part.

Plasma sources provide the foundational technology underpinning many of today's frontline industries. Examples include plasma processing for semiconductors and microprocessors, large area electronics for displays and solar panels, coatings for the packaging and automotive industries, treatment of architectural glass, etc. There is a constant demand for larger and more uniform plasma sources, which are running up against the limits of conventional capacitive and inductive RF plasma reactors, due to standing wave effects and the asymptotic impedance of large area reactors. The inherent properties of resonant network antennas can overcome these limitations because of their spatially distributed internal resonances, and real impedance which circumvents the high voltages and currents of monolithic capacitive or inductive reactors.

This book therefore aims to show that resonant RF networks are versatile, alternative plasma sources with numerous potential applications in plasma processing, helicon sources, plasma thrusters, etc [1]. Resonant antennas have been

doi:10.1088/978-0-7503-5296-3ch1

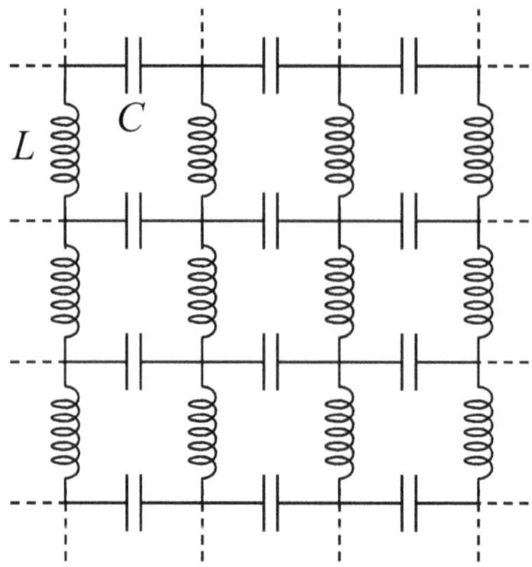

Figure 1.1. One type of a two-dimensional regular mesh of inductive and capacitive elements which can be extended in either dimension.

employed increasingly by the plasma industry in recent years, and the theory has now developed alongside the technological applications to the extent that it is timely to document the progress here, to aid antenna design for future RF plasma sources.

1.1 Resonant network antennas…

From its obvious similarity to a round bird cage, the cylindrical resonant network shown on the right in figure 1.2 is also known as a birdcage antenna. In this example of a high-pass, closed network, the top and bottom end rings are made up of series capacitors, capacitance C, joined by legs (the bars of the birdcage) of self inductance L. We note, with reference to section 1.2.1, that if any capacitor is removed, the birdcage becomes an open network.

Birdcage antennas are well known in nuclear magnetic resonance (NMR) excitation, used for magnetic resonance imaging (MRI), where a uniform transverse RF magnetic field is generated by RF currents, as explained in chapter 8. In MRI the oscillating magnetic field is used to excite the nuclear spins within a sample such as human tissue; in this book, the oscillating magnetic field is instead used to induce electric fields to excite a plasma. The electrical properties of MRI antennas have been extensively described in the book *Electromagnetic Analysis and Design in Magnetic Resonance Imaging* by J-M Jin [3] for the non-dissipative (lossless) case. This book extends the theory to dissipative (lossy) antennas in general, and to inductively coupled plasma sources in particular.

In passing, we address one important point here before it raises a doubt in the reader's mind: The leg of a birdcage is a straight conductor, so why, in the equivalent circuit of figure 1.2, is it represented as a coil although there is no loop to calculate magnetic flux linkage? Moreover, how to attribute a self inductance L to individual leg segments in a convoluted circuit? The answer is that the contemporary

Figure 1.2. Left: a classical birdcage with two end rings and many legs. Right: the equivalent electrical circuit of a 16-leg, high-pass birdcage antenna; the legs have an effective self inductance *L*, and the end rings are made up of capacitors *C*. The RF input and ground connections are on opposite points of the top end ring in this case. This is an example of a closed resonant network. (Adapted from [2] with permission. Copyright 2013 IOP Publishing.)

undergraduate concept of self inductance must be replaced by an earlier, more general, but apparently long-forgotten, alternative concept called 'self partial inductance' [4]. This will be explained in chapter 4. For the present, it suffices to accept that each leg can be considered as an inductor of self inductance *L*, symbolized by a coil icon, even though the leg itself has no loop nor coil structure.

Where the capacitors are concerned, there is no such ambiguity in figure 1.2 because they are discrete high power capacitor components, or lumped element assemblies of high-*Q* capacitors, as shown in the RF plasma sources of section 1.2 and described in section 13.2. The question of spatially distributed inductance and capacitance does not arise until the antenna legs become long enough to behave as transmission lines, as described in chapters 2 and 7.

A planar network is made by unwrapping a birdcage antenna to form a flat ladder structure as shown in figure 1.3. The birdcage end rings then become the stringers ('stiles' in the UK) of the ladder whose ends are in open circuit. The planar antenna can also be considered as a segment of a ladder filter, where the stringers consist of series capacitors joined by parallel inductors. For the purposes of this book, we choose to call the ladder inductors 'legs' instead of 'rungs' because they have the same function and position as the legs of the birdcage. In both antennas, the positions of the RF input and ground connections, in combination with the values of *L*, *C*, and RF frequency, are chosen to optimize various properties of the antenna

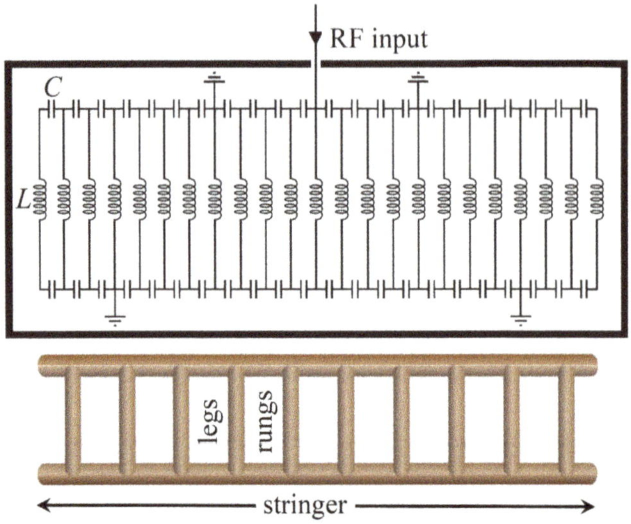

Figure 1.3. Ladders have two stringers joined by rungs, which we will call legs in this book. In the equivalent electrical circuit of the high-pass ladder antenna, the stringers are made up of capacitors *C*, and the legs are inductors *L*. This is an example of a high-pass, open resonant network.

such as its input impedance, and the spatial distribution of the generated electromagnetic field. These relations are described in the following chapters.

The particular arrangement of series capacitors and parallel inductors in both figures constitutes a high-pass filter because low frequencies are 'short circuited' via the parallel inductors, whereas the capacitors present a low impedance for the passage of high frequencies. Throughout this book we will consider high-pass networks for our birdcage and ladder antennas, although other configurations are envisaged easily (low-pass, hybrid, two-dimensional, etc) and some of these are mentioned as future applications in chapter 14.

1.2 …for radio-frequency plasma sources

Resonant network antennas are well suited for inductively coupled plasma (ICP) sources and for helicon wave excitation. Both of these RF plasma types require excitation by RF electric fields which in turn require magnetic induction by high RF currents in the source [5–7]. Strong currents occur within a resonant network antenna, whose mechanical structure can be specifically designed to withstand internal oscillation of high currents, whilst minimizing ohmic losses. In this way, the external power supply, power cables, and matching circuit are spared from excessive load currents and voltages. Indeed, the dominantly real input impedance of a resonant antenna is much closer to the characteristic impedance of cables and generator output impedance, typically 50 Ω. This is one significant advantage compared to the effective 'open circuit' infinite impedance of capacitively coupled reactors having opposing unconnected electrodes, or the effective 'short-circuit' zero impedance of conventional inductive or helicon antennas, which are entirely built from coiled wire or interconnected metal strips.

Another related advantage is that the spatially distributed, multiple resonant modes of network antennas avoid the extremely high voltages which occur at the input terminals of large area spiral coils or long solenoids. Indeed, the inductance of conventional ICP sources increases strongly with size, leading to very high voltages (typically 10^4 V) even for moderate RF power, with concomitant problems of arcing and energetic ion sputtering damage across high voltage sheaths caused by plasma capacitive coupling. In contrast, the dominantly real impedance of resonant antennas means that they can be scaled to large size while maintaining voltages below 1 kV for several kW of RF power. These aspects are developed further in chapter 13, as well as in part II of the book.

For motivation, we briefly present two examples of high RF power applications at the Swiss Plasma Center (SPC) of the EPFL—one for cylindrical geometry, and the other for large area planar sources. The theory is developed in the subsequent chapters.

1.2.1 Birdcage antenna helicon source

The first concerns a 2 m long helicon linear device, figure 1.4(a), called the Resonant Antenna Ion Device (RAID), operating in steady state up to 10 kW RF power at an industrial standard frequency of 13.56 MHz [8–15]. The red glow that can just be seen from the cylinder on the left is the plasma emission filtered through the walls of the water-cooled ceramic tube; it is not red hot. The plasma is generated by a birdcage antenna inside the cubic metal box visible inside the leftmost coil, mounted around the ceramic tube. An axial magnetic field enables helicon waves to propagate from the birdcage antenna, sustaining an intensely bright plasma column. Figure 1.4(b) shows the visible light emission from a typical hydrogen plasma discharge. This photograph was taken from the end of the plasma column through

(a) (b)

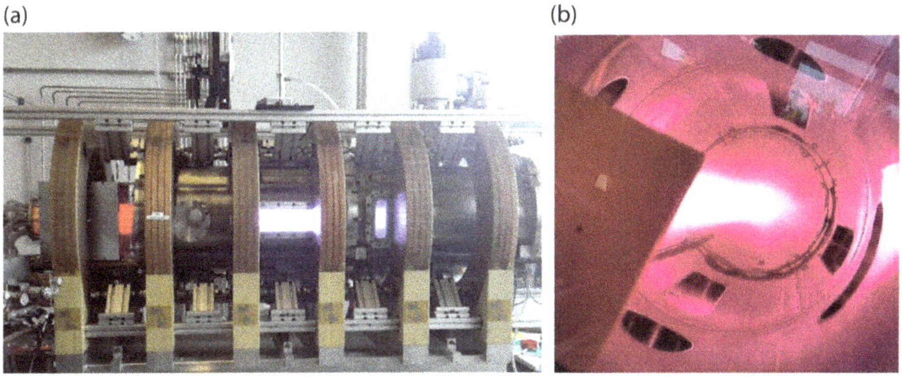

Figure 1.4. Photographs of the RAID experiment at the Swiss Plasma Center of the Ecole Polytechnique Fédérale de Lausanne. (a) The RAID vacuum vessel is surrounded by six water-cooled coils. (b) Downstream view of the visible light emission from a typical plasma column in hydrogen gas. The target is partially visible in the foreground.

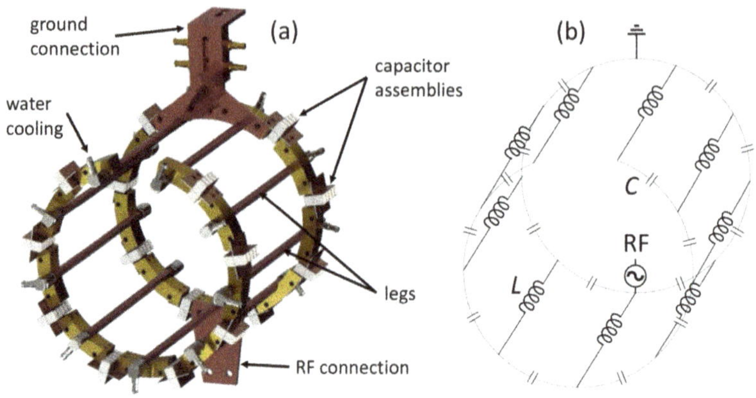

Figure 1.5. (a) Engineering scale drawing of the RAID birdcage antenna, showing copper legs, and capacitor assemblies mounted on brass connectors in the end rings. (b) Equivalent circuit of the open birdcage, which is distinguished by the missing capacitor in one end ring, in contrast to the continuous end rings in figure 1.2. Birdcage antennas are suitable for generating plasma in a cylindrical volume. (Reproduced with permission from [15]. Copyright 2021 IOP Publishing.)

an off-axis view port. Prospective applications include ion thrusters, or as a negative ion source for neutral beam heating of nuclear fusion devices such as tokamaks. In the mechanical drawing and equivalent circuit, figure 1.5, note that this birdcage is an open network due to the break in an end ring. In fact, birdcage antennas can be closed networks (figure 1.2) or open networks (figure 1.5), whereas planar antennas can only be open networks.

1.2.2 Ladder antenna inductive source

Plasma processes are used for the industrial production of solar cells, flat panel displays, large area electronics, semiconductors on silicon wafers, roll-to-roll food packaging, and surface treatment of architectural glass, to name just a few applications. One important aspect lies in the possibility of large area plasma processing for the treatment of larger single substrates, or for larger batches of work pieces [18]. Figure 1.6 shows a 1.7 m-diagonal planar inductive plasma source which operates at 15 kW RF power at 13.56 MHz in the SPC [16]. The whole ladder antenna assembly is mounted inside a grounded housing filled with a dielectric. This is then placed inside a vacuum chamber where the antenna surface is protected from the plasma by glass tiles (not shown); technological details are given in chapters 7 and 13. Figure 1.7 shows the individual inductance L and capacitor C elements of the ladder antenna.

1.3 Evolution of the antenna design

In the following chapters, we will essentially focus on regular networks, constituted by repetition of an elementary mesh. At the most basic level, this electrical system is analytically described in terms of uncoupled, purely reactive lumped elements, i.e. capacitors and inductors. This analysis, although over-simplified, serves to introduce

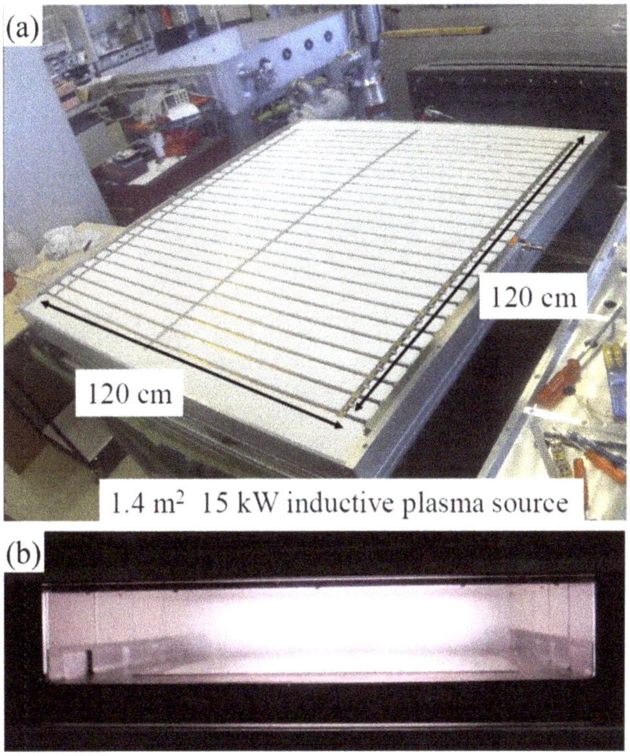

Figure 1.6. (a) Photograph of a large area ladder antenna inductive plasma source embedded in a dielectric foam. Its vacuum chamber can be seen in the background [16]. (b) Photograph of the plasma in a smaller planar reactor, width 37 cm [17].

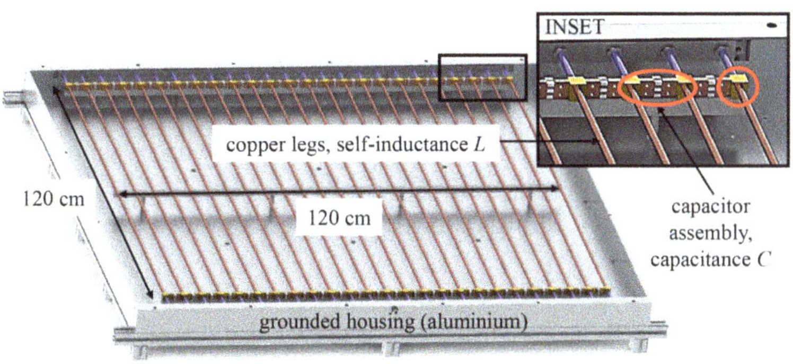

Figure 1.7. Engineering scale drawing of the ladder antenna whose equivalent circuit is similar to figure 1.3. The capacitor assembly is shown in the inset. Ladder antennas are suitable for generating large area plasma.

the notions of resonance and normal modes. Building on this ground level, three further important steps have to be taken into account to reach a complete model of the system as a resonant ICP, summarized in figure 1.8 as follows:

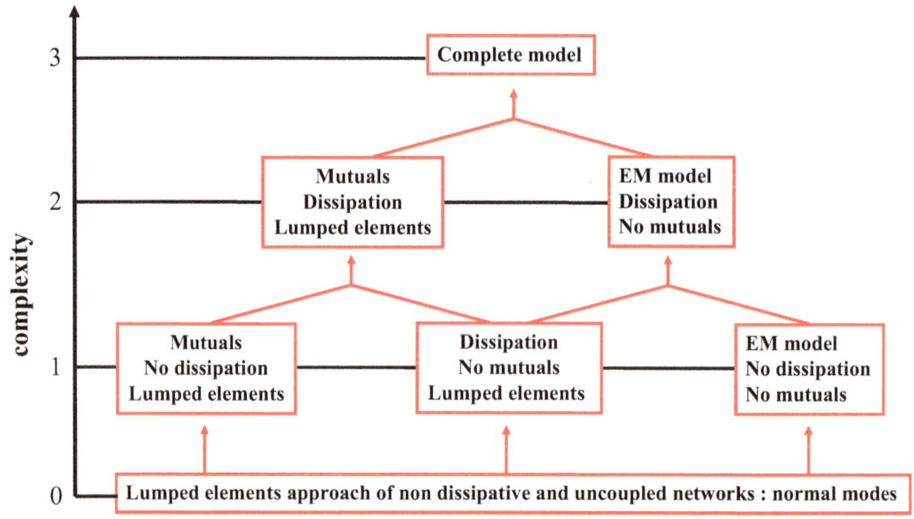

Figure 1.8. Development of the theory and models for resonant network antennas. 'EM' means 'electromagnetic', for when the antenna size becomes comparable to the RF wavelength.

- The mutual inductance between the network conductors themselves, and between the network and its environment (e.g. metallic housing) must be calculated in order to properly predict the resonance frequencies of the system. Technically, this is of prime importance as most of the time in RF plasma systems the excitation frequency is fixed, and as such it is not possible to tune it in order to make it match a particular resonance frequency of the antenna. Therefore, in practice, the system is designed so that the desired resonant mode occurs at the RF generator frequency by properly choosing the value of the antenna capacitors. Furthermore, when operating as a plasma source, the network is also coupled to the plasma by mutual induction, which must also be incorporated into the theoretical model.
- Another important aspect is dissipation, due in a part to resistance of the antenna components, but essentially associated with plasma loading. The introduction of dissipation into the various models is necessary to quantify the system input impedance, and hence all the currents, voltages, and power levels.
- For large antennas and/or for high frequencies, the lumped element approximation is no longer valid, and the spatial variation of current and voltage waves propagating along the antenna elements must also be taken into account.

1.4 Why use resonant network antennas?

For convenient reference, the advantages of and interest in resonant network antennas are gathered here at the outset, to avoid searching through each chapter for the various motivations.

1. Intense currents, appropriate for RF inductively coupled plasma (part II) and helicon plasma sources (part III), circulate internally around the robust mechanical framework of a resonant network antenna.
2. The normal mode resonances induce a spatial distribution of adjacent current loops, better suited for uniform plasma induction compared to the single current loop of monolithic coil sources (chapters 3 and 6).
3. The dominantly real impedance at the antenna input means that the currents and voltages in the RF matching circuit are much smaller than within the resonant antenna itself—in contrast to conventional reactive-impedance sources—thereby reducing the required current and voltage ratings, and power loss, of the matching components (chapters 2, 5, and 13). For the same reason, the RF power supply and matching circuit can be conveniently distanced from the antenna plasma source by a long coaxial cable, without overheating the cable (chapter 13).
4. Resonant network antennas are versatile: their design can be adapted for high or low pressures, small or large sizes, surface or volume plasma sources, and one-, two-, or three-dimensional configurations (chapters 6, 7, 9, 13, and 14).

1.5 Outline of the book

The first three parts of the book are structured in order of general mathematical and physical complexity as follows:

- Part I: resonant network antennas without plasma.
- Part II: resonant network antennas in non-magnetized plasma.
- Part III: resonant network antennas in magnetized plasma.

Furthermore, remember that we mostly concentrate on high-pass configurations of ladder and birdcage resonant antennas (figure 1.9). From this point of view, the book is roughly (but not rigorously) split first into the Cartesian planar geometry of ladder antennas until chapter 8, whereupon the dominant geometry becomes the cylindrical treatment of birdcage antennas. Part IV mixes all of these aspects in a description of antenna technology and future developments. Where relevant, applications are presented at the end of each chapter.

The complete model at the pinnacle of figure 1.8 can be reached via arbitrary different pathways through ascending levels of complexity. In this book, we have chosen to develop the various aspects of resonant network antennas in the following sequence of chapters:

- **Part I** begins with definitions, and revision of the most basic concepts of parallel resonant circuits in chapter 2. Loop inductance is defined and the limits of the lumped element approximation are quantified.
- Chapter 3 presents an analytical model of uncoupled networks in a vacuum, using a lumped element (discrete component) approach to derive the normal modes of the resonant network. This gives a basic understanding of the main

Figure 1.9. 'A ladder approach to a closed birdcage.' (Illustration by Alex Howling. Copyright 2023 Alex Howling.)

RF properties of resonant networks and a rough estimate of their resonance frequencies.

- The mutual coupling between antenna elements requires the concept of mutual partial inductance, which is conceptually different from the familiar loop inductance. The basics are explained in chapter 4, along with a matrix analysis of mutually coupled, dissipative components. This description of resonant network antennas in vacuum yields an accurate estimate of resonance frequencies and load impedance, which completes part I.
- **Part II** explains why low pressure RF plasmas are used, and the basics of inductive and capacitive coupling to non-magnetized plasma in chapter 5. Chapter 6 introduces the complex image method (CIM) to account for the inductive coupling due to mutual partial inductance with a plasma slab.
- Chapter 7 is the first chapter to account for electromagnetic (EM) standing wave effects in large and/or high frequency inductively coupled plasma sources, where the lumped element approximation is no longer appropriate. Using a multi-transmission line (MTL) model, the complete model at the apex of figure 1.8 is attained for large area planar plasma sources.
- The following chapters 8 (analytical) and 9 (experimental) take up the analogous challenge for birdcage antennas which are inductively coupled to cylindrical plasma, to conclude part II.
- **Part III** adds a static uniform magnetic field to the previous configurations. This gives rise to the diverse plasma waves in magnetized plasma in chapter 10, of which whistlers are of principal interest in uniform plasma at RF frequencies.
- Chapter 11 describes helicon modes, which are confined whistler waves, in a magnetized cylindrical column. Chapter 12 compares the theory with measurements on birdcage antennas. It is also demonstrated how helicon waves can be launched by planar resonant antennas, to close part III.

- **Part IV** describes technological aspects of resonant network antennas in chapter 13, and considers alternative configurations for future RF plasma sources in chapter 14. The appendices explain basic plasma physics results and give more details of methods and theorems. Finally, to give a head start to the reader, a link to programs used to produce many of the book's figures is given in the last appendix K.

References

[1] Guittienne Ph 2010 Apparatus for large area plasma processing *Patent* WO2010092433

[2] Hollenstein Ch, Guittienne Ph and Howling A A 2013 Resonant RF network antennas for large-area and large-volume inductively coupled plasma sources *Plasma Sources Sci. Technol.* **22** 055021

[3] Jin J-M 1998 *Electromagnetic Analysis and Design in Magnetic Resonance Imaging* (Boca Raton, FL: CRC Press)

[4] Paul C R 2010 *Inductance: Loop and Partial* (Hoboken, NJ: Wiley)

[5] Lieberman M A and Lichtenberg A J 2005 *Principles of Plasma Discharges and Materials Processing* (Hoboken, NJ: Wiley) 2nd edn

[6] Chabert P and Braithwaite N J 2011 *Physics of Radio-Frequency Plasmas* (Cambridge: Cambridge University Press)

[7] Chen F F 2016 *Introduction to Plasma Physics and Controlled Fusion* 3rd edn (Cham: Springer)

[8] Furno I, Agnello R, Fantz U, Howling A A, Jacquier R, Marini C, Plyushchev G, Guittienne Ph and Simonin A 2017 Helicon wave-generated plasmas for negative ion beams for fusion *EPJ Web Conf.* **157** 03014

[9] Agnello R *et al* 2018 Cavity ring-down spectroscopy to measure negative ion density in a helicon plasma source for fusion neutral beams *Rev. Sci. Instrum.* **89** 103504

[10] Agnello R *et al* 2020 Negative ion characterization in a helicon plasma source for fusion neutral beams by cavity ring-down spectroscopy and Langmuir probe laser photodetachment *Nucl. Fusion* **60** 026007

[11] Agnello R, Andrebe Y, Arnichand H, Blanchard P, De Kerchove T, Furno I, Howling A A, Jacquier R and Sublet A 2020 Application of Thomson scattering to helicon plasma sources *J. Plasma Phys.* **86** 905860306

[12] Agnello R 2020 Negative hydrogen ions in a helicon plasma source *PhD Thesis* no. 7817, Ecole Polytechnique Federale Lausanne (EPFL), Switzerland

[13] Marini C *et al* 2017 Spectroscopic characterization of H_2 and D_2 helicon plasmas generated by a resonant antenna for neutral beam applications in fusion *Nucl. Fusion* **57** 036024

[14] Furno I *et al* 2023 Helicon volume production of H^- and D^- using a resonant birdcage antenna on RAID *Physics and Applications of Hydrogen Negative Ion Sources* ed M Bacal (Cham: Springer Nature) (Springer Series on Atomic, Optical, and Plasma Physics) ch 9

[15] Guittienne Ph, Jacquier R, Pouradier Duteil B, Howling A A, Agnello R and Furno I 2021 Helicon wave plasma generated by a resonant birdcage antenna: magnetic field measurements and analysis in the RAID linear device *Plasma Sources Sci. Technol.* **30** 075023

[16] Guittienne Ph, Jacquier R, Howling A A and Furno I 2017 Electromagnetic, complex image model of a large area RF resonant antenna as inductive plasma source *Plasma Sources Sci. Technol.* **26** 035010

[17] Demolon P, Guittienne Ph, Howling A A, Jost S, Jacquier R and Furno I 2018 RF bias to suppress post-oxidation of μc-Si:H films deposited by inductively-coupled plasma using a planar RF resonant antenna *Vacuum* **147** 58
[18] Guittienne Ph, Fayet P, Larrieu J, Howling A A and Hollenstein Ch 2012 Plasma generation with a resonant planar antenna *55th Ann. SVC (Surface Vacuum Coaters) Technical Conf. (Santa Clara, CA, 28 April–3 May)*

Part I

Resonant network antennas without plasma

IOP Publishing

Resonant Network Antennas for Radio-Frequency Plasma Sources

Theory, technology and applications

Philippe Guittienne, Alan Howling and Ivo Furno

Chapter 2

Introduction to resonant circuits

In this first part, resonant network antennas are analysed in the absence of plasma, which, for all intents and purposes, means in a vacuum or in air.

Readers who are not familiar with RF systems may find this short chapter a useful introduction to the basic concepts needed in the analysis of resonant networks [1], especially for the next two chapters 3 and 4. As can be seen in the equivalent circuits of figures 1.2, 1.3, and 1.5, these networks are essentially combinations of L, C parallel components, therefore, we first revise the basics of parallel resonant circuits. Beyond this, section 2.3 analyses the relation between the lumped element approximation and transmission lines. This is necessary to define the maximum length of antenna legs before the lumped element approximation breaks down and instead an electromagnetic treatment is required (see chapter 7).

Useful background to alternating current theory can be found in standard textbooks such as B I Bleaney and B Bleaney *Electricity and Magnetism* [2]. First of all, we now define the conventions to be used self-consistently in all chapters.

2.1 Definitions and conventions

2.1.1 Angular frequency and wavenumber

Without exception in this book, the antennas are excited by RF power generators at a single radio frequency f, usually an Industrial, Scientific, and Medical (ISM) frequency such as 13.56 MHz or a harmonic such as 40.68 MHz. Hence, every field, voltage, and current oscillates at a single angular frequency $\omega = 2\pi f$ rad s^{-1}, provided that any non-linear response of the plasma can be neglected. For self-consistency throughout, we take every sinusoidally oscillating parameter to vary as

$$\text{time dependence} \propto e^{-j\omega t}, \tag{2.1}$$

doi:10.1088/978-0-7503-5296-3ch2

where, by convention, the complex exponential notation means that the real part of the expression is taken as the measurable quantity. We choose the symbol $j = \sqrt{-1}$ to avoid any confusion between $i = \sqrt{-1}$ and electrical current i.

As a reminder, it is customary to express each variable's magnitude and phase ϕ by a complex amplitude called a phasor. For example, the voltage phasor $V = |V|e^{j\phi}$ and the instantaneous measurement of voltage, $V(t)_{\text{measured}}$, are related by

$$V(t)^{\text{measured}} = \text{Re}\,[|V|\,e^{j\phi}e^{-j\omega t}] = \text{Re}\,[Ve^{-j\omega t}]. \tag{2.2}$$

In linear theory, equations can be written directly in terms of phasors because, for discrete components, the common factor $e^{-j\omega t}$ cancels throughout. The common factor $e^{-j\omega t}$ is re-inserted before taking the real part of the final result. The measured voltage amplitude is the amplitude of the phasor, $(V.\ V^*)^{\frac{1}{2}}$, where V^* is the complex conjugate of V. Note that the phasor description applies only to linear combinations of parameters; the time-dependent electrical power[1] $P(t)$, on the other hand, is a non-linear combination and must be described using the real values directly, i.e. $P(t) = \text{Re}(V)\text{Re}(I)$.

2.1.2 Loop self inductance

We briefly revise the contemporary meaning of self inductance as taught in undergraduate courses for several generations of students [2, 3], thereby including the authors and essentially all of the readers. A lumped element inductor is a compact coil, solenoid, or loop of wire. When carrying a current I (A), the inductor generates a magnetic field density $\mathbf{B}$ (T) in the space surrounding the current loop. A magnetic flux $\phi = \int_S \mathbf{B} \cdot d\mathbf{S}$ links (or threads) its own loop winding(s), where S is an area bounded by the current loop [2]. The total magnetic flux linkage, $\Phi = n\phi$ (Wb $\equiv$ Tm^2), is proportional to the integer number of loop turns $n \geqslant 1$. There are various definitions of loop self inductance, which are all self-consistent, and one of these is

$$L^{\text{loop}} = \Phi/I \quad (\text{WbA}^{-1}) \equiv (\text{H}). \tag{2.3}$$

Loop self inductance, in units of H (henry), is the constant of proportionality between a coil's magnetic flux linkage and its own current which generates it—hence the use of the word 'self'. Ferromagnetic materials are not used in the resonant network antennas of this book, hence the inductance is independent of the values of the flux and current, and depends only on the dimension of the loop and its shape, i.e. inductance is a purely geometric property [3].

Loop self inductance in this book will always have the explicit suffix 'loop'. This is to distinguish it from self partial inductance L, whose unstated 'partial' label remains implicit because various other suffixes and subscripts will need to be added for partial inductances in chapter 4. This distinction is important because, historically, confusion arose in the literature due to a lack of explicit definitions of the two

[1] The time-averaged electrical power can be written as $\frac{1}{2}\,\text{Re}(V.\ I^*)$; see appendix D.2.1 and (D.20).

types of inductance, as clearly pointed out in the recent (2010) book *Inductance: Loop and Partial* by C R Paul [3].

For a lumped element inductor, Faraday's law can be expressed as

$$V = -\frac{d\Phi}{dt} = -L^{\text{loop}}\frac{dI}{dt} = j\omega L^{\text{loop}} I, \tag{2.4}$$

hence the complex impedance of a lumped element's loop self inductance is the conventional expression $Z_L = j\omega L^{\text{loop}}$. The distinction between lumped element impedance and spatially distributed impedance will be quantified in section 2.3.

2.2 Parallel resonant circuits

We begin with the most simple case of a parallel resonant circuit with no ohmic losses, and then go on to consider the effect of a small resistance associated with the inductor.

2.2.1 The lossless *LC* parallel circuit

Consider the parallel assembly of two complex lumped element impedances Z_1 and Z_2 in figure 2.1. Although this is a trivial example, let us consider the possible current oscillation in this system. According to Ohm's law and the conservation of current (Kirchoff's first law) we must have:

$$\begin{aligned} V_B - V_A &= Z_1 I_1 \\ V_B - V_A &= Z_2 I_2 \\ I_1 &= -I_2, \end{aligned} \tag{2.5}$$

where V_A and V_B are a consequence of the hypothetical current; they are not externally imposed voltages. This system then has a unique solution under the condition

$$Z_1 = -Z_2. \tag{2.6}$$

If (2.6) is satisfied, a current can oscillate in the parallel assembly, and it is the only possibility for this isolated system. The oscillating current is said to be the unique normal mode of this structure, as a trivial example of normal modes to be defined in chapter 3.

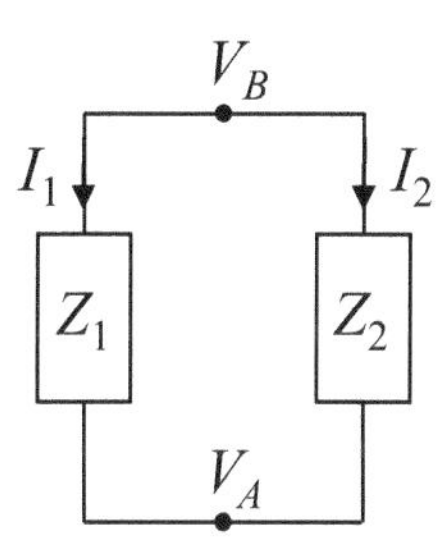

Figure 2.1. An isolated circuit made up of two parallel impedances Z_1 and Z_2.

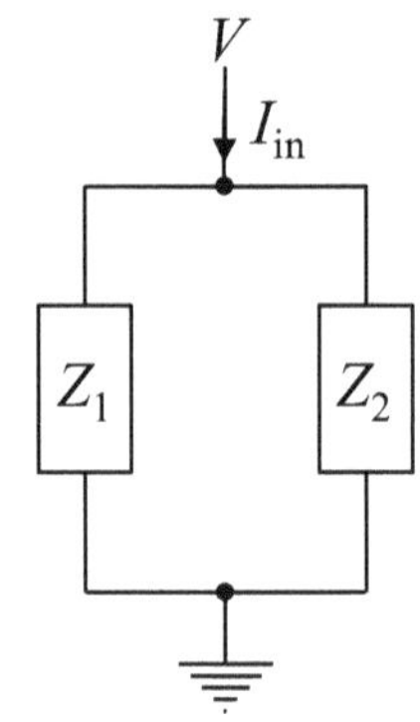

Figure 2.2. A parallel circuit with a voltage source.

Let us now apply a voltage source V to the parallel assembly, as shown in figure 2.2. To estimate the input current I_{in} we express the input impedance of the system, Z_{in}, as

$$Z_{\text{in}} = \frac{V}{I_{\text{in}}} = \frac{Z_1 Z_2}{Z_1 + Z_2}. \tag{2.7}$$

According to this expression, if (2.6) is satisfied, the parallel assembly input impedance is infinite, which means, for a finite voltage source, that the input current I_{in} drops to zero. Alternatively, this is used as a tuner circuit to obtain a high voltage signal for an infinitesimal aerial current in a radio receiver.

The condition $Z_1 = -Z_2$ implies opposite signs for both the real and imaginary parts of the two impedances. As the real part of an impedance cannot be negative, a normal mode can only exist in non-dissipative systems, with zero real impedance, constituted by purely reactive impedances. This applies also to normal modes on resonant antennas in chapter 3.

We now consider ideal electrical components. For time harmonic variations $e^{-j\omega t}$, capacitors and inductors are components which have purely imaginary impedances of opposite signs: $Z_C = \frac{1}{j\omega C} = -\frac{j}{\omega C}$ for a capacitor of capacitance C, and $Z_L = j\omega L^{\text{loop}}$ for an inductor of inductance L^{loop}. Then a parallel $L \| C$ assembly can present a normal mode under the condition $Z_L = -Z_C$, so that

$$j\omega_0 L^{\text{loop}} = -\frac{1}{j\omega_0 C} \quad \Rightarrow \quad \omega_0 = 2\pi f = \frac{1}{\sqrt{L^{\text{loop}} C}}. \tag{2.8}$$

This specific frequency ω_0 at which the system behaves like an ideal oscillator is called the resonance frequency, or the normal mode frequency, of the parallel $L \| C$ circuit. In the ideal case, infinite current oscillates at the normal mode frequency within the resonant circuit, with zero input current $I_{\text{in}} = 0$. This behaviour extends to normal modes on resonant network antennas in chapter 3.

2.2.2 The lossy *LC* parallel circuit

In reality, purely reactive capacitors or inductors do not exist and there is always some dissipation due to component resistance. Therefore, we now consider the circuit in figure 2.3 where a resistance R has been introduced uniquely into the inductor branch, because real inductors are generally far more lossy than capacitors. This $RL||C$ arrangement is a standard model for R, L, C circuits, but other variations are possible.

Intuitively, the quality factor Q of an inductor in a resonant circuit can be defined as the ratio of its reactive impedance at the resonance frequency, to its resistance [2],

$$Q = \omega_0 L^{\text{loop}}/R, \tag{2.9}$$

because the higher the value of Q, the better the approximation to an ideal lossless inductor. For typical RF coils, $Q = 100$–200 [2]. We shall see below that this quality factor recurs often in the analysis of the $RL||C$ circuit.

The parallel assembly input impedance is now an unwieldy expression:

$$Z_{\text{in}} = \frac{Z_1 Z_2}{Z_1 + Z_2} = \frac{R + j\omega[L^{\text{loop}} - CR^2 - (\omega L^{\text{loop}})^2 C]}{1 + \omega^2 C[CR^2 + L^{\text{loop}}(\omega^2 L^{\text{loop}} C - 2)]}. \tag{2.10}$$

In contrast to the lossless impedances, the denominator of this expression can no longer vanish, and strictly speaking, no normal mode can be found in a parallel $RL||C$ circuit. But it is physically reasonable that in the limit of small R, i.e. high Q, this $RL||C$ circuit should behave as a slightly perturbed ideal $L||C$ circuit.

Given the parallel arrangement of the impedances, it is algebraically more convenient to calculate the complex input admittance $Y_{\text{in}} = 1/Z_{\text{in}}$ as the sum of the individual branch admittances:

$$Y_{\text{in}} = j\omega C + \frac{1}{R + j\omega L^{\text{loop}}} = j\omega C + \frac{R - j\omega L^{\text{loop}}}{R^2 + (\omega L^{\text{loop}})^2}. \tag{2.11}$$

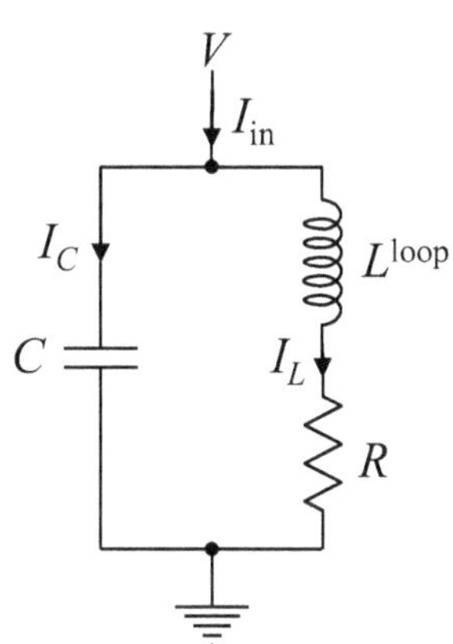

Figure 2.3. A parallel circuit with a voltage source and a lossy inductor.

A conventional definition of resonance for $RL\|C$ circuits is the point for which the imaginary part of the input impedance and admittance is zero [2]. Hence, using (2.10) or (2.11), we have

$$R^2 + (\omega_0' L^{\mathrm{loop}})^2 = L^{\mathrm{loop}}/C, \tag{2.12}$$

which defines the resonance frequency ω_0' for a parallel $RL\|C$ assembly as

$$\omega_0' = \sqrt{\frac{1}{L^{\mathrm{loop}}C}}\sqrt{1 - \frac{R^2C}{L^{\mathrm{loop}}}} = \omega_0\sqrt{1 - \frac{1}{Q^2}}. \tag{2.13}$$

From this definition, the lossy $RL\|C$ resonance frequency ω_0' is slightly below the lossless $L\|C$ resonance frequency ω_0, although the fractional downshift,

$$\frac{d\omega_0'}{\omega_0} = \frac{df_0'}{f_0} \approx -\frac{1}{2Q^2}, \tag{2.14}$$

is less than 0.005% for typical RF inductors with $Q > 100$. The input admittance at this $RL\|C$ resonance frequency (where the imaginary part cancels out by definition) is a pure conductance whose value, directly from (2.11) and (2.12), is

$$Y_{\mathrm{in}}^{\mathrm{res}} = \frac{R}{R^2 + (\omega_0' L^{\mathrm{loop}})^2} = \frac{RC}{L^{\mathrm{loop}}}. \tag{2.15}$$

Consequently, the input impedance at resonance is a pure resistance,

$$Z_{\mathrm{in}}^{\mathrm{res}} = \frac{1}{Y_{\mathrm{in}}^{\mathrm{res}}} = \frac{L^{\mathrm{loop}}}{RC}. \tag{2.16}$$

These are effectively the minimum admittance and the maximum impedance, respectively, whose exact values occur at a frequency in the narrow range between ω_0' and ω_0. Hence we have the counter-intuitive result that input impedance falls when the resistance R rises. Later on, in section 6.4.1, this will help to understand why the antenna input impedance falls when it ignites a dissipative plasma by inductive coupling.

The ratios of the currents at resonance are readily calculated for small R: For a given applied voltage V, the input current I_{in} is at its minimum value $I_{\mathrm{in}} = VRC/L^{\mathrm{loop}}$; the capacitor current amplitude $I_C \simeq \omega_0 CV$, and the inductor current amplitude is $I_L \simeq V/(\omega_0 L^{\mathrm{loop}})$. Using the definitions (2.8) and (2.9) for ω_0 and Q shows directly that the magnitude of the resonance currents in the capacitor and the inductor are equal, and larger than the input current by a factor Q, namely

$$|I_C/I_{\mathrm{in}}| \simeq |I_L/I_{\mathrm{in}}| \simeq Q. \tag{2.17}$$

Hence Q is the 'magnification factor' for the oscillating current in the inductor and capacitor branches, compared to the current in the external circuit. This demonstrates the statement in the Introduction, section 1.2: 'Strong currents oscillate within a resonant network antenna … (so that) the external power supply, power cables, and matching circuit are spared from excessive load currents and voltages.' This is an extremely useful property for an ICP source, where strong internal RF currents generate strong magnetic fields for efficient inductive coupling, because the antenna can be designed as a robust mechanical structure to withstand high currents with only minimal ohmic losses.

Looking ahead, it will be useful to have an approximate expression for Y_{in} near resonance in order to compare with the behaviour of resonant networks around their normal mode frequencies; for future reference, this comparison is made in section 3.5.6. This expression is obtained by writing $\omega = \omega_0 + \delta\omega$ in (2.11), expanding, and retaining terms of order $\delta\omega$. After some algebra, the expansion yields [2]

$$Y_{in} \simeq \frac{R}{\omega^2 L^2}\left[1 + 2jQ\frac{\delta\omega}{\omega_0}\right]. \tag{2.18}$$

Additional important results can now be deduced, showing the usefulness of the quality factor Q [2].

- When $\frac{\delta\omega}{\omega_0} = \pm\frac{1}{2Q}$, the admittance real and imaginary parts are of equal magnitude. Therefore, the admittance magnitude rises by a factor $\sqrt{2}$ from its resonance minimum $\frac{1}{Q^2 R}$, and its phase angle changes from 0 to $\pm\frac{\pi}{4}$. The impedance magnitude $|Z_{in}|$ falls by a factor $\sqrt{2}$.
- These are also known as the half power points. For a constant amplitude of input current, $|I_L| \propto |V_{in}| \propto |Z_{in}|$ for the narrow range of frequency $\pm\delta\omega$. Hence the dissipated power $|I_L|^2 R \propto |Z_{in}|^2$, which falls by a factor 2 at $\omega_0 \pm \delta\omega$. The frequency bandwidth, Δf, between these two half power points is the full width at half maximum (FWHM), $\Delta f = 2\delta f$, so that

$$Q = \frac{\omega_0}{2\delta\omega} = \frac{\omega_0}{\Delta\omega} = \frac{f_0}{\Delta f}, \tag{2.19}$$

  and $1/Q$ defines the fractional FWHM bandwidth of the $RL||C$ circuit. Note that the resonance frequency downshift (2.14) due to circuit losses is smaller than the FWHM bandwidth by another factor Q.
- The input impedance at resonance (2.16) is maximum and purely real; it can also be written as $Z_{in}^{res} = \frac{Q}{\omega_0 C}$. The dissipated ohmic power at resonance is therefore $P_{res} = I_{in}^2 Q/(\omega_0 C)$ or $P_{res} = V^2 \omega_0 C/Q$. Hence, for a circuit whose resonance frequency and capacitance is invariable, the power is determined by the Q value, for a given input current or voltage.
- Finally, the time-averaged stored energy is $\frac{1}{2}L^{loop}I^2 + \frac{1}{2}CV^2 = \frac{1}{2}L^{loop}I_0^2$ at resonance, and the mean dissipated power is $RI_0^2/2$ [2]. Their ratio is L^{loop}/R, hence

$$Q = \frac{\omega_0 L^{loop}}{R} = \omega_0\frac{\text{stored energy}}{\text{power dissipation}} = 2\pi\frac{\text{stored energy}}{\text{dissipated energy per cycle}}, \tag{2.20}$$

since the cycle period is $2\pi/\omega_0$. This general relation is useful for defining Q in more complicated resonant circuits when individual values of L, C, and R are not identifiable [1] (see also section 3.6).

For a given pair of L^{loop} and C, the difference between the $RL\|C$ circuit behaviour and an ideal $L\|C$ oscillator will increase with the resistance R until the quality factor becomes so low, at $Q < \frac{1}{2}$, that the circuit is over-damped and ceases to oscillate. When perturbed from its equilibrium steady-state it returns to it by exponential decay.

A worked example of a lossy parallel resonant circuit

By way of a worked example of an $RL\|C$ circuit, we use an arbitrary set of values $C = 1$ nF, $L^{\text{loop}} = 253.3$ nH, and $R = 1\ \Omega$, with L, C chosen so that, for a lossless $L\|C$ circuit, $f_0 = 1/(2\pi\sqrt{L^{\text{loop}}C}) = 10$ MHz. Figure 2.4 shows plots of the magnitude, real part, and imaginary part of the input impedance and admittance, as a

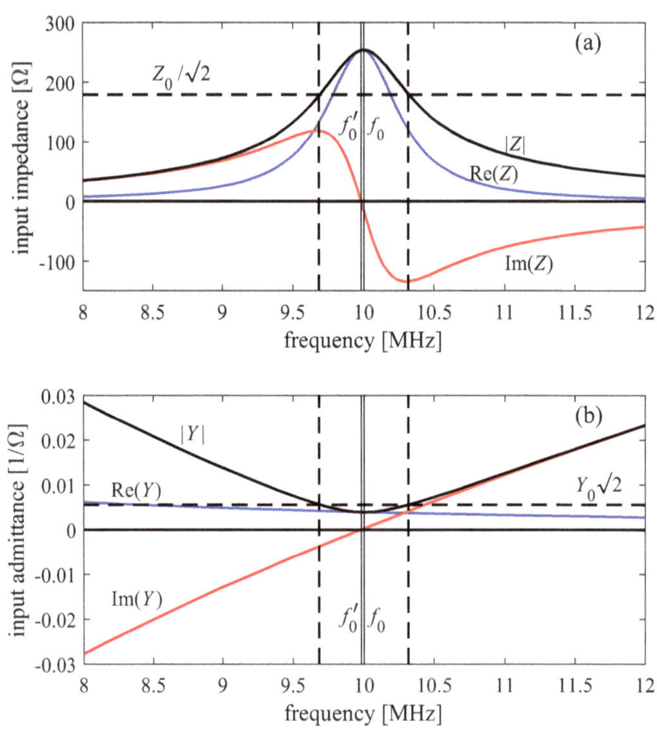

Figure 2.4. (a) The input impedance and (b) the input admittance of the lossy parallel resonant circuit in figure 2.3 as a function of frequency. The magnitude, and real and imaginary parts are shown. $f_0 = 10$ MHz and $f_0' = 9.98$ MHz mark the resonance frequencies of the lossless and lossy circuits, respectively. The vertical dashed lines mark the FWHM bandwidth limits.

function of frequency. For comparison with the figure, we now estimate various properties of this $RL||C$ circuit using the preceding definitions.

From (2.9) we have $Q = 15.9$, a very poor quality inductor which is chosen intentionally to show the influence of its resistance. Nevertheless, using (2.13), the resonance frequency $f_0' = 9.98$ MHz is only 20 kHz below the lossless resonance frequency, and the solid vertical lines at f_0 and f_0' are almost indistinguishable in figure 2.4. Using (2.16), the maximum input impedance of 253.3 Ω occurs extremely close to the frequency of zero imaginary impedance $\text{Im}(Z_{\text{in}}) = 0$.

From (2.18), the half-power points, where impedance (admittance) falls (rises) by a factor $\sqrt{2}$, occur at $\pm\frac{f_0}{2Q} = 0.314$ MHz either side of f_0 (the vertical dashed lines in figure 2.4). Hence the FWHM bandwidth is 0.628 MHz and the fractional bandwidth is 6.28%; an unimpressive selectivity for this low-Q circuit. The half-power points are also at the crossing point for equal real- and imaginary-magnitude parts. To the eye, the frequency dependence of the impedance in (a) is more evocative of a symmetrical resonance behaviour than that of the admittance in (b), hence the impedance will be used for all subsequent figures of resonant circuits, even though the algebra is more convenient for the admittance.

The current amplitudes in the system are shown in figure 2.5(a) for $V_{\text{in}} = 1$ V fixed amplitude, hence $|I_{\text{in}}| \propto |Y_{\text{in}}|$ for this case. The minimum input current, $I_{\text{in}} = V/253.3 = 0.0039$ A, occurs at $f_0' = 9.98$ MHz where the current amplitudes,

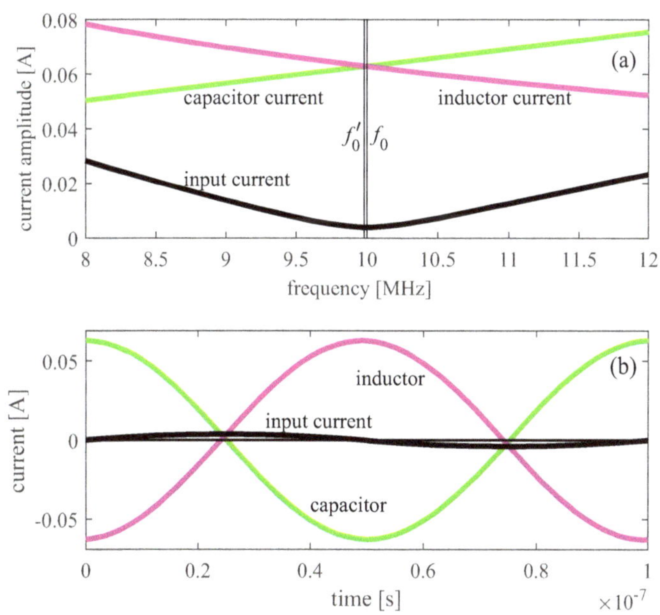

Figure 2.5. (a) Amplitudes of the input current I_{in}, and the currents in the inductor and capacitor branches of the lossy parallel resonant circuit in figure 2.3, as a function of frequency for a fixed input voltage of 1 V. (b) Time dependence, at resonance, of the input current I_{in}, and the currents in the L^{loop} and C branches of the lossy parallel resonant circuit in figure 2.3.

$QI_{\text{in}} = 0.063$ A, are equal. Figure 2.5(b) shows the temporal evolution of these currents at 9.98 MHz. The currents in the L^{loop} and C branches oscillate back-and-forth almost, although not exactly, in anti-phase. The inductor current is, in fact, phase shifted with regard to the anti-phase of the capacitor current by a small angle $\tan\theta \simeq \theta = 1/Q$ rad.

In conclusion, even for this low-Q circuit, it appears that we essentially recover the characteristics of the resonating $L\|C$ at almost the same frequency, namely, a minimum input current for a given source voltage (i.e. maximum input impedance) and equal amplitude of anti-phase currents in the two branches. These properties are not met at exactly the same value of the excitation frequency, but at very similar values between 10 MHz resonance frequency of the non-dissipative $L\|C$ circuit, and 9.98 MHz of the dissipative $RL\|C$ circuit.

2.3 From lumped element inductor to transmission line

In the previous section, each impedance was called a lumped element because its impedance acts over a negligible length, $\Delta z \to 0$, so it is inserted into the circuit effectively at a single point. This is because commercial capacitors have large-area, closely spaced electrodes packed into a small volume; and inductors, such as a coil or short solenoid, have many windings in a compact space.

In the first chapter, the legs of birdcage and ladder antennas were introduced with no discussion as to the consequences of their length. In this section, we will show what approximations are made by assuming lumped element inductance for the antenna legs. We also revise the basics of waves of current and voltage on a transmission line.

For this, we will use the illustrative example in figure 2.6 where we compare a conventional lumped-element inductor in (a), with a single-turn wire loop inductor in (b), especially designed as a long straight thin loop. A current in each circuit causes a flux linkage Φ through its area, associated with its particular lumped-element loop inductance L^{loop}. By construction, the thin loop in (b) can equivalently be considered as a bifilar transmission line, length l, terminated by a short circuit of width $s \ll l$, and area $l \times s$ threaded by the flux Φ. Our task is to calculate the impedance Z, and ask how long the loop can be for it still to be treated as a lumped element inductor.

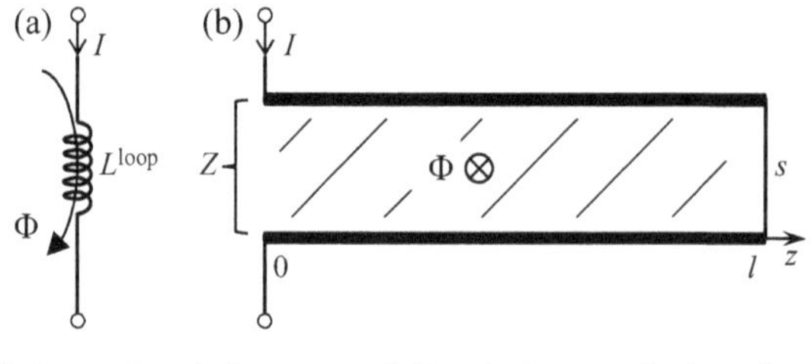

Figure 2.6. (a) A lumped element loop inductance and (b) a single-turn wire loop, impedance Z, represented as a bifilar transmission line, length l, separation $s \ll l$, terminated by a short circuit. Φ is the flux linkage in (a) or (b).

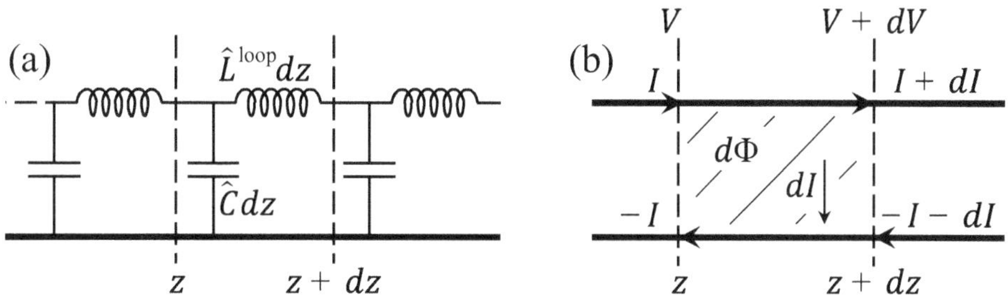

Figure 2.7. Two representations of a transmission line (a) as a chain of infinitesimal sections with loop inductance per unit length, $\hat{L}^{\rm loop}$, and capacitance per unit length, $\hat{C}$, and (b) one infinitesimal section showing the current flow and voltage between conductors.

Alternating current flow along a transmission line creates time-varying magnetic flux and induced voltage (due to inductance) between the conductors. The crucial difference introduced by the transmission line description, in contrast to a lumped element inductor, is that capacitance also exists between two conductors. An alternating voltage wave associated with the AC current now causes time-varying electrical charges and capacitive current flow between the conductors.

The classical analysis of a transmission line [2] considers an infinitesimal section of the lossless conductors of length dz, as shown in figure 2.7. In this context, the bifilar line is characterized by its capacitance per unit length, $\hat{C}$ F m^{-1}, and its loop inductance per unit length, $\hat{L}^{\rm loop}$ H m^{-1}. Dissipation due to the conductor resistivity and dielectric losses will be accounted for in chapter 7. Kirchoff's laws and the definitions of capacitance and inductance applied to the infinitesimal section in figure 2.7(b) lead directly to $dV = -(\hat{L}^{\rm loop}dz)\frac{\partial I}{\partial t}$ and $dI = -(\hat{C}dz)\frac{\partial V}{\partial t}$, hence

$$\frac{dV}{dz} = j\omega\hat{L}^{\rm loop}\ I, \tag{2.21}$$

$$\frac{dI}{dz} = j\omega\hat{C}\ V. \tag{2.22}$$

By differentiating either equation with respect to z and substituting the other, we arrive at

$$\frac{d^2V}{dz^2} = -\omega^2\hat{L}^{\rm loop}\hat{C}\ V, \tag{2.23}$$

which is the wave equation for V, and similarly for I. The wave equation has a general solution of the form:

$$\begin{aligned} V &= A^+e^{jkz} + A^-e^{-jkz}, \\ Z_cI &= A^+e^{jkz} - A^-e^{-jkz}, \end{aligned} \tag{2.24}$$

Figure 2.8. 'An expansion of small arguments.' (Illustration by Alex Howling. Copyright 2023 Alex Howling.)

where A^+ and A^- are the phasor voltage amplitudes of waves $e^{j(kz-\omega t)}$ and $e^{j(-kz-\omega t)}$ travelling along $+z$ and $-z$, respectively; $k = \omega(\hat{L}^{\text{loop}}\hat{C})^{\frac{1}{2}}$ is the wavenumber, and $Z_c = (\hat{L}^{\text{loop}}/\hat{C})^{\frac{1}{2}}$ is called the characteristic impedance of the line, which is purely real. Note that the wave voltages superpose (addition), whereas the oppositely flowing currents subtract. The short-circuit boundary condition $V = 0$ at $z = l$ gives the relation between the amplitudes $A^+e^{jkl} + A^-e^{-jkl} = 0$. Finally, we can find the input impedance Z of the single-turn loop in figure 2.6(b) by setting $z = 0$ and taking the ratio of V and I to obtain

$$Z = \left(\frac{V}{I}\right)_{z=0} = jZ_c \tan(kl) = jZ_c \tan\left(\frac{2\pi l}{\lambda}\right), \tag{2.25}$$

by means of the trigonometric identity $\tan\theta \equiv -j(e^{j\theta} - e^{-j\theta})/(e^{j\theta} + e^{-j\theta})$.

The first important result is that if the loop is shorter than a quarter of a wavelength, $l < \lambda/4$, its impedance Z is purely reactive and positive, i.e. the loop behaves as an inductance. Moreover, by expansion of small arguments (figure 2.8), $\tan\theta = \theta(1 + \frac{\theta^2}{3} + \frac{2\theta^4}{15} + \cdots)$, and substitution for Z_c and k from above, we can write that

$$Z \simeq jZ_c kl = j\omega\left(\hat{L}^{\text{loop}}l\right), \quad \text{for} \quad kl \ll 1, \tag{2.26}$$

so the single-turn loop inductance behaves exactly as a lumped element inductance $L^{\text{loop}} = (\hat{L}^{\text{loop}}l)$ to second order in the small quantity $kl = \frac{2\pi l}{\lambda}$. The condition $kl \ll 1$ simply means $l \ll \frac{\lambda}{2\pi}$, i.e. $l \ll \lambda/10$, to all intents and purposes.

The results for a short-circuited transmission line, acting as a thin, lossless wire loop, are resumed using (2.25) and (2.26) as follows:

1. For $l < \lambda/10$, the loop acts like a lumped element inductor, with inductance approximately proportional to it length, $L^{\text{loop}} = (\hat{L}^{\text{loop}} l)$, which means that there is no significant influence from the line's capacitance. The fractional underestimate of the effective inductance increases from 0 in the limit of $l \ll \lambda$, up to +15% at $l = \lambda/10$. Two numerical examples at 13.56 MHz for the small antenna ($l = 0.2$ m) in chapters 3, 4, 6, and the large antenna ($l = 1.2$ m) in chapter 7, underestimate the effective inductance by 0.1% and 4% respectively. Clearly, the long antenna is more in need of an electromagnetic analysis which includes the leg capacitance per unit length.
2. For $l < \lambda/4$, the loop acts like a pure inductor, but whose effective inductance increases non-linearly with length because of the contribution from the line's intrinsic capacitance per unit length, $\hat{C}$.
3. For $l = \lambda/4$, the loop acts as a lossless resonant circuit with infinite input impedance ($\tan\frac{\pi}{2} \to \infty$) because of the resonance with the line capacitance resulting in a quarter-wavelength standing wave on the line.
4. For $\lambda/4 < l < \lambda/2$, the line behaves as a capacitance, being purely reactive and negative.

This completes the quantification of the transition from lumped element inductor to transmission line. The results are general, independent of the conductor shapes, although the specific case of a bifilar transmission line[2] will serve again in the comparison between loop and partial inductances of chapter 4.

To answer the question in chapter 1 and at the beginning of this chapter, namely 'What is the maximum length of antenna legs before the lumped element approximation breaks down?', we find that the antenna dimensions must be at least an order of magnitude smaller than the RF free-space wavelength, $\lambda_{\text{rf}} = c/f$ (see also appendix D.1.1). The wavelength at 13.56 MHz is 22.1 m, therefore beyond 2 m length, an electromagnetic treatment becomes necessary, although we will see in section 7.6.5 that coupling with a plasma makes this limit even more stringent. The full electromagnetic treatment of a large-area plasma reactor and a high-frequency inductive antenna probe is given in chapter 7, following on from these basic notions.

References

[1] Jin J-M 1998 *Electromagnetic Analysis and Design in Magnetic Resonance Imaging* (Boca Raton, FL: CRC Press)

[2] Bleaney B I and Bleaney B 1976 *Electricity and Magnetism* 3rd edn (London: Oxford University Press) W.1 p

[3] Paul C R 2010 *Inductance: Loop and Partial* (Hoboken, NJ: Wiley)

[2] For completeness' sake, a bifilar transmission line in vacuum has an inductance per unit length $\hat{L}^{\text{loop}} = \frac{\mu_0}{\pi}\ln\frac{s}{a}$, and a capacitance per unit length $\hat{C} = \pi\varepsilon_0/\ln\frac{s}{a}$, where s is the wire separation and a is the wire radius.

Chapter 3

Normal modes and dissipative networks

For the next step up in complexity from a parallel resonant circuit, this chapter presents an example of a planar resonant network antenna designed as an industrial prototype [1–4]. The aim is to perform an analytical analysis, as far as possible, of this real network without plasma, in order to compare with its measured resonance frequencies and distributions of current and voltage. We will find that an analytical solution is adequate for open networks, with good accuracy except for the precise values of the resonance frequencies; for the latter, the mutual partial inductances must be computed, as will be shown in chapter 4. The algebra is cumbersome, but the principal result is that the antenna input impedance can be described by a parallel resonant equivalent circuit close to each resonance frequency [5].

The experimental antenna is first described in section 3.1 to present the real circuit elements and connection configurations which will need to be taken into account. The first step in the analysis is to approximate the antenna as a lossless circuit in order to define the normal modes of the system in section 3.4. Next, a novel general solution is derived in section 3.5 for the currents in the driven dissipative network which includes component resistances. The final section, 3.5.6, shows that the input impedance of the complete antenna network can be approximated by a simple parallel resonant equivalent circuit in the neighbourhood of each resonance frequency. This is a convenient analytical aid for designing antennas with a desired resonance frequency and input impedance, as discussed for some applications.

3.1 Experimental set-up of the ladder antenna

To give a concrete example, figure 3.1 shows a labelled photograph of the principal antenna components, with the corresponding schematic circuit in figure 3.2.

The network consists of $N = 23$ water-cooled tubular copper legs, overall length $l_b = 20$ cm, external radius $a = 0.3$ cm, with parallel axes spaced 2.5 cm apart, for a

doi:10.1088/978-0-7503-5296-3ch3

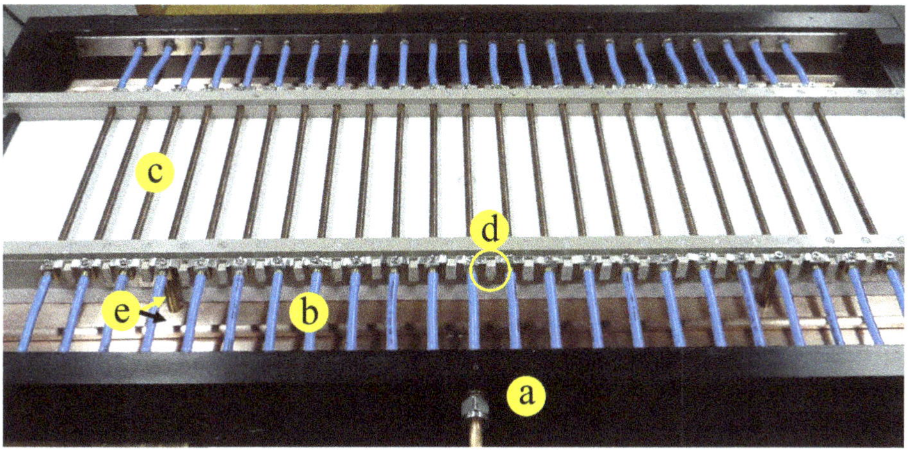

Figure 3.1. Photograph of a planar resonant antenna with 23 parallel legs, measuring 55 cm from left to right, and 22 cm across. The black external frame (a) carries cooling water which is injected via blue plastic tubes (b) through every tubular copper leg (c) of the antenna. The legs are joined together at each end by a capacitor (d) which is a stacked assembly of four high quality American Technical Ceramics (ATC) [6] ceramic capacitors. The antenna sits above a water-cooled copper baseplate ((e): black arrow) which provides a grounded reference plane for connector stubs to the antenna ((e): yellow arrow) and for the RF type N coaxial connector below the baseplate. For high power plasma operation (chapter 6), the whole assembly is embedded in thermally conducting silicone (not yet applied in the antenna shown here).

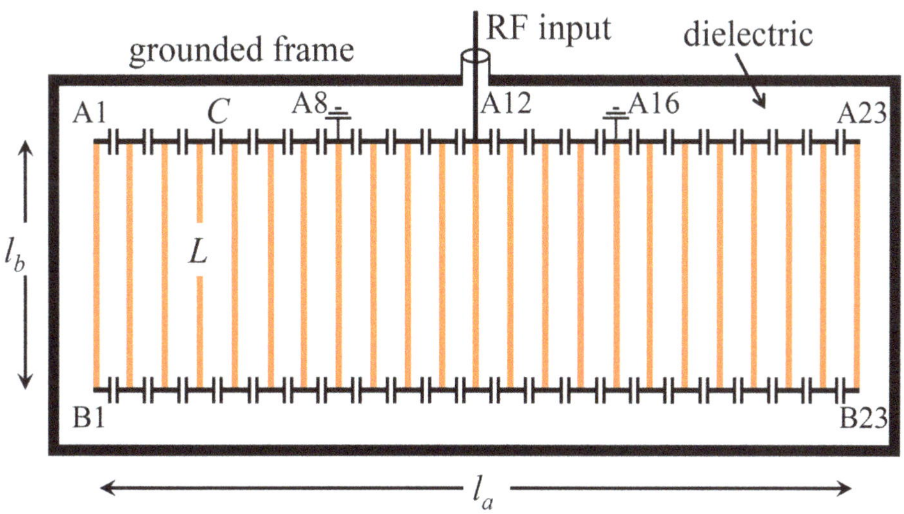

Figure 3.2. Schematic of the planar RF antenna showing the antenna copper bars of self inductance L, the capacitors C, the RF input connection at node number A12, and the ground connections at nodes A8 and A16, to promote mode m = 6; see section 3.4.

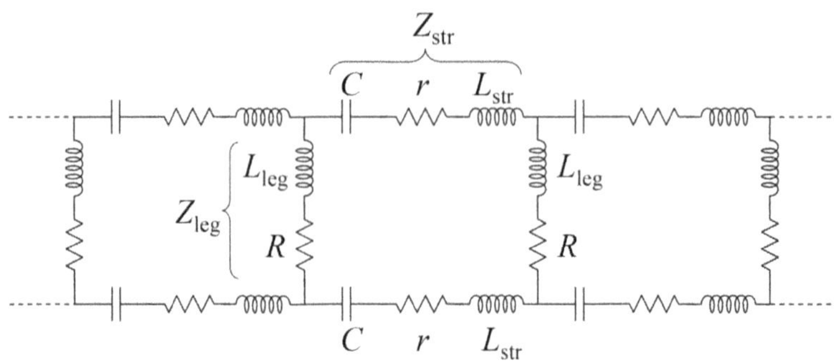

Figure 3.3. High-pass dissipative ladder network made up of identical meshes, showing the leg self partial inductances L_{leg} with resistance R, linked by stringer segments of capacitance C having short connections of self partial inductance L_{str} and resistance r. (Adapted with permission from [5]. Copyright 2014 IOP Publishing.)

total stringer length of $l_a = 55$ cm. For these short lengths, a lumped element model is appropriate for the leg self partial inductance L around 13.56 MHz. By the same token, capacitive coupling between the copper legs is negligible. Accounting for the RF current path along the skin depth of the conductors, the effective leg length is $l_{leg} = 19$ cm (see section 4.4.1). The legs are connected at both ends by stringers which consist of 2.6 nF ceramic capacitor assemblies (see section 13.2) with copper strip connectors $l_{str} = 1.9$ cm long and $w = 0.6$ cm wide. In figure 3.1, the antenna assembly is set in an open top metal housing with the axes of the antenna leg network 5.5 cm above the grounded baseplate. The RF input power is centrally connected at node A12 in figure 3.2, with internal ground connections at A8 and A16. The screen of the RF input coaxial cable is the only ground connection to the external RF power supply and matching circuit, to avoid any possibility of spurious current in ground loops.

Finally, to ensure that the analytical model is as general as possible, the equivalent circuit of the real antenna in figure 3.3 includes the small effective self partial inductance of the stringer segments, L_{str} calculated (chapter 4) to be $\simeq$ 10.5 nH, as well as $L_{leg} \simeq 146$ nH and their respective resistances r and R. To a first order approximation for the model of a dissipative antenna, therefore, the stringer impedances are

$$Z_{str} = -j/(\omega C) + j\omega L_{str} + r, \tag{3.1}$$

and the leg impedances are

$$Z_{leg} = R + j\omega L_{leg}. \tag{3.2}$$

Note that here we simply use self partial inductance with no distinction with respect to conventional inductance; this is valid in the present context where no mutual partial inductance is considered. This completes the circuit description of the antenna in figures 3.1, 3.2, and 3.3.

To experimentally characterize the RF properties of the antenna in air (i.e. without plasma), its input impedance spectrum was measured by a vector network

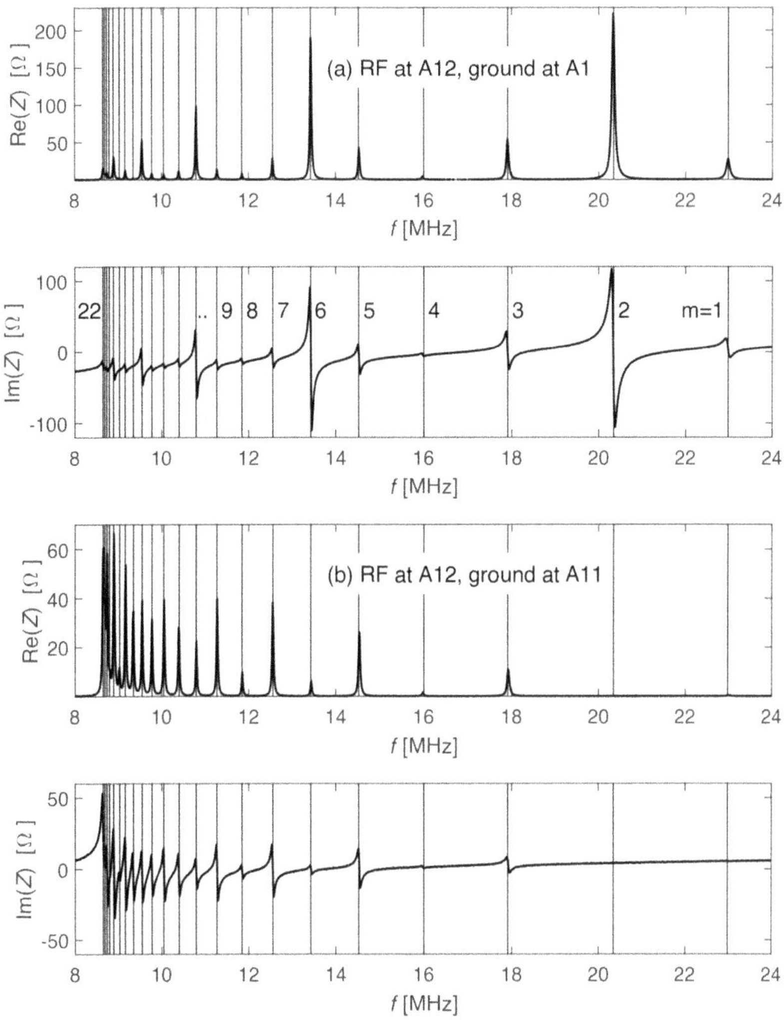

Figure 3.4. Real and imaginary parts of the network impedance measured by a vector network analyser. Referring to figure 3.2, the RF input is centrally placed at node A12, but here there is only one ground connection which is either (a) at the end of the antenna on node A1; or (b) on the node A11, adjacent to the RF input. The mode numbers m = 1–22 were identified by counting downwards from the highest resonance frequency (m = 1), and by verifying the normal mode structures (section 3.4.4).

analyser (Rohde & Schwarz ZVL 9 kHz–6 GHz) for various ground connection positions, with the RF input fixed centrally at A12. Two examples are shown in figure 3.4 for the ground at node A1 in (a), and for the ground position at A11 in (b). The real and imaginary parts of the input impedance show the complexity of the network frequency response, but qualitative comparison with figure 2.4(a) suggests a sequence of parallel resonances at different resonance frequencies; a mathematical confirmation of this is derived in section 3.5. Note also that the resonance frequencies are apparently identical for the two RF connection positions—only the relative amplitudes are changed. These impedance spectra will be deciphered step by step in this chapter and in chapter 4. Further measurements used a variable frequency RF generator to excite the antenna while the spatial distribution of network node voltages A_n and B_n were measured referenced to the ground plane, and a magnetic pickup probe monitored the relative distribution of currents in the legs. These measurements will be shown below in section 3.4.4 for comparison with models.

3.2 Introduction to normal modes

A general definition of a normal mode is an oscillation of a lossless system in which all parameters oscillate sinusoidally with the same frequency and with a fixed phase relation. This free (lossless) motion occurs at fixed frequencies known as the natural, normal mode frequencies, or resonance frequencies. A normal mode can exist alone or superposed with others; the most general oscillation of a system is a superposition of its normal modes (by excitation at multiple frequencies, or an impulse such as for the strings of a piano or guitar). The modes are 'normal' in the sense that they can be excited independently of each other, i.e. excitation of one mode will not excite another mode. In mathematical terms, the normal modes are orthogonal.

In the case of lossless resonant network antennas, a normal mode[1] 'm' is represented as a spatial distribution of oscillating current and voltage which occurs at its specific resonance frequency ω_{m}. A trivial example was given for a lossless $L\|C$ parallel resonant circuit in section 2.2.1. The calculation of normal modes for closed or open antennas is distinguished only by the boundary conditions inherent to their structure (chapter 9). In the lossless antennas described by J-M Jin [7], the voltage and current distribution of an excited normal mode is independent of where any infinitesimal excitation is applied. However, this will not be the case when dissipation occurs because of power transmission along the antenna away from the RF power source connection. In contrast to lossless antennas, it is then necessary to account for current injection by the RF power source, ohmic losses in the antenna components, and power transmitted to an inductively coupled plasma. The following treatment of normal modes will therefore be adapted to pave the way for the dissipative analysis [5].

3.3 General solution for the network currents

As a first step, figure 3.5 shows an isolated, electrically floating segment of a ladder network of N legs with generalized stringer impedances Z_{str} and leg impedances Z_{leg}. There is no consideration of voltage input nor ground points for the moment. I_n, J_n, and K_n are, respectively, the current phasors in the legs, the top stringer, and the bottom stringer, while A_n and B_n denote the voltage phasors at the top and bottom stringer nodes labelled by $n = 1, 2, \ldots, N$. Ladder networks are conventionally analysed using a circulating current description [7], where the circulating current i_n

[1] We use roman (normal) font for the mode number m to distinguish it from the azimuthal periodicity m conventionally used in cylindrical coordinates, for example, in chapter 8.

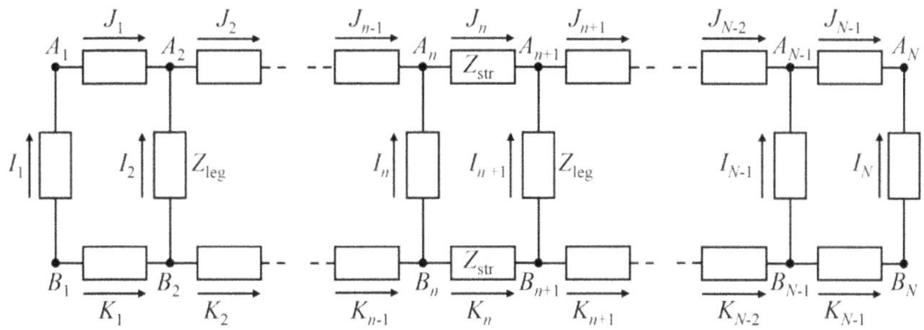

Figure 3.5. A ladder network segment with N legs made up of generalized stringer impedances Z_{str} and leg impedances Z_{leg}. A and B are the node voltages in the top and bottom stringers; I, J, and K are, respectively, the currents in the legs, the top stringer, and the bottom stringer. (Adapted with permission from [5]. Copyright 2014 IOP Publishing.)

in each section (not shown) is given by $i_n = J_n = -K_n$, and $I_n = i_n - i_{n-1}$. However, as mentioned at the end of section 3.2, to treat the general case of a dissipative driven antenna in section 3.5, at some point we will have to introduce the RF input current I_{rf}, driven by the antenna RF power supply, into the circuit analysis. This injection of RF current means that $J_n = -K_n$ will not hold at every node junction on the antenna, and so the circulating current description cannot be used for the dissipative driven antenna; only for lossless antennas. For consistent notation throughout this work, we will therefore use independent currents K_n and J_n instead of circulating currents i_n. A derivation of the normal mode frequencies and current distributions is now reviewed briefly to introduce the notation and to give a reference point of comparison for the general solution for dissipative antennas in section 3.5.

For the currents and impedances defined in the nth elementary mesh of figure 3.5, the general set of equations for the circuit analysis is given by Kirchhoff's and Ohm's laws for the currents and voltages,

$$J_n = J_{n-1} + I_n, \tag{3.3}$$

$$K_n = K_{n-1} - I_n, \tag{3.4}$$

$$Z_{leg}I_n + Z_{str}J_n - Z_{leg}I_{n+1} - Z_{str}K_n = 0. \tag{3.5}$$

A recurrence relation for the leg currents I_n is obtained by subtracting the voltage equation for the $(n-1)$th mesh from the nth mesh voltage equation in (3.5), and eliminating J_n and K_n using (3.3) and (3.4) to obtain

$$I_{n+1} - 2\left(1 + \frac{Z_{str}}{Z_{leg}}\right)I_n + I_{n-1} = 0. \tag{3.6}$$

The conventional solution is $I_n = a_1(\Gamma_1)^n + a_2(\Gamma_2)^n$, where Γ_1 and Γ_2 are the roots of the quadratic characteristic equation which can be arranged as

$$\frac{\Gamma + 1/\Gamma}{2} = \left(1 + \frac{Z_{str}}{Z_{leg}}\right). \tag{3.7}$$

Recognizing that these roots are reciprocal, we write $\Gamma_1 = e^{\gamma}$ and $\Gamma_2 = e^{-\gamma}$, whereupon the characteristic equation becomes

$$\cosh\gamma = \left(1 + \frac{Z_{\text{str}}}{Z_{\text{leg}}}\right), \tag{3.8}$$

and the general solution for the antenna leg currents is therefore

$$I_n = a_1 e^{\gamma n} + a_2 e^{-\gamma n}, \tag{3.9}$$

where a_1 and a_2 are complex constants to be determined from boundary conditions. The solution given by (3.8) and (3.9) holds for any impedances Z_{str} and Z_{leg}, and is therefore valid for dissipative and lossless antennas alike, and furthermore, can also be applied to low-pass networks or any hybrid combination of lumped element impedances. For harmonic currents with $e^{-j\omega t}$ time dependence, the leg currents are $I_n e^{-j\omega t}$ so that the general solution is seen to consist of a forward travelling wave $a_1 \exp(\gamma n - j\omega t)$ and a backward travelling wave $a_2 \exp(-\gamma n - j\omega t)$, where $\gamma = \alpha + j\beta$ is the (complex) propagation constant, e^{α} is the phasor amplitude ratio per section, and β is the phase change per section (α and β are both real). For comparison with waves on the continuous transmission line in section 2.3, β, the phase change per section along the antenna, takes the place of the wavenumber k, which can be seen as the phase change per unit distance along a transmission line, $d\phi/dz = 2\pi/\lambda = k$.

3.3.1 Characteristic equation on lossless networks

Normal modes apply to lossless networks, therefore in figure 3.3 we consider a non-dissipative ladder antenna with purely imaginary impedances $Z_{\text{str}} = 1/(j\omega C) + j\omega L_{\text{str}}$ and $Z_{\text{leg}} = j\omega L_{\text{leg}}$. The impedance ratio in (3.8) is now purely real, so the propagation constant corresponds to $\gamma = j\beta$ in the characteristic equation for lossless networks (3.8):

$$\cosh\gamma = \cos\beta = 1 + \frac{L_{\text{str}}}{L_{\text{leg}}} - \frac{1}{\omega^2 L_{\text{leg}} C}. \tag{3.10}$$

The network normal mode frequency ω is found below by re-arranging this characteristic equation, and finding β using the appropriate boundary conditions. For this special case of a lossless antenna, there is no injection of RF current to consider, so $J_n = -K_n$ for all n. Substituting into equations (3.3) to (3.5) gives the same recurrence relation as for I_n. Hence, for normal modes on open or closed lossless networks, the stringer currents are given by

$$J_n = -K_n = b_1 e^{j\beta n} + b_2 e^{-j\beta n}, \tag{3.11}$$

where b_1 and b_2, as well as β, depend on the antenna boundary conditions. The connection between $a_{1,2}$ in (3.9) and $b_{1,2}$ in (3.11) is given by the recurrence relation (3.3).

3.4 Normal mode solution for open networks

Every ladder resonant network necessarily has an open configuration, whereas birdcage antennas can be open or closed (see section 1.2.1). Conventional birdcages have a closed configuration, so a discussion of closed networks is postponed until chapter 9 on birdcage antennas, section 9.2.

3.4.1 Normal mode frequencies on open networks

All of the preceding equations apply to networks in general, whether open or closed. To determine the remaining unknown values of β, b_1, and b_2, it is now necessary to apply the boundary conditions specific to the network in hand. With reference to chapter 1, we recall that ladder antennas, and birdcage antennas missing at least one capacitor, are both open networks because their extremities are on open circuit.

Considering figure 3.5, obvious boundary conditions at both ends of the ladder would be $I_1 = J_1$, and $I_N = -J_{N-1}$. However, these can equivalently be seen as zero current in virtual additional stringers added at $n = 0$ and $n = N$ (not shown in the figure), i.e. $J_0 = 0$ and $J_N = 0$, respectively, which is mathematically more straightforward. Substituting $J_0 = 0$ in (3.11), we have $b_1 = -b_2$, and the currents J_n in the non-dissipative ladder network are therefore proportional to $(e^{j\beta n} - e^{-j\beta n})$, i.e. proportional to $\sin(n\beta)$. The finite length condition imposed by $J_N = 0$ further implies that $\sin(N\beta) = 0$, i.e. $N\beta = \mathrm{m}\pi$, where $\mathrm{m} = 1, 2, \ldots, N-1$ because there is no excited current ($J_n = 0$ for all n) for $\mathrm{m} = 0$ and $\mathrm{m} = N$ [7]. Substituting $\beta = \frac{\mathrm{m}\pi}{N}$ into (3.10), leads to the existence of normal mode frequencies [7]

$$\cos\frac{\mathrm{m}\pi}{N} = 1 + \frac{L_{\mathrm{str}}}{L_{\mathrm{leg}}} - \frac{1}{\omega_{\mathrm{m}}^2 L_{\mathrm{leg}} C},$$
$$\omega_{\mathrm{m}} = \frac{1}{\sqrt{C\left(L_{\mathrm{str}} + 2L_{\mathrm{leg}}\sin^2\left(\frac{\mathrm{m}\pi}{2N}\right)\right)}}, \quad \mathrm{m} = 1, 2, \ldots, (N-1), \tag{3.12}$$

where the trigonometric identity $\cos\theta \equiv 1 - 2\sin^2\frac{\theta}{2}$ has been used for simplification, and m is defined as the 'mode number' of the open network[2].

Normal modes are the network analog of the $L\|C$ parallel resonant circuit's trivial single mode in chapter 2. To stretch an analogy perhaps too far, the simplest network corresponds to a single two-leg segment, $N = 2$, which also has a single mode $\mathrm{m} = 1, \ldots, N-1 = 1$. Substituting $N = 2$, $\mathrm{m} = 1$ in (3.12) gives $\omega_1 = 1/\sqrt{L_{\mathrm{leg}}C}$, the same as the trivial $L\|C$ case, provided that L_{str} is removed. For any value of N, note that mode $\mathrm{m} = 1$ has the highest frequency, and $\mathrm{m} = N-1$ has the lowest, with the normal mode frequencies crowding towards the low frequency limit as m increases (the order would be reversed for a low-pass network).

[2] If the ladder network is wrapped around to form a *closed* birdcage (closed by adding a supplementary stringer capacitor between the first and last legs), then the mode number m and the azimuthal periodicty *m* are, in fact, the same in this case; see chapter 9 and footnote 1 here.

There is no relation here to harmonics which are multiples of a fundamental frequency; in the present case, each mode m is a normal mode independent of the others.

3.4.2 Normal mode currents on open networks

The spatial distribution of the stringer currents $J_n = -K_n$, already imposed by the above boundary conditions, is

$$J_n = -K_n \propto \sin\left(n\frac{\mathrm{m}\pi}{N}\right), \qquad \mathrm{m} = 1, 2, \ldots, (N-1), \tag{3.13}$$

where the phase shift per section $\beta = \frac{\mathrm{m}\pi}{N}$ is proportional to the mode number m. This is analogous to the normal modes of a finite length of string, pinned at both ends, which supports an integral number of half-wavelengths. Using this familiar example, the amplitude of a standing wave on the string is proportional to $\sin\left(x\frac{2\pi}{\lambda}\right)$, where x is the distance along the string, and λ is the standing wave wavelength. Equating with $\sin\left(n\frac{\mathrm{m}\pi}{N}\right)$, the wavelength in terms of the number of legs is $\lambda = \frac{2N}{\mathrm{m}}$. Equivalently, the number of wavelengths across the antenna is $\frac{N}{\lambda} = \frac{\mathrm{m}}{2}$, i.e. an integral number of half-wavelengths, where the number equals the mode number, m.

The normal mode current phasor I_n in each leg, using (3.13) in (3.3) and $(\sin\theta - \sin\phi) \equiv 2\cos\frac{1}{2}(\theta+\phi)\sin\frac{1}{2}(\theta-\phi)$, is given by [7]

$$I_n \propto 2\sin\left(\frac{\mathrm{m}\pi}{2N}\right)\cos\left[\left(n-\frac{1}{2}\right)\frac{\mathrm{m}\pi}{N}\right], \qquad \mathrm{m} = 1, 2, \ldots, (N-1). \tag{3.14}$$

The spatial variation of the leg current phasor, for $n = 1$ to N across the antenna, depends only on the cos term in (3.14), and is shown in figure 3.6 for m = 1, 2, and 6.

Figure 3.6. The fixed spatial distribution of leg current phasor amplitudes I_n on a 23-leg ladder antenna for the normal modes m = 1 (black diamonds), m = 2 (red circles), and m = 6 (blue squares), calculated using (3.14). The leg currents oscillate in phase at frequency ω_m, bounded by the amplitude. The symbols mark the current amplitude in each leg; the cosine lines guide the eye to indicate the mode structure [2]. The number of wavelengths across the antenna is m/2.

To recap so far, a N-leg lossless open network presents $(N - 1)$ normal modes [7] whose current amplitudes have sinusoidal spatial distributions and which oscillate in phase. There is one normal mode current distribution for each normal mode frequency; this may seem obvious, but it is not the case for degenerate modes on closed networks, see section 9.2. The number of wavelengths across an open network is m/2. The currents in the legs for the m = 1 mode constitute a half wavelength over the antenna length, while three wavelengths occur for the mode m = 6.

3.4.3 Normal mode voltages on open networks

The node voltages are useful to gauge the relative level of excitation of different normal modes depending on the particular location of the RF and ground connectors. They can be calculated beginning with Ohm's law across each impedance in the top and bottom stringers, referring to figure 3.5:

$$A_n - A_{n+1} = Z_{\text{str}} J_n, \tag{3.15}$$

$$B_n - B_{n+1} = Z_{\text{str}} K_n. \tag{3.16}$$

The sum of (3.15) and (3.16) is $(A_n + B_n) - (A_{n+1} + B_{n+1}) = Z_{\text{str}}(J_n + K_n) = 0$ for normal modes (section 3.4.1), whence the sum of voltages at the nth node is constant:

$$B_n + A_n = V_0, \tag{3.17}$$

where V_0 is a constant amplitude (remember that the time oscillation $e^{-j\omega t}$ is implicit for linear combinations of phasors from section 2.1.1). The remaining Ohm's law relation is across the legs,

$$B_n - A_n = j\omega_{\text{m}} L_{\text{leg}} I_n, \tag{3.18}$$

giving a pair of simultaneous equations whose solution is in terms of I_n (3.14) and constants. The latter can be defined by choosing a pair of nodes on the top stringer in figure 3.5 to be the RF feed-point, $A_{N_f} = V_{\text{rf}}$, and a ground connection, $A_{N_g} = 0$, respectively. Note that, although we have defined these RF connection nodes, the antenna can still be treated as a single network because no input/output currents (no dissipation) are to be considered here. Finally, the solution for the normal mode voltage phasors, when excited at a normal mode frequency ω_{m}, is as follows:

$$A_n = \frac{V_{\text{rf}}}{D}\left(\cos\left[\left(n - \frac{1}{2}\right)\frac{\text{m}\pi}{N}\right] - \cos\left[\left(N_g - \frac{1}{2}\right)\frac{\text{m}\pi}{N}\right]\right), \tag{3.19}$$

$$B_n = -\frac{V_{\text{rf}}}{D}\left(\cos\left[\left(n - \frac{1}{2}\right)\frac{\text{m}\pi}{N}\right] + \cos\left[\left(N_g - \frac{1}{2}\right)\frac{\text{m}\pi}{N}\right]\right), \tag{3.20}$$

where D is a denominator term dependent on the RF and ground locations,

$$D = \cos\left[\left(N_f - \frac{1}{2}\right)\frac{\text{m}\pi}{N}\right] - \cos\left[\left(N_g - \frac{1}{2}\right)\frac{\text{m}\pi}{N}\right]. \tag{3.21}$$

The time-dependent node voltages are given by multiplying the voltage phasors by $e^{-j\omega_{\mathrm{m}}t}$ and taking the real part, i.e. $A(n, t) = \mathrm{Re}\,[A_n e^{-j\omega_{\mathrm{m}}t}] = A_n \cos(\omega_{\mathrm{m}}t)$ and similarly for $B(n, t)$.

The constants of proportionality for the currents in (3.13) and (3.14) are calculable only now, because we chose first to define the input voltage instead of an input current. Using (3.18), the corresponding leg and stringer current phasor amplitudes are

$$I_n = \frac{2jV_{\mathrm{rf}}}{D\omega_{\mathrm{m}}L_{\mathrm{leg}}} \cos\left[\left(n - \frac{1}{2}\right)\frac{\mathrm{m}\pi}{N}\right], \tag{3.22}$$

$$J_n = -K_n = \frac{jV_{\mathrm{rf}}}{D\omega_{\mathrm{m}}L_{\mathrm{leg}}} \frac{\sin\left(n\frac{\mathrm{m}\pi}{N}\right)}{\sin\left(\frac{\mathrm{m}\pi}{2N}\right)}. \tag{3.23}$$

The time-dependent leg and stringer currents are again given by multiplying the voltage phasors by $e^{-j\omega_{\mathrm{m}}t}$ and taking the real part, but there is a phase change of $\pi/2$ relative to the voltages due to the factor j, hence

$$I(n, t) = \frac{2V_{\mathrm{rf}}}{D\omega_{\mathrm{m}}L_{\mathrm{leg}}} \cos\left[\left(n - \frac{1}{2}\right)\frac{\mathrm{m}\pi}{N}\right] \sin(\omega_{\mathrm{m}}t), \tag{3.24}$$

$$J(n, t) = -K(n, t) = \frac{V_{\mathrm{rf}}}{D\omega_{\mathrm{m}}L_{\mathrm{leg}}} \frac{\sin\left(n\frac{\mathrm{m}\pi}{N}\right)}{\sin\left(\frac{\mathrm{m}\pi}{2N}\right)} \sin(\omega_{\mathrm{m}}t). \tag{3.25}$$

The normal modes are now completely defined in terms of their node voltages and component currents [2].

3.4.4 Comparison with measurements on the ladder antenna

We are now in a position to compare measurements of the real antenna with the normal mode solution. To measure the antenna voltages and currents, a variable frequency, low power RF generator was used to excite different modes of the antenna; for example, for the measurements in figure 3.7, the mode m = 6 was excited at 13.423 MHz. The node voltage amplitudes A_n and B_n were measured one-by-one on the antenna in air, before filling with a dielectric; see section 13.4.2. A high impedance 10× oscilloscope probe was used, whose ground reference was placed at the closest possible point on the metal ground plane to minimize stray inductance errors due to the loop area of the probe wires. The relative current in the legs was measured by a magnetic pickup probe consisting of a Rogowski coil connected to a 50 Ω oscilloscope input. The Rogowski was wound on a half-ring former, designed to fit snugly around the leg tubes to ensure a reproducible measurement position from leg to leg. The probe was carefully screened to avoid pickup from neighbouring legs.

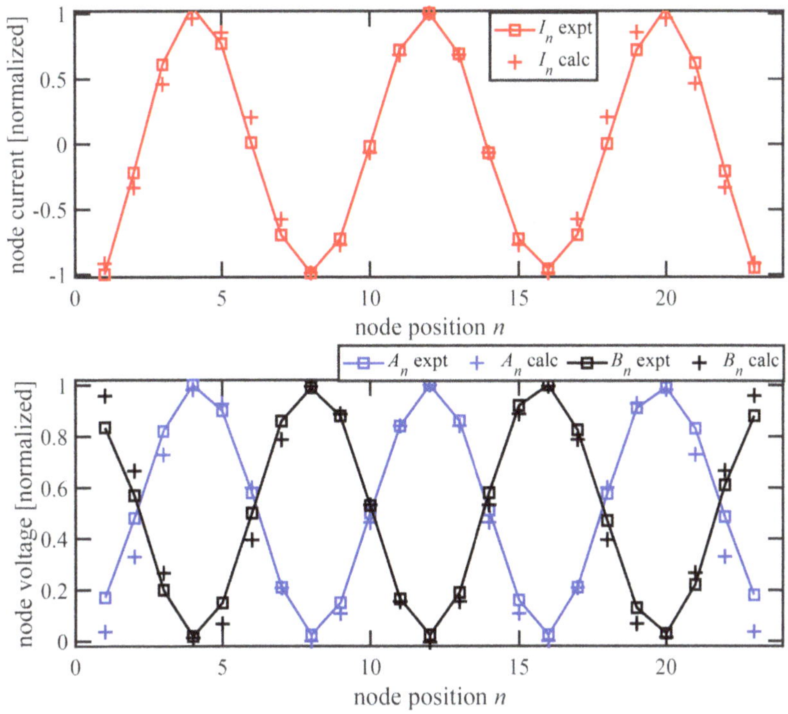

Figure 3.7. Top: Measured (red line and squares) and calculated (red crosses) leg currents divided by the central leg current in the 23 legs of the planar RF antenna for mode m = 6. Bottom: Measured (blue line and squares) and calculated (blue crosses) A_n top node voltages divided by the voltage $A12$ at the RF input. For the B_n bottom node voltages, the measured (black line and squares) and calculated (black crosses) voltages are for the same conditions. Ground nodes on A8 and A16. Compare also with figure 4.18.

Figure 3.7 shows good correspondence between the measured leg currents and the calculated normal mode current distribution (3.14) for the antenna excited in the mode m = 6 at 13.423 MHz. Small discrepancies are observed in the outer parts of the antenna, which are attributed to the neglect of the mutual inductance between the antenna elements; the outer elements have fewer neighbours than the central elements, which renders the inductive coupling dependent on position along the network. Fair agreement, with discrepancies for the outer legs, is also visible in figure 3.7 for the measured and calculated voltages (3.19) and (3.20) at the A_n and B_n nodes. As expected from the normal mode calculations, these current and voltage distributions are characterized by three wavelengths over the antenna length and are symmetrical about the central leg. Mutual partial inductance is fully accounted for in chapter 4, where the discrepancies are mostly eliminated; see figure 4.18.

Two impedance spectra of the antenna measured by a network analyser are shown in figure 3.4. From section 2.2.2, the resonance frequencies are defined by zero imaginary impedance, and very closely approximated by the peaks in the real impedance. From normal mode theory, the expected number of resonance frequencies is $(N - 1) = 22$. Figure 3.4(a) shows predominantly high frequency, low-m

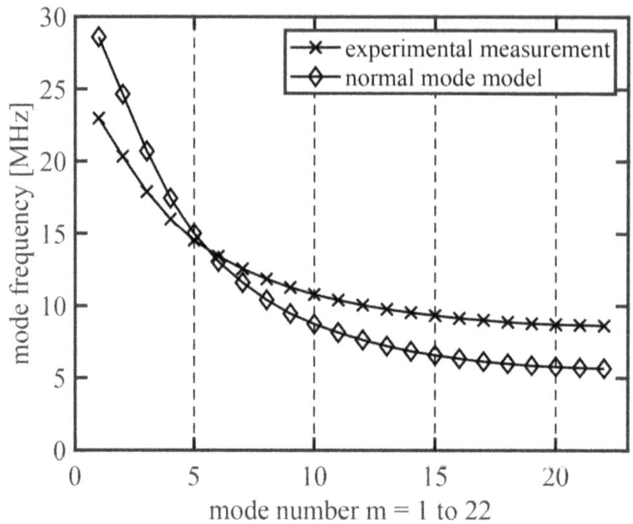

Figure 3.8. The normal mode frequencies for m = 1 − 22, calculated using (3.12) for the 23-leg resonant network, compared with the resonance frequencies measured from the impedance spectra, shown by the set of 22 vertical lines in figure 3.4. See also figure 4.21.

modes because the RF and ground connections are widely spaced (A12 and A1), thereby principally exciting long wavelength modes. At the other extreme, the spectrum in figure 3.4(b) is dominated by modes crowding to the low frequency limit of high-m modes, because short wavelengths are preferentially excited by the closely spaced adjacent RF and ground connections, A12 and A11. Thus, by acquiring the impedance spectra for a range of RF-to-ground node spacing, all 22 mode frequencies can be unambiguously identified and accurately measured. In figure 3.8, the measured mode frequencies are compared with the normal mode frequencies (3.12). The frequency trends with m are qualitatively similar, but the agreement is far worse than for the current and voltage mode spatial structures in figure 3.7. This is because the resonance frequencies are determined by electrical parameters as yet unaccounted for, such as the mutual inductance between the antenna elements themselves, and with image currents induced in the grounded baseplate [7], whereas the mode structure is principally defined by the antenna's physical geometry. The net effect is that the measured frequencies in figure 3.8 are 'compressed' into a smaller range of frequency. Mutual inductance and the precise correction to the calculated mode frequencies will be readdressed more successfully in chapter 4, figure 4.21.

The selective excitation of low-m or high-m modes in figure 3.4 was obtained by changing the ground node position A_{N_g} from $A1$ to $A11$. The influence of the excitation positions on normal mode voltage amplitude can also be seen mathematically in (3.19)–(3.21). Unsurprisingly, the measured impedance for each mode also depends on the configuration connection. The m = 6 mode of figure 3.7 has a particularly high real input impedance peak in figure 3.9(a) at 13.423 MHz (near to the industrial ISM frequency 13.56 MHz, by design choice of the capacitor value)

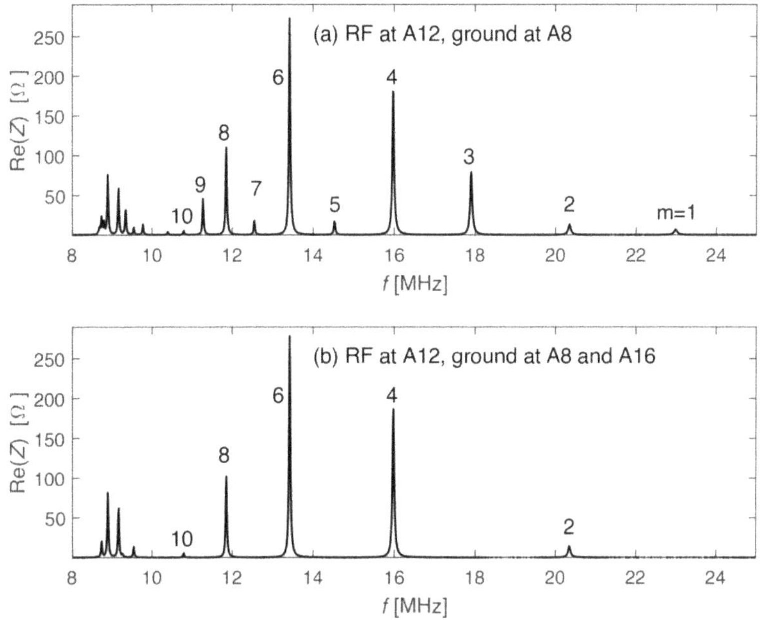

Figure 3.9. Real part of the network impedance measured with the RF input placed centrally at node A12, and the ground connection (a) only at node A8; (b) at nodes A8 and A16 (refer to figure 3.2). The mode m = 6 at 13.423 MHz is better isolated and stabilized by the imposition of its specific mode structure symmetry by the double grounding A8 and A16 in (b), because all odd-m modes are suppressed. See also figure 4.20.

for the ground node at A8 and the RF central node, A12. The RF input current is therefore particularly low for m = 6; this is crucial for suppressing perturbations to the uniformity of normal mode current and voltage distributions, because the normal mode condition implies zero input current. The asymmetric position of the single ground node A8, with respect to the central RF input in figure 3.9(a), means that low amplitude odd-m modes appear in the impedance spectrum. However, in figure 3.7, we see that A16 is also a zero-voltage node by symmetry of mode m = 6. By grounding A16 as well as A8, the odd modes of resonance are forbidden by symmetry arguments, as shown in figure 3.9(b). Additional ground connections, such as B4 and B20 [2] in figure 3.7, over-constrain the system and bring no further advantage of uniformity nor stability. Experimentally, then, it is found that the best results are obtained for a central RF input with two grounds symmetrically positioned on the same side of the antenna (side *A*), or on the opposite side (side *B*). Optimal mode stability could be important for maintaining uniform plasma, because the induced currents will depend on the electron density and temperature profiles. This, then, explains the experimental RF and ground configuration chosen in figures 3.1 and 3.2.

Clearly, the measured impedances cannot be compared with the normal mode model because it assumes ideal lossless components, zero input current, and hence infinite impedance at resonance (see section 2.2.1). A first comparison with the measured impedance values will depend on a dissipative model in section 3.5.

3.5 Dissipative networks: Helyssen plasma sources

In the next step towards a model of a real antenna, we now generalize the normal mode model by adding dissipation [5] (figure 3.10). In contrast to normal modes on lossless networks [7], we define a Helyssen planar resonant antenna as a driven, dissipative ladder network with arbitrary feed-point positions for the RF power connections as shown in figure 3.11.

A real antenna dissipates power due to the electrical resistance of its components. Furthermore, when used as a plasma source, power injected into the antenna is partially transferred to the plasma by inductive coupling to the dissipative medium. The question then arises as to whether the normal mode properties of non-dissipative antennas are preserved to efficiently maintain a stable, uniform plasma source. Plasma coupling is postponed until chapter 6. In the following, only the resistances R and r are responsible for the power dissipation, and the lumped element impedances are now $Z_{\rm str} = r - j/(\omega C) + j\omega L_{\rm str}$ (3.1), and $Z_{\rm leg} = R + j\omega L_{\rm leg}$ (3.2), as in figure 3.3.

This section considers the following points for antenna design:

1. An estimation of perturbations to the spatial uniformity of currents and voltages due to the injected RF driving current (section 3.5.2).
2. Calculation of the antenna input impedance spectrum in terms of the antenna components $L_{\rm leg}$, R, C, $L_{\rm str}$, r and the RF connection node positions N_f and N_g (section 3.5.4).
3. The possibility of describing the dissipative antenna using an equivalent circuit with analytical expressions for the equivalent circuit components (section 3.5.6).
4. The optimum design of the connection configuration to optimize the antenna input impedance (section 3.5.7).

3.5.1 Currents in a dissipative antenna with arbitrary RF connections

Because of current injection $I_{\rm rf}$ from the RF power supply in figure 3.11, a single solution for the whole antenna, in the style of the normal modes, can no longer be

Figure 3.10. 'Normal dissipated mode.' (Illustration by Alex Howling. Copyright 2023 Alex Howling.)

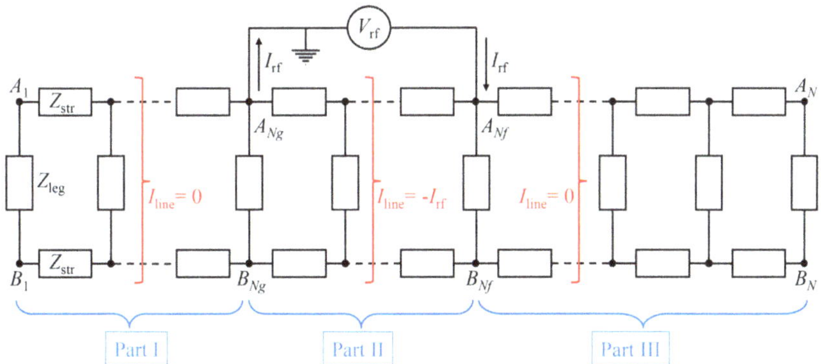

Figure 3.11. Schematic circuit of a Helyssen driven ladder antenna showing the three parts I, II, and III with their net current I_{line} along the ladder. The driving current I_{rf} from the RF power supply is fed into the antenna via node $n = N_f$ with the return path via node N_g. Node N_g is arbitrarily defined as ground to be consistent with sections 3.1 and 3.4. The boundaries between the three parts are defined by the nodes N_g and N_f, where $N_g < N_f$ in this work.

found. Instead, separate solutions have to be determined for each part of the antenna, respecting the boundary conditions at the nodes of RF current injection, N_f and N_g. The symbols for stringer voltages, and stringer and leg currents, are the same as for figure 3.5.

For a given antenna part in figure 3.11, the total net current along the line from left to right is $I_{line} = J_n + K_n$. Adding (3.3) and (3.4), by current continuity, we now have

$$I_{line} = J_n + K_n = J_{n-1} + K_{n-1}, \tag{3.26}$$

which must be constant along a line segment if not interrupted by a current injection node. For current continuity, referring to figure 3.11, $I_{line} = 0$ in parts I and III because of their open circuit boundary condition, the same as for normal modes. However, $I_{line} = -I_{rf}$ in part II of the line which carries the RF antenna input current.

For antenna impedance calculations, it will also be necessary to consider the RF voltages A_n and B_n at the antenna nodes in figure 3.11. Recallng the previous equations (3.15) to (3.18) for convenience, the voltages across the antenna impedances, by Ohm's law, are

$$A_n - A_{n+1} = Z_{str} J_n, \tag{3.27}$$

$$B_n - B_{n+1} = Z_{str} K_n, \tag{3.28}$$

$$B_n - A_n = Z_{leg} I_n. \tag{3.29}$$

The sum of (3.27) and (3.28) is $(A_n + B_n) - (A_{n+1} + B_{n+1}) = Z_{str}(J_n + K_n) = Z_{str} I_{line}$ from which the sum of voltages at the nth node can be written as

$$B_n + A_n = V_0 - n Z_{str} I_{line}, \tag{3.30}$$

where V_0 is a different constant amplitude phasor for each part. Using (3.29) and (3.30), the node voltages in terms of the dissipative antenna leg currents I_n are

$$A_n = \left[V_0 - nZ_{\rm str}I_{\rm line} - Z_{\rm leg}I_n \right] / 2, \tag{3.31}$$

$$B_n = \left[V_0 - nZ_{\rm str}I_{\rm line} + Z_{\rm leg}I_n \right] / 2, \tag{3.32}$$

where $I_{\rm line} = 0$ for parts I and III, and $I_{\rm line} = -I_{\rm rf}$ in part II as above. The stringer currents are then found using (3.27) and (3.28):

$$J_n = \left[I_{\rm line} + \frac{Z_{\rm leg}}{Z_{\rm str}}(I_{n+1} - I_n) \right] \Big/ 2, \tag{3.33}$$

$$K_n = \left[I_{\rm linc} - \frac{Z_{\rm leg}}{Z_{\rm str}}(I_{n+1} - I_n) \right] \Big/ 2, \tag{3.34}$$

where it can be seen that, in part II of the antenna, the oscillating RF current contribution $I_{\rm line} = -I_{\rm rf}$ is shared equally between the two stringers.

3.5.2 The leg currents in the three parts of the dissipative antenna

The three solutions for the currents in the dissipative antenna parts are obtained by calculating the constants a_1 and a_2 for the currents I_n in (3.9) for each part of the antenna, where $\gamma = \alpha + j\beta$ is the (complex) propagation constant (3.8). These six constants are found using the boundary conditions at both ends of the ladder (as in section 3.4) and by respecting current continuity between the different parts at nodes $n = N_g$ and $n = N_f$. After tedious algebra, the leg currents for the general configuration of figure 3.11 are

$$I_n^{\rm I} = I_{\rm rf}F_{(\gamma,N)}[\cosh g_1 + \cosh g_{2a} - \cosh g_3 - \cosh g_{4a}], \tag{3.35}$$

$$I_n^{\rm II} = I_{\rm rf}F_{(\gamma,N)}[\cosh g_1 + \cosh g_{2b} - \cosh g_3 - \cosh g_{4a}], \tag{3.36}$$

$$I_n^{\rm III} = I_{\rm rf}F_{(\gamma,N)}[\cosh g_1 + \cosh g_{2b} - \cosh g_3 - \cosh g_{4b}], \tag{3.37}$$

$$F_{(\gamma,N)} = \frac{\tanh(\gamma/2)}{2\sinh(\gamma N)}, \tag{3.38}$$

where the arguments of the cosh functions

$$\begin{aligned} g_1 &= \gamma(n - 1 + N_g - N), \\ g_{2a} &= \gamma(n - N_g + N), \\ g_{2b} &= \gamma(n - N_g - N), \\ g_3 &= \gamma(n - 1 + N_f - N), \\ g_{4a} &= \gamma(n - N_f + N), \\ g_{4b} &= \gamma(n - N_f - N), \end{aligned} \tag{3.39}$$

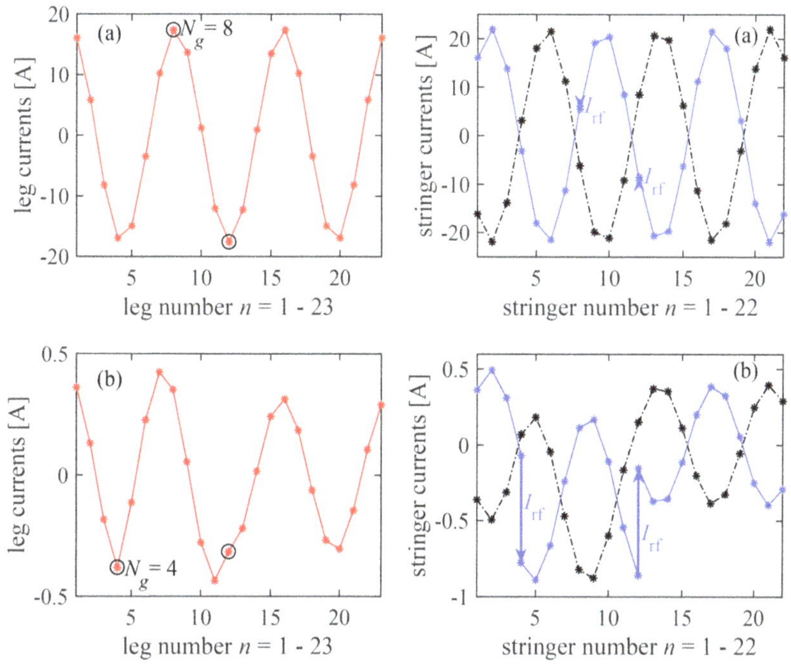

Figure 3.12. Driven RF currents in a dissipative antenna calculated for RF connection points $N_f = 12$ and (a) $N_g = 8$ (corresponding to figure 3.7), or (b) $N_g = 4$. Left: leg currents I_n, where the RF connection positions are marked by circles. Right: stringer currents, full blue line J_n (top stringer), dashed black line K_n (bottom stringer), and the discontinuities due to the RF current injection into the top stringer are labelled by I_{rf}. The current distribution is more uniform in (a) because the RF current is in spatial phase with the m = 6 normal mode, and consequently of much smaller amplitude than in (b). Representative experimental antenna component values: $N = 23$ legs, $L_{leg} = 143$ nH, $R = 36$ mΩ, $C = 2.6$ nF, $L_{str} = 7.9$ nH, $r = 2$ mΩ. Mode m = 6, at time $\exp(-j\omega t) = \exp(-j\pi/4)$ and RF current amplitude $|I_{rf}| = 1$ A. (Adapted with permission from [5]. Copyright 2014 IOP Publishing.)

represent the influence of the chosen RF connection node pair (N_g, N_f). From (3.35) to (3.37) it can be seen that the current flow in each element of a dissipative antenna is proportional to the driving current I_{rf}, varies non-linearly with γ (via $F_{(\gamma,N)}$), and depends on the RF connection positions. The importance of these three parameters for the antenna input impedance, the antenna currents, and their relative phases is discussed in sections 3.5.3 to 3.5.7.

Figure 3.12 shows the calculated current distribution in the legs and stringers of a dissipative antenna corresponding to the experimental arrangement in section 3.1. Two cases are shown: (a) for feed-points $(N_g, N_f) = (8, 12)$ where the antenna resonance currents are much larger than I_{rf}; and (b) for feed-points $(N_g, N_f) = (4, 12)$ where I_{rf} is comparable to the antenna currents. The vertical arrows in (b) mark the differences in the J_n currents at nodes N_g and N_f which arise to respect current continuity with the RF current injection. From the point of view of current uniformity, it is clearly advantageous to arrange for the RF input current to be as small as possible compared to the antenna currents. In this way, the current distribution approaches the uniform normal mode situation where the input current is zero. This is logically achieved by connecting the RF generator to the current anti-nodes of the mth normal mode structure corresponding to the generator's frequency, $\omega = \omega_{\mathrm{m}}$. In this way, the driving RF current is in spatio-temporal resonance with the corresponding normal mode of the antenna.

3.5.3 Driven resonance currents in the limit of weak dissipation

It is instructive to compare the leg currents in (3.35) to (3.37), at resonance, with the normal mode current distribution. In the limit of weak dissipation, $\alpha \to 0$, we should expect to find the same expression as (3.22).

For the limit of small attenuation per section, $|\alpha| \ll 1$ from section 3.3, the numerator of $F_{(\gamma,N)}$ in (3.38) is $\tanh(\gamma/2) \simeq \tanh(j\beta/2) = j\tan(\beta/2)$. In the denominator, $\sinh(\gamma N) = \sinh(\alpha N + j\beta N)$ can be expanded as $\sinh(\alpha N)\cos(\beta N) + j\cosh(\alpha N)\sin(\beta N)$. For normal mode resonances, $\sin(\beta N) = \sin(\mathrm{m}\pi) = 0$, so we have $\sinh(\gamma N) \simeq \alpha N(-1)^{\mathrm{m}}$ for $\alpha N \ll 1$. One expression for $F_{(\gamma,N)}$ at resonance, in the limit of weak dissipation, is therefore

$$F^{\mathrm{res}}_{(\gamma,\,N)} \simeq \frac{j\tan\left(\dfrac{\mathrm{m}\pi}{2N}\right)}{2(-1)^{\mathrm{m}}\alpha N}. \tag{3.40}$$

Patient expansion of the sums of the four cosh terms in (3.35) to (3.37) in the limit of small α gives $2D(-1)^{\mathrm{m}}\cos\left[\left(n-\frac{1}{2}\right)\frac{\mathrm{m}\pi}{N}\right]$ for all three of the leg currents, where D is the same as (3.21) for normal modes.

Finally, at resonance, in the limit of weak dissipation $\alpha N \ll 1$, the leg current distributions (3.35) to (3.37) in all three parts of the antenna tend towards the same limit:

$$\left(\frac{I_n}{I_{\mathrm{rf}}}\right)_{\alpha N \ll 1} \simeq \left(\frac{D}{j\alpha N}\right)\tan\left(\frac{\mathrm{m}\pi}{2N}\right)\cos\left[\left(n-\frac{1}{2}\right)\frac{\mathrm{m}\pi}{N}\right], \tag{3.41}$$

$$D = \cos\left[\left(N_f - \frac{1}{2}\right)\frac{\mathrm{m}\pi}{N}\right] - \cos\left[\left(N_g - \frac{1}{2}\right)\frac{\mathrm{m}\pi}{N}\right]. \tag{3.42}$$

The first bracket of (3.41) shows that the RF driving current I_{rf} is in phase quadrature ($j = e^{i\pi/2}$) with respect to the leg currents I_n. The inverse dependence on α means that the RF driving current becomes negligible compared to the leg currents, $|I_{\mathrm{rf}}| \ll |I_n|$, in the limit of a lossless antenna, i.e. a non-zero resonance current persists as the driving current tends to zero. Therefore, very large resonance currents circulate within the antenna's internal structure for only a small RF input current. This is an ideal property of an RF source for plasma inductive coupling distributed over large areas or volumes, and is a major motivation for using resonant network antennas—see also the first paragraph of the book, and sections 1.2, 1.4, 2.2.2, and 5.5. High resonance currents are consistent with normal mode operation, and the cosine term in (3.41) shows that the uniform normal mode current distribution (3.14) is recovered for a weakly dissipative antenna. Also, the amplitude of the spatial distribution of leg currents is uniform along all three parts of the antenna in the normal mode limit $|\alpha| \rightarrow 0$ because the driving current $I_{\rm rf}$ becomes negligible compared with I_n.

The factor D in (3.42) and (3.21) accounts for the choices of mode number m and of RF current injection connections (N_g, N_f); it is the same as for normal modes (3.21). When the injection points coincide with a maximum and a minimum in the leg currents, as for $(N_g, N_f) = (8, 12)$ in figure 3.12(a), the RF excitation is efficiently coupled to the antenna resonance current distribution, then $|I_n/I_{\rm rf}| \gg 1$, and $|D| = 1.99$ is close to its maximum of 2. Conversely, when the RF connections positions do not correspond to the normal mode spatial variation of the leg currents, as for $(N_g, N_f) = (4, 12)$ in figure 3.12(b), we have $|D| = 0.037$ close to its minimum value, zero. The RF injected current is then significant compared to the leg currents, so the current distribution becomes lopsided across the antenna. The importance of D for the antenna input impedance is described further in section 3.5.7.

3.5.4 Input impedance of a dissipative antenna

The input impedance of a Helyssen antenna determines the required level of driven RF current and voltage for a given power injection, and defines the impedance matching conditions required between the antenna and the RF power source, see chapter 13.

In figure 3.11, the RF voltage across the input nodes is given by $A_{N_f} - A_{N_g} = V_{\rm rf}$. Substituting $n = N_f$, then $n = N_g$, in (3.31), the voltage difference gives

$$V_{\rm rf} = \left[Z_{\rm str}(N_f - N_g)I_{\rm rf} - Z_{\rm leg}(I_{N_f} - I_{N_g})\right]/2, \tag{3.43}$$

because $I_{\rm line} = -I_{\rm rf}$ in part II of the antenna. The exact expression for the antenna input impedance, $Z_{\rm in} = V_{\rm rf}/I_{\rm rf}$, for any level of dissipation, follows directly using (3.36) to give

$$Z_{\rm in} = \frac{Z_{\rm str}}{2}(N_f - N_g) + Z_{\rm leg}F_{(\gamma,N)}G_{(\gamma,N_g,N_f,N)}, \tag{3.44}$$

where $F_{(\gamma,N)}$ is given in (3.38), and $G_{(\gamma,N_g,N_f,N)}$ is a form factor which accounts for the influence of the general configuration of the RF connection positions:

$$\begin{aligned} G_{(\gamma,N_g,N_f,N)} = {} & (\cosh\gamma(N - 2N_f + 1) + \cosh\gamma(N - 2N_g + 1))/2 \\ & + \cosh\gamma N - \cosh\gamma(N - N_g - N_f + 1) - \cosh\gamma(N + N_g - N_f). \end{aligned} \tag{3.45}$$

Physically, the first term of the impedance expression (3.44) is the parallel combination of the two rows of $(N_f - N_g)$ impedances $Z_{\rm str}$ between the RF driving current connections; this is dominantly capacitive. The second term, proportional to $Z_{\rm leg}$, is periodic in mode number and driving frequency due to its dependence on γ in $F_{(\gamma,N)}$ and $G_{(\gamma,N_g,N_f,N)}$. These terms are responsible for the series of peaks shown in figure 3.13. Strictly, instead of presenting normal modes, we can now say that the

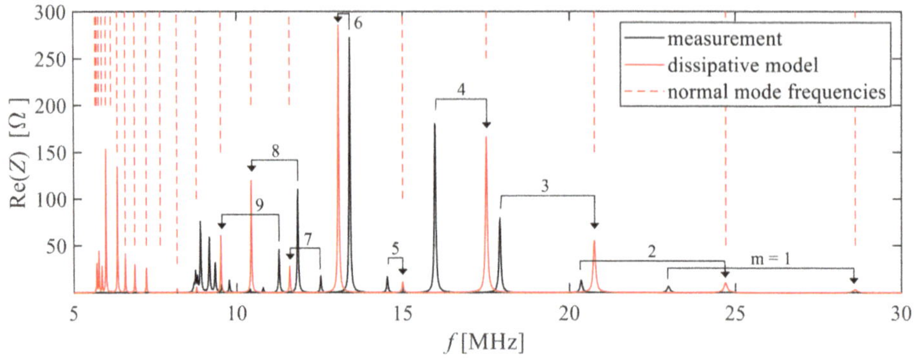

Figure 3.13. Real part of the input impedance of a resonant ladder antenna. Black curve: measured with the network analyser in figure 3.9(a) for dominant excitation of mode m = 6. Red curve: calculated using (3.44) for the same RF connections $(N_g, N_f) = (8, 12)$. Peaks with the same mode number are bracketed for m = 1 − 9. Each bracket arrow indicates the frequency shift from the measured mode to the calculated mode. The red dashed lines indicate the normal mode frequencies according to (3.12), which are seen to correspond closely to the dissipative mode frequencies. Model parameters: $L_{\text{leg}} = 146$ nH, $R = 14.2$ mΩ, $C = 2.6$ nF, $L_{\text{str}} = 10.5$ nH, $r = 9.2$ mΩ. The resistances R and r are given for mode m = 6 at 13.4 MHz, and scale with the frequency-dependent skin depth in the metal components.

antenna exhibits a set of driven resonant modes associated with peaks in its real input impedance. The analytical input impedance Z_{in} in (3.44) is exact for any values of component impedances Z_{str}, Z_{leg}, and hence for any level of dissipation, and for a continuous range of frequencies.

In figure 3.13, the calculated analytical impedance (3.44) is confronted with the network analyser measurements previously presented in figure 3.9(a). The whole of the model spectrum is calculated from the single set of five parameters in the figure caption. Qualitatively, the correspondence of amplitude and width is good for most pairs of peaks, except for the group with the highest mode numbers. The large frequency shifts indicated by the brackets are addressed in the next section 3.5.5.

3.5.5 Resonance frequencies of dissipative antennas

In the previous sections 3.5.2 to 3.5.4, solutions for the dissipative antenna have been derived in terms of the complex propagation constant $\gamma = \alpha + j\beta$. To progress to circuit modelling, we now wish to calculate the frequencies of the driven resonant modes in terms of the circuit elements. For this, we return to the characteristic equation (3.8) and adapt it for the dissipative antenna, using $Z_{\text{str}} = r + \frac{1}{j\omega C} + j\omega L_{\text{str}}$ and $Z_{\text{leg}} = R + j\omega L_{\text{leg}}$. After some algebraic manipulation (appendix A.1), the modified characteristic equation is now

$$\cosh\gamma \simeq \left[1 + \frac{L_{\text{str}}}{L_{\text{leg}}} - \frac{1}{\omega^2 L_{\text{leg}} C}\right] - j\left[\frac{R'}{\omega^3 L_{\text{leg}}^2 C}\right], \tag{3.46}$$

$$R' \simeq R\left[1 + \omega^2 L_{\text{str}} C\left(\frac{rL_{\text{leg}}}{RL_{\text{str}}} - 1\right)\right] = R(1 - \omega^2 CL_{\text{str}}) + r\omega^2 CL_{\text{leg}}, \tag{3.47}$$

where R' is the combined effective resistance of R and r, and products of small resistances have been neglected. Note that $R' = R$, the leg resistance, if the stringer residual inductance L_{str} is negligible. By expanding $\cosh\gamma = \cosh(\alpha + j\beta) = (\cosh\alpha)(\cos\beta) + j(\sinh\alpha)(\sin\beta)$ and equating real and imaginary parts, we obtain

$$\cos\beta \simeq \left[1 + \frac{L_{\text{str}}}{L_{\text{leg}}} - \frac{1}{\omega^2 L_{\text{leg}} C}\right], \tag{3.48}$$

$$\alpha \sin\beta \simeq -\frac{R'}{\omega^3 L_{\text{leg}}^2 C}, \tag{3.49}$$

assuming $|\alpha| \ll 1$ so that $\cosh\alpha \approx 1$. The finite length condition for open networks, $J_N = 0$, still applies, so $\sin(N\beta) = 0$ and $N\beta = \text{m}\pi$, where $\text{m} = 1, 2, \ldots, (N - 1)$ [7] as in section 3.4. Substituting $\beta = \frac{\text{m}\pi}{N}$ into (3.48), and recognizing the trigonometric identity $(1 - \cos x) \equiv 2\sin^2(\frac{x}{2})$, leads to identical expressions for the resonant mode frequencies as in (3.12)

$$\omega_{\text{m}} \simeq \frac{1}{\sqrt{C\left(L_{\text{str}} + L_{\text{leg}}\left(1 - \cos\left(\frac{m\pi}{N}\right)\right)\right)}}, \quad \text{m} = 1, 2, \ldots, (N - 1), \tag{3.50}$$

$$\simeq \frac{1}{\sqrt{C\left(L_{\text{str}} + 2L_{\text{leg}}\sin^2\left(\frac{\text{m}\pi}{2N}\right)\right)}}, \quad \text{m} = 1, 2, \ldots, (N - 1), \tag{3.51}$$

to first order in the small quantity α, hence the impedance peaks of the weakly dissipative antenna are at almost the same frequencies as the normal mode frequencies (3.12) of the lossless antenna, as shown in figure 3.13. The frequency shifts in figure 3.13 are therefore essentially the same as the discrepancy between the two curves in figure 3.8, and the inclusion of R and r does not alter the resonance frequencies to first order in α. This could have been been suspected from (2.14), because ohmic losses shift the resonance frequency of parallel resonant circuits by only a very small fraction; a similar observation has now been shown to apply to resonant networks as well. This would be a useful property, for a hypothetical case of weak, purely resistive plasma loading, because a fixed frequency RF generator could be used to drive the same antenna resonance before and after plasma ignition. However, we will see later in part II—figure 6.11, for example—that plasma loading also changes the antenna inductance values, and hence the resonance frequencies also.

3.5.6 Parallel resonant equivalent circuit for dissipative antennas

In the previous section 3.5.5, the resonance frequencies were calculated in terms of circuit parameters. We now wish to find an equivalent circuit for the input impedance of the dissipative antenna. In the neighbourhood of a resonance frequency, each peak in figure 3.13 apparently has a form similar to the parallel resonance behaviour of a lossy parallel resonant circuit in section 2.2.2. A model of the antenna impedance would help to understand how the individual electrical components influence the input impedance instead of the rather obscure exact expression (3.44).

In contrast to the discrete frequencies, $\omega_{\rm m}$, and infinite impedance of normal modes on lossless networks, the introduction of resistance results in a finite impedance over a frequency continuum. To interpret the antenna impedance behaviour close to resonance, we consider an approximation for $Z_{\rm in}$ by expansion of (3.44) in terms of a small deviation $\delta\omega$ from a resonance frequency $\omega_{\rm m}$. To spare the reader this fastidious task, the expansion is fully developed in appendix A.2. It is found that the input impedance (3.44) of the dissipative antenna in the neighbourhood of the mth mode resonance can be represented by an impedance $Z_{\rm str}(N_f - N_g)/2$ in series with a parallel resonant equivalent circuit as shown in figure 3.14. To be clear, the imaginary part of the parallel resonant equivalent circuit is zero at resonance, by definition, but now the total input impedance of the antenna includes a series impedance $Z_{\rm str}(N_f - N_g)/2$ which has an imaginary component, even at resonance. In figure 3.14, the equivalent circuit parameters around mode m are as follows:

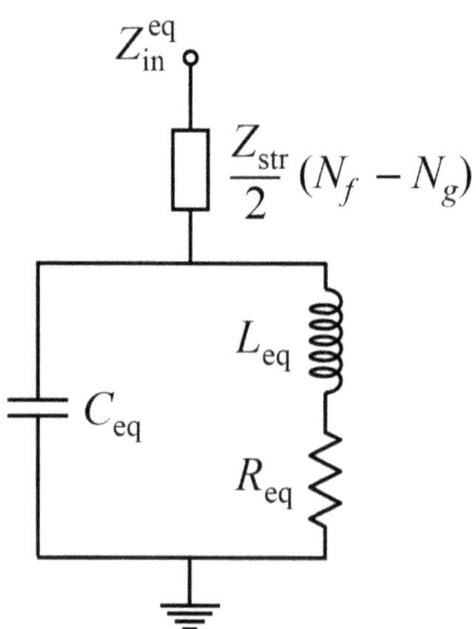

Figure 3.14. An equivalent circuit, in the neighbourhood of a mode resonance, for the dissipative antenna shown in figure 3.11 with the components of figure 3.3. Using (3.52) to (3.57), this circuit accounts for all the antenna components $L_{\rm leg}$, R, C, $L_{\rm str}$, r, the mode number m, the number of legs N, and the general RF connection positions (N_g, N_f). (Reproduced with permission from [5]. Copyright 2014 IOP Publishing.)

$$Z_{\text{in}}^{\text{eq}} \simeq \frac{Z_{\text{str}}}{2}(N_f - N_g) + \frac{1}{Y_{\text{eq}}}, \tag{3.52}$$

$$Y_{\text{eq}} = \frac{1}{\text{fac}} \cdot \frac{R'}{\omega_{\text{m}}^2 L_{\text{leg}}^2}\left[1 + 2jQ\frac{d\omega}{\omega_{\text{m}}}\right] = \frac{R_{\text{eq}}}{\omega_{\text{m}}^2 L_{\text{eq}}^2}\left[1 + 2jQ\frac{\delta\omega}{\omega_{\text{m}}}\right], \tag{3.53}$$

$$\text{where} \quad Q = \omega_{\text{m}} L_{\text{leg}}/R' = \omega_{\text{m}} L_{\text{eq}}/R_{\text{eq}}, \tag{3.54}$$

$$R_{\text{eq}} = \text{fac} \cdot R', \tag{3.55}$$

$$L_{\text{eq}} = \text{fac} \cdot L_{\text{leg}}, \tag{3.56}$$

$$C_{\text{eq}} = C\left[1 - \cos\left(\frac{\text{m}}{\pi N}\right)\right] \cdot \frac{1}{(1 - \omega_{\text{m}}^2 L_{\text{str}} C)} \cdot \frac{1}{\text{fac}}, \tag{3.57}$$

$$\text{fac} = (1 - \omega_{\text{m}}^2 L_{\text{str}} C)\frac{D^2}{2N}, \tag{3.58}$$

$$D = \cos\left[\left(N_f - \frac{1}{2}\right)\frac{\text{m}\pi}{N}\right] - \cos\left[\left(N_g - \frac{1}{2}\right)\frac{\text{m}\pi}{N}\right]. \tag{3.59}$$

The admittance Y_{eq} in (3.53) for a parallel resonant $RL\|C$ circuit is recognized from (2.18) in the introduction to resonant circuits, section 2.2, where $\delta\omega$ is now the frequency shift around the mode resonance frequency ω_{m}. R' is the combined resistance of the leg and stringer values, R and r, respectively, weighted by the network inductances and capacitance, given by (3.47). The values of R_{eq}, L_{eq}, C_{eq} can be determined for any mode number m by using the antenna individual component values L_{leg}, R, C, L_{str}, r, the mode number m, the number of legs N, and the RF connection positions (N_g, N_f) to calculate D in (3.42). By choice of C_{eq}, the angular frequency at resonance, $\omega_{\text{m}} = 1/\sqrt{L_{\text{eq}}C_{\text{eq}}}$, is the same expression as for the normal mode resonance in (3.50) or (3.51). This simple equivalent circuit for a complicated network means that we can profit from our previous results for $RL\|C$ circuits in chapter 2, section 2.2.2.

Before this, however, we note that the series impedance $Z_{\text{str}}(N_f - N_g)/2$ in (3.52) is principally due to two strings of $(N_f - N_g)$ impedances Z_{str} in parallel. These are principally capacitive, and add only a small downward shift to the total imaginary impedance; strictly, this should be accounted for before the parallel resonance definition of zero imaginary part is applied, although it makes little difference. Fortunately, the peak in the magnitude of $Z_{\text{in}}^{\text{eq}}$ (or its real part) marks the resonance frequency unambiguously. The real input impedance at resonance, when $\delta\omega = 0$ in (3.53), of the equivalent circuit is

$$Z_{\text{res}}^{\text{eq}} = \text{fac} \cdot \frac{\omega_{\text{m}}^2 L_{\text{leg}}^2}{R'}. \tag{3.60}$$

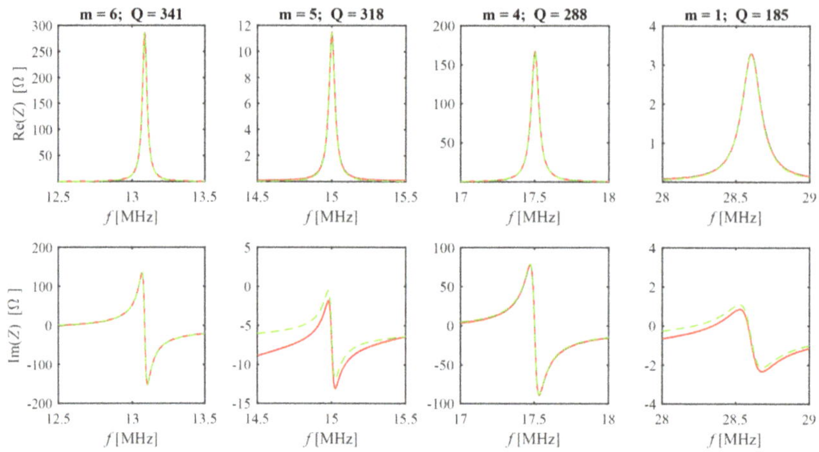

Figure 3.15. Comparison of the exact impedance calculation taken from figure 3.13 (red lines) and the equivalent circuit approximation (3.52) (superposed green dashed lines), for modes m = 6, 5, 4, 1 over a 1 MHz range for each mode. The calculated curves are practically indistinguishable for the high impedance, well-isolated modes m = 4 and 6. Same circuit parameters as for figure 3.13. Note the large difference in the ordinate scales.

The peak impedance is therefore strongly influenced by the term $0 \leqslant D^2 \leqslant 4$ in (3.58) which depends on m, N, and the RF connection positions (N_g, N_f) in equations (3.59), (3.42), and (3.21). This is the main reason for the large variation in input impedance as m varies across the mode spectrum in figures 3.4, 3.9, and 3.13.

The quality factor Q of the equivalent circuit represents the Q factor of the whole network. Surprisingly, from (2.9) and (3.54), Q is simply the quality factor of a single leg inductance L_{leg} in series with the effective resistance R', at the mode frequency ω_{m}. Therefore, Q is independent of the RF connection positions which have such a strong influence on the input impedance via (3.58). We note in passing that the apparently small values of $r < R$ and $L_{\text{str}} < L_{\text{leg}}$ have a significant effect and must be accounted for in the equation (3.47) for R', in order to obtain an accurate fit across the mode spectrum [7]. Moreover, the square-root-frequency dependence of R and r due to the skin depth is included. The increase of R' with frequency is the reason for the visible decrease in Q with frequency in figure 3.13. Note that the general expression for Q, (2.20), would only be necessary if the equivalent parallel resonant circuit parameters were not known [7].

In figure 3.15, the exact calculated modes m = 1, 4, 5, 6 from figure 3.13 are compared with the equivalent circuit expression (3.52) for the real and imaginary parts. The equivalent circuit approximation is practically indistinguishable from the exact impedance (3.44) for the high impedance, isolated modes m = 4 and 6, using the same parameters as in figure 3.13. The single expression (3.44) is valid across the whole spectrum, whereas the equivalent circuit impedance (3.52) applies only in the neighbourhood of its individual, isolated mth mode; this explains the apparent discrepancy in the imaginary part for the intermediate peak m = 5, which is small enough to be influenced by the peak at m = 6. The Q values around 300 correspond to very high quality factors for the RF resonant network without plasma, and explain the accuracy of the equivalent circuit approximations.

3.5.7 Optimization of the input impedance for antenna design

The antenna input impedance is influenced by the mode number m and the RF connection positions (N_g, N_f). A high input impedance is necessary to minimize the RF injection current for a given power so that the antenna current sinusoidal distribution is as uniform as possible along the antenna, approaching a normal mode distribution. Figure 3.16 is a contour plot of the antenna real input impedance Re (Z_{in}) in the (N_g, N_f) plane for mode m = 6. This is a graphical demonstration of the impedance modulation which is approximately proportional to D^2. Once the number of legs N and the mode number m have been chosen, the value of D^2 depends only on the antenna connection positions (N_g, N_f). For each mode, the connection positions can be chosen to obtain a value close to the maximum, $D^2 \simeq 4$. For example, in figure 3.16, $D^2 = 3.96$ for (N_g, N_f) = (8, 12) for which the real input

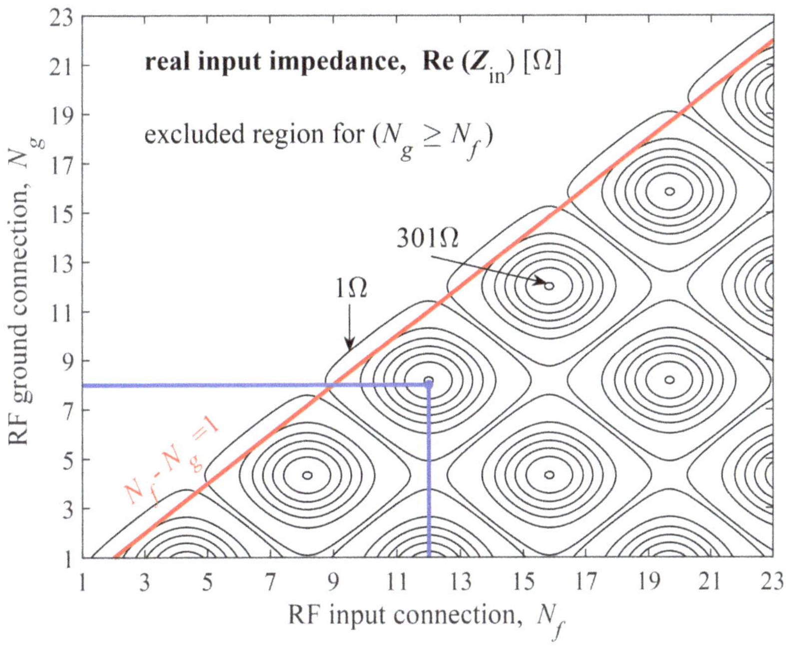

Figure 3.16. Contour plot of the real part of the antenna input impedance, Re (Z_{in}), using (3.44) for artificial continuum ranges of the RF connection N_f, and a single ground connection $N_g < N_f$ on the same stringer. The applicable values of real input impedance correspond uniquely to integer values of N_g and N_f. RF connections (N_g, N_f) = (8, 12), at the intersection of the blue lines, correspond to an optimum input impedance for minimum input current and good current uniformity. The red line $N_f - N_g = 1$ corresponds to RF connections across a single capacitor. The excluded region $N_g \geqslant N_f$ is an arbitrary choice determined with reference to figure 3.11. The contours descend from 301 Ω (the innermost circles) down to 1 Ω, in 50 Ω intervals. Mode m = 6, for the same antenna component values as in figure 3.12. (Adapted with permission from [5]. Copyright 2014 IOP Publishing.)

impedance is close to the maximum, but $D^2 = 0.0014$ for $(N_g, N_f) = (4, 12)$ for which the real input impedance is only 0.13 Ω. The choice of N_g and N_f is therefore a design parameter for the real input impedance of the driven resonant antenna, to minimize the externally injected RF input current compared to the internal circulating resonance current, and thus to ensure uniformity of the current across the whole network. Finally, for stability of an m = 1 mode, the RF and ground connections should be spaced far apart on the corresponding nodes, whilst still respecting $D^2 \approx 4$. The input impedance then includes some imaginary component due to the parallel combination of stringer impedances (see section 3.5.6), but this only affects the matching condition.

The model in this chapter has concentrated on ladder antennas, which are open networks by definition. However, exactly the same results apply also to open birdcage resonant antennas, because they satisfy the same end-boundary conditions. The habitual practice for birdcage antennas is to connect the RF power across a single capacitor [7–9]. In the present context (although birdcages are usually operated in mode m = 1), this corresponds to the straight line $N_f - N_g = 1$, superposed in red in figure 3.16. It can be seen that a high input impedance is not achieved for any single-capacitor position; this could overload the capacitor, as discussed in section 9.2.4. In contrast, the RAID open birdcage in chapter 12 shows an optimal design for mode m = 1 with diametrically opposed RF connections.

Further advantages of the dominantly real input impedance of resonant network antennas are discussed in section 5.5, in comparison with other large area plasma sources. Finally, the design of the RF matching circuit is considered in section 13.1.

3.6 Application: frequency resolution of MRI antennas

To anticipate chapter 9, if the ladder antenna is wrapped around to form a cylinder, this creates an open birdcage antenna provided that the end nodes are not fully connected, as in figure 1.5. The birdcage antenna is well known from nuclear magnetic resonance measurements (NMR) used for magnetic resonance imaging (MRI) [7], where the antennas operate using the m = 1 mode to obtain a homogeneous oscillating magnetic field (see sections 8.2 and 8.3). High frequency resolution is critical to MRI performance, and this is determined by the quality factor $Q = \frac{\omega 1}{\Delta\omega}$, where ω_1 is the m = 1 mode frequency, and $\Delta\omega$ denotes the FWHM bandwidth, from section 2.2.2, where Q depends on the effective resistance of the antenna. J-M Jin [7] describes MRI antenna design, although resistive components were not considered explicitly, so a general expression for Q in (2.20) was used in absence of an analytical definition. Driven birdcage resonators with losses were considered by A Novikov [10] using low impedance voltage sources applied to single mesh elements. R J Pascone *et al* [11] and J Tropp [12] treated the birdcage resonator as a lossy transmission line. Other authors [13, 14] represented the dissipation as an effective coil resistance.

The Q value of an open birdcage can now be estimated analytically using the parallel resonant equivalent circuit of the ladder antenna developed in section 3.5.6. Putting m = 1 in (3.54) gives

$$Q = \omega_1 L_{\text{leg}}/R', \tag{3.61}$$

where R' is given in terms of the antenna component values using (3.47). MRI antennas are usually closed birdcage antennas (section 9.2), and mutual inductance is neglected in the equivalent circuit approach, therefore we do not aim here to accurately reproduce the mode frequency. Consequently, we can simplify R' by setting $L_{\text{str}} = 0$ in (3.47) to obtain

$$\begin{aligned} Q &\approx \frac{\omega_1 L_{\text{leg}}}{R + r\omega_1^2 C L_{\text{leg}}} \\ &\approx \frac{\omega_1 L_{\text{leg}}}{R + r\Big/\left[\frac{L_{\text{str}}}{L_{\text{leg}}} + 2\left(\frac{\pi}{2N}\right)^2\right]}, \end{aligned} \tag{3.62}$$

where the small angle approximation $\sin\frac{\pi}{2N} \approx \frac{\pi}{2N}$ has been used in (3.51), because $N=9$ is the smallest number of legs for reasonable uniformity of the magnetic field (see section 8.3.3). This small factor means that it is not justifiable to neglect L_{str} in the normal mode frequency expression (3.51) to substitute for ω_1^2 in the denominator. To achieve high Q, (3.62) shows the importance of minimizing not only the leg resistance R, but also the stringer resistance r by using low-loss capacitors (see section 13.2). Perhaps surprisingly, a non-negligible stringer inductance L_{str} also helps to increase Q. Despite the insight afforded by (3.62), a more precise value is easily calculated using (3.61) directly[3]. Hence, the MRI frequency resolution can be specifically designed, optimized, and predicted analytically from the known values of the antenna components.

3.7 Chapter summary

An experimental and theoretical analysis of Helyssen resonant antennas has been presented to aid antenna design for its development as a plasma source for industrial applications. Analytical expressions are given for the mode resonance frequencies, the antenna currents, the Q value, and the input impedance. These apply for arbitrary antenna impedances, for any level of dissipation, and for general RF connection positions. The antenna can be represented by a simple equivalent circuit which includes a parallel resonant circuit.

The analytical solution is adequate for open networks, needing only the 5 circuit components shown in figure 3.3, and quantified in the caption of figure 3.13, to model the whole mode spectrum from m = 1 to m = 22. Comparison with network analyser measurements in figure 3.13 shows good accuracy for the spatial distribution of voltage and current modes, reasonable estimates of Q and the input impedance, but imprecise values of the resonance frequencies. For the latter, the mutual partial inductances must be included, as will be shown in the next chapter 4.

[3] For example, (3.61) gives the precise value $Q = 185$ for the ladder resonant antenna impedance in figure 3.15 for m = 1, whereas the approximation (3.62) gives $Q = 166$.

To design a planar antenna for an inductively coupled plasma source, it was shown that a high input impedance is necessary to maintain the current amplitude uniformity of normal modes. This can be achieved by optimal choice of RF connection positions for a given mode number. The dominantly real input impedance near to antenna resonance avoids the problem of strong reactive currents and voltages in the matching box and RF power connections found in conventional large area plasma sources.

The equivalent circuit model with analytical solutions is the farthest we can go before resorting to digital computation. Analytical solutions enable physical interpretation and present a reasonable agreement with experimental measurements, but higher precision in the following chapters will require numerical inversion of matrices.

References

[1] Guittienne Ph, Fayet P, Larrieu J, Howling A A and Hollenstein Ch 2012 Plasma generation with a resonant planar antenna *55th Annual SVC (Surface Vacuum Coaters) Technical Conf. (Santa Clara, CA, 28 April–3 May)*

[2] Guittienne Ph, Lecoultre S, Fayet P, Larrieu J, Howling A A and Hollenstein Ch 2012 Resonant planar antenna as an inductive plasma source *J. Appl. Phys.* **111** 083305

[3] Lecoultre S, Guittienne Ph, Howling A A, Fayet P and Hollenstein Ch 2012 Plasma generation by inductive coupling with a planar resonant RF network antenna *J. Phys. D: Appl. Phys.* **45** 082001

[4] Lecoultre S, Guittienne Ph, Howling A A, Fayet P and Ch Hollenstein 2012 Corrigendum: plasma generation by inductive coupling with a planar resonant RF network antenna *J. Phys. D: Appl. Phys.* **45** 409502

[5] Guittienne Ph, Howling A A and Hollenstein Ch 2014 Analysis of resonant planar dissipative network antennas for RF inductively coupled plasma sources *Plasma Sources Sci. Technol.* **23** 015006

[6] ATC American Technical Ceramics https://rfs.kyocera-avx.com/

[7] Jin J-M 1998 *Electromagnetic Analysis and Design in Magnetic Resonance Imaging* (Boca Raton, FL: CRC Press)

[8] Guittienne Ph, Chevalier E and Hollenstein Ch 2005 Towards an optimal antenna for helicon waves excitation *J. Appl. Phys.* **98** 083304

[9] Romano F *et al* 2020 RF helicon-based inductive plasma thruster (IPT) design for an atmosphere-breathing electric propulsion system (ABEP) *Acta Astronaut.* **176** 476

[10] Novikov A 2011 Advanced theory of driven birdcage resonator with losses for biomedical magnetic resonance imaging and spectroscopy *Magn. Reson. Imaging* **29** 260

[11] Pascone R J, Garcia B J, Fitzgerald T M, Vullo T, Zipagan R and Cahill P T 1991 Generalized electrical analysis of low-pass and high-pass birdcage resonators *Magn. Reson. Imaging* **9** 395

[12] Tropp J 2002 Dissipation, resistance, and rational impedance matching for TEM and birdcage resonators *Concepts Magn. Reson.* **15** 177

[13] Hoult D I and Richards R E 1976 The signal-to-noise ratio of the nuclear magnetic resonance experiment *J. Magn. Reson.* **24** 71

[14] Hayes C E, Edelstein W A, Schenck J F, Mueller O M and Eash M 1985 An efficient, highly homogeneous radiofrequency coil for whole-body NMR imaging at 1.5 T *J. Magn. Reson.* **63** 622

IOP Publishing

Resonant Network Antennas for Radio-Frequency Plasma Sources

Theory, technology and applications

Philippe Guittienne, Alan Howling and Ivo Furno

Chapter 4

Partial inductance and the matrix model

In the previous chapter, chapter 3, the self inductances of the network legs and stringers were treated, for convenience, in the same way as conventional loop self inductances. Now in this chapter we will explain why they are, in fact, self and mutual *partial* inductances, why this is necessary, what the physical interpretation is, what the difference with respect to loop self inductance is, and how to calculate partial inductances. Based on this, the measured driven resonance frequencies and input impedance of a resonant network antenna will be calculated precisely using a matrix treatment. The theory and model are still purely analytical, but now the network solutions involve matrix inversion, and so must be calculated using a computer.

At the outset, we restrict our discussion in this chapter to only dimensions that are small compared to the RF vacuum wavelength, $l \ll \lambda$. Therefore, according to section 2.3, we will be concerned here only with lumped element inductors, whether loop or partial. The contrasting case of transmission lines involves distributed inductance and also distributed capacitance, and will be postponed until chapter 7.

4.1 A brief history of inductance: loop and partial

The reference text for this chapter is the book by C R Paul [1], published as recently as 2010, titled *Inductance: Loop and Partial*. As described by Paul [2], the concept of inductance pre-dates Maxwell in 1873, but the distinction between partial and loop inductance was mostly lost some time after the 1920s.

In the absence of non-linear magnetic (ferromagnetic) materials, inductance is independent of current and depends only on the circuit's geometry. There are two fundamental approaches to inductance calculations. The contemporary approach, already defined in section 2.1.2, is to calculate the *loop inductance* from the magnetic flux linking a closed current circuit [1]. This represents the whole circuit by a single

doi:10.1088/978-0-7503-5296-3ch4

loop self inductance; it is convenient when the circuit geometry is sufficiently simple and well-defined for the magnetic flux to be calculated, such as for a coil or a solenoid. However, when a current path is difficult to define because parts of the circuit are coupled to wires carrying different currents, it is not possible to calculate a unique loop inductance. A resonant network antenna is a good example of a multi-loop, multi-connector circuit [3].

On the other hand, the *partial inductances* [1] of all the contiguous elements of a circuit can be calculated separately and combined in an impedance matrix, which includes all the mutual partial inductances, to give the complete circuit inductance. These two different methods date from the origins of electromagnetic theory (for a review see [4]) up to modern works [1].

Fundamental partial inductance calculations were presented, for example, by Rosa [5] in 1907 where the concept 'partial' was apparently self-evident and not even mentioned; nonetheless, the integrals used are incomprehensible in terms of modern-day 'loop' inductance (see section 4.4.2). Grover [6] only briefly mentions 'partial', presumably because the dichotomy between the general partial inductances and the modern-day loop inductances did not exist at the time of publication in 1946. It is as if the different epochs implicitly considered inductance to be either self-evidently partial, or self-evidently loop, and therefore not naming either definition explicitly. Needless to say, it is bewildering to read books and papers from both epochs without this prior knowledge.

According to Paul [1], partial inductance is not covered in any undergraduate course on electrical engineering despite its growing importance in the design of semiconductors. His recent book acknowledges A E Ruehli of the IBM T J Watson Research Center who wrote a seminal paper on partial inductance in 1972 [3]. Their interest in partial inductance is driven by the need to understand stray and 'unintentional' inductance, to improve the design of high-speed digital systems.

In a different application, Testa [7] uses partial inductance for the exact calculation of the inductance of magnetic field diagnostic coils for tokamaks; the analytical method is so accurate that numerical simulation would bring no improvement in precision.

Our interest in partial inductance is different from both Paul and Testa. We want to model the coupling, due to mutual partial inductance, between the elements in complicated networks of L, C macroscopic components where neighbouring conductors all carry different currents. The aim is to accurately design resonant network antennas with the required driven resonance frequencies, input impedance, and normal mode structures having uniform voltage and current amplitudes.

4.2 Can the self inductance of a wire be measured?

In the experimental set-up of section 3.1, we mentioned that the values of $L_{\mathrm{leg}} \simeq 146$ nH and $L_{\mathrm{str}} \simeq 10.5$ nH were calculated (the formulas are given in section 4.5.2). The leg and stringer inductances were not directly measured for reasons which will become clear by carrying out a thought experiment sketched in figure 4.1. The difference between measuring a lumped element loop self inductance, and measuring the self inductance

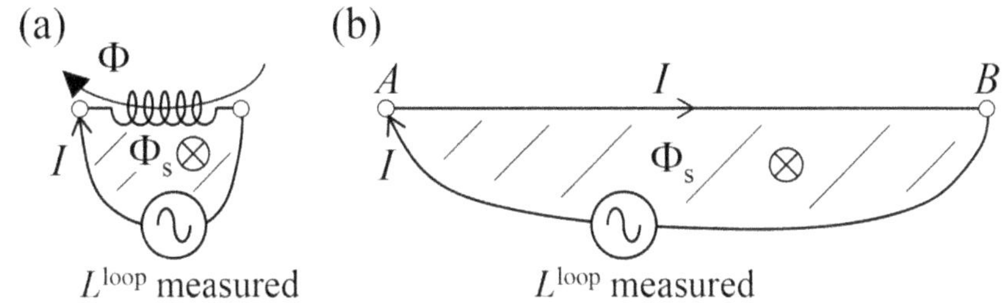

Figure 4.1. (a) A wirewound inductor several millimetres across, linking flux Φ for a current I, connected to an inductance bridge measuring the circuit loop self inductance L^{loop}. The connecting wires link a flux $\Phi_S \ll \Phi$ due to the small stray inductance. (b) An antenna leg AB represented by a wire 20 cm long, carrying a current I. The inductance bridge measures a loop inductance L^{loop} which is entirely due to the 'stray' flux Φ_S.

of a short wire, will help to illustrate why partial inductance is a necessary and useful concept [3]. The reader will forgive the rather pedantic approach, which is intended to expose any preconceptions.

4.2.1 Measurement of a coil's self inductance

One definition of loop inductance was already touched upon in section 2.1.2. Figure 4.1(a) represents a lumped element RF commercial wirewound coil or solenoid inductor, taken off the shelf. Its inductance value is printed on it and could typically be anything from, say, 1 μH to several mH. Smaller inductances down to several nH can be purchased, but these are usually sub-millimetric surface mounted devices (SMD). The most convenient way to measure the loop inductance of an inductor such as a coil or solenoid is to use an inductance bridge, vector impedance analyser, or LCR meter. Oscilloscopes can be used in more indirect methods.

In figure 4.1(a), the inductor is shown connected to the inductance meter because a current-carrying component cannot properly be considered in isolation, only as part of a closed circuit [1]. The measurement device passes a small alternating current I through the inductor, at the frequency $f = \frac{\omega}{2\pi}$ of interest, and analyses the voltage and relative phase at its own terminals to determine the total reactance X and resistance R of the entire circuit complex impedance, $Z = R + jX$. Provided that capacitance can be neglected[1], the measured loop inductance L^{loop} is directly deduced using $X = \omega L^{loop}$. The closed circuit of the inductor and the connection wires causes a small magnetic flux $\Phi_S \ll \Phi$ in the enclosed area. This represents a small inductance, called the stray (or parasitic, or 'unintentional' [1]) inductance, which is unavoidable, and it is desirable to minimize it by using connecting wires as short as possible. This is easy to do here because the compact coil is physically small, so the stray inductance will be of the order of nH or less. The stray inductance will add to, or subtract from, the coil loop inductance, depending on its winding

[1] We assume that the measurement frequency is far below the self-resonance frequency of the coil, at which the small capacitance between windings resonates with its inductance.

direction, but this makes little difference because the coil inductance is orders of magnitude larger owing to its multiple windings.

Finally, therefore, in the case of a lumped element coil inductor, there is no ambiguity as to the value of its inductance L^{loop}, nor as to the location of the loop inductance in the circuit, because the flux ix highly concentrated at the small coil. This is why the standalone inductor component can have its value printed on it, independently of its use in arbitrary different circuits. This was the implicit assumption for inductors with loop self inductance L^{loop} in chapter 2.

4.2.2 Measurement of a straight wire's self inductance

Figure 4.1(b) shows the wire *AB* whose self inductance we wish to measure. As before, the inductance meter connection is made via two connecting wires[2]. There are now several differences compared to the coil in figure 4.1(a):

- Unlike the coil, the wire *AB* has no loop self inductance of its own (despite the drawings in figures 1.2, 1.3, 1.5(b), or 3.3, for example).
- The connecting wires presumably have inductance properties comparable to the wire itself, and are even longer than the wire being measured.
- The measured inductance L^{loop} is entirely due to the so-called 'stray' loop self inductance, whose single turn area depends on the routing of the connecting wires.

In conclusion, the dilemma is that the measured L^{loop} is not a property of the wire *AB* alone, and furthermore, it is not even clear where L^{loop} should be attributed in the circuit path—at the terminals of the inductance bridge? Or somewhere on the wire *AB*? Or somehow distributed among *AB* and the connecting wires? Evidently, there is a problem in using loop self inductance to characterize simple wires, and so this conventional concept will not be adequate for describing the inductance shared among all the *N* legs in a resonant network antenna. This abortive example is motivation for a more fundamental alternative approach to inductance, called partial inductance, in the next section, 4.3.

4.3 Definition of partial inductance

Following [1], we give a step-by-step description to define the self partial inductance and mutual partial inductance between wires (legs and stringers) that make up a resonant network antenna.

---

[2] It might be argued that the straight wire could be bent into a circle, or a hairpin shape (section 2.3), or any shape in between, to eliminate connecting wires and their associated stray inductance. But this then becomes a loop self inductance measurement of some sort of a single-turn coil instead of the straight wire. Inductance depends on geometry (dimensions, shape, and relative orientation [1]), so the straight wire and the bent wire are different cases with different inductance values.

4.3.1 Step 1: introduce the magnetic vector potential A

We will concentrate on the simplest example which is still relevant to a ladder antenna: a rectangular loop of current filaments in figure 4.2(a), carrying a circulating current I. Our trivial first step is to substitute the magnetic flux density $\mathbf{B}$ by the magnetic vector potential $\mathbf{A}$, using $\mathbf{B} = \nabla \times \mathbf{A}$ [8]. The total magnetic flux linking the rectangular current loop, Φ, is given by

$$\Phi = \int_S \mathbf{B} \cdot d\mathbf{S} = \int_S (\nabla \times \mathbf{A}) \cdot d\mathbf{S}, \tag{4.1}$$

where $\mathbf{B}$ and $\mathbf{A}$ apply over the whole loop area S, and $d\mathbf{S}$ is a vector normal to the surface of an elemental area dS. The loop self inductance $L^{\mathrm{loop}} = \Phi/I$ is self-consistent with our initial definition in section 2.1.2. At the risk of repetition, the problem with this loop definition is that it is not known where to situate, or how to distribute, the single inductance value of L^{loop} on or around the loop perimeter—should it be positioned on one of the sides, or shared equally between the sides, or proportional to the side length, etc?

4.3.2 Step 2: convert to a contour integral around the loop

The essential second step shown in figure 4.2(b) is to convert the surface integral of $(\nabla \times \mathbf{A})$ into a line integral around the loop contour, c, by using Stokes' theorem:

$$\Phi = \int_S (\nabla \times \mathbf{A}) \cdot d\mathbf{S} = \oint_c \mathbf{A}_c^{\mathrm{net}} \cdot d\mathbf{l} = \oint_c A_{c\|}^{\mathrm{net}}\, dl, \tag{4.2}$$

where $\mathbf{A}_c^{\mathrm{net}}$ is the net total magnetic vector potential at each point around the contour c, sourced from the circulating current I in the four sides, and $d\mathbf{l}$ is an elemental contour length vector. Their dot product means that only the component of $\mathbf{A}_c^{\mathrm{net}}$ parallel to the contour, namely $A_{c\|}^{\mathrm{net}}$, actually contributes to the integral.

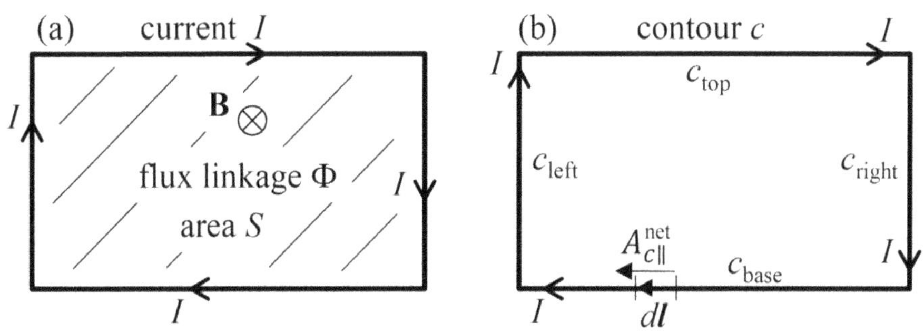

Figure 4.2. (a) A rectangular loop of area S, whose circulating current I produces a magnetic flux density $\mathbf{B}$, causing a magnetic flux linkage Φ in the loop; see (4.1). (b) The same loop, with perimeter contour c divided into four sides, showing a general elemental vector length $d\mathbf{l}$, and the parallel component $A_{c\|}^{\mathrm{net}}$ of the net total magnetic potential vector $\mathbf{A}_c^{\mathrm{net}}$ on the contour c, sourced from the current I on the four sides of the loop; see (4.2).

4.3.3 Step 3: decomposition of the contour

Our first glimpse of the meaning of partial inductance comes from splitting, or decomposing, the contour integral (4.2) into four parts corresponding to the four sides of the rectangle in figure 4.3:

$$\begin{aligned}\Phi &= \oint_c A_{c\|}^{\text{net}}\, dl \\ &= \int_{c_{\text{left}}} A_{\text{left}\|}^{\text{net}}\, dl + \int_{c_{\text{top}}} A_{\text{top}\|}^{\text{net}}\, dl + \int_{c_{\text{right}}} A_{\text{right}\|}^{\text{net}}\, dl + \int_{c_{\text{base}}} A_{\text{base}\|}^{\text{net}}\, dl.\end{aligned} \tag{4.3}$$

Notice that the complete, closed contour integral symbol $\oint_c$ is replaced by four partial, line integrals. We remember that loop self inductance is $L^{\text{loop}} = \Phi/I$ from our initial definition in section 2.1.2, and also in step 4.3.1. By analogy, we can attribute a partial contribution to the loop self inductance from each segment of the loop by dividing by the current. For example, the partial loop self inductance of the left side of the rectangular loop is

$$L_{\text{left}}^{\text{net}} = \frac{\int_{c_{\text{left}}} A_{\text{left}\|}^{\text{net}}\, dl}{I}, \tag{4.4}$$

which can be called the net partial inductance [9] of the left side. Thus, (4.3) can be written as

$$L^{\text{loop}} = L_{\text{left}}^{\text{net}} + L_{\text{top}}^{\text{net}} + L_{\text{right}}^{\text{net}} + L_{\text{base}}^{\text{net}}. \tag{4.5}$$

4.3.4 Step 4: magnetic vector potential due to a line current

However, we are not yet out of the woods, because each $A_{c\|}^{\text{net}}$ component, such as $A_{\text{left}\|}^{\text{net}}$, is a net sum superposition of magnetic vector potentials sourced by all four sides of the rectangle. Fortunately, for current in straight wires, there is a simple

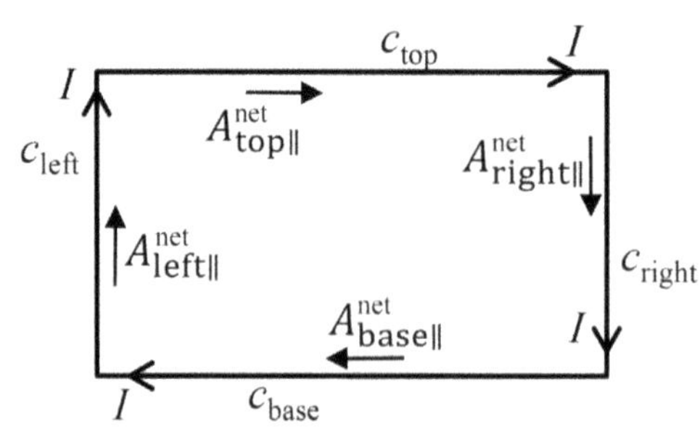

Figure 4.3. The parallel components of $A_{c\|}^{\text{net}}$, in figure 4.2(b), labelled individually around the rectangular contour c.

expression for the magnetic vector potential, which we derive now, to help in the following steps.

Starting from Maxwell's equation $\nabla \times \mathbf{H} = \mathbf{J}$, in the absence of electric displacement current (a quasistatic approximation—specifically, the magneto-quasistatic approximation [10] in appendix D.1.1—consistent with the lumped element assumption [3]), in a non-ferromagnetic medium ($\mu_r = 1$), gives

$$\nabla \times \mathbf{B} = \mu_0 \mathbf{J}. \tag{4.6}$$

We again substitute the magnetic vector potential:

$$\nabla \times (\nabla \times \mathbf{A}) = \mu_0 \mathbf{J}. \tag{4.7}$$

Using the vector identity $\nabla \times (\nabla \times \mathbf{a}) \equiv \nabla(\nabla \cdot \mathbf{a}) - \nabla^2 \mathbf{a}$, and with $\nabla \cdot \mathbf{A} = 0$ for the Coulomb gauge, we have

$$\nabla^2 \mathbf{A} = -\mu_0 \mathbf{J}. \tag{4.8}$$

For spatially distributed current density in the same equation (E.6), this is related to the diffusion equation (E.8) for magnetic vector potential in the same magneto-quasistatic approximation. However, here we are concerned with current filaments, and we proceed using Green's functions as follows.

Each vector component of (4.8) is of a similar form to the Poisson equation, so using Green's functions for fields vanishing at infinity, the solution is

$$\mathbf{A} = \frac{\mu_0}{4\pi} \int_V \frac{\mathbf{J}\, dV}{R}, \tag{4.9}$$

where R is the distance between the point where $\mathbf{A}$ is determined, and the elemental volume dV containing the current density $\mathbf{J}$. For a line current, we have $\mathbf{J}\, dV = I\, d\mathbf{l}$, where $d\mathbf{l}$ is the vector elemental length along the current direction, as in figures 4.2(b) and 4.4. For a straight line current segment of length l, along a unit vector $\hat{\mathbf{z}}$, we finally have

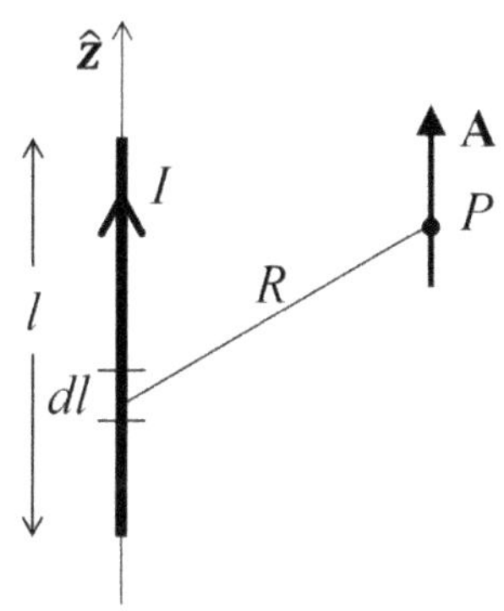

Figure 4.4. The magnetic vector potential $\mathbf{A}$ at a point P, sourced by a straight line current segment length l, current I, along a unit vector $\hat{\mathbf{z}}$, according to (4.10). Note that $\mathbf{A}$ is always parallel to its source current.

$$\mathbf{A} = \left(\frac{\mu_0 I}{4\pi}\right)\left[\int_l \frac{dl}{R}\right]\hat{\mathbf{z}}, \tag{4.10}$$

which scales with a purely geometrical integral factor $\left[\int_l \frac{dl}{R}\right]$, because the current I can only be constant along the segment.

With reference to (4.10) and figure 4.4, note that the magnetic vector potential $\mathbf{A}$ at a given point P is always, and everywhere, parallel (and proportional) to the current. This holds true for all points in space surrounding the finite-length wire, even up to its ends, so the calculations remain exact despite fringing fields which are usually the bugbears of electromagnetic field estimates. This property does not depend on two-dimensional symmetry arguments—for example, for an infinite wire along $\hat{\mathbf{z}}$, where the only component is A_z. This remarkable parallelism is extremely practical for interpreting the contour integrals in the next steps and beyond.

4.3.5 Step 5: partial contributions to the magnetic vector potential

Our next task is to understand how the currents on the different sides of the rectangle contribute to the net magnetic vector potential at a point on the contour. To take an example, we wish to know how to calculate the integrand $A^{\text{net}}_{\text{left}\|}$ in (4.4) and figure 4.3. Using the parallelism result from step 4.3.4, figure 4.5 is intended to show the partial contributions of the $\mathbf{A}$ vectors of the four currents, acting at one point on the left side. Clearly, the partial magnetic vector potentials from the top and base currents, $\mathbf{A}_{\text{top}}$ and $\mathbf{A}_{\text{base}}$, respectively, have zero contribution to the parallel resultant of the dot product in (4.2), $A_{\text{left}\|}$, because they are perpendicular to the left side contour c_{left}. Equating the parallel components then, we immediately have, at every point along the left side:

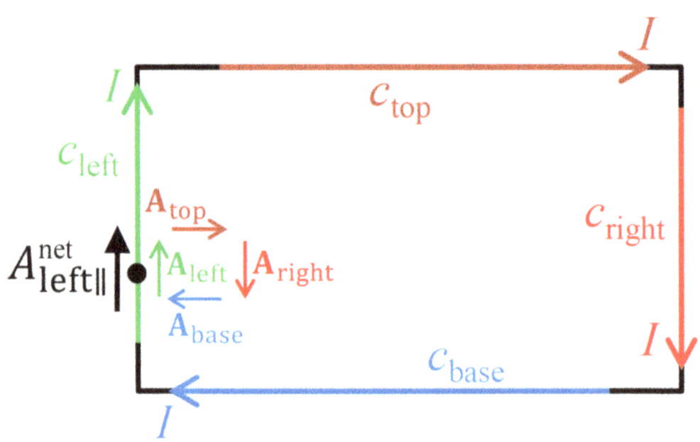

Figure 4.5. The magnetic vector potentials $\mathbf{A}_{\text{top}}$, $\mathbf{A}_{\text{right}}$, $\mathbf{A}_{\text{base}}$, and $\mathbf{A}_{\text{left}}$, which are sourced uniquely by the currents on their respective sides, are to be understood as partial contributions superposed at the same single point (the black dot) on the left side. These vectors and their source currents I on the four sides are colour coded on-line to aid identification. The net sum of only the parallel components make up the resultant parallel component $A^{\text{net}}_{\text{left}\|}$ at that point.

$$A_{\text{left}\|}^{\text{net}} = A_{\text{left}\|} - A_{\text{right}\|}, \tag{4.11}$$

respecting the opposite sign of the partial vector components because of the opposite direction of their anti-parallel source currents on the left and right sides of the rectangle in figure 4.5 (note that parallel currents would give an additive contribution in (4.11)). Therefore, only the currents in the parallel sides make partial contributions to $A_{\text{left}\|}$, and perpendicular currents do not contribute at all. This is an especially useful simplification for orthogonal networks such as ladder antennas, because we will see that stringers do not contribute to leg inductance, and legs do not contribute to stringer inductance. This allows one independent inductance calculation for the set of legs, and a separate calculation for the set of stringers.

4.3.6 Step 6: self partial inductance and mutual partial inductance

Direct integration of (4.11) along the left side contour c_{left} yields

$$\int_{c_{\text{left}}} A_{\text{left}\|}^{\text{net}}\, dl = \int_{c_{\text{left}}} A_{\text{left}\|}\, dl - \int_{c_{\text{left}}} A_{\text{right}\|}\, dl. \tag{4.12}$$

Substituting into (4.4), the net partial inductance contributed by the left side is

$$L_{\text{left}}^{\text{net}} = \frac{\int_{c_{\text{left}}} A_{\text{left}\|}^{\text{net}}\, dl}{I} = \frac{\int_{c_{\text{left}}} A_{\text{left}\|}\, dl}{I} - \frac{\int_{c_{\text{left}}} A_{\text{right}\|}\, dl}{I}. \tag{4.13}$$

The first contribution to the net partial inductance $L_{\text{left}}^{\text{net}}$ in (4.13) concerns only its magnetic vector potential component $A_{\text{left}\|}$, which is sourced by its own current I in the left side. Hence, for the left side, this is called the

$$\text{self partial inductance, } L_{\text{left}} = \frac{\int_{c_{\text{left}}} A_{\text{left}\|}\, dl}{I}, \tag{4.14}$$

where the label 'partial' in L_{left} is implicit throughout this book to avoid redundancy.

The second contribution to the net partial inductance $L_{\text{left}}^{\text{net}}$ in (4.13) concerns the magnetic vector potential component $A_{\text{right}\|}$, which is sourced by the current I in a different wire (the right-hand wire). Hence, this is called the mutual partial inductance between the left and right wires:

$$\text{mutual partial inductance, } M_{\text{left/right}} = \frac{\int_{c_{\text{left}}} A_{\text{right}\|}\, dl}{I}, \tag{4.15}$$

where $M_{\text{left/right}}$ is subtracted in (4.13), in order to account for the opposite direction of I in the right wire. Again, the label 'partial' in $M_{\text{left/right}}$ is implicit. Generally, mutual partial inductance is always reciprocal, i.e. $M_{\text{left/right}} = M_{\text{right/left}}$.

To recapitulate (4.13), the net partial inductance for the left side is

$$L_{\text{left}}^{\text{net}} = L_{\text{left}} - M_{\text{left/right}}, \tag{4.16}$$

which is the algebraic sum of its self partial inductance and the mutual partial inductance.

By extension, we can express the loop self inductance L^{loop} in (4.5) entirely in terms of its self partial, and mutual partial, inductances, by summing the net inductances around the contour as in section 4.3.3. This is summarized in (4.17) and pictorially in figure 4.6.

$$\begin{aligned} L^{\text{loop}} &= L_{\text{left}}^{\text{net}} + L_{\text{top}}^{\text{net}} + L_{\text{right}}^{\text{net}} + L_{\text{base}}^{\text{net}} \\ &= \left(L_{\text{left}} - M_{\text{left/right}}\right) + \left(L_{\text{top}} - M_{\text{top/base}}\right) \\ &\quad + \left(L_{\text{right}} - M_{\text{right/left}}\right) + \left(L_{\text{base}} - M_{\text{base/top}}\right) \\ &= \left(L_{\text{left}} + L_{\text{top}} + L_{\text{right}} + L_{\text{base}}\right) - 2\left(M_{\text{left/right}} + M_{\text{top/base}}\right). \end{aligned} \tag{4.17}$$

Before calculating explicit expressions for the self and mutual partial inductances, we summarize some of their useful properties here:

- The self and mutual partial inductances of any circuit carrying a current I can always be summed, respecting signs, to find the loop inductance. But the inverse is not true: partial inductances cannot be deduced from the loop inductance alone. Hence, the theory of partial inductance is the more fundamental concept. Loop inductance is just a special case of the more general theory of partial inductance [1, 9].
- The self partial inductance L of a wire is an intrinsic property of the wire, in a similar way that loop self inductance is an intrinsic property of a coil inductor. Stated bluntly, with reference to section 4.2.1, its L value could be printed on it; it depends only on the wire's physical dimensions (some correction must be made if the effective length of the current path in the wire is altered by connection to other components).

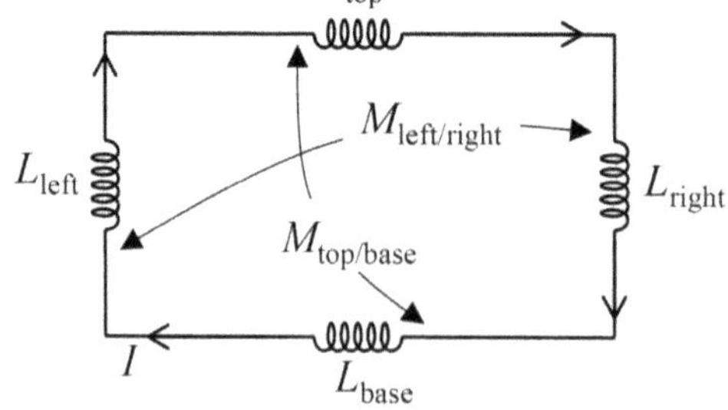

Figure 4.6. An equivalent circuit of the self partial inductances L, and mutual partial inductances M, which fully characterize the loop inductance L^{loop} of a rectangular current loop.

- Conversely, the mutual partial inductance M of a pair of wires depends on their spatial disposition with respect to each other. If there are several conductors, there is an M value associated with each of the other conductors. Mutual inductance is usually not important between loop inductances unless, of course, they are arranged as a transformer, where it is a lumped, or loop, mutual inductance.
- Calculations of self partial inductance in section 4.4 will further demonstrate that there is no equivalent concept in the conventional picture of loop self inductance. To be explicit, the question: 'What is the self inductance of a length of wire?' has no answer and, indeed, no meaning in the contemporary loop model of self inductance, because there is no clearly defined flux linkage. This was alluded to in the introduction, section 1.1.
- The net partial inductance L^{net} of a wire in a circuit with current I in every component, is the algebraic sum of its self partial inductance L and all of its mutual partial inductances, $\sum M_i$. This concept resolves the apparent dilemma in section 4.2.2, because we now see that the effective inductance of a wire in a circuit depends not only on itself, but also on the rest of the circuit, in quantifiable ways. Furthermore, a net partial inductance can be attributed to each and every element around the circuit—this is not possible in the loop inductance concept. The net partial inductance has no meaning for an isolated element, because any current-carrying element must be part of a closed circuit.
- Looking at figure 4.6, we can well imagine why the legs and stringers in chapter 3, specifically figure 3.3, were non-rigorously represented as lumped element loop self inductances. They were treated effectively as coil inductors, even though they are straight wires. The same applies to figures 1.2 and 1.3. We now know that the legs and stringers should be considered as partial inductances, and that the neglect of their significant mutual partial inductances is responsible for the errors in the calculated resonance frequencies in figures 3.8 and 3.13. This is a critical point for source design because the network resonance spectrum must be accurately matched to the frequency of the RF generator (usually 13.56 MHz or, often, its third harmonic, 40.68 MHz).
- In a ladder antenna, all of the legs are perpendicular to all of the stringers, and hence the leg partial inductances are independent of the stringer partial inductances (section 4.3.5).
- To answer the question posed by the title of section 4.2, partial inductance can be measured by judicious juxtaposition of differential voltage probes, as described in specialized books on electromagnetic compatibility (EMC), and using $\Delta V = j\omega L^{\text{net}} I$ [9]. A E Ruehli also shows how partial inductance can be measured using long measurement wires which are perpendicular to the plane of a planar circuit to be measured [3].
- From section 4.3.4, the magnetic vector potentials, and therefore partial inductances, depend on relatively simple geometric integrands. The integrals

are lengthy, but straightforward, as shown in section 4.4.1. Exact values can be calculated which account precisely for fringing fields.

- There is a 'dot convention' [1] which is used in circuit diagrams to indicate the polarity of induced voltages due to mutual (loop) inductance between coils, accounting for their relative orientation(s). However, straight wire inductances have no such ambiguity because they have no windings: the mutual partial inductance flux is additive for parallel currents, and subtractive for anti-parallel currents (see section 4.5.2). The set of I_n leg currents, and the set of J_n, K_n currents, are each consistently drawn in a parallel manner (see figure 3.5 and later also figure 4.13), so the mutual partial inductances are all additive in the algebra of (4.31), and there is no need here to use the dot convention explicitly in figure 4.13.

4.4 Calculation of the partial inductance of wires

Figure 4.7 now shows the thickness of the wires, radius a, in the rectangular loop. An important point is that at RF frequencies, the skin depth in copper wires is so small (66 μm at 1 MHz) that the RF current flows effectively only along the conductor surfaces [8]; all electric fields and currents are screened from penetrating into the wire. Consequently, there is no current nor magnetic flux within the conductors, and so no internal self inductance [1]. All of the magnetic flux sourced by the skin current occurs only in space outside the wire itself, and so the external self inductance must be calculated within the inner contour shown in figure 4.7 [1]. The skin current is assumed to be uniformly distributed around the wire circumference (proximity effects due to neighbouring currents are ignored for $a \ll w, l$ [1]), so by symmetry, the wire current is taken to be effectively on its axis for the calculation of magnetic vector potential and magnetic flux density.

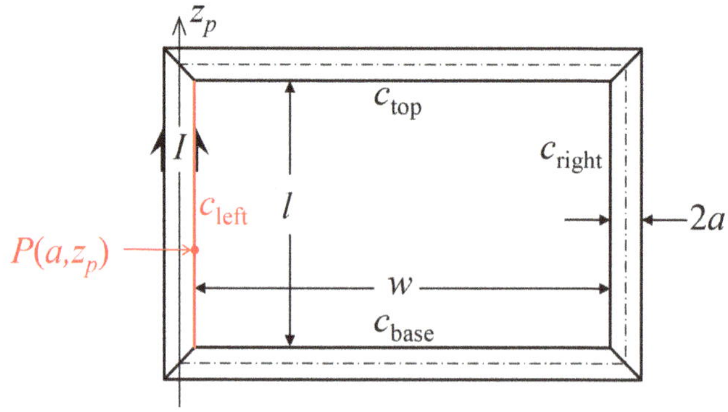

Figure 4.7. Another schematic of the rectangular loop of wire, but this time showing the thickness of the wire, radius a. The dashed rectangle marks the axis of the wires. The split arrow denotes the RF current I flowing in the skin depth along the surface of the wires. The magnetic flux links the circuit entirely within the inner contour, length l, width w. The magnetic vector potential is to be integrated around this inner contour, marked by c_{left}, c_{top}, c_{right}, and c_{base} [1]. The self partial inductance of the left wire is found by integrating its $A_{\mathrm{left}\|} = A_z(a, z_p)$ component at point $P(a, z_p)$ along the c_{left} contour (shown in red on-line) along $\Delta z_p = l$.

It is instructive to follow the calculation of the self partial inductance of a wire. This can be done in two ways: in section 4.4.1 by direct integration of the magnetic vector potential along the wire surface using (4.14); or by construction of a flux surface in section 4.4.2. Both methods require a two-stage integration.

4.4.1 Partial inductance by integration of the magnetic vector potential

All of the preceding explanation of partial inductance was developed purely in terms of magnetic vector potentials, so it seems logical to pursue the integration of (4.14) and (4.15) to obtain the self and mutual partial inductances. First, we adapt figure 4.4 in figure 4.8 to show the co-ordinates of interest for the integration. By Ampère's law, the magnetic field for $r \geqslant a$ due to an axially symmetric skin current at $r = a$ is identical to the field of the same current on the axis. The contour segment represents c_{left} in figure 4.7. To calculate $\mathbf{A}$ at the point $P(a,\, z_p)$, we integrate (4.10) along the wire from $z_p = -l/2$ to $l/2$:

$$\begin{aligned} A_z(a,\, z_p) &= \left(\frac{\mu_0 I}{4\pi}\right) \int_{-l/2}^{l/2} \frac{dz}{R}, \\ &= \left(\frac{\mu_0 I}{4\pi}\right) \int_{-l/2}^{l/2} \frac{dz}{\sqrt{(z_p - z)^2 + a^2}}, \\ &= \frac{\mu_0 I}{4\pi}\left[\sinh^{-1}\left(\frac{z_p + l/2}{a}\right) - \sinh^{-1}\left(\frac{z_p - l/2}{a}\right)\right], \end{aligned} \tag{4.18}$$

using Pythagoras' theorem for R, and a standard integral $\int dx/\sqrt{x^2 + a^2} = \sinh^{-1}(x/a)$. The *faux problème* of apparent divergence for an infinitely long wire, $l \to \infty$, is given short shrift in section 8.5.1.

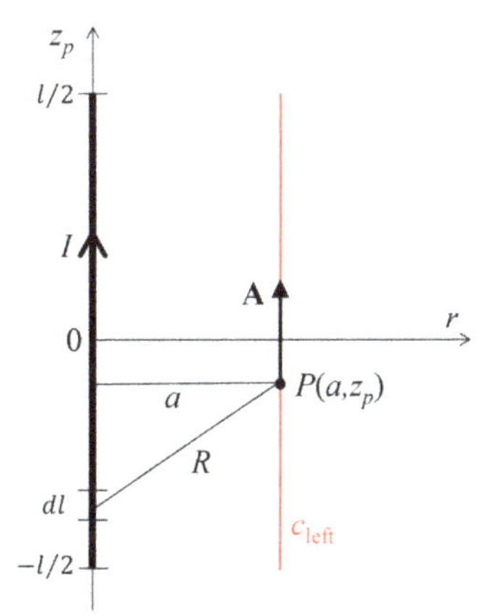

Figure 4.8. Magnetic vector potential $\mathbf{A}$ at a fixed point $P(a,\, z_p)$, according to (4.10) and figure 4.4. The wire of length l, radius a, is positioned centrally about the origin, with its effective current, by axial symmetry, along the z-axis [1]. c_{left} (in red) corresponds to the left contour in figure 4.7.

The second stage is to substitute (4.18) into the formula for the self partial inductance (4.14). To do this, the point $P(a,\ z_p)$ moves along the contour c_{left} for $-l/2 \leqslant z_p \leqslant l/2$, and $A_{\text{left}\|}$ corresponds to the left wire inner surface, i.e. $A_{\text{left}\|} = A_z(a,\ z_p)$ at point $P(a,\ z_p)$ in figure 4.7. The self partial inductance of the left wire is [1]

$$
\begin{aligned}
L_{\text{left}} &= \frac{\int_{c_{\text{left}}} A_{\text{left}\|}\, dl}{I} = \frac{\int_{c_{\text{left}}} A_z(a,\ z_p)\, dl}{I}, \\
&= \frac{\mu_0}{4\pi}\int_{-l/2}^{l/2}\left[\sinh^{-1}\left(\frac{z_p + l/2}{a}\right) - \sinh^{-1}\left(\frac{z_p - l/2}{a}\right)\right] dz_p, \\
&= 2\frac{\mu_0}{4\pi}\int_{\lambda=0}^{l}\left(\sinh^{-1}\frac{\lambda}{a}\right) d\lambda, \\
&= \frac{\mu_0}{2\pi}\left[\lambda \sinh^{-1}\frac{\lambda}{a}\right]_0^l - \frac{\mu_0}{2\pi}\int_0^l \frac{\lambda}{\sqrt{\lambda^2 + a^2}}\, d\lambda, \\
&= \frac{\mu_0}{2\pi}\left[l \sinh^{-1}\frac{l}{a}\right] - \frac{\mu_0}{2\pi}\left[\sqrt{\lambda^2 + a^2}\right]_0^l, \\
&= \frac{\mu_0}{2\pi}\left(l \sinh^{-1}\frac{l}{a} - \sqrt{l^2 + a^2} + a\right), \\
L_{\text{left}} &= \frac{\mu_0}{2\pi} l\left(\sinh^{-1}\frac{l}{a} - \sqrt{1 + \frac{a^2}{l^2}} + \frac{a}{l}\right),
\end{aligned}
\tag{4.19}
$$

$$
L_{\text{left}} = \frac{\mu_0}{2\pi} l\left[\ln\left(\frac{l}{a} + \sqrt{\frac{l^2}{a^2} + 1}\right) - \sqrt{1 + \frac{a^2}{l^2}} + \frac{a}{l}\right], \tag{4.20}
$$

using, successively, $\lambda = z_p \pm l/2$, that $\sinh^{-1}x$ is an odd function, integration by parts, and the identity $\sinh^{-1}\frac{l}{a} = \ln\left(\frac{l}{a} + \sqrt{\frac{l^2}{a^2} + 1}\right)$. This is an exact expression for the self partial inductance of a cylindrical wire, length l, radius a, which accounts precisely for fringing fields [1, 3, 5, 6]. There is no comparable term in loop inductance models.

For the leg dimensions in section 3.1, namely l = 19 cm and a = 0.3 cm, expressions (4.19) and (4.20) each give L_{leg} = 146.6 nH. For the case $a \ll l$ in (4.20), a simpler approximation is often used [11–13],

$$
L_{\text{left}} \approx \frac{\mu_0}{2\pi} l\left[\ln\left(\frac{2l}{a}\right) - 1\right]. \tag{4.21}
$$

This value is $L_{\text{leg}} \approx$ 146.0 nH, as in section 3.1, which is smaller than the exact value by only 0.4%, and $a/l = 0.016 \ll 1$, consistent with this good approximation.

Note that the physical length of the wire c_{left} in figure 4.7 could be anything from l, to $l + 4a$, depending on the details of the wire's connection to the adjacent wires.

For example, in the experimental set-up, section 3.1, the legs are 20 cm long, but the best agreement was found for $l = 19$ cm. This 'electrical length' is shorter than the physical length by $\Delta l = 1$ cm $= 3.\dot{3}a$ (which is in the physically reasonable range $0 < \Delta l < 4a$) because the RF skin current takes the most direct internal short cut via the stringer strip connectors.

Finally, the mutual partial inductance in (4.15) involves the magnetic vector potential of the right wire, integrated along the same contour as before, c_{left}. In figure 4.7, we see that the right wire current axis is at a distance $(w + a)$ from c_{left}, instead of a. Hence the exact mutual partial inductance is simply obtained by replacing a, in (4.19) or (4.20), by $s = (w + a)$:

$$M_{\text{left/right}} = \frac{\mu_0}{2\pi} l \left[\ln\left(\frac{l}{s} + \sqrt{\frac{l^2}{s^2} + 1} \right) - \sqrt{1 + \frac{s^2}{l^2}} + \frac{s}{l} \right], \tag{4.22}$$

Again, the approximation

$$M_{\text{left/right}} \approx \frac{\mu_0}{2\pi} l \left[\ln\left(\frac{2l}{w} \right) - 1 \right], \tag{4.23}$$

is often used when $a \ll w \ll l$.

For the ladder antenna in section 3.1, the leg axis spacing is $(w + 2a) = 2.5$ cm, therefore $s = (w + a) = 2.2$ cm. We calculate that the mutual partial inductance between neighbouring legs is $M_{\text{leg/leg}} = 74.5$ nH. Therefore the net partial inductance of the legs is significantly different, by approximately 50%, compared to the 146 nH self partial inductance used in chapter 3. It is not surprising that the mode frequency estimations were strongly in error in figures 3.8 and 3.13.

4.4.2 Partial inductance by integration of the magnetic flux density

The calculation of partial inductance by direct integration of **A** in section 4.4.1 was the logical approach, although the results are rather unsatisfyingly buried in algebra. A more physical image of partial inductance will now be constructed [1], using the parallelism picture in figure 4.4, but there will be no escaping a two-stage integration.

We start from the same equation (4.14) as in section 4.4.1, namely

$$\text{self partial inductance, } L_{\text{left}} = \frac{\int_{c_{\text{left}}} A_{\text{left}\parallel}\, dl}{I}. \tag{4.24}$$

But this time, by 'reverse engineering' Stokes' theorem in (4.2), we search for a convenient surface integral which is equivalent to the contour integral in (4.24). This desired surface is constructed in figure 4.9, and is justified as follows. As in section 4.3, the magnetic flux surface integral $\int_S \mathbf{B} \cdot d\mathbf{S} = \oint_c \mathbf{A} \cdot d\mathbf{l}$, the line integral around the whole contour of the dashed rectangle. Taking advantage of the parallelism

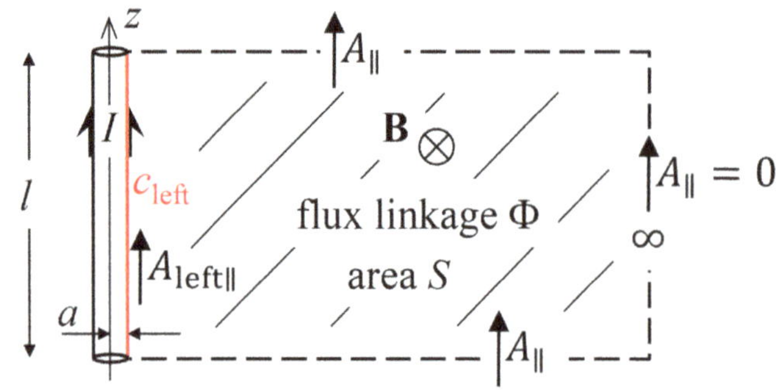

Figure 4.9. Calculation of the self partial inductance L of a wire: The magnetic flux Φ, linking the dashed rectangular area from the wire surface to infinity, is mathematically equivalent to the required line integral of $A_{\text{left}\|}$ along the wire surface.

result in figure 4.4, we see that $\int_c \mathbf{A} \cdot d\mathbf{l}$ is zero for the top and bottom sides. Furthermore, because the right contour is at infinity, the magnetic vector potential on the right is zero. Hence, the magnetic flux integral $\Phi = \int_S \mathbf{B} \cdot d\mathbf{S} = \oint_c \mathbf{A} \cdot d\mathbf{l}$ in figure 4.9 is equivalent to $\int_{c_{\text{left}}} A_{\text{left}\|}\, dl$ alone. It directly follows from (4.24) that [1]

> the self partial inductance L of a line segment is mathematically equivalent to the magnetic flux between the current segment and infinity, divided by its current.

It is to be understood that the area is projected perpendicularly outwards from the wire, so that **A** is perpendicular to the lateral sides of the rectangle. The magnetic flux density **B** can be calculated by integrating the Biot–Savart law along the line current segment, and the flux follows from the area integral of **B**. Suffice it to say that the expression for L_{left} is identical to (4.19) and (4.20) [1].

It is important to realize that this area integral of **B** is not some sort of a flux calculation for the loop self inductance of a finite-length wire! The loop inductance concept requires the flux linkage to be integrated over the area within a current loop, but no return current path is defined here. Even if it were assumed that current returns via some path at infinity, then the flux would have to be integrated over all space, not only the rectangle in figure 4.9, which leads to divergence.

The mathematical construct of figure 4.9 is used implicitly by E B Rosa [5], F W Grover [6], and others. The fact is that **B** is non-zero all around the wire, including fringing fields near the ends, and so the real magnetic flux also exists outside of the artificial integration zone. This is why the integration limits employed by Rosa *et al* are so confusing if erroneously interpreted as a loop inductance calculation on the basis of $L^{\text{loop}} = \Phi/I = \int_S \mathbf{B} \cdot d\mathbf{S}/I$. The mathematical construction corresponds to the real physical case only for infinite parallel wires, where there are no ends and therefore no fringing fields.

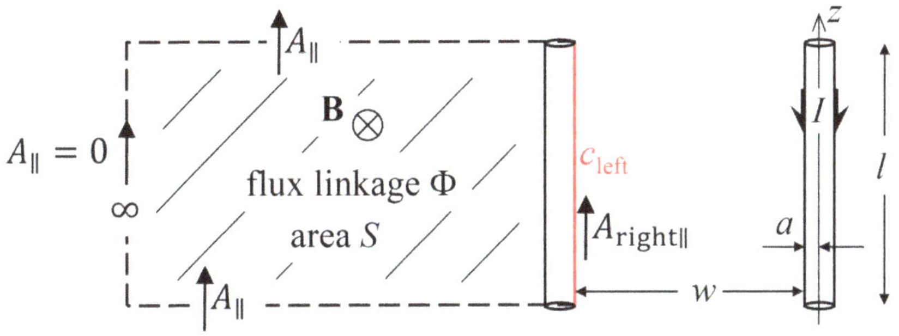

Figure 4.10. Calculation of the mutual partial inductance $M_{\text{left/right}}$ between two wires: The magnetic flux Φ, linking the dashed rectangular area from the left wire to infinity, is mathematically equivalent to the line integral of $A_{\text{right}||}$, along the left wire surface, due to the current in the right wire. The orientation is the same as in section 4.3 for reference.

Mutual partial inductance is obtained by using the same construction as above, but now with reference to figure 4.10. The magnetic flux integral $\Phi = \int_S \mathbf{B} \cdot d\mathbf{S} = \oint_c \mathbf{A} \cdot d\mathbf{l}$ in figure 4.10 is equivalent to $\int_{c_{\text{left}}} A_{\text{right}||}\, dl$ alone. It directly follows from (4.15) that:

> the mutual partial inductance $M_{1/2} = M_{2/1}$ between two line segments is mathematically equivalent to the magnetic flux, due to unit current in one segment, between the other segment and infinity on its far side.

It is to be understood that the area is projected perpendicularly outwards from the second wire, so that $\mathbf{A}$ is perpendicular to the lateral sides of the rectangle. The mutual partial flux adds to, or subtracts from, the self partial flux according to whether the currents are parallel, or anti-parallel, respectively.

The combination of sections 4.4.1 and 4.4.2 gives the best of both worlds; the exact analytical expressions, and a physical interpretation, for calculating partial inductance. There is a third, related method called the Neumann integral [1] which also shows that inductance is purely a geometrical construct depending on dimensions, shape, and relative orientation [1]. The inductance unit henry (H) comes from the factor μ_0 (H m^{-1}) multiplied by the integral whose overall dimension is length (m).

4.5 Relevant special cases of partial inductance

We will consider two example cases of partial inductance to prepare the reader before plunging into the matrix of resonant networks.

4.5.1 Parallel wires with anti-parallel currents

To draw a comparison with more familiar textbook expressions for (loop) self inductance, we consider two parallel wires. Figure 4.11(a) shows two short ($l \ll \lambda$) wire segments as connecting wires between other parts of some arbitrary circuit

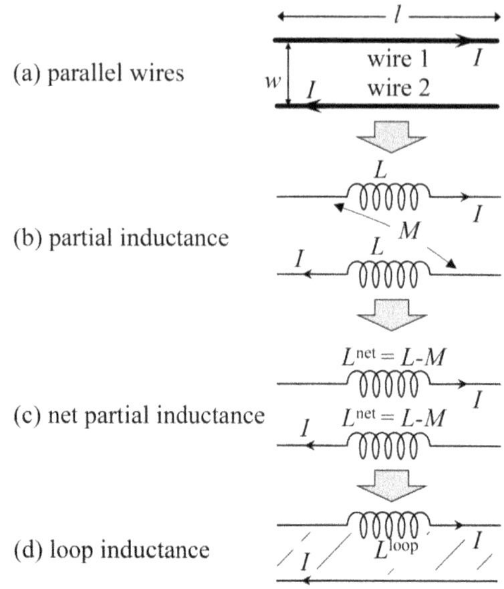

Figure 4.11. Four equivalent descriptions of short, go-and-return connecting wires, radius a. (a) Two parallel wires, length l, spacing w; (b) in terms of self partial inductances L, and their mutual partial inductance M; (c) the partial inductances combined into net partial inductances L^{net}; and (d) the loop self inductance picture.

(not shown). This figure recalls the lumped element bifilar inductance in section 2.3. Exact values for the self partial inductance L and mutual partial inductance M of each wire in figure 4.11(b) are given in (4.20) and (4.22), respectively; approximations are in (4.21) and (4.23).

In figure 4.11(c) the net partial inductance of each identical wire is $L^{\text{net}} = L - M$, and the total inductance summed around the wires, which is the loop self inductance in figure 4.11(d), is $L^{\text{loop}} = 2L^{\text{net}} = 2L - 2M$. Let us first calculate this value using the approximate expressions (4.21) and (4.23):

$$\begin{aligned} L^{\text{loop}} &\approx 2\frac{\mu_0}{2\pi}l\left[\ln\left(\frac{2l}{a}\right) - 1\right] - 2\frac{\mu_0}{2\pi}l\left[\ln\left(\frac{2l}{w}\right) - 1\right], \\ &\approx 2\frac{\mu_0}{2\pi}l\left[\ln\left(\frac{2l}{a}\cdot\frac{w}{2l}\right)\right], \\ &\approx \frac{\mu_0}{\pi}l\left[\ln\left(\frac{w}{a}\right)\right]. \end{aligned} \tag{4.25}$$

The loop self inductance per unit length is therefore

$$\hat{L}^{\text{loop}} = \frac{L^{\text{loop}}}{l} \approx \frac{\mu_0}{\pi}\left[\ln\left(\frac{w}{a}\right)\right], \tag{4.26}$$

which is the same classical result[3] as in standard textbooks [8] for the (loop) self inductance per unit length of a two-wire transmission line of infinite length, wire radii a separated by distance w. We repeat, nonetheless, that it is not clear where, or how, to place the loop inductance in figure 4.11(d).

It is reassuring to be able to reproduce this classical result, but partial inductances are even more versatile [14]. For example, we can calculate the exact inductance of the finite-length wires in figure 4.11(a) by using, this time, the exact expression (4.22) for the mutual partial inductance, instead of the approximation (4.23) which is tantamount to assuming infinite wires ($w \ll l$). To give a convenient numerical example, we choose the wire length to be the same as the effective spacing, $l = w$ in (4.22), and with very thin wires, $a \ll l$, so that the approximation (4.21) for L is accurate. Now, the total inductance is

$$\begin{aligned}
2L^{\rm net} &= 2L - 2M, \\
&= 2\frac{\mu_0}{2\pi}l\left[\ln\left(\frac{2l}{a}\right) - 1\right] - 2\frac{\mu_0}{2\pi}l\left[\ln\left(\frac{l}{l} + \sqrt{\frac{l^2}{l^2} + 1}\right) - \sqrt{1 + \frac{l^2}{l^2}} + \frac{l}{l}\right], \\
&= \frac{\mu_0}{\pi}l\left[\ln\left(\frac{2l}{a}\right) - 1\right] - \frac{\mu_0}{\pi}l\left[\ln\left(1 + \sqrt{2}\right) - \sqrt{2} + 1\right], \qquad (4.27) \\
&= \frac{\mu_0}{\pi}l\left[\ln\left(\frac{l}{a}\right) + \ln 2 - \ln\left(1 + \sqrt{2}\right) + \sqrt{2} - 2\right], \\
&= \frac{\mu_0}{\pi}l\left[\ln\left(\frac{w}{a}\right) - 0.774\right],
\end{aligned}$$

substituting $w = l$ in the last line. Compared to the value for the infinite line (4.25), the term -0.774 represents the exact effect of fringing fields and end effects which are not accounted for in the classical expression [14].

On a practical note, narrow spacing (small w) of the wires increases M, thereby reducing $L^{\rm net}$ and $L^{\rm loop}$, which is a common technique to diminish stray inductance in go-and-return connecting wires.

4.5.2 Mutual partial inductance of offset parallel wires

Leg mutual partial inductances

Mutual partial inductance between the legs is the simplest case in figure 4.12(a) because the expressions are given directly by (4.22) and (4.23) for parallel wires. The mutual partial inductance M_{nq} between two parallel wires n and q is always positive, but the sign of the mutual partial *flux* contribution of wire q, $M_{nq}I_q$, to the self partial flux of wire n, L_nI_n, depends on the relative direction of the currents: the mutual partial flux is additive for the case of parallel currents, and subtractive for anti-parallel currents in sections 4.3 and 4.5.1. This can be understood by inspection of the flux

[3] For an infinite wire, $B = \frac{\mu_0 I}{2\pi r}$. The flux per unit length from a across to w is $\hat{\Phi} = \int_a^w B\,dr = \frac{\mu_0 I}{2\pi}\ln\frac{w}{a}$. Summing the flux from both wires, the loop self inductance per unit length is $\hat{L}^{\rm loop} = 2\hat{\Phi}/I = \frac{\mu_0}{\pi}\ln\frac{w}{a}$.

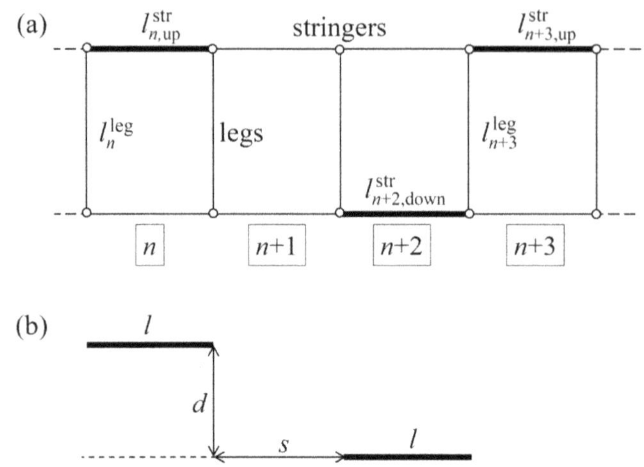

Figure 4.12. (a) Schematic of four sections of a ladder antenna, n to $(n + 3)$, with some stringers marked to show the various orientations necessary for mutual partial inductance calculations. (b) Geometry of parallel, offset filaments of equal length l, distance d apart and offset by s, for calculation of the mutual partial inductances.

diagrams in section 4.4.2. The set of I_n leg currents are consistently drawn in a parallel manner (see figure 3.5 and later also figure 4.13), therefore the leg mutual partial flux contributions, $M_{nq}I_q$, are all additive in section 4.7. Fortunately, section 4.3.5 finds that stringers, being perpendicular to legs, do not contribute to leg inductance. This is the only situation for which conductors are locally de-coupled [3].

Stringer mutual partial inductances

Figure 4.12(a) shows that stringer mutual inductance is more complicated than for legs. For example, the stringer labelled $l^{str}_{n,\,up}$ has in-line mutual partial inductance with every upper stringer such as $l^{str}_{n+3,\,up}$. Not only this, but $l^{str}_{n,\,up}$ also has mutual partial inductance with every lower stringer such as $l^{str}_{n+2,\,down}$. These mutual partial inductances can be calculated using the same integrating procedure for magnetic vector potential as in section 4.4.1. General expressions[4] are given by Grover [6] and demonstrated by Paul [1]. Figure 4.12(b) shows the simplest geometry which is sufficiently general to describe all of our mutual partial inductances used for ladder antennas. The pair of conductors have always the same length l, and the leg tubes, or stringer strips, are replaced by

[4] For unequal-length parallel offset filaments in figure 4.12(b), with lengths l and m, Paul [1] derives

$$M = \frac{\mu_0}{4\pi}\left[z_2 \sinh^{-1}\frac{z_2}{d} - z_1 \sinh^{-1}\frac{z_1}{d} - (z_2 - m)\sinh^{-1}\frac{(z_2 - m)}{d} + (z_1 - m)\sinh^{-1}\frac{(z_1 - m)}{d} - \sqrt{z_2^2 + d^2} + \sqrt{z_1^2 + d^2} + \sqrt{(z_2 - m)^2 + d^2} - \sqrt{(z_1 - m)^2 + d^2}\right],$$

where $z_1 = m + s$ and $z_2 = m + s + l$. This is identical to Grover [6] when the notation is changed to $s \to \delta$, $z_1 \to \gamma$, $z_2 \to \alpha$, $(z_1 - m) \to \delta$, and $(z_2 - m) \to \beta$.

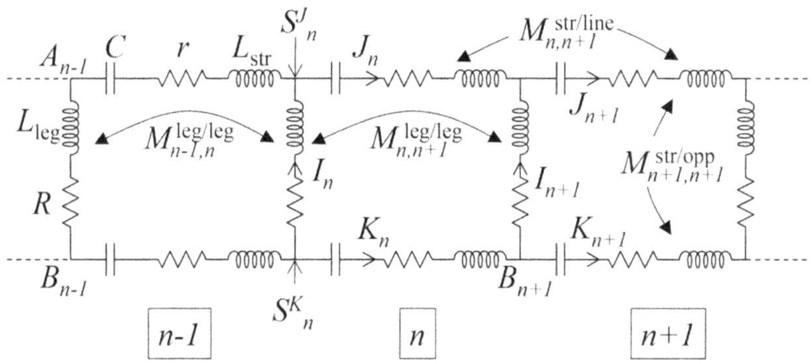

Figure 4.13. The high-pass dissipative ladder network of figure 3.3 re-drawn to include mutual partial inductances between the legs and stringers. The antenna network shows meshes $\{\ldots, n-1, n, n+1, \ldots\}$, with leg self partial inductances L_{leg} and resistance R, linked by stringers with capacitors C and short connectors of self partial inductance L_{str} and combined resistance r. The arrows represent coupling by mutual partial inductances. I_n are the leg currents, J_n and K_n are, respectively, the upper and lower stringer currents. S_n^J and S_n^K are the source currents, and A_n and B_n the voltages, at the upper and lower nodes, respectively. Mutual partial inductances are shown between nearest neighbours only, for clarity. (Adapted with permission from [12]. Copyright 2015 IOP Publishing.)

filaments coincident with their axis (except for self partial inductance—see below). The mutual partial inductance of the filament pair in figure 4.12(b) is

$$M = \frac{\mu_0}{4\pi}\left[(2l+s)\sinh^{-1}\frac{(2l+s)}{d} - 2(l+s)\sinh^{-1}\frac{(l+s)}{d} + s\sinh^{-1}\frac{s}{d}\right.$$
$$\left. - \sqrt{(2l+s)^2+d^2} + 2\sqrt{(l+s)^2+d^2} - \sqrt{s^2+d^2}\right]. \tag{4.28}$$

This gives a good approximation for separation d much larger than the conductor cross-section width. For in-line stringers: $l = l^{\text{str}}$, $d = 0$. For stringers on opposite sides of the antenna: $l = l^{\text{str}}$, $d = l^{\text{leg}}$. In both cases, s is an appropriate multiple (positive or negative) of l^{str}.

Self partial inductances

As just mentioned, the conductors in figure 4.12(b) are considered as filaments with their current placed on an effective axis [6]. This is adequate for mutual partial inductance where the pairs of conductors are widely spaced. However, self partial inductance, which is a special case of mutual partial inductance [1, 6], depends sensitively on the spatial distribution of current in the conductor because of the intimate proximity of the current density and its self-generated magnetic field. The self partial inductance therefore depends on the shape of the conductor cross-section, not forgetting the skin effect which concentrates the RF current on the conductor's perimeter. The experimental ladder antenna of section 3.1 consists of hollow cylindrical tubes for the legs, and strip connectors for the capacitor assemblies in the stringers. The following approximations were used.

For the cylindrical legs length l, radius a, as used in [1, 6, 11] and section 4.4.1:

$$L_{\text{leg}} \approx \frac{\mu_0}{2\pi} l_{\text{leg}} \left[\ln \left(2l_{\text{leg}}/a \right) - 1 \right]. \tag{4.29}$$

For a stringer strip length l, width b, as used in [6, 12, 15].

$$L_{\text{str}} \approx \frac{\mu_0}{2\pi} l_{\text{str}} \left[\ln \left(2l_{\text{str}}/b \right) + \frac{1}{2} \right]. \tag{4.30}$$

These analytical expressions for mutual and self partial inductances are employed in programs whose link is given in appendix K. Self partial inductance for other cross-sectional shapes are given in [1, 3, 6]. Failing an analytical expression, self or mutual partial inductance can be computed numerically [15].

4.6 Antenna equivalent circuit including mutual partial inductance

The antenna equivalent circuit of figure 3.3 can now be improved by accounting for the mutual inductances between all of the parallel elements, as shown in figure 4.13. The antenna circuit in figure 4.13 is still a balanced high-pass filter ladder network, and the lumped element equivalent circuit remains a valid approximation provided that the antenna dimensions are small compared to the wavelength of the RF excitation (22 m in vacuum at 13.56 MHz). The legs have self partial inductance L_{leg} determined by their geometry (see 4.5.2), with a small resistance R estimated from their skin depth resistance in copper. The high-quality-factor capacitor assemblies C in the stringers have copper strips linking the legs, having self partial inductance L_{str} and resistance r given by the skin depth resistance plus the effective series resistance (ESR) of the capacitor assemblies.

4.7 Mutual partial inductance matrix equations

In contrast to the previous analysis of the planar resonant antenna [16, 17] in chapter 3, the mutual partial inductances between all antenna elements are accounted for in this section. The resulting impedance matrix model (figure 4.14) is used to calculate the input impedance of the antenna, and the distribution of currents and voltages for any symmetrical arrangement of RF power feeding and ground connections.

The values of the resistances, capacitances, and self partial inductances are independent of their position in the circuit of figure 3.3, but the mutual partial inductances of the legs and the stringers do depend on their location in the network of figure 4.13. This is because their inductive coupling is different according to the number, position, and currents of their neighbours. For example, current I_q in leg q couples a magnetic flux $M_{nq}^{\text{leg/leg}} I_q$ with leg n, where $M_{nq}^{\text{leg/leg}}$ is their mutual partial inductance which is a purely geometric quantity [1, 3, 11]. By the principle of superposition, this flux is independent of fields caused by other currents. The current I_q itself may be influenced by the circuit response to coupling with other circuits, but its flux coupled with leg n is always $M_{nq}^{\text{leg/leg}} I_q$, independent of fluxes produced elsewhere. The total flux coupled to leg n, due to currents in all of the legs, is

Figure 4.14. 'The Matrix model.' (Illustration by Alex Howling. Copyright 2023 Alex Howling.)

therefore $\sum_{q=1}^{N} M_{nq}^{\mathrm{leg/leg}} I_q$. Note that $M_{nn}^{\mathrm{leg/leg}} \equiv L_{\mathrm{leg}}$ is the self partial inductance of a leg [1, 6]. In chapter 3, all of the mutual partial inductances were neglected, $M_{nq}^{\mathrm{leg/leg}} = 0$ for $n \neq q$. We recall also that the set of I_n leg currents in figure 4.13 are all drawn parallel, so that the leg mutual partial fluxes are additive.

Because the net partial inductance of each element depends on its position $\{1, 2, \ldots, n, \ldots, N\}$ in the network, and a different current I_q in each element, a matrix approach will be necessary to calculate the antenna impedance. An advantage of the partial inductance approach [1, 3] is that the impedance matrix of any network can be built up from the mutual partial inductances between pairs of elements, which are easily calculated.

To define the impedance matrices, we first consider Ohm's law for the nth leg, referring to figure 4.13, where currents are taken to be time harmonic as usual. Applying the superposition principle for all currents and voltages,

$$B_n - A_n = RI_n + j\omega \sum_{q=1}^{N} M_{nq}^{\mathrm{leg/leg}} I_q, \tag{4.31}$$

where $j\omega M_{nq}^{\mathrm{leg/leg}} I_q$ is the voltage induced in leg n by a current I_q in leg q. This is the equivalent of the Ohm's law in (3.18) for normal modes, and (3.29) for the dissipative antenna, where mutual partial inductance was neglected ($q = n$ only).

Equivalently, Ohm's law in (4.31) can be expressed as a matrix equation for all the N legs using $N \times N$ matrices for the impedances, and column vectors for the voltages, $\mathbf{A}$, $\mathbf{B}$, and currents $\mathbf{I}$:

$$\mathbf{B} - \mathbf{A} = (\bar{R} + j\omega \bar{M}_{\mathrm{leg/leg}})\mathbf{I}, \tag{4.32}$$

where $\bar{R} = R\bar{I}$ is the leg resistance multiplied by the $N \times N$ identity matrix $\bar{I}$, and the mutual partial inductance matrix $\bar{M}_{\text{leg/leg}}$ is the $N \times N$ matrix for $M_{nq}^{\text{leg/leg}}$, as described in appendix B. The impedance matrix of the antenna legs is directly

$$\bar{Z}_{\text{leg}} = R\bar{I} + j\omega\bar{M}_{\text{leg/leg}}. \tag{4.33}$$

The stringer impedance matrix $\bar{Z}_{\text{str}}$ is derived similarly in appendix B; it is independent of the leg impedance matrix because the legs and stringers are perpendicular, therefore their mutual partial inductance is de-coupled [3]. Also, the set of J_n, K_n currents in figure 4.13 are all drawn parallel, so that all of the stringer mutual partial inductance terms are additive.

The required solutions for the antenna electrical properties are found using Kirchoff's laws, similarly to the analytical solution in section 3.5, except that now matrix equations are used because of the mutual inductance terms. The matrix solution is derived in appendix B, where the antenna currents, voltages, and input impedance are determined in terms of the impedance matrices $\bar{Z}_{\text{leg}}$ and $\bar{Z}_{\text{str}}$, along with the source current vectors $\mathbf{S}^J$ and $\mathbf{S}^K$. For example, the leg current distribution is given by

$$\mathbf{I} = \left(\bar{U}\,\bar{Z}_{\text{str}}^{-1}\,\bar{U}^T\,\bar{Z}_{\text{leg}} + 2\bar{I}\right)^{-1}(\mathbf{S}^K - \mathbf{S}^J), \tag{4.34}$$

where the matrix $\bar{U}$ is defined in appendix B, and the antenna input impedance is

$$Z_{\text{in}} = A_{\text{RFnode}}/S_{\text{RFnode}}^J, \tag{4.35}$$

where n= 'RFnode' refers to the node of the RF input connection, which was $n = 12$ (node A12) for all experiments presented here. Although the matrix solution can be expressed analytically, the values must be computed numerically because of the difficulty of matrix inversion and calculation of all the mutual inductances. Appendix K gives a link to the programs used for this calculation.

4.7.1 Effect of a planar screen on the mutual partial inductance matrix

In the last step before this mutual partial inductance model can be compared with measurements on a real antenna, the effect of image currents in the antenna baseplate screen must also be accounted for. The ladder antennas in this book are usually mounted above a copper baseplate, for example, in figure 3.2, which acts as a reference ground plane and screens the antenna fields from the external environment (and vice versa). Consequently, the inductive coupling of the antenna with its current image in the screen must be included. Fortunately, this is a relatively simple exercise if the screen is assumed to be an infinite, perfectly conducting plane.

First, consider a straight wire distance h above an infinite, perfectly conducting ground plane as shown in figure 4.15. According to the method of images, the magnetic field above the screen, produced by the induced current profile in the screen, is identical to the field produced by an image of the source current reflected in the screen [1, 2]. The image current is in the opposite direction, parallel and at an equal distance h below the screen. The mutual inductance between the source current in the wire and the induced current in the ideal screen can be calculated directly using the mutual partial inductance of parallel wires, as previously. The mutual partial inductance of an antenna leg with its own image in figure 4.16 at a distance $2h_{\text{screen}}$ is directly

$$M^{\text{leg/screen}} \approx \frac{\mu_0}{2\pi} l \left[\ln\left(\frac{l}{h_{\text{screen}}} \right) - 1 \right], \tag{4.36}$$

where the approximation is good for $h_{\text{screen}} \ll l$. The validity of using image currents to calculate mutual partial inductance is considered in more detail in section 8.5.1. Note that each leg is mutually coupled to all the leg image currents, not only its own image. The mutual partial inductance matrix due to all the leg image currents, $\bar{M}_{\text{leg/screen}}$, can be calculated using the general expression (4.28) in section 4.5.2. It then suffices to replace the antenna mutual inductance matrix $\bar{M}_{\text{leg/leg}}$, as used throughout this section 4.7 up to now, by $(\bar{M}_{\text{leg/leg}} - \bar{M}_{\text{leg/screen}})$, and similarly for the stringers. This is a particularly useful property of mutual partial inductances, where modifications and refinements can simply be retrospectively added, or 'plugged in', to the overall mutual partial inductance matrix. Figure 4.17 shows this operation

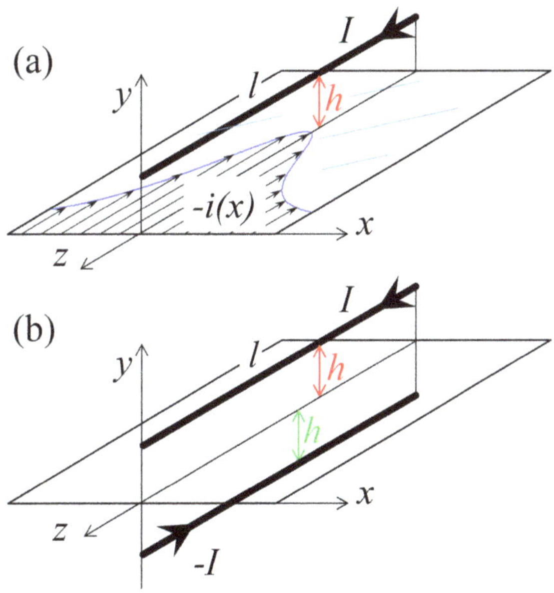

Figure 4.15. (a) Perspective view of a linear current source element length l, at a distance h (shown in red) above an infinite, perfectly conducting ground plane. The profile of the induced surface current on the plane is shown schematically. The integral along x of the current per unit width, $\int_{-\infty}^{\infty} - i(x)\, dx = -I$, is equal and opposite to the source current I. (b) To illustrate the method of images, the same source current is shown, with its equal and opposite image current at a distance h (shown in green) below the screen. (Adapted with permission from [11]. Copyright 2015 IOP Publishing.)

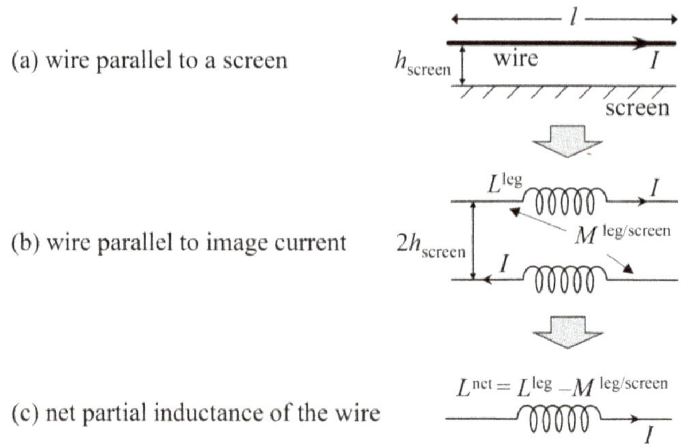

Figure 4.16. (a) The antenna leg parallel to a perfectly conducting screen, at distance h_{screen}, (b) the equivalent circuit of the wire with its own image current mutual partial inductance $M^{leg/screen}$, and (c) the net partial inductance of the wire due to the presence of the screen.

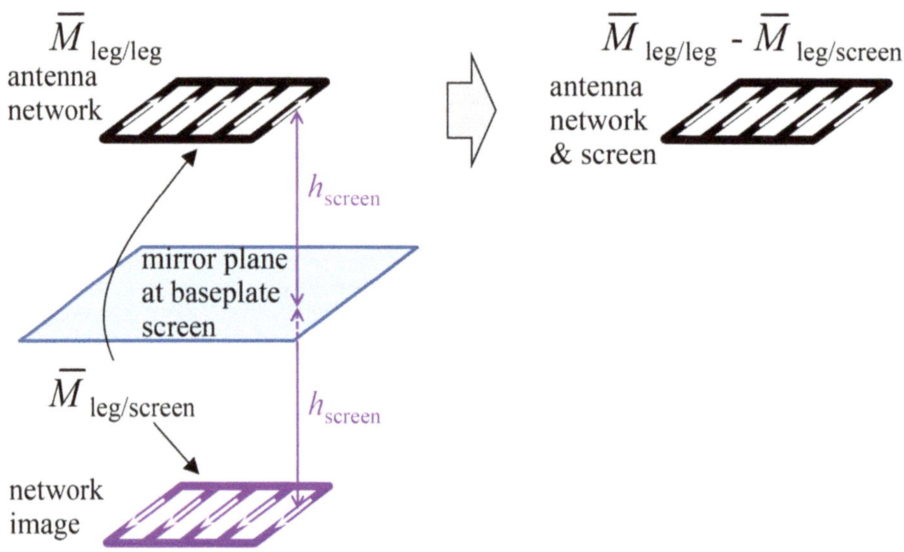

Figure 4.17. Left hand side: Schematic of a ladder antenna network, with mutual partial inductance $\bar{M}_{leg/leg}$ for the currents shown schematically in its legs, above a perfectly conducting, infinite screen. The network image has mutual partial inductance $\bar{M}_{leg/screen}$ with the antenna legs, with equal, but opposite, image currents. Right-hand side: The combined effective mutual partial inductance of the antenna and screen is therefore $(\bar{M}_{leg/leg} - \bar{M}_{leg/screen})$.

pictorially. The inductive coupling with plasma will also be treated in this fashion in part II, but for now, the calculation of Z_{in}—still without plasma but now including the screen—can finally be compared with the experimental measurements in vacuum already described in chapter 3.

4.8 Experiment and theory for an antenna without plasma

The validity of the mutual partial inductance matrix model is now investigated by comparison with experiments in air (no plasma) for the resonant network antenna, which include the baseplate screen.

4.8.1 Voltage distribution of the network modes

Comparison of voltage node measurements with the mutual inductance solution for antenna voltages shows excellent agreement in figure 4.18. Following (2.2) for the convention of comparison with measurements, the normalized voltage distribution is $\mathrm{Re}(A_n/A_{12})$, to respect the phase relation between the complex phasor amplitudes A_n and A_{12}. The solution given by the analytical approach from section 3.5, figure 3.7(b), is good although not as accurate; this confirms that the mode structure on the antenna is not strongly perturbed by the mutual inductance between antenna elements and the screen. When the excitation frequency is deliberately set far off resonance, even the distortion of the voltage distribution is reproduced in detail by the mutual inductance matrix model, see figure 4.19.

4.8.2 Matrix model of the input impedance without plasma

Accounting for the mutual partial inductances, the measured impedance spectra in figure 4.20 can be so accurately reproduced that the experimental data and the matrix model calculations are largely superposed to the point of being

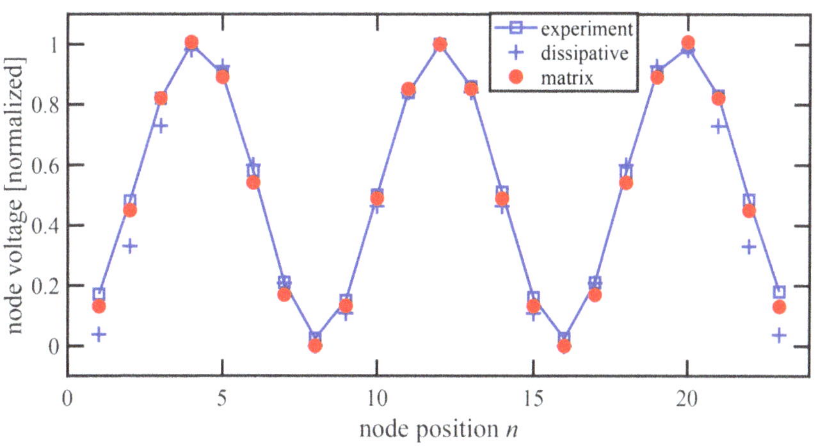

Figure 4.18. Normalized voltage distribution, $\mathrm{Re}(A_n/A_{12})$, for a m = 6 mode across the 23-leg network, $n = 1, 2, \ldots, 23$. Measurements ('experiment' blue line and squares) are compared with the analytical solution for a dissipative antenna ('dissipative' blue crosses) calculated in figure 3.7, and with the matrix model for mutual inductances ('matrix' red circles). The RF input, at 13.38 MHz, is at A12 and ground connections at A8 and A16. The matrix model shows even better agreement with experiment than the dissipative model, especially for the outer legs. (Adapted with permission from [12]. Copyright 2015 IOP Publishing.)

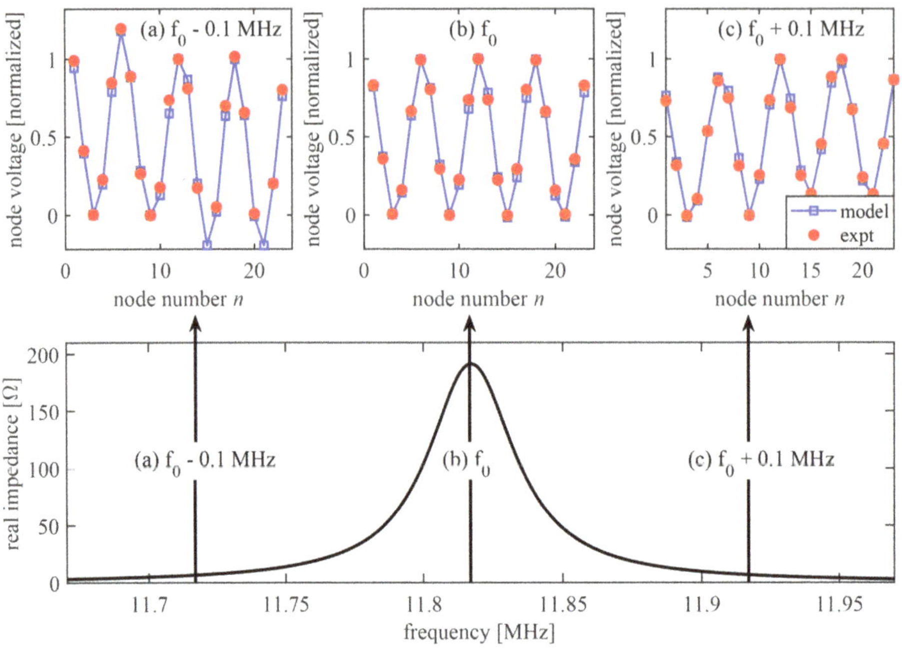

Figure 4.19. Comparison between experimental measurements (blue line and squares) and matrix calculations (red circles) of normalized node voltage distributions, $\mathrm{Re}(A_n/A_{12})$, for a m = 8 mode measured (a) 0.1 MHz below, (b) at, and (c) 0.1 MHz above, the resonance frequency f_0 = 11.817 MHz. To exaggerate the voltage asymmetry, only a single ground node connection was used, at node $n = 9$. The corresponding frequencies are shown in the bottom figure on the impedance spectrum. The matrix model faithfully reproduces the intentionally distorted measurements. (Adapted with permission from [12]. Copyright 2015 IOP Publishing.)

indistinguishable[5]. Compared to the analytical calculations for a simple dissipative model in figure 3.13, this is a convincing validation of the mutual partial inductance model. The detailed agreement in frequency and complex impedance across all the modes in the spectrum is all the more impressive when it is realized that the matrix model uses only the known physical dimensions of the antenna elements and the given capacitance value. The only adjusted parameters are the effective electrical length (19 cm) of the legs (physically 20 cm long), shortened due to the stringer connectors as explained in section 4.4.1, and a fine tuning of the stringer self partial inductance whose expression (4.30) is only approximate.

4.8.3 Matrix model for the mode frequencies without plasma

In view of the spectacular impedance correspondence in figure 4.20, a comparison of the measured and calculated frequencies is somewhat redundant. Nevertheless, it is shown in figure 4.21 to emphasize the strong effect of mutual partial inductance

[5] To obtain this precision, it is necessary to model the mutual partial inductance of each stringer's geometry as two metal strips with the capacitor assembly inserted between them.

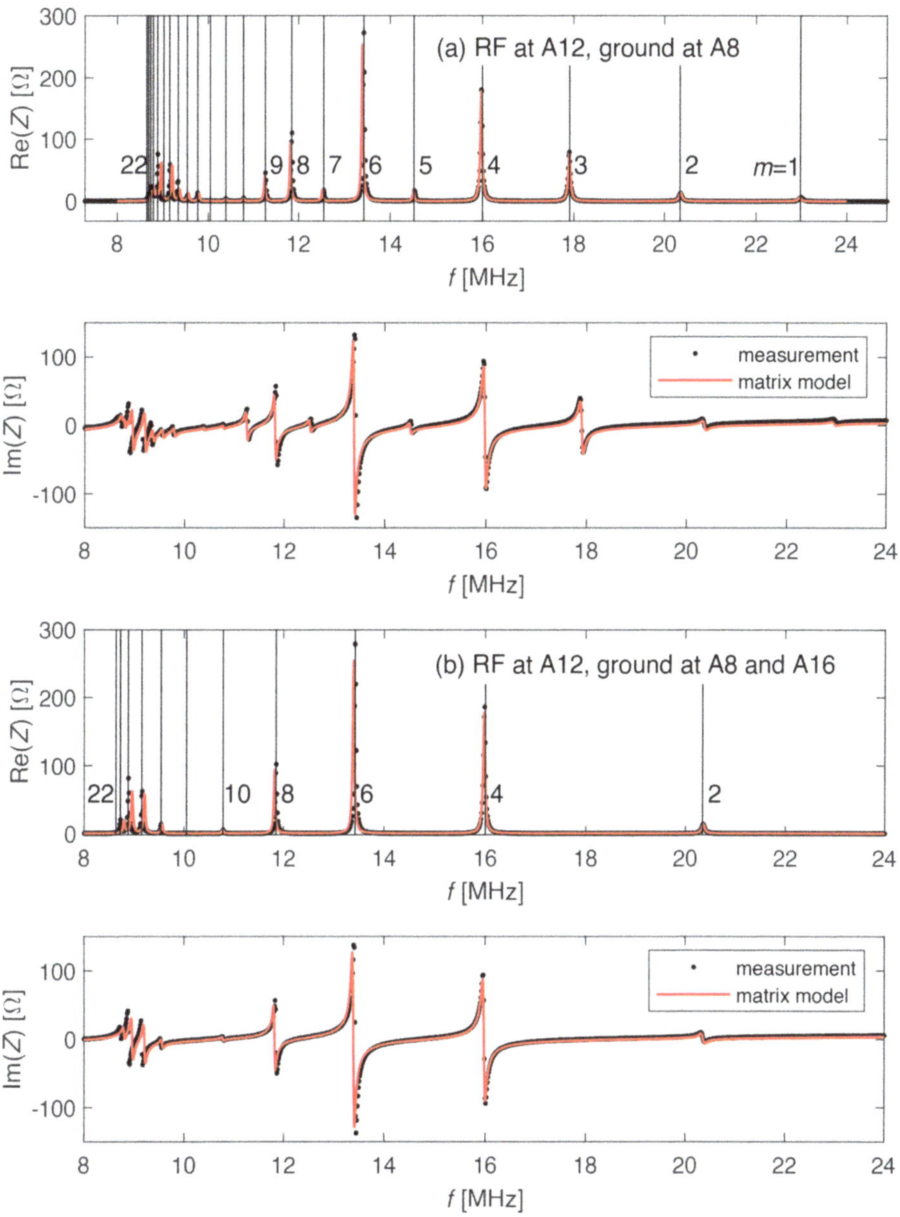

Figure 4.20. Comparison between measurement (black dots), and the mutual inductance matrix model (red line), for the ladder antenna input impedance, real and imaginary parts. RF connection at A12, with (a) ground connection at A8, and (b) ground at A8 and A16, as in figure 3.9. The network analyser data point interval is 11.9 kHz. Vertical lines mark the mode frequencies m = 1 to 22, the same as in figure 3.4.

when compared to the analytical dissipative model in figure 3.8. The introduction of the mutual partial inductances is clearly necessary [15] to obtain good agreement with experiment, and this was lacking in section 3.5 and earlier work on resonant

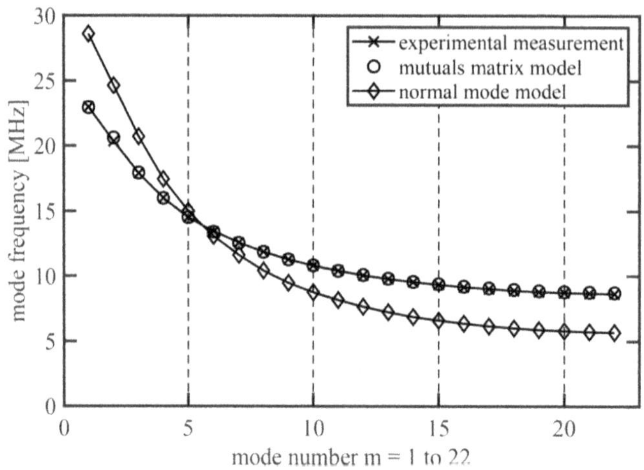

Figure 4.21. Measurement of the network resonance frequencies (×) compared with the mutual partial inductance matrix model (○), and the analytical dissipative model of section 3.5 (◇), figure 3.8. The matrix calculations, for the parameters of figure 4.20, are almost superposed with the experimental values, showing the success of the mutual inductance model. (Adapted with permission from [12]. Copyright 2015 IOP Publishing.)

network antennas [16–20]. Concerning the mode frequency shift shown in figures 3.8, 3.13, and 4.21, this is certainly due to the asymmetry of the mutual inductance contribution from the different elements across the network: legs and stringers in the outer segments of the antenna have less inductive coupling than the central elements, simply because they have fewer neighbours. However, we do not have an intuitive, physical account of the mode frequency 'compression' and pivot near mode m = 6, relative to the normal mode model, at the time of writing.

4.9 Conclusions for part I

The analytical theory for dissipative networks in chapter 3 showed that the driven resonances of ladder antennas behave as equivalent R, L, C parallel circuits as described in chapter 2. Analytical expressions give fair estimates for the input impedance, but not for the resonance frequencies.

In contrast, the excellent agreement of the matrix model with impedance measurements, based only on the network's geometry and capacitance, lends confidence to the model for mutual partial inductive coupling between the antenna elements. Building on this assurance, the partial inductance method will be extended to include plasma inductive coupling in part II 'Resonant network antennas in non-magnetized plasma'.

References

[1] Paul C R 2010 *Inductance: Loop and Partial* (Hoboken, NJ: Wiley)

[2] Paul C R 2006 *Introduction to Electromagnetic Compatibility* 2nd edn (Hoboken, NJ: Wiley)

[3] Ruehli A E 1972 Inductance calculations in a complex integrated circuit environment *IBM J. Res. Dev.* **16** 470–81
[4] Leferink F B J 1995 Inductance calculations: methods and equations *IEEE Proc. Int. Symp. Electromagnetic Compatibility (Atlanta, GA, 14–18 August)* p 16
[5] Rosa E B 1907-1908 The self and mutual inductances of linear conductors *Bull. Natl Bur. Stand.* **4** 80:301
[6] Grover F W 1962 *Inductance Calculations: Working Formulas and tables* (New York: Dover)
[7] Testa Dthe EUROfusion MST1 team and the TCV team *et al* 2019 LTCC magnetic sensors at EPFL and TCV: lessons learnt for ITER *Fusion Eng. Des.* **146** 1553
[8] Bleaney B I and Bleaney B 1976 *Electricity and Magnetism* (London: Oxford University Press) 3rd edn p 1
[9] Ott H W 2009 *Electromagnetic Compatibility Engineering* (Hoboken, NJ: Wiley)
[10] Haus H A and Melcher J R 2022 Electromagnetic Fields and Energy *Massachusetts Institute of Technology: MIT OpenCourseWare* http://ocw.mit.edu (Accessed: 22 April 2022)
[11] Howling A A, Guittienne Ph, Jacquier R and Furno I 2015 Complex image method for RF antenna-plasma inductive coupling calculation in planar geometry. Part I: basic concepts *Plasma Sources Sci. Technol.* **24** 065014
[12] Guittienne Ph, Jacquier R, Howling A A and Furno I 2015 Complex image method for RF antenna-plasma inductive coupling calculation in planar geometry. Part II: measurements on a resonant network *Plasma Sources Sci. Technol.* **24** 065015
[13] Guittienne Ph, Jacquier R, Howling A A and Furno I 2017 Electromagnetic, complex image model of a large area RF resonant antenna as inductive plasma source *Plasma Sources Sci. Technol.* **26** 035010
[14] Paul C R 2010 Partial inductance *IEEE EMC Society Newsletter,* Summer 2010 **226** 34
[15] Jin J-M 1998 *Electromagnetic Analysis and Design in Magnetic Resonance Imaging* (Boca Raton, FL: CRC Press)
[16] Guittienne Ph, Lecoultre S, Fayet P, Larrieu J, Howling A A and Ch Hollenstein 2012 Resonant planar antenna as an inductive plasma source *J. Appl. Phys.* **111** 083305
[17] Guittienne Ph, Howling A A and Ch Hollenstein 2014 Analysis of resonant planar dissipative network antennas for RF inductively coupled plasma sources *Plasma Sources Sci. Technol.* **23** 015006
[18] Lecoultre S, Guittienne Ph, Howling A A, Fayet P and Hollenstein Ch 2012 Plasma generation by inductive coupling with a planar resonant RF network antenna *J. Phys. D: Appl. Phys.* **45** 082001
[19] Lecoultre S, Guittienne Ph, Howling A A, Fayet P and Hollenstein Ch 2012 Corrigendum: plasma generation by inductive coupling with a planar resonant RF network antenna *J. Phys. D: Appl. Phys.* **45** 409502
[20] Hollenstein Ch, Guittienne Ph and Howling A A 2013 Resonant RF network antennas for large-area and large-volume inductively coupled plasma sources *Plasma Sources Sci. Technol.* **22** 055021

Part II

Resonant network antennas in non-magnetized plasma

Chapter 5

Introduction to inductively coupled plasma

Plasma enters the book here for the first time, and is in every chapter from here on. This introduction covers some basic generalities such as the mechanism of power transfer to the plasma, why radio frequency (RF) power supplies are often employed, and why plasmas are usually at low gas pressure (10^{-1}–10^{1} Pa). Part II concentrates on non-magnetized plasma, for which there are two main reactor types introduced in this chapter: capacitively coupled plasmas (CCP) and inductively coupled plasmas (ICP). Resonant network antennas fall into the second category, so conventional solenoid ICP reactors will be briefly reviewed here for a point of comparison. Magnetized plasmas for wave-sustained sources will be postponed until part III.

5.1 RF plasma generalities

The second part of the book title mentions RF plasma sources, so we should explain why we use plasma, and why we specify RF.

5.1.1 Why use plasma?

Electrical power is transferred from a generator to a plasma via electric fields accelerating electrons in a gas. The electrons give up their acquired kinetic energy to the neutral gas particles through collisions. These electron–molecule collisions can be elastic—where total kinetic energy is conserved—or inelastic, where electron energy is transferred to internal energy of the atoms or molecules. From appendix C.1, the small mass m_e of electrons means that their elastic collisions transfer only a small fraction ($\sim\frac{m_e}{M} \ll 1$) of their kinetic energy to neutrals of mass M, so electrons do not efficiently heat the gas at low pressure (see also 5.1.3).

On the other hand, in inelastic collisions, electrons can efficiently transfer $\frac{M}{m_e + M} \sim 100\%$ of their kinetic energy to molecular internal energy (see appendix C.2) such as ro-vibrational or electronic excitation, dissociation, attachment (and polymerization [1]), ionization, etc, depending on the electron energy and the chemical properties of the molecule. The inelastic electron–molecule collisions count as the useful work in plasma chemistry; the final chemical products obtained depend on the various inelastic collision cross-sections and the electron temperature, as well as on secondary reactions among the products themselves.

Moreover, note that acceleration through a potential difference of only 1 V imparts an energy of 1 eV $=1.6 \cdot 10^{-19}$ J to an electron, corresponding to the mean energy of a thermal distribution of electrons at temperature $T_e = \frac{e}{k_B} \approx 11'600$ K, where k_B is Boltzmann's constant (strictly speaking, continuous power transfer is due to the phase randomizing effect of collisions with respect to the oscillating electric field; see appendix D.2.1). Particle balance between ionization source rate and surface loss fluxes constrains the electron temperature to a narrow range [2] around 3 eV $\sim 35 \cdot 10^3$ K. To obtain similar gas reactions by means of conventional thermal chemistry would entail heating the gas to many thousands of degrees. These last three paragraphs, then, contain the basic phenomena motivating the use of plasma, which can be resumed as follows:

> Plasma electrons can cause high temperature chemistry
> without strongly heating the gas.

This exotic plasma chemistry offers unique opportunities in plasma processing, but also brings particular challenges, touched upon in section 9.3.2. In contrast to electrons, ions have practically the same mass as the neutrals, hence their energy transfer fraction in head-on elastic collisions is ≈ 1. This means that the ions thermalize efficiently with the gas, which itself is in thermal contact with the walls. In general, therefore, the ion temperature is comparable to the gas and wall temperature. In conclusion, for low gas pressure (see section 5.1.3), we have a non-equilibrium plasma with electrons much hotter than the ions, $T_e \gg T_i$.

5.1.2 Why use radio frequency?

So many plasma generators use RF power that the choice of RF could be taken for granted, but it is preferable to consider why. Following on from section 5.1.1, it is clear that it is advantageous to transfer electrical energy to the electrons—which do the useful plasma chemistry—but not to the ions, which otherwise waste power and overheat the apparatus.

Electrons have low inertia and so can be accelerated by high frequency electric fields, whereas ions have high inertia, and this can be used to discriminate against them. The frequency responses are quantified by the plasma frequency for electrons, f_{pe}, and ions, f_{pi} (taking $^{40}Ar^+$ ions for example) [2–4]:

$$f_{\mathrm{pe}} = \frac{1}{2\pi}\sqrt{\frac{n_{e0}e^2}{\varepsilon_0 m_e}}, \tag{5.1}$$

$$f_{\mathrm{pi}} = \frac{1}{2\pi}\sqrt{\frac{n_{e0}e^2}{\varepsilon_0 40 m_p}}, \tag{5.2}$$

$$\frac{f_{\mathrm{pe}}}{f_{\mathrm{pi}}} = \sqrt{\frac{40 m_p}{m_e}} \approx 271, \tag{5.3}$$

where m_p is the proton mass and n_{e0} is the electron density (equal to the plasma density, assuming singly charged ions). The 271 ratio of electron and ion plasma frequencies leaves a wide window for an intermediate frequency to selectively accelerate electrons, but not ions. The plasma density is roughly determined by power balance [2], and $n_{e0} \sim 10^{16}\ \mathrm{m}^{-3}$ is a representative value in plasma processing. Using this, we find $f_{\mathrm{pi}} \sim 3$ MHz and $f_{\mathrm{pe}} \sim 0.9$ GHz, which covers the RF spectrum from the high frequency (HF) band (3–30 MHz), through the very high frequency (VHF) band (30–300 MHz) and above. The geometric mean of these two plasma frequencies is roughly 50 MHz.

The Industrial, Scientific, and Medical (ISM) frequencies are reserved in the electromagnetic spectrum for high power applications, and RF generators and matching circuits are widely available, especially for 13.56 MHz and its lower harmonics 27.12 MHz, 40.68 MHz, etc. In practice, standing wave effects, and stray capacitance and inductance begin to cause problems towards 100 MHz, and so these low to mid-range harmonics are usually preferred. These ISM commercial RF generators often satisfy

$$f_{\mathrm{pi}} < f_{\mathrm{rf}} \ll f_{\mathrm{pe}}, \tag{5.4}$$

which is suitable for selectively driving electron currents, whereas ions can only respond to time-averaged electric fields. This is the main reason for using ISM radio frequency generators.

5.1.3 Why use low pressure?

From section 5.1.1, the electrons are inefficient at heating the gas, but if the pressure rises above, say, 10^4 Pa, the electron mean free path falls to the extent that too few electrons gain sufficient energy for inelastic collision thresholds. The frequency of elastic collisions rises with the pressure and electron energy is dissipated in strong heating of the gas, eventually becoming a thermal plasma ($T_e \sim T_i \sim T_{\mathrm{gas}}$). Thermionic emission and arcing will eventually occur if metal electrodes are exposed to the plasma. On the other hand, if the pressure is below, say, 10^{-1} Pa, the electron mean free path becomes comparable to the reactor dimensions, ionization collisions become rare, and the plasma cannot be maintained.

In practice, therefore, a plasma is most efficiently sustained at a mid-range gas pressure where there are at least several electron–neutral mean free paths, λ_{mfp}, within the reactor characteristic dimension, $d > 10^1\lambda_{\mathrm{mfp}}$, and yet the mean free path is long enough for electrons to reach significant energy ($qE\lambda_{\mathrm{mfp}} \sim \varepsilon_i$, the gas ionization energy). This reasoning qualitatively resembles the Paschen law for minimum DC breakdown voltage typically occurring around 10^0 mbar·cm or 10^2 Pa·cm (although RF breakdown itself is different from DC breakdown). Anticipating section 5.2, the reactor characteristic dimension is usually a few centimeters for the electrode gap of CCP reactors, or the skin depth of ICP reactors; this therefore fixes the optimum pressure roughly at 10^1 Pa. Broad limits around this value suggest 10^0 to 10^2 Pa, to give a conservative operational pressure range. This tallies with the low pressure values generally used in RF plasma reactors. In this book, only sections 9.3.1 and 13.5 concern higher pressures.

5.2 RF plasma sources in non-magnetized plasma

So far, the questions were applicable to RF plasmas in general. Now we will briefly compare two common types of RF plasma reactors which do not have externally applied DC magnetic fields, namely: RF capacitively coupled plasma (CCP), and RF inductively coupled plasma (ICP) [5].

5.2.1 Capacitively coupled RF reactors: the *E*-mode

In an RF CCP, figure 5.1, an RF peak-to-peak voltage V_{pp} of typically 10^2 V is applied between two opposing electrodes in a low pressure gas. The (slowly) oscillating quasi-electrostatic electric field $\tilde{E}$ perpendicular to the electrodes drives the electron current, hence the term '*E*-mode', corresponding to the electro-quasistatic approximation in appendix D.1.1. The electric field lines close outside the plasma via the RF external circuit [8]. The oscillating magnetic field generated by this current generally has a negligible influence (but see section 5.2.3). The electron current is intercepted by the surfaces and a negative surface charge with a positively charged adjacent layer (sheath) forms to brake the electrons and equalize the

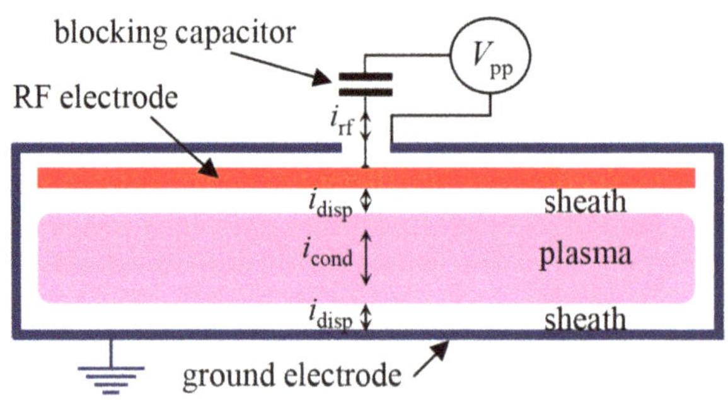

Figure 5.1. Schematic of a capacitively coupled plasma (CCP). The RF generator's peak-to-peak voltage, V_{pp} on the RF electrode, drives a current i_{rf} around the circuit. Through the sheaths, i_{disp} is principally a displacement current; through the plasma, i_{cond} is a conduction current carried by the electrons. All these RF currents are equal, by continuity. The blocking capacitor is to prevent DC current flow, and DC bias appears on the RF electrode for asymmetric electrode areas [6, 7].

time-averaged ion and electron fluxes to the walls. Consequently, the time-averaged plasma potential is positive with respect to the surfaces. The resulting perpendicular (normal) electric field into the plasma is mostly screened over a distance of a few Debye lengths from the wall, which defines the sheath width (usually a few millimetres).

In a steady state, electron ionization of the gas compensates for the electron and ion loss fluxes to the surfaces. The RF current is carried by oscillating electrons in the plasma bulk, and displacement current in the sheaths ensures current continuity to the external circuit. This displacement current can be extended through dielectric layers on the RF electrodes, therefore an RF CCP can operate with exposed electrodes, or with coated, insulated electrodes (in contrast to DC discharges). The strong, time-averaged E-field in the sheaths causes ions to impact the substrate with high energy. Their maximum energy is typically $\sim eV_{pp}/4$ for symmetric electrodes, or up to $eV_{pp}/2$ on the small electrode of strongly asymmetric electrode areas [6, 7]. This ion bombardment energy can be damaging for electrode surfaces, or can be useful for etching, depending on the application. It can be difficult to control the ion energy independently of the RF power and the plasma density [2].

5.2.2 Inductively coupled RF reactors: the *H* -mode

In contrast, the RF current in an inductively coupled plasma (ICP) is driven by a rotational (vortex, or solenoidal) RF electric field which is induced, obligatorily through a dielectric wall or window, by an oscillating magnetic field. The (slowly) oscillating quasistatic magnetic field $\tilde{H}$ is generated by an RF current in a neighbouring primary circuit, see figure 5.2. This is called the '*H*-mode', referring to the magnetic field strength *H*, and it corresponds to the magneto-quasistatic approximation in appendix D.1.1. The RF electron current oscillates around a

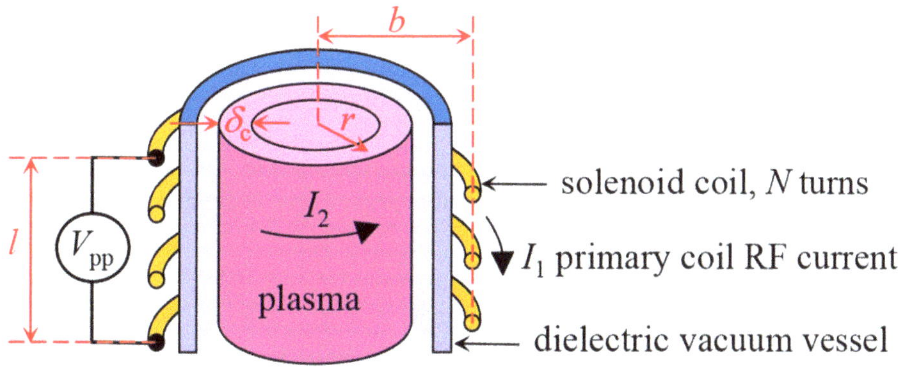

Figure 5.2. Schematic section of a cylindrical, inductively coupled plasma (ICP) with RF current I_1 in the primary circuit solenoidal coil, radius *b*, length *l*, number of turns *N*. The solenoid is usually in air outside a cylindrical dielectric wall which forms the vacuum vessel for the low pressure plasma. I_1 induces an opposing (by Lenz's law) current I_2 in the plasma which acts as a single turn, air-core (i.e. no magnetic material), secondary circuit. The inductively coupled plasma current, with effective radius *r*, flows in a surface layer of radial thickness δ_c which is the collisional skin depth for penetration of the induced electric field [9].

closed loop in the plasma, without interruption by a surface. Therefore there are no sheaths interrupting the electron current, and the induced electric field, $E = \rho_p j$, is equal to the Ohm's law product of the local electron current density j and the plasma electrical resistivity ρ_p which is non-zero due to electron collisions in the gas. In short, the RF current source is the primary circuit, and the plasma behaves as a single turn, resistive secondary circuit; the pair form a loosely coupled, air-core transformer. The absence of high voltage sheaths means that ion energy to the walls can, theoretically, be as low as the floating potential, 20–30 V, although some unavoidable capacitive coupling can still damage the dielectric window by sputtering [10] if the coil voltage is high (see also section 5.5). Independent control to obtain higher ion energies can be achieved using an independent RF biased substrate as described in section 6.5.

A prerequisite of inductive coupling is that there must be sufficient electron density to carry the induced current. Until breakdown occurs, this cannot be the case, so ICP ignition depends initially on *E*-mode operation to generate a low density plasma due to high voltage across the source coil. Once the plasma density is high enough to carry induced current and absorb sufficient RF power, the *E*-mode becomes screened out by the induced currents, and the discharge is sustained by the *H*-mode currents [2]. This process is called the '*E* to *H* transition' [10, 11].

On a particular point of understanding, we note that the *H*-mode-induced electric field can persist and penetrate into the plasma over skin depth distances much longer than the Debye-length sheath of the *E*-mode. Because of the nature of the *H*-mode closed-loop, circulating current $j = E/\rho_p$ does not lead to charge accumulation on any surfaces, only a relative (oscillating) drift velocity of the electron fluid relative to a stationary ion background; the plasma remains quasi-neutral at all places and times.

Solenoids for cylindrical reactors in figure 5.2, and flat spiral coils for planar sources (section 6.1), are the usual designs for ICP reactors [2]. Resonant network antennas—the subject of this book—provide analogous alternatives for ICP sources. Birdcage resonant antennas are suitable for cylindrical reactors instead of solenoids. Ladder resonant antennas provide planar sources instead of flat spiral coils. The ladder resonant antenna is particularly interesting because much larger planar reactors can be designed; see section 5.5 and chapters 6 and 7. Consequently, from now on, this book will concentrate on resonant network antennas for RF plasma sources.

5.2.3 Large-scale RF reactors: the *EM*-mode

Once again, the implicit assumption in this chapter has been that the reactor dimensions are much smaller than the RF wavelength, so that lumped element circuit descriptions, and quasistatic approximations apply. For both CCP and ICP, the RF field–plasma interaction therefore manifests a quasi-steady-state nature, because the wavelength of the electromagnetic field is much longer than the plasma dimensions [8]. Plasma reactors have been divided into *E*-mode, section 5.2.1, or *H*-mode, section 5.2.2. However, through standing wave effects etc, large CCP reactors partly exhibit H-mode behaviour as well as E-mode [3, 12, 13], and large ICP reactors partly show some E-mode behaviour as well as H-mode [2, 3, 14, 15]; see chapter 7. General solutions of Maxwell's equations would always account for both E- and H-modes in a self-consistent electromagnetic (EM) treatment. We could call this the EM-mode, where the plasma is an electromagnetically coupled plasma (EMCP).

5.3 Skin depth in inductively coupled plasma

As stated in section 5.1.2, the RF excitation frequency is much less than the electron plasma frequency f_{pe}. This means that the ICP electromagnetic induction field incident on the plasma surface is always cut off; it cannot propagate through an unmagnetized plasma as if it were a conventional dielectric (see appendix D.5.4).

5.3.1 Collisional skin depth

In collisional plasmas, $\omega \ll \nu, \omega_{\mathrm{pe}}$, where ν is the electron–molecule collision frequency introduced in (5.5). The cut off occurs by $1/e$ damping of the wave amplitude over the 'collisional skin depth' δ_c. This skin depth damping is, at one and the same time, the source of ICP heating and the cause of ICP non-uniformity, because the induced electrical power dissipation can only occur inwards from the edge of the plasma volume, whether cylindrical as in figure 5.2, or planar as in appendix D.5.2. In practice, the plasma radius (cylindrical) or thickness (planar) should be designed to optimize the volume ratio [working plasma volume]:[vacuum vessel volume] for the plasma reactor, accounting for the skin depth limitation. Intuitively, if the skin depth is too small, or too large, compared to the solenoid radius r, the power efficiency falls away. The optimal power efficiency lies between these extremes, and for a collisional, quasi-uniform cylindrical plasma, the optimum is $\delta_c \approx 0.57r$ [2], demonstrating the link between reactor design and plasma properties. The collisional skin depth (D.42)

$$\begin{aligned} \delta_c &\approx \sqrt{\frac{2}{\omega\mu_0\sigma_{\mathrm{dc}}}} = \sqrt{\frac{1}{\pi f_{\mathrm{rf}}\mu_0\sigma_{\mathrm{dc}}}} = \sqrt{\frac{m_e\nu}{\pi f_{\mathrm{rf}}\mu_0 n_{e0}q^2}}, \\ \text{where}\quad \nu &\approx \frac{c_e^{\mathrm{th}}}{\lambda_{\mathrm{mfp}}} \approx n_{\mathrm{gas}}\sigma_{\mathrm{en}}\sqrt{\frac{8k_{\mathrm{B}}T_e}{\pi m_e}} = \frac{p}{k_{\mathrm{B}}T_{\mathrm{gas}}}\sigma_{\mathrm{en}}\sqrt{\frac{8k_{\mathrm{B}}T_e}{\pi m_e}}, \end{aligned} \tag{5.5}$$

depends on only a few common experimental parameters, namely RF frequency f_{rf}, gas pressure p, and electron density n_{e0}, so we can see that ICP reactors mostly have similar dimensions. The symbols introduced in (5.5) are the electron–neutral gas kinetic collision frequency ν, mean free path λ_{mfp}, electron mean thermal speed c_e^{th}, gas number density n_{gas}, electron–neutral collision cross-section $\sigma_{\mathrm{en}} \sim 10^{-19}\ \mathrm{m}^2$, electron temperature $T_e \sim 3$ eV $\sim$35 000 K, and gas temperature $T_{\mathrm{gas}} \sim 400$ K. Taking arbitrary but representative values for the three experimental parameters $f_{\mathrm{rf}} = 13.56$ MHz, $p = 30$ Pa, $n_{e0} = 10^{17}\ \mathrm{m}^{-3}$, we find $\nu/\omega = 7$ (i.e. collisional

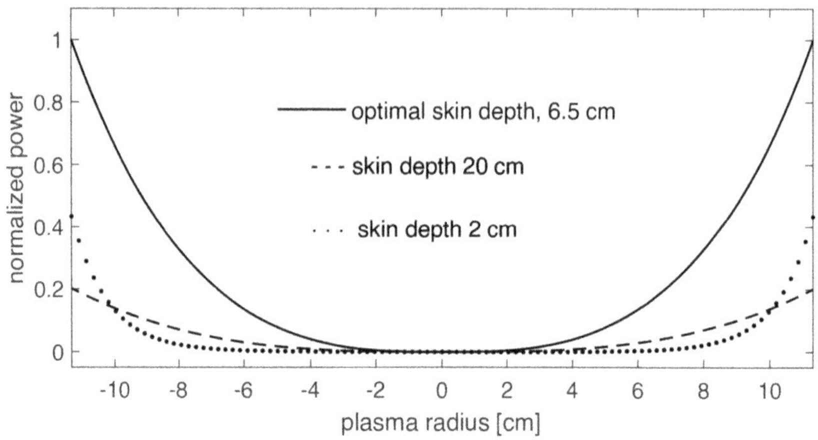

Figure 5.3. Calculated radial profile (solid line) of the ohmic power, normalized to its maximum value for the optimum skin depth, across a cylindrical uniform plasma of 23 cm diameter in a solenoidal ICP reactor, figure 5.2. The plasma parameters correspond to the maximum power transfer in section 5.3.1, for which the electron density is $n_{e0} = 10^{17}$ m^{-3} and the optimum collisional skin depth is $\delta_c = 0.57r = 6.5$ cm. Two bracketing cases are shown for comparison: $n_{e0} = 10^{16}$ m^{-3} with skin depth $\delta_c = 20$ cm (dashed line); and $n_{e0} = 10^{18}$ m^{-3} with skin depth $\delta_c = 2$ cm (dotted line). The ohmic power density in a solenoidal ICP is always non-uniform because of the skin depth effect.

conditions, for high pressure in section 5.1.3), and $\delta_c = 6.5$ cm, for which an optimal reactor diameter would be $2r = 2\delta_c/0.57 \approx 23$ cm.

Figure 5.3 shows the calculated radial profile[1] of the RF inductively coupled power deposition across this plasma diameter, assuming uniform plasma, for the optimal skin depth and two other values. Relying on inward plasma diffusion, this power deposition profile might be acceptable for uniform plasma processing of an individual, modest-sized silicon wafer of 20 cm diameter [10]. Note, however, that there is no question of designing a wider, solenoidal, multi-wafer batch reactor with these ICP properties. In short, the skin depth is an inherent cause of non-uniformity in the solenoidal ICP reactor design of figure 5.2, and is a serious limitation for scaling up to larger substrates.

5.3.2 Collisionless skin depth

For a collisionless plasma (very low pressure, $\nu \ll \omega_{\mathrm{rf}}$), the electron oscillations are not interrupted by collisions with gas neutrals. When transverse electromagnetic radiation impinges on the plasma surface, the electrons oscillate freely in the electric field and re-radiate, reflecting the electromagnetic field. Cut off in this case occurs because the field penetrates only as evanescent damping, with no propagation and no power dissipation (because no collisions), to a $1/e$ depth called the 'collisionless

[1] The radial radiated power per unit area is given by the Poynting vector $S_r = \frac{1}{2}\,\mathrm{Re}\left(E_\phi H_z^*\right)$, where $E_\phi = H_{z0}\frac{k}{\sigma_{\mathrm{dc}}}\frac{J_1(kr)}{J_0(kR)}$, $H_z = H_{z0}\frac{J_0(kr)}{J_0(kR)}$, and $k = -j^{1/2}\sqrt{2}/\delta_c$ [2]. See also section 8.4.1 for birdcage antennas with $m = 1$ azimuthal periodicity.

skin depth' $\delta_{nc} = c/\omega_{pe}$ (see appendices D.5.3 and D.5.4). This is the distance travelled by light during a fraction $\frac{1}{2\pi}$ of an electron plasma frequency period, $2\pi/\omega_{pe}$. In this case, the only experimental parameter is the electron density n_{e0}. Taking the same value as in section 5.3.1, $n_{e0} = 10^{17}$ m^{-3}, we have $\delta_{nc} = 1.7$ cm. For these parameters, the separate collisionless conditions $\nu \ll \omega_{rf}$ and $\delta_{nc} \ll \lambda_{mfp}$ are both satisfied for gas pressures much less than 3 Pa. There is another skin depth—the anomalous skin depth—due to stochastic heating, where electrons move in and out of the skin depth in a time shorter than the RF or collision periods [2].

5.4 Transformer model for inductively coupled plasma

For the design and operation of ICP sources, it is important to understand the coupling to the plasma, because the plasma loading affects the primary circuit impedance, and therefore the impedance matching and the power transfer efficiency [2]. We will see that the coupling is defined by the source–plasma mutual inductance, because this is the mechanism by which the primary and secondary circuits interact with each other.

The transformer model [3, 9, 16], using discrete (lumped) circuit elements to represent self-inductances and mutual inductance with the plasma, is often used to model solenoid and coil ICP sources, including spiral coils [17]. For simple geometries such as a solenoid, the loop self inductance of an ICP source primary circuit can be calculated unambiguously using standard formulas [2, 9, 18]. The loop self inductance of the plasma and the source–plasma loop mutual inductance can likewise be estimated by making reasonable assumptions about the plasma current geometry. In this way, the electrical properties of an inductive low pressure RF plasma can be analysed by considering the plasma to be a one-turn secondary of an air-core transformer [9]. However, the transformer model approach is less straightforward in other cases, such as plane spiral RF antennas [17, 19–21] or resonant network antennas, where the loop inductances of the source and plasma are difficult and/or impractical to determine. Nevertheless, we now work through the transformer model for a solenoid ICP because we will need to put other approaches—such as image models—into context for comparison (see section 5.4.3 and chapter 6).

5.4.1 Loop inductances for a solenoid ICP reactor

Referring to figure 5.2, we have an N-turn primary solenoid surrounding a single turn, air-core ($\mu_r = 1$) plasma secondary solenoid. The plasma sheet current radius r is a weighted value averaged over the radial current distribution. The magnetic flux density within the primary solenoid, $B_1 = \mu_0 N I_1/l$ [2, 18], is uniform, to all intents and purposes, over all its internal area πb^2. The magnetic flux linkage with the N turns of the primary due to its own current is therefore $\Phi_1 = N \cdot \pi b^2 \cdot \mu_0 N I_1/l$, hence the primary loop self inductance is

$$L_1^{\text{loop}} = \frac{\Phi_1}{I_1} = \mu_0 \pi b^2 N^2 / l. \tag{5.6}$$

Similarly for the plasma sheet current (single turn, $N = 1$), radius r and same length l, its magnetic flux density is $B_2 = \mu_0 I_2 / l$ over its area πr^2. The magnetic flux linkage of the plasma secondary due to its own current is $\Phi_2 = \pi r^2 \cdot \mu_0 I_2 / l$. Hence, the plasma secondary loop self inductance is

$$L_2^{\text{loop}} = \frac{\Phi_2}{I_2} = \mu_0 \pi r^2 / l. \tag{5.7}$$

The flux linked with the plasma area due to the primary current I_1 is $\Phi_{12} = B_1 \cdot \pi r^2 \cdot 1$, hence the loop mutual inductance is

$$M_{12}^{\text{loop}} = \frac{\Phi_{12}}{I_1} = \mu_0 \pi r^2 N / l. \tag{5.8}$$

Equivalently, the flux linked with the primary due to the plasma secondary current I_2 is $\Phi_{21} = B_2 \cdot \pi r^2 \cdot N$, hence the loop mutual inductance is

$$M_{21}^{\text{loop}} = \frac{\Phi_{21}}{I_2} = \mu_0 \pi r^2 N / l. \tag{5.9}$$

We have $M_{12}^{\text{loop}} = M_{21}^{\text{loop}}$ as expected. This loop mutual inductance defines the coupling between the primary coil and the plasma secondary.

5.4.2 Transformer model for a solenoid ICP reactor

In figure 5.4, the ICP source is a primary solenoid circuit with current I_1, voltage V_1, resistance R_1, and loop self inductance L_1^{loop} given by (5.6) in section 5.4.1. The plasma loop self inductance L_2^{loop} is determined by the geometry of the plasma current path, given by (5.7). This current path also has plasma impedance $Z_{\text{pl}} = R_2(1 - j\omega/\nu)$, where R_2 is the plasma current path resistance, and $-jR_2\omega/\nu = -j\omega L_p$ is due to the L_p plasma inductance contribution from electron inertia which follows from the plasma complex conductivity in (D.16) [2, 9]. The plasma resistance can be estimated from the cylindrical shell, radius r, of skin thickness δ_c [2, 18], using $R_2 = \frac{2\pi r}{\sigma_{\text{dc}} l \delta_c}$. Applying Kirchoff's law to the primary and secondary circuits gives

$$V_1 = R_1 I_1 + j\omega L_1^{\text{loop}} I_1 - j\omega M_{12}^{\text{loop}} I_2, \tag{5.10}$$

$$0 = Z_{\text{pl}} I_2 + j\omega L_2^{\text{loop}} I_2 - j\omega M_{12}^{\text{loop}} I_1. \tag{5.11}$$

Substituting for I_2 from (5.11) in (5.10) gives the primary impedance $Z_1 = V_1 / I_1$ as follows:

$$Z_1 = R_1 + j\omega L_1^{\text{loop}} + \frac{(\omega M_{12}^{\text{loop}})^2}{Z_{\text{pl}} + j\omega L_2^{\text{loop}}}. \tag{5.12}$$

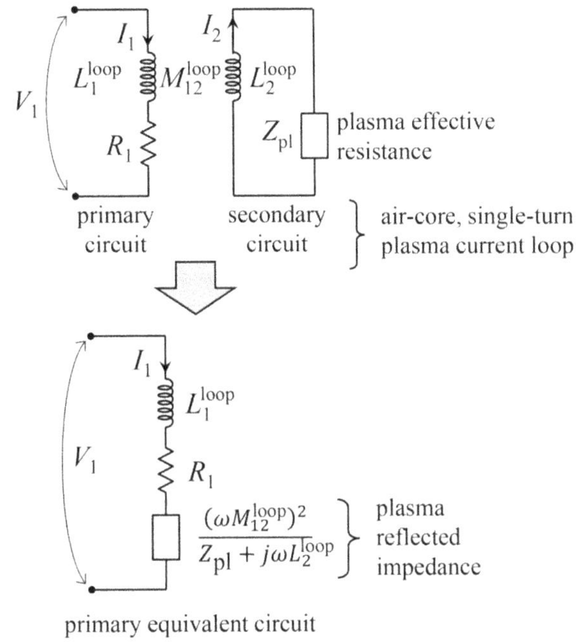

Figure 5.4. The electrical circuit of the solenoid ICP source in figure 5.2 showing the primary circuit and an inductively coupled plasma (secondary circuit). The equivalent circuit of the primary is transformed by inductive coupling to the plasma, effectively adding a 'reflected impedance'. The loop inductance expressions are given in (5.6)–(5.9).

In figure 5.4, the secondary circuit impedance has been transformed into its series equivalent impedance (the reflected impedance) in terms of the primary circuit current. We note that the reflected impedance is generally complex; it causes a decrease in the total inductance (concomitant with Lenz's law), and an increase in resistance due to transfer of the power dissipation in the secondary circuit.

The accuracy of the transformer method suffers from the approximations made for the plasma loop self and mutual inductance in section 5.4.1: the magnetic field within the plasma column is, in fact, not uniform because of the skin depth effect and more sophisticated weighted averages would be required. The required mutual inductance can be estimated in more self-consistent ways, for example, by calculating the electromagnetic time-harmonic fields radiated by the primary coil, and the influence at this coil of the fields scattered back by currents induced in a resistive cylindrical core [22]. Nevertheless, the transformer model serves to demonstrate how plasma loading of the source depends on the mutual inductance M_{12}^{loop} between the source and the plasma.

5.4.3 Comparison between transformer and image models for ICP

The aim here is to make a bridge between the familiar transformer model and the lesser-known partial inductance image model, by revisiting the example of a wire

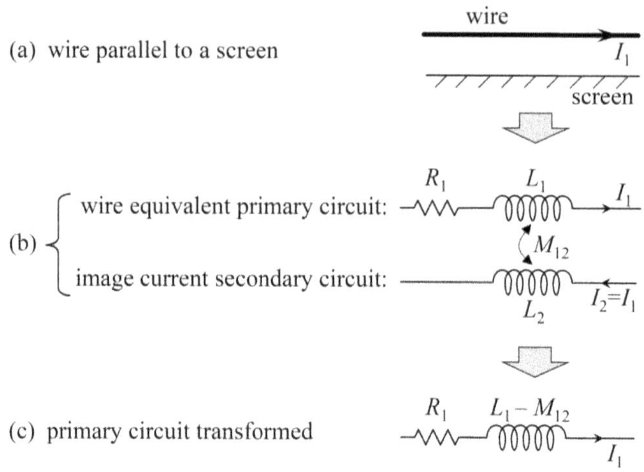

Figure 5.5. (a) The wire and screen image current of figure 4.16 recast as a transformer example in (b): the wire as a primary circuit (resistance R_1, self partial inductance L_1); and the perfectly conducting screen image as a secondary circuit (no resistance, self partial inductance L_2), coupled by their mutual partial inductance M_{12}. The image current in the screen, I_2, is equal and opposite to the wire current, I_1. (c) The equivalent primary circuit of the wire, transformed by coupling to the ideal screen. The transformation entails the subtraction of the mutual partial inductance.

current coupled to its image current. We have already met the mutual partial inductance between a straight leg primary circuit and an induced current in a perfectly conducting screen in section 4.7.1, figures 4.15 and 4.16. The partial inductance concept is useful to separate the total (loop) inductance of a wire/image system into effective primary and secondary transformer circuits coupled by their mutual partial inductance. In figure 5.5, the self partial inductance L_1 of the wire and its resistance R_1 represent the primary circuit, carrying current I_1; the same as for the transformer model. The self partial inductance, L_2, of the image current in the perfectly conducting screen ($R_2 = 0$) represents the secondary circuit carrying the induced current I_2. Finally, the mutual partial inductance M_{12}, from (4.36), represents the mutual inductance between the primary and secondary circuits. The voltage across the primary circuit is therefore

$$V_1 = R_1 I_1 + j\omega L_1 I_1 - j\omega M_{12} I_2, \tag{5.13}$$

exactly the same as for the transformer model, (5.10). For the transformer model, the next step would be to substitute for the current I_2 in terms of I_1 and the impedances in the secondary circuit [2, 9], as resumed in section 5.4.2. However, in the terminology of 'go-and-return circuits' (figure 4.11) and 'ideal mirror image' methods (figures 4.15, 4.16, and 5.6), it is implicit that the induced current in the secondary circuit here is equal and opposite to the primary circuit source current. This is inherent in the symmetry of the source and its image, contrary to the transformer where the geometry and number of turns are different between the primary coil and the plasma secondary turn. Hence $I_2 = I_1$ in (5.13) and figure 5.5,

Figure 5.6. 'The ideal mirror image method.' (Illustration by Alex Howling. Copyright 2023 Alex Howling.)

respecting the opposite directions of the currents, and so the equation for the secondary (5.11) is redundant here. Substitution for I_2 in (5.13) simply gives

$$V_1 = R_1 I_1 + j\omega L_1 I_1 - j\omega M_{12} I_1. \tag{5.14}$$

The impedance V_1/I_1 of the transformed, or coupled, primary circuit in figure 5.5(b) is therefore directly

$$Z_1 = V_1/I_1 = R_1 + j\omega(L_1 - M_{12}), \tag{5.15}$$

where $(L_1 - M_{12})$ is the net partial inductance of the primary circuit including the reflected impedance of the screen image current. The reflected impedance of the secondary circuit transformed into the primary circuit is therefore equal to the mutual partial impedance, because the currents in the primary and secondary circuits are equal; this is a consequence of the image description.

This is all well and good for a perfectly conducting screen, but ideally we would like an image model which gives the mutual partial inductance in presence of a plasma—this is the subject of the next chapter, chapter 6.

5.5 Prohibitively high voltages in large area ICP

The development of inductively coupled plasma (ICP) sources was motivated by the requirement of high densities, in the order of 10^{18} m^{-3}, and high dissociation rates [10]. The first inductive plasmas for relatively large area processes were generated by spiral inductive couplers; see figure 5.7. Electrical problems arise, however, for diameters above 20–30 cm due to the high RF voltages required at the feeding of the inductor to sustain the high plasma density [10]. This is because the input impedance of an inductive source is dominantly reactive; consequently the power factor [18] is low, and high voltages are needed even for moderate RF power dissipation. The same applies to any monolithic ICP reactor, i.e. where the inductive element is a single element. Problems include arcing inside and/or outside of the vacuum vessel, and high energy ion bombardment across high voltage sheaths at the dielectric walls, causing plasma contamination and wall erosion. Slotted Faraday screens—inside or outside the dielectric wall—may be necessary to reduce the E-mode capacitive

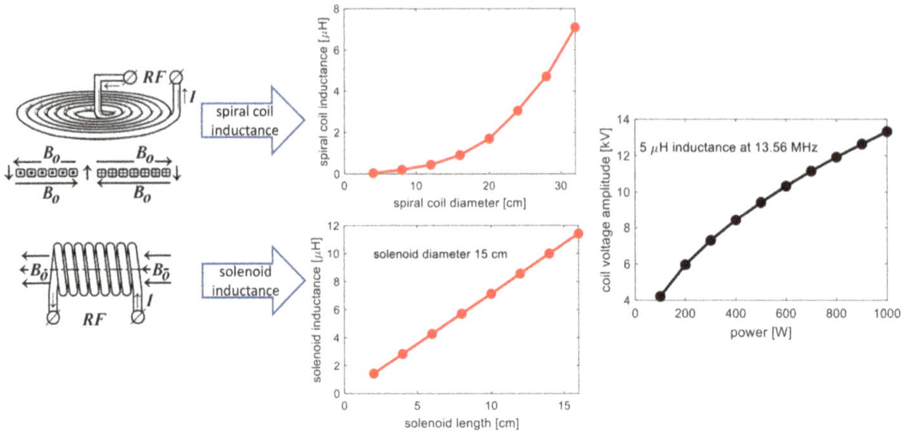

Figure 5.7. The scaling of inductance with diameter for a spiral coil, and with length for a solenoid coil. For a typical coil inductance of 5 μH at 13.56 MHz, the coil voltage amplitude exceeds 13 kV for 1 kW input power.

coupling of the high voltage coil to the plasma. There are also many alternative large area designs, briefly reviewed in section 7.1. Arrays of multiple smaller ICPs [23, 24] or helicon sources are another possibility provided that uniform impedance and power dissipation can be guaranteed [25].

For the large coil example in figure 5.7, the reactive impedance of a typical coil inductance of $L = 5\,\mu$H at 13.56 MHz is $\omega L = 426\,\Omega$. Assuming a constant reflected real impedance of, say, $R \sim 2\ \Omega$ due to plasma loading, this gives a power factor $\cos\phi = \frac{R}{Z_L} \approx \frac{R}{\omega} = \frac{2}{426}$ so that $\phi = 89.7°$, which is close to 90°. The impedance is dominantly imaginary, i.e. reactive. Using Ohm's law, $P = I_{\rm rms}^2 R = \frac{V_{\rm rms}^2}{Z^2} R$, so $V_{\rm rms} = \sqrt{PZ^2/R}$ gives the square root power dependence for voltage in figure 5.7. For $P = 1$ kW, $V_{\rm rms} = \sqrt{1000 \cdot 426^2/2} = 9.5$ kV for which the voltage amplitude is 13.3 kV, an extremely high and dangerous RF input voltage for an industrial plasma reactor.

In contrast, a resonant network antenna generates multiple *internal* mode resonances of currents and voltages which are spatially distributed along the antenna, according to the chosen mode structure. The dominantly real input impedance of resonant antennas is independent of its size. This means that they can be scaled to large size and high RF power, maintaining input voltage below 1 kV for several kW of RF power, so Faraday screens are not necessary. Figure 5.8 summarizes the resonant network approach. At resonance, the input impedance is purely real, $R = 115\ \Omega$ (the small downshift in imaginary impedance is due to the series capacitance described in section 3.5.6). Ohm's law applies directly so that $P = \frac{V_{\rm rms}^2}{R}$, and $V_{\rm rms} = \sqrt{PR}$ gives the square root power dependence for voltage in figure 5.8. For $P = 1$ kW, $V_{\rm rms} = \sqrt{1000 \cdot 115} = 339$ V for which the RF voltage amplitude is 480 V, a perfectly manageable voltage, almost 30 times less than for a large coil reactor at the same RF power. Further comparisons of matching between solenoid ICPs and resonant network antennas are made in chapter 13, for example, section 13.3.4.

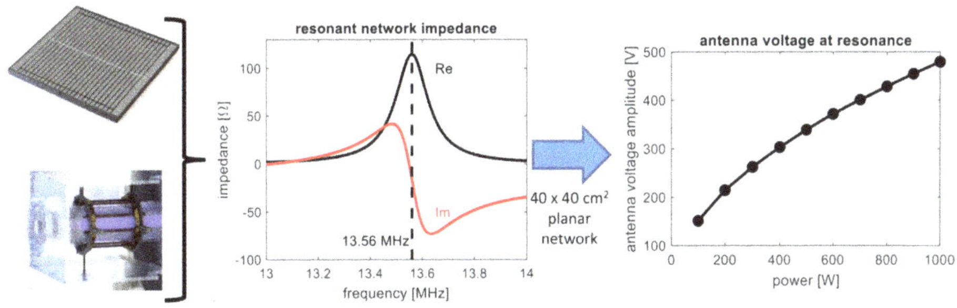

Figure 5.8. Two high power resonant networks, represented by ladder and birdcage antennas, and their resonance input impedance at 13.56 MHz, which is dominantly real. The corresponding voltage/power relation is shown for a planar network example. The voltage amplitude is below 500 V for 1 kW input power.

5.6 Chapter summary for the introduction to ICP

The motivation for RF plasma and some basics of inductive coupling are introduced, notably the plasma skin depth and the transformer model. The mutual inductance which couples the source and the induced plasma current is the key parameter for estimating the effect of plasma loading on the source impedance and voltage.

Two main difficulties which prevent upscaling to larger ICP reactors are the non-uniform skin depth profile, and the dangerous voltage across the inductance of big RF coils. These drawbacks are part and parcel of monolithic ICP reactors having a single large source.

In the next chapters, we focus on new, versatile solutions afforded by resonant network antennas. A 'complex image method' will be introduced to account for the self-consistent mutual partial inductance between antenna and plasma.

References

[1] Howling A A, Descoeudres A and Hollenstein Ch 2020 Multiple dehydrogenation reactions of negative ions in low pressure silane plasma chemistry *Plasma Sources Sci. Technol.* **29** 105015

[2] Lieberman M A and Lichtenberg A J 2005 *Principles of Plasma Discharges and Materials Processing* 2nd edn (Hoboken, NJ: Wiley)

[3] Chabert P and Braithwaite N 2011 J *Physics of Radio-Frequency Plasmas* (Cambridge: Cambridge University Press)

[4] Chen F F 2016 *Introduction to Plasma Physics and Controlled Fusion* 3rd edn (Cham: Springer)

[5] Chabert P, Tsankov T V and Czarnetzki U 2021 Foundations of capacitive and inductive radio-frequency discharges *Plasma Sources Sci. Technol.* **30** 024001

[6] Köhler K, Coburn J W, Horne D E, Kay E and Keller J H 1985 Plasma potentials of 13.56-MHz RF argon glow discharges in a planar system *J. Appl. Phys.* **57** 59

[7] Köhler K, Horne D E and Coburn J W 1985 Frequency dependence of ion bombardment of grounded surfaces in RF argon glow discharges in a planar system *J. Appl. Phys.* **58** 3350

[8] Godyak V A 1986 *Soviet Radio Frequency Discharge Research* (Falls Church, VA: Delphic Associates)

[9] Piejak R B, Godyak V A and Alexandrovich B M 1992 A simple analysis of an inductive RF discharge *Plasma Sources Sci. Technol.* **1** 179
[10] Hopwood J 1992 Review of inductively coupled plasmas for plasma processing *Plasma Sources Sci. Technol.* **1** 109
[11] Lee H-C 2018 Review of inductively coupled plasmas: nano-applications and bistable hysteresis physics *Appl. Phys. Rev.* **5** 011108
[12] Lieberman M A, Booth J P, Chabert P, Rax J M and Turner M M 2002 Standing wave and skin effects in large-area, high-frequency capacitive discharges *Plasma Sources Sci. Technol.* **11** 283
[13] Sansonnens L, Howling A A and Hollenstein Ch 2006 Electromagnetic field nonuniformities in large area, high-frequency capacitive plasma reactors, including electrode asymmetry effects *Plasma Sources Sci. Technol.* **15** 302
[14] Guittienne Ph, Jacquier R, Howling A A and Furno I 2017 Electromagnetic, complex image model of a large area RF resonant antenna as inductive plasma source *Plasma Sources Sci. Technol.* **26** 035010
[15] Kawamura E, Graves D B and Lieberman M A 2011 Fast 2D hybrid fluid-analytical simulation of inductive/capacitive discharges *Plasma Sources Sci. Technol.* **20** 035009
[16] 2008 *Advanced Plasma Technology* ed R d'Agostino, P Favia, Y Kawai, H Ikegami, N Sato and F Arefi-Khonsari (Weinheim: Wiley)
[17] Gudmundsson J T and Lieberman M A 1998 Magnetic induction and plasma impedance in a planar inductive discharge *Plasma Sources Sci. Technol.* **7** 83
[18] Bleaney B I and Bleaney B 1976 *Electricity and Magnetism* (London: Oxford University Press) 3rd edn p 1
[19] Guittienne Ph, Lecoultre S, Fayet P, Larrieu J, Howling A A and Ch Hollenstein 2012 Resonant planar antenna as an inductive plasma source *J. Appl. Phys.* **111** 083305
[20] M-Fayoumi El I and Jones I R 1998 The electromagnetic basis of the transformer model for an inductively coupled RF plasma source *Plasma Sources Sci. Technol.* **7** 179
[21] Meyer J A, Mau R and Wendt A E 1996 Plasma properties determined with induction loop probes in a planar inductively coupled plasma source *J. Appl. Phys.* **79** 1298
[22] Zaman A J M, Long S A and Gardner C G 1981 Impedance of a loop surrounding a conducting cylinder *IEEE Trans. Instrum. Meas.* **IM-30** 41
[23] Godyak V and Chung C W 2006 Distributed ferromagnetic inductively coupled plasma as an alternative plasma processing tool *Jpn. J. Appl. Phys.* **45** 8035
[24] Ahr P, Tsankov T V, Kuhfeld J and Czarnetzki U 2018 Inductively coupled array (INCA) discharge *Plasma Sources Sci. Technol.* **27** 105010
[25] Lee J-W, Lee Y-S, Chang H-Y and An S-H 2014 On the possibility of the multiple inductively coupled plasma and helicon plasma sources for large-area processes *Phys. Plasmas* **21** 083502

IOP Publishing

Resonant Network Antennas for Radio-Frequency Plasma Sources
Theory, technology and applications
Philippe Guittienne, Alan Howling and Ivo Furno

Chapter 6

Inductive coupling using plane plasma sources

Inductively coupled plasma (ICP) sources are widely used in many applications, particularly in the crucial industry of plasma processing for semiconductors. Spiral (also called pancake, or stove-top) coil antennas are often used, although alternative plasma sources for larger and more uniform reactors would always be welcome. This chapter briefly reviews conventional RF planar ICP sources [1, 2] before comparing with the plasma performance of a ladder resonant network antenna.

The ladder resonant ICP antenna is shown to exhibit an *E*- to *H*-mode transition, typical of ICP sources. As further proof, the plasma density patterns for the *E*- or *H*-mode correspond, respectively, to electro-quasistatic or magneto-quasistatic electric fields, as expected. The plasma uniformity obtained is influenced by diffusion within the induced normal mode structure described in chapter 3; the latter can be determined by design, according to the desired uniformity, by choosing the leg spacing and/or the mode number m. In principle, the resonant ladder area can be extended to arbitrary dimensions (see also chapter 7) and shapes (chapter 14).

For reactor design, it is helpful to be able to model the RF plasma coupling to the resonant network. The complex image method gives the inductive coupling in terms of the complex mutual partial inductance between the plasma and the antenna. In this chapter and appendix E, analytical expressions are obtained for the inductance and resistance of the ICP source, and the calculated impedance spectrum due to plasma loading agrees well with measurement.

6.1 Introduction to planar ICP sources

The overwhelming majority of planar ICP sources are flat spiral coils, therefore it behooves us to make a brief review of this important source type for imminent comparison with planar resonant ladder antennas. Literature reviews can be found

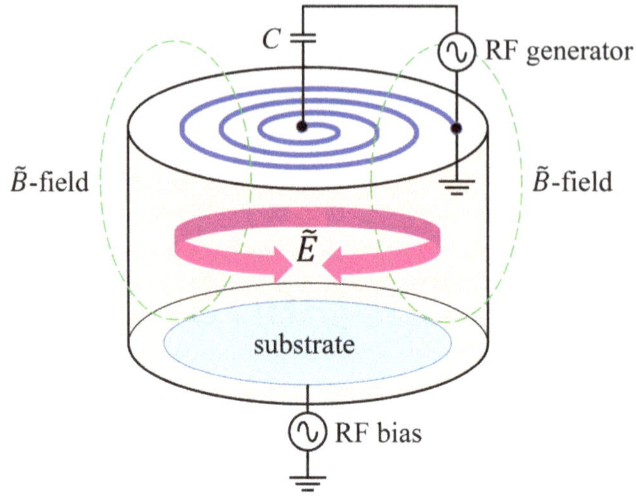

Figure 6.1. Schematic of a planar ICP with a spiral coil driven by an RF generator via an effective capacitor C. The oscillating current creates a $\tilde{B}$-field which induces an oscillating azimuthal electric field $\tilde{E}$ to generate the RF plasma. Ion bombardment of the substrate can be enhanced by a separate RF bias supply (see also section 6.5).

Figure 6.2. 'Contours of current density in a spiral coil ICP reactor at phase $\omega t = 0.3\pi$.' (Illustration by Alex Howling. Copyright 2023 Alex Howling. Adapted with permission from figure 11 in [7]. Copyright 1998 IOP Publishing.)

from 1992 [3] to many contemporary works such as [4, 5], to cite just two. Figure 6.1 shows the basic elements of an RF planar spiral coil ICP reactor. The power dissipation from induced fields in the plasma is confined to a few skin depths below the spiral coil, similarly to the edge of a solenoidal reactor in figure 5.3, except that the field profiles are more complex in the spiral case [6, 7]; see figure 6.2 for an example. The capacitor C in figure 6.1 symbolizes a matching system for resonant excitation of the coil. As mentioned in section 5.5, very high voltage occurs across the single large coil inductance. This monolithic resonant antenna is in contrast to the resonant network antennas of this book, because the network's mode structure excites an internal array of spatially distributed resonances. Its impedance is dominantly real at the network input terminals, so for a given RF input power, the network voltages are consequently much lower and easier to handle than for the single planar coil.

These sources typically exhibit characteristic E- to H-mode transitions [3, 5], as already mentioned in section 5.2.2, and described below in detail for the ladder resonant ICP antenna in section 6.3.1. Non-uniformity is an intrinsic property of spiral coils because of the symmetry requirement of a null in the induced azimuthal electric field on axis [7–9]. This inherent non-uniformity is mitigated by diffusion which is especially effective at low pressures, for example, below 1 Pa. The uniformity can be ameliorated using special coil designs [10], so magnetic fields [11] are generally not needed [4] although multipolar magnetic cusp confinement (for example, an array of bar magnets on the walls) can play a role by slowing electron diffusion losses [1, 3, 12]. Two examples are multi-spiral coils carrying different RF currents, and cone-spiral coils [4, 13]. Note, however, that any coil winding scheme that distances part of the coil from the plasma reduces the mutual partial inductance and thereby the coupling efficiency. Uniformity limitations mean that spiral coil reactors are generally intended as single-wafer plasma processing reactors [1, 3, 10, 12]. The many other types of specifically large-scale planar antenna sources will be covered in chapter 7.

Other variations on this theme include the 'ladder antenna' [14] (not to be confused with the resonant network ladder antenna), developed as an alternative configuration to increase the processed area—its structure of parallel inductive legs can diminish the reactive input impedance, although it remains primarily reactive nevertheless. Uniform plasma glow was obtained by choosing RF input and ground points for which all currents paths have equal lengths and impedance. Another example, among many, employs crossed internal oscillating current sheets consisting of two orthogonal sets of eight wires [15]; the uniformity is improved compared with spiral coils.

As previously detailed in section 5.5, a general drawback of non-resonant antennas for plasma processing lies in the fact that their input impedance is principally reactive, which implies very high input voltage and ion bombardment [3], extremely low power factor, limited current in the segments, and therefore limited plasma density. The matching circuit of a reactive-impedance system can often limit the RF power, because of large circulating currents and large ohmic dissipation in the RF skin depth of all the conducting surfaces. Furthermore, as the inductance increases with size, the tuning capacitor to maintain the RF resonance frequency has to be decreased to values comparable to stray capacitance (pF), with attendant uncertainties [3].

6.2 Experimental set-up for an ICP ladder resonant antenna

In this chapter, we present a planar, resonant network RF plasma source operating at the ISM frequency of 13.56 MHz, suitable for both large area processing, and high electron density. It can be designed in principle up to large sizes by adding inductive copper legs linked together by capacitors, in contrast to sources where the capacitance is coupled towards the plasma and not in series with the bar inductances [16]. This experimental antenna plasma reactor was tested in argon up to 2 kW for possible plasma processing applications [17, 18]; one of these applications is described in section 6.5. The antenna itself was already fully described in section 3.1 and figure 3.2. It was characterized in air, on the bench, by chapters 3 and 4. A supplementary end-on view is added here in figure 6.3. We now proceed to the design and implementation of a plasma reactor using this antenna. The antenna is embedded in a polyimide foam dielectric to prevent spurious plasma ignition within the antenna itself, with a protective glass cover and a grounded metal baseplate as shown schematically in figure 6.4; see section 13.4.2 for a technological description. The copper tubes are water cooled to prevent excessive heating of the antenna components such as the capacitors. Note that the principal heat load on the antenna is due to the plasma itself, not the ohmic heating of the antenna RF currents in the bars. Note also that the comparatively low voltages in a resonant network

Figure 6.3. Photograph of a small antenna (55 cm × 22 cm) for operation in mode m = 6 at 13.56 MHz, shown without the dielectric filling. See figure 3.1 for a full description.

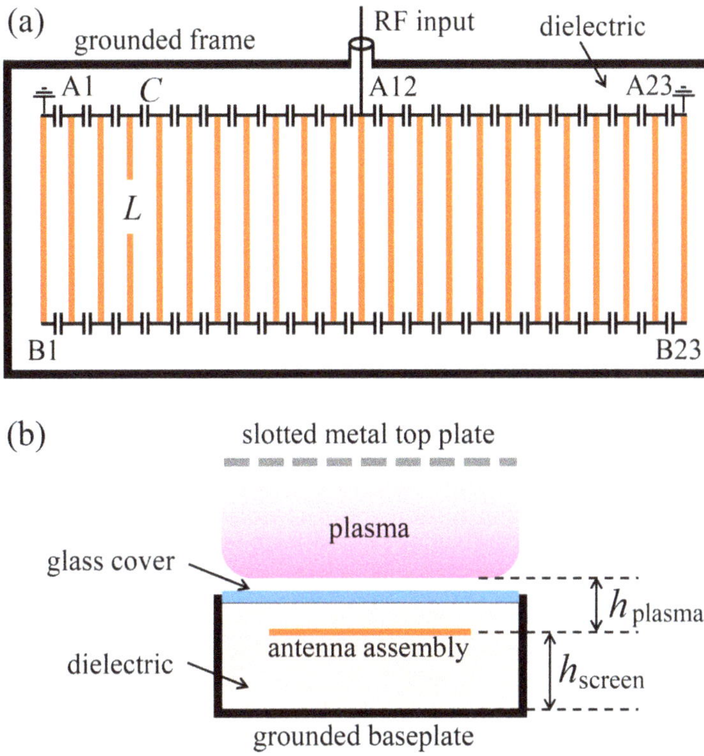

Figure 6.4. (a) Schematic top view of the ladder antenna assembly as in figure 3.2, but grounded at nodes A1 and A23. (b) Schematic end view of the antenna showing the grounded metal baseplate distance $h_{screen} = 5.5$ cm below the antenna; the antenna assembly embedded in a polyimide foam dielectric to a level 0.6 cm above the leg axes, and the foam surface which is protected by a 0.3 cm thick glass cover. The plasma lower surface boundary is at height h_{plasma} above the antenna assembly, accounting for the effective vacuum sheath thickness. The plasma slab is bounded by a slotted metal top plate 8 cm above the glass surface. The whole system is placed within a vacuum vessel. (Adapted with permission from [19]. Copyright 2015 IOP Publishing.)

antenna (see section 5.5) mean that no Faraday shield is necessary, because the ion bombardment across weak capacitive sheaths is of corresponding low energy. This assembly constitutes the plasma source module which is entirely placed within a vacuum vessel and connected via 30 cm of 50 Ω coaxial cable inside the vacuum vessel to a vacuum feed-through. This is possible thanks to the impedance properties of the resonant network, because the reflected power in the connecting cable between the external matchbox and the internal antenna is sufficiently low. In contrast, for a conventional ICP, whose impedance is highly reactive, the matching box must be placed directly on the antenna to minimize the ohmic power loss and heating from the very high reflected RF current—see sections 13.3.4 and 13.3.5. Finally, because the antenna is placed in the vacuum vessel, the plasma source module is directly exposed to the low pressure plasma ('immersed inductive coupler' [3]). Therefore, it does not need a thick dielectric window to withstand atmospheric pressure over the large area of the antenna. Mechanical construction and matching circuits for resonant network antennas are described in chapter 13.

An insulating substrate could be placed directly on the antenna, with a gas showerhead placed opposite for uniform processing with reactive gases [20, 21]. Alternatively, any substrate could be positioned as an upper boundary of the plasma as shown in figure 6.4(b), in which case a gas showerhead could conceivably be incorporated into the antenna assembly itself. The upper boundary in the present case was a slotted metal plate for access by a Langmuir probe array, and there were no sidewalls to facilitate access for a microwave diagnostic. Measurements were made using argon in this geometry [22], although an industrial design for plasma deposition in reactive gases would preferably use a closed, showerhead reactor for fast equilibration of the plasma [23].

The mode m = 6 was chosen for a detailed study of the antenna performance, because its frequency is distant from neighbouring modes in the input impedance spectrum (figure 4.20), and because it performs well in terms of plasma uniformity because of the adjacent proximity of its current nodes, as discussed in section 6.3. The RF input power is centrally connected at node A12 in figure 6.4, with ground connections at extreme ends A1 and A23 (for modes m/2 odd, such as m = 6 here), or B1 and B23 (for modes m/2 even; not shown here) to ensure symmetrical power feeding [19]. This can be understood from figure 3.7 for the example of m = 6, where the possible A_n ground nodes are at A1, A8, A16, and A23. Ground nodes A8, or {A8, A16} by symmetry, were employed in chapters 3 and 4, whereas ground nodes {A1, A23} are chosen for the high power applications in this chapter. In the context of input impedance optimization in figure 3.16, the (ground, RF) connection node sets $(N_g, N_f) = (8, 12)$ or $(1, 12)$ are essentially equivalent. The two sets of experimental parameters for the plasma in this section were either 1 Pa pressure of argon with 50–500 W delivered power at 13.56 MHz for mode m = 6; or a frequency range from 10 to 23 MHz at a constant delivered RF power of 200 W to excite modes m = 2, 4, 6, 8, 10.

6.2.1 Plasma diagnostics

The two-dimensional plasma density profile was mapped using the ion saturated currents measured by a linear array of 15 Langmuir probes traversing the slotted plate. The probe pins are made of steel wires 0.4 mm diameter and 4 mm long (the differences between the areas A (m^2) of the pins are estimated to be less than 10%). Scanning this array of biased probes along the length of the antenna gives the topography of the ion saturation current, I_{sat} (A) [1]

$$I_{sat} \approx 0.61 q n_{e0} A \sqrt{\frac{k_B T_e}{m_e}}, \tag{6.1}$$

where q (C) is the magnitude of the electron charge, k_B (J K^{-1}) is Boltzmann's constant, and the factor $e^{-1/2} \approx 0.61$ depends on the specifics of the plasma sheath model used [24, 25]. The electron density n_{e0} (m^{-3}) is roughly proportional to I_{sat}, assuming a relatively narrow range of electron temperature T_e (K) variation in (6.1).

Only a small perturbation of the plasma is observed when moving the whole probe system. Note that the plasma parameters are very stable and the reactor can be operated in steady state for hours at a time, as required for any industrial plasma processing. This is particularly convenient for long-duration diagnostic measurements such as a Langmuir probe scan; we will see in later chapters that sensitive laser measurements can be acquired reliably during 6 hours or more. The absolute electron density, line-averaged along the antenna width, was measured by a 33 GHz microwave interferometer via horns on either side and a PTFE focusing lens [26], with macor blocks arranged in the vacuum chamber to absorb stray radiation and minimize reflections.

The plasma-coupled antenna was electrically characterized by measuring its input impedance at the vacuum feed-through, by means of a wideband current/voltage/phase probe (Z-Scan™ [27] from Advanced Energy), as a function of frequency for several modes. The same probe was used to measure the RF power delivered to the antenna (forward power minus reflected power). The power transfer efficiency for a given mode can be estimated from the changes in mode resonance impedance and frequency due to plasma, using an approximate method described in section 9.2.6. It is important to remember that the network analyzer impedance measurements for chapters 3 and 4 were made directly at the antenna input, in air, on the bench. In contrast, for the plasma set-up reported in this chapter, the antenna was connected via 30 cm of 50 Ω coaxial cable inside the vacuum vessel to the vacuum feed-through on the vacuum side. For a consistent comparison here, the vacuum impedance measured by a network analyzer, and the impedance measured during plasma, were both measured at the vacuum feed-through on the atmosphere side. The measured impedance Z_{mes} therefore corresponds to the antenna input impedance Z_{in} transformed by a transmission line of length l = 30 cm and characteristic impedance Z_{c} = 50 Ω. The relation is [28]

$$Z_{\mathrm{mes}} = Z_{\mathrm{c}} \frac{Z_{\mathrm{in}} + jZ_{\mathrm{c}} \tan(2\pi l/\lambda)}{Z_{\mathrm{c}} + jZ_{\mathrm{in}} \tan(2\pi l/\lambda)}, \tag{6.2}$$

where $\lambda = c/(f\sqrt{\varepsilon_r})$ is the RF wavelength in the cable dielectric of relative permittivity ε_r, assuming a lossless cable.

6.3 Plasma performance of an ICP ladder resonant antenna

For an antenna operating in mode m = 6 at 13.56 MHz, argon plasmas could be ignited and sustained with RF power as low as 10 W and up to about 2000 W at pressures between 0.1 and 10 Pa. Figure 6.5 shows an example of the electron density distribution for an argon plasma at 1000 W measured over a plane in the middle of the 8 cm gap between the antenna and the floating metal slotted plate. Here, the electron density is peaked in the centre and falls off abruptly near the edges, typical for a diffusion profile. The electron density measured by microwave interferometry was greater than 10^{17} m^{-3}. Contour lines shown in figure 6.5 are plotted in intervals of 10% in electron density and show that it is symmetrical in both dimensions and presents a non-uniformity of less than ±5% in the central region (±0.2 m along Ox

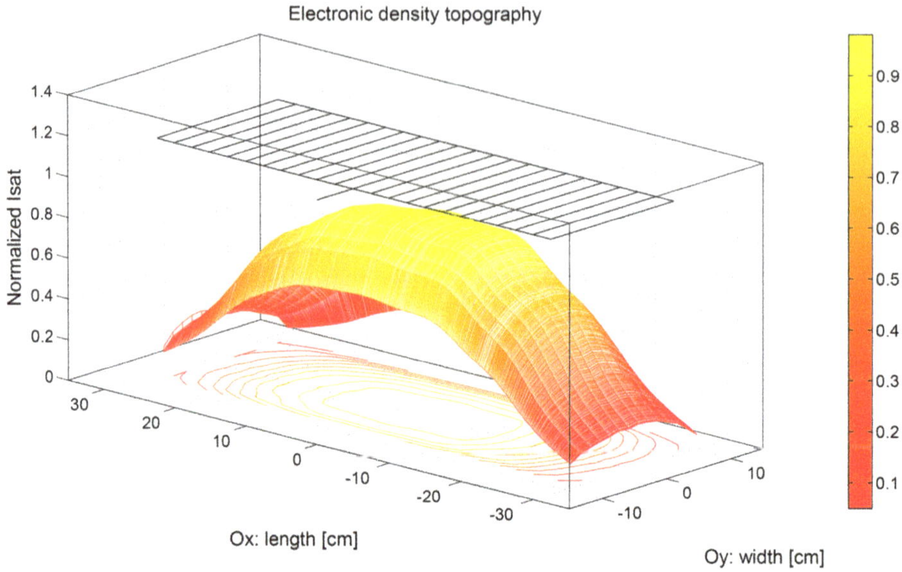

Figure 6.5. Normalized electron density surface plot obtained by scanning Langmuir probes at about 4 cm above the antenna and at a power of 1000 W. The Ar flux is 10 sccm and the pressure 1 Pa. The slotted, floating metal substrate is placed about 8 cm above the RF antenna. The lines in the contour plot are spaced at 10% intervals. A schematic of the antenna is drawn to show the relative dimensions. (Reproduced with permission from [29]. Copyright 2012 AIP Publishing.)

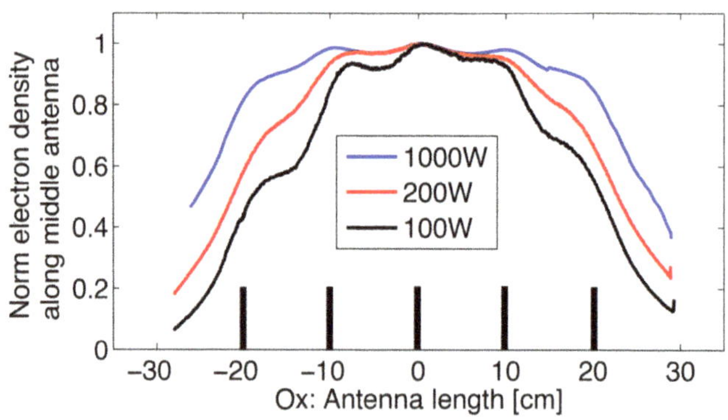

Figure 6.6. Normalized electron density profiles measured along the RF antenna length for 100, 200, and 1000 W at 4 cm above the antenna. The argon pressure is 0.7 Pa and the flowrate is 10 sccm. The black line segments indicate the node positions of the legs with maximal currents for mode m = 6, i.e. leg numbers 4, 8, 12, 16, and 20, as in figure 3.7. (Reproduced with permission from [29]. Copyright 2012 AIP Publishing.)

and ±0.08 m along Oy). Although the antenna is inductively coupled to the plasma, the image of the antenna current distribution is scarcely visible in the plasma density —this is related to the low gas pressure which favours diffusion of the plasma species within the plasma volume.

The electron density measured along the middle of the antenna is given in figure 6.6. These profiles broaden with increasing power when passing from 100 W and 200 W to 1000 W. The black line segments indicate the position of the legs with maximal currents for mode m = 6, and these correspond to the local density maxima, as expected for inductive coupling. Note that the shoulders of electron density near the antenna ends at ±20 cm rise progressively to approach values similar to the three central maxima near the RF input, showing that the plasma becomes more and more uniform with increasing power. This indicates that the plasma ionization rate begins to dominate the diffusive losses at the edges, extending the high electron density from the centre of the antenna to the edges.

6.3.1 Transition from *E*-mode to *H* -mode

The transition from the capacitive *E*-mode to the inductive *H*-mode was monitored by gradually increasing the RF power and measuring the electron density along the antenna width using the microwave interferometer [29]. Figure 6.7 shows the evolution of the electron density from 30 W to 2000 W. A clear jump by more than one order of magnitude in electron density is observed around 60 W for an argon pressure of 1 Pa. In the inductive mode, the electron density increases monotonically with power and since no saturation is observed, even higher electron densities should be reached above 2 kW.

As discussed in section 5.2, the supposition is that capacitive coupling (*E*-mode) from the high voltage points on the antenna first ignites a plasma. Then, as the antenna input power is raised, currents induced in the plasma finally cause a transition to an inductively coupled plasma (*H*-mode) [22, 30–34]. To put this interpretation on a firmer footing, the mode transition was investigated in more

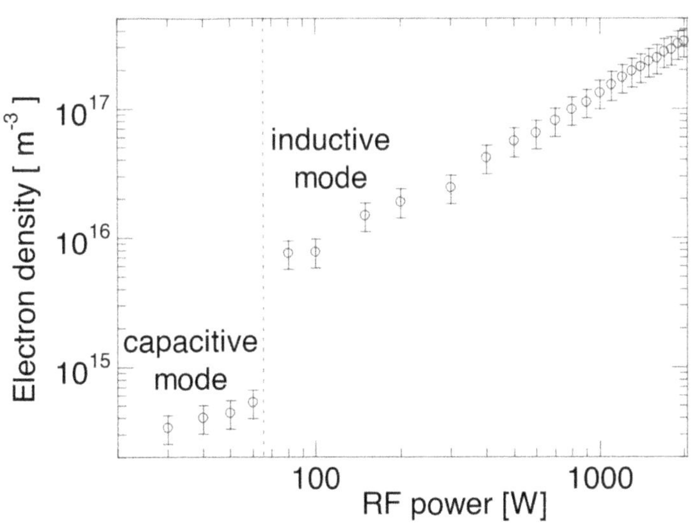

Figure 6.7. Maximum electron density measured by microwave interferometry in the Ar plasma for RF power increasing from 30 W to 2000 W, shown on a log/log scale. The Ar pressure is 1 Pa and the flowrate is 30 sccm. The transition from capacitive to inductive coupling occurs at around 60 W. (Reproduced with permission from [29]. Copyright 2012 AIP Publishing.)

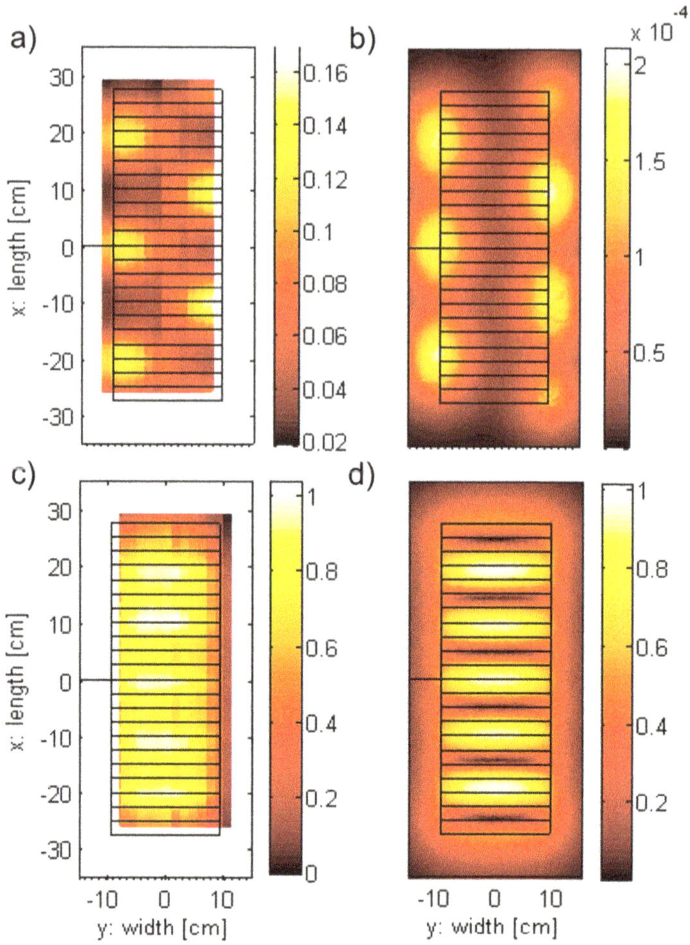

Figure 6.8. Images (a) and (c) present measured electron density profiles for mode m = 6 at 20 W and 1000 W, respectively, in a plane 2 cm above the antenna. The Ar flowrate is 50 sccm and the pressure is 2 Pa. Images (b) and (d) present calculated electric field magnitude profiles in the same plane, determined in the first case (b) by the voltages at each node, and in the second case (d) also by the induced electric field due to the currents in each segment. (Reproduced with permission from [29]. Copyright 2012 AIP Publishing.)

detail. Figures 6.8(a) and (c) present two measured electron density profiles obtained at RF powers of 20 W and 1000 W, respectively, with the probe array located 2 cm above the antenna. The argon flowrate and pressure were held constant at 50 sccm and 2 Pa. The plasma profiles are clearly not the same, indicating that the plasma couples in two different ways with the antenna at low and high power. These measurements are compared in figures 6.8(b) and (d) with two calculated electric field magnitude profiles at the same position using a COMSOL numerical simulation [35]. In a first calculation, figure 6.8(b), the physical parameters defining the RF antenna are its normal mode voltages according to (3.19) and (3.20). The electric field is then calculated in a plane parallel to the antenna at the same position as the measurement, assuming the electro-quasistatic approximation (E-mode operation) appropriate to the electric fields in absence of strong currents:

$$\mathbf{E} \approx -\nabla V. \tag{6.3}$$

In the second calculation, figure 6.8(d), the antenna currents were also taken into account in the calculation, using Ohm's law and the impedance of each segment of the antenna. Thereby, an additional contribution to the electric field inside the plasma is added due to electromagnetic induction. The Coulomb gauge consistent with a solution of Maxwell's equations gives

$$\mathbf{E} = -\nabla V - \frac{\partial \mathbf{A}}{\partial t}, \tag{6.4}$$

as shown in section E.1, where $\mathbf{A}$ is the magnetic vector potential defined by $\mathbf{B} = \nabla \times \mathbf{A}$. The magneto-quasistatic approximation consists of neglecting ∇V in the presence of strong currents, to obtain

$$\mathbf{E} \approx -\frac{\partial \mathbf{A}}{\partial t}, \tag{6.5}$$

which corresponds to H-mode operation. Quasistatic approximations are appropriate for these reactor dimensions which are much smaller than the RF wavelength (see appendix D.1.1).

Comparison between the measurements and the simulations demonstrates that the plasma generated by the antenna at low power (figures 6.8(a) and (b)) is capacitively coupled with the antenna electro-quasistatic potentials (E-mode). Both the measured electron density and the calculated electric field magnitude show five areas of higher density located above the nodes where the RF voltages are maximal, that is to say nodes A4, B8, A12, B16, and A20 as shown in figure 3.7(b).

At higher RF power (figures 6.8(c) and (d)), a transition occurs into an inductive mode (H-mode), in which the plasma couples with the RF currents in the antenna; compare figures 3.7(a) and 6.6. This time, the electron density is strongly increased in the areas above the legs carrying the maximum currents. The normalization of the scales for both inductive (H-mode) profiles, figures 6.8(c) and (d), makes it clear that the electron density and the electric field are much larger than for the capacitive (E-mode) profiles of figures 6.8(a) and (b). Figure 6.8 therefore presents a graphic demonstration of the E- to H-mode transition.

6.3.2 Plasma uniformity for spiral coil and ladder resonant antenna

Planar ICPs are naturally suited for plasma processing of flat substrates, such as silicon wafers or glass substrates, and for roll-to-roll treatment. The plasma current induced by a monolithic spiral planar coil is inherently non-uniform, although better than for solenoidal coils [8]. By radial symmetry, both are fated to have a single global null in azimuthal electric field in the centre, as mentioned above in section 6.1, and shown here in figure 6.9(a) and (b). The stronger non-uniformity for the 40 cm coil is due to a standing wave node, hence zero current density, at the position of the black ring in

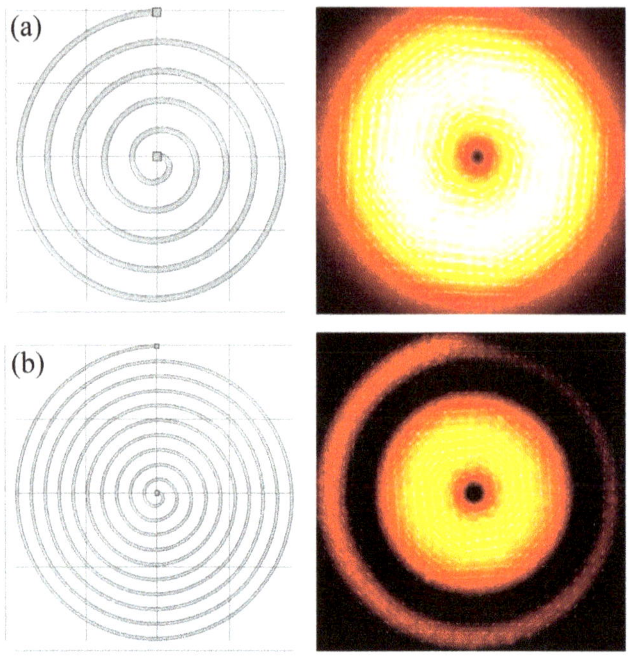

Figure 6.9. Calculated current density induced in a plasma by a spiral coil at 13.56 MHz. (a) 20 cm diameter, five-turn spiral; (b) 40 cm diameter, ten-turn spiral. The current density is zero on axis in both cases.

figure 6.9(b). If the spiral coil is loosely wound, the source-to-substrate gap may have to be widened so that diffusion sufficiently smears out the coil image appearing on the processed substrate. However, the chamber volume is thereby increased, and the diffusive loss to the taller sidewall causes its loss perturbation to extend inwards and encroach on the substrate area, as estimated in section 6.3.3.

On the other hand, the multiple normal nodes of ladder resonant antennas in figure 6.10 are internal resonances of its structure which are spatially distributed across the antenna area, so it does not suffer from the universal single hollow centre of a planar spiral antenna. In fact, the mode m is akin to having m adjacent planar coils because of the m cells of circulating current shown in the figure. In contrast to the spiral coil in figure 6.9, the ladder's mode structure can be designed flexibly by choosing the number of legs, the leg spacing, and, especially, the mode number m. Furthermore, the resonant antenna can be extended to arbitrary dimensions (chapter 7) and shapes (section 14.2). The plasma uniformity is further improved by diffusion within the induced normal mode structure, in the same way as for the spiral coil, but more effectively because of the shorter distances between the m nodes.

6.3.3 Diffusion profiles close to reactor sidewalls

The mode structure and reactor dimensions can be designed so that, in principle, any arbitrary degree of uniformity can be obtained, within the constraints of the diffusive loss density profile at the sidewalls. Note that a reactor which completely encloses the plasma on all sides, with no dead gas volume, is the

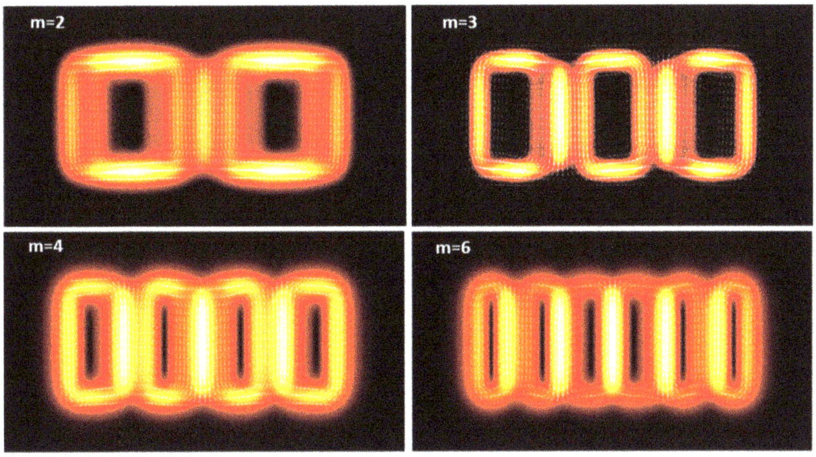

Figure 6.10. Calculated current density loops, circulating in a plane plasma, induced by a ladder resonant antenna for mode numbers m = 2, 3, 4, or 6. The number of internal resonances, spatially distributed over the antenna length, can be chosen according to the normal mode number m.

optimal design for the fastest possible plasma-chemical equilibration time after plasma ignition [23]. The mathematical expressions for the flow and diffusive profiles near the sidewall edge of two-dimensional rectangular channel flow [20, 36] show that the dominant perturbation terms in the infinite series vary as $\exp(-\pi\zeta/D)$, where ζ is the lateral distance inwards from the sidewall, and D is the sidewall height, equal to the source-to-substrate gap. The e-folding length for edge non-uniformity is therefore D/π. For a tolerance of $e^{-3} \sim 5\%$ non-uniformity, the outer border of width $3D/\pi \sim D$ is therefore unusable, non-uniform substrate area, which is motivation to use as small a gap as possible. Edge non-uniformity can thus be seen as the penalty for depending on a wide source-to-substrate gap for diffusion to average out coil or antenna images on the substrate. Note that the uniformity of the plasma density profile in figure 6.5 is considerably worse than this prediction because of the open-sided construction (for the microwave diagnostic access), which is incompatible with the mathematical assumption of flow of uniform plasma in a two-dimensional rectangular channel.

6.4 Induced currents in the plasma: the complex image method

6.4.1 The impedance loading spectrum when inductively coupled to plasma

Figure 6.11(a) shows the spectrum of the antenna input impedance magnitude, measured at the vacuum feed-through for modes m = 2, 4, 6, 8, 10 in a vacuum, and with plasma in the H-mode, at 200 W. The input impedance in a vacuum (i.e. without plasma) is conveniently measured using a network analyser. We note in passing that the resonance peaks are not as symmetrical as in the measurements of figures 3.4 or 4.20, mainly because of the impedance transformation by the internal coaxial cable described by (6.2). The input impedance measurements in presence of plasma were more problematic because of the necessary high voltages involved.

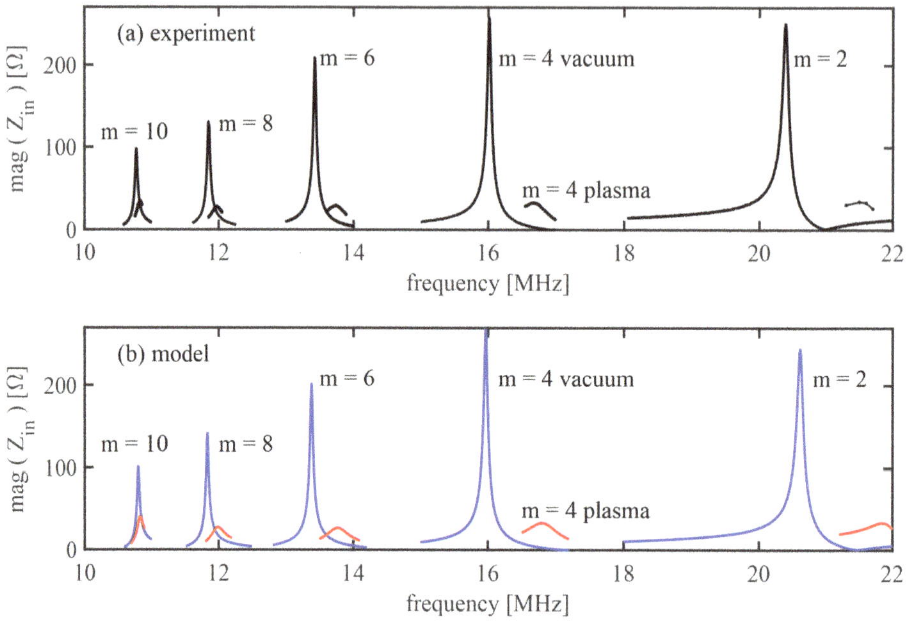

Figure 6.11. (a) Measurement of the antenna input impedance magnitude, $|Z_{\rm in}|$, for the mode impedance spectra in a vacuum, and with plasma, for modes m = 10, 8, 6, 4, 2. (b) Calculations of the corresponding mode impedance spectra by means of the impedance matrix model, in a vacuum (blue lines) and with plasma (red lines). The mutual partial inductance due to plasma coupling with the antenna was estimated using the complex image method [37]. Experimental parameters are Ar pressure 1 Pa and 200 W nominal constant delivered power. The RF input is centrally connected at A12 as in figure 6.4(a) with symmetric ground connections at A1 and A23 to suppress the odd-m modes. (Adapted with permission from [19]. Copyright 2015 IOP Publishing.)

The plasma measurements were carried out using a wideband current/voltage/phase probe (section 6.2.1) and a variable frequency RF generator continually adjusted for a nominal constant delivered power of 200 W. Due to the practical difficulty of maintaining the nominal RF power and the impedance matching conditions when strongly off resonance, the plasma measurements were made only for a limited phase range of $-\pi/4$ to $+\pi/4$ radians across each of the mode resonances.

By inspection of the change in measured impedance due to plasma coupling, shown in figure 6.11(a), three qualitative observations are immediately apparent:

1. Each mode impedance falls strongly due to the power inductively coupled to the plasma; for mode m = 6, the measured impedance at resonance falls by a factor ~7.1 from 208 Ω to 29.3 Ω. This (perhaps counter-intuitive) result is explained just after equation (2.16), where the input impedance at resonance of a $RL\|C$ circuit, $Z_{\rm in}^{\rm res}$, is *inversely* proportional to its inductor resistance R. Remember from figure 3.14 that each antenna resonant mode can be considered as an equivalent $RL\|C$ circuit.

2. Each mode frequency increases noticeably due to plasma coupling, because the induced eddy currents, by Lenz' law, act to oppose the magnetic flux changes caused by the antenna primary circuit. From the transformer model in figures 5.4 and 5.5, the mutual partial inductance subtracts from the primary inductance. The reduction in antenna inductance causes the mode frequency of the equivalent $RL||C$ circuit to rise, according to the basic relation (2.8), or, more accurately, (2.13). The frequency of mode m = 6 increases from 13.43 MHz to 13.74 MHz; an upshift of 2.3%. Supposing that the capacitance is unchanged, this would suggest that the equivalent circuit inductance decreases by about 4.5%.
3. The mode resonances are significantly broader in presence of plasma coupling. The quality factor Q (see section 2.2.2) of the antenna equivalent $RL||C$ circuit is diminished because of the power dissipation coupled into the plasma, whose reflected impedance adds a resistive component. Taking mode m = 6 again as an example, we measure $Q = 336$ in a vacuum; a high quality factor for an RF component as expected for the low-resistance copper tube leg inductors and high-Q RF capacitors (section 13.2). Plasma coupling reduces the quality factor to $Q = 52.8$, a fall of factor ~6.4. Note that, although the Q value has fallen strongly, it is still more than 100 times above the non-oscillating condition called overdamping, at $Q = \frac{1}{2}$. Using equations (2.16) and (2.9), we see that $Z_{\text{in}}^{\text{res}} = \omega_0^2 L^2/R$ and $Q = \omega_0 L/R$, therefore the respective ratios of 7.1 and 6.4 in points 1 and 3 here are commensurate with a similar increase in the equivalent circuit resistance R, given that $\omega_0 L$ in point 2 does not change by more than a few percent.

Power transfer efficiency

As mentioned in section 6.2.1, the delivered RF power (forward power minus reflected power) can be measured by means of a current/voltage/phase probe module, or by individual high-frequency voltage and current probes, placed after the matching box as near as possible to the antenna input; this measurement is feasible owing to the almost-real input impedance of the antenna [22]. In this way, the power measurement does not include the unknown power losses in the matching box itself, which is a significant error in power measurements of conventional antennas with highly reactive input impedance. Nevertheless, a fraction of the measured power is still lost in the antenna circuit itself, with the remainder dissipated as useful power in the plasma. The power transfer efficiency is intuitively defined as the plasma power divided by the total RF power delivered to the antenna.

The power transfer efficiency for a given mode can be estimated from the changes in mode resonance impedance and frequency due to plasma, using an approximate method described in section 9.2.6. Taking advantage of the $RL||C$ equivalent circuit description for each mode, the power transfer efficiency η is given by

$$\eta = 1 - \frac{\omega_{\text{plasma}}^2 \cdot Z_{\text{plasma}}^{\text{res}}}{\omega_{\text{vac}}^2 \cdot Z_{\text{vac}}^{\text{res}}} = 1 - \frac{13.74^2 \times 29.3}{13.43^2 \times 208} \approx 0.85, \tag{6.6}$$

where the numerical values are taken from points 1 and 2 here above for the mode m = 6. The expression (6.6) is suited to resonant network ICP, and so is different from the conventional subtractive method [38]. The non-optimized high power efficiency of 85% is principally related to the strong reduction ratio of the vacuum loading peak down to the plasma loading peak in figure 6.11(a). This high efficiency is typical also of conventional ICP sources operated in noble gases, although high power ICP sources can have η as low as ~50%, before optimization, with 76% of the ohmic losses being due to currents induced in a Faraday screen [38].

The challenge now is to be able to understand and to model the antenna input impedance in presence of plasma.

6.4.2 Inductive coupling of straight wires to plane plasma—the complex image method

By analogy with antenna coupling to a perfectly conducting screen in sections 4.7.1 and 5.4.3, we need to find the effective mutual partial inductance between the antenna and the inductively coupled plasma. Fortunately, the complex image method gives a ready-made approximation for the case of straight wires parallel to a resistive slab, and the mathematical demonstration is given in appendix E. The complex image method can be resumed as follows:

- With reference to figure 6.12(a), the source current I induces an eddy current distribution, $-j(x, y)$, in the plasma. By current conservation, the cross-section surface integral of this plasma eddy current density is equal and opposite to the source current I; namely, $\int_{-\infty}^{\infty}\int_{-\infty}^{0}(-j_z(x, y))\,dx\,dy = -I$. This is consistent with the image concept of an equal and opposite image current.
- Appendix E shows that the magnetic field in the medium of the source I (which is above the plasma, in air, in this case), due to the induced eddy currents in the plasma, is approximately equal to the magnetic field of an image current $-I$ reflected in a mirror plane which is located at the complex skin depth $p = (1 - j)\delta/2$ below the plasma surface [39–48]—see figure 6.12(b) and figure E.2 for a graphic description. Note, therefore, in the complex image method, the image current is real; it is the image distance which is complex. There is nothing conceptually wrong with a complex depth because images are, in any case, a mathematical construct, not a physical manifestation [44, 49]. The name 'complex image method' is therefore rather a misnomer, because the image current itself is still real and equal and opposite to the source current—which is helpful for net partial inductance calculation of the reflected impedance (see section 5.4.3)—it is only the hypothetical image distance which is complex, not the image current itself.
- The required mutual partial inductance between the source current I and the plasma is given directly by the mutual partial inductance expression for two parallel wires, length l, separated by a complex distance $2h_{\text{plasma}} = 2(h + p)$, namely

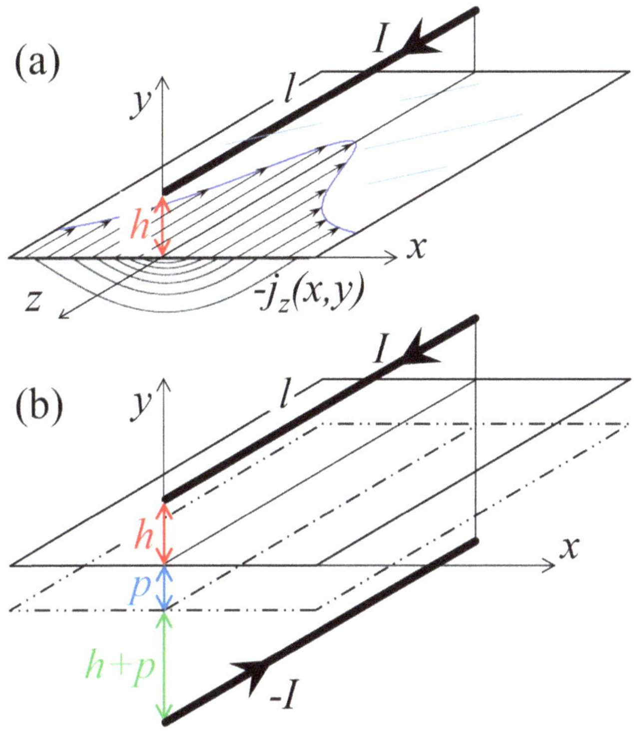

Figure 6.12. (a) Perspective view of a linear current source element length l, parallel to and at distance h (shown in red) above the surface of a plasma whose complex skin depth is $p = (1 - j)\delta/2$. The surface and depth profiles of the induced current density $-j_z(x, y)$ in the plasma are shown schematically. The cross-section surface integral of this plasma eddy current density is equal and opposite to the source current I; namely, $\int_{-\infty}^{\infty}\int_{-\infty}^{0}(-j_z(x, y))\,dx\,dy = -I$. (b) To illustrate the complex image method, the effective mirror surface (dash-dotted line boundary) is at the complex skin depth distance p (blue) below the plasma surface (continuous line boundary). Hence the equal and opposite image current $-I$ is at a complex distance $h + p$ (green) below the mirror surface. Finally, the image current $-I$ is therefore at a total (complex) distance of $2(h + p)$ below the current source I. See also figure E.2. (Adapted with permission from [37]. Copyright 2015 IOP Publishing.)

$$M^{\text{leg/plasma}} \approx \frac{\mu_0}{2\pi} l\left[\ln\left(\frac{l}{h_{\text{plasma}}}\right) - 1\right], \tag{6.7}$$

by analogy with (4.36). This is the great advantage of the complex image method, because it replaces complicated integrals [6, 7, 50] (see appendix E) by a simple, approximate, closed-form[1] analytical solution. Also, it does not involve an arbitrary assumption about the supposed effective depth of the plasma current [6].

[1] An expression is said to be a closed-form solution if it solves a given problem in terms of functions and mathematical operations from a given generally accepted set.

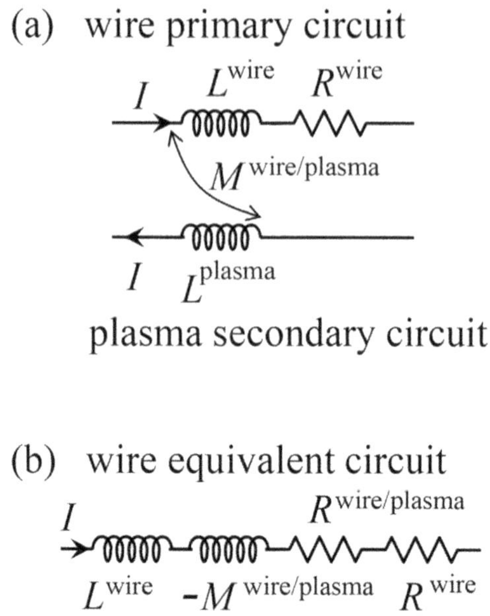

Figure 6.13. (a) The wire as a primary circuit, and the plasma as a secondary circuit, coupled by their complex mutual partial inductance. The image current in the plasma is equal and opposite to the wire current. (b) The equivalent circuit of the wire, transformed by coupling to the plasma, which reduces the primary net inductance and also increases its resistance. (Adapted with permission from [37]. Copyright 2015 IOP Publishing.)

- The net partial inductance of the source wire, in the presence of its equal and opposite plasma image current, is then $L^{\text{net}} = L^{\text{leg}} - M^{\text{leg/plasma}}$, as shown in figure 6.13 by analogy with figure 4.16. It is important to note that $h_{\text{plasma}} = (h + p)$ is a complex quantity, therefore the mutual partial inductance $M^{\text{leg/plasma}}$ in (6.7) is also complex. This brings a self-consistent dissipative component to the mutual partial inductance, representing the coupling to the resistive plasma as shown in section 6.4.3 below.

6.4.3 Complex image model for the loading spectrum

By analogy with figure 4.16, section 5.4.3, and (5.15) in particular, the impedance of the primary, transformed by inductive coupling to the plasma, is

$$Z^{\text{wire/plasma}} = R^{\text{wire}} + j\omega(L^{\text{wire}} - M^{\text{wire/plasma}}), \tag{6.8}$$

where the only difference compared with (5.15) is that the mutual partial inductance, from (6.7), is now complex, $M^{\text{wire/plasma}} = \text{Re}\,[M^{\text{wire/plasma}}] + j\,\text{Im}\,[M^{\text{wire/plasma}}]$. This introduces a resistive component into the reflected impedance (i.e. the mutual partial impedance), which accounts for the ohmic dissipation by the inductively coupled eddy currents in the plasma. With reference to figure 6.13(b), this can be made explicit by substituting and re-arranging (6.8) as

$$\begin{aligned} Z^{\text{wire/plasma}} &= R^{\text{wire}} + \text{Im}\,[\omega M^{\text{wire/plasma}}] + j\omega(L^{\text{wire}} - \text{Re}\,[M^{\text{wire/plasma}}]), \\ &= (R^{\text{wire}} + R^{\text{wire/plasma}}) + j\omega(L^{\text{wire}} - \text{Re}\,[M^{\text{wire/plasma}}]). \end{aligned} \tag{6.9}$$

The measured primary resistance therefore increases by $R^{\text{wire/plasma}} = \text{Im}\,[\omega M^{\text{wire/plasma}}]$, and the measured primary inductance decreases by $\text{Re}\,[M_{12}^{\text{wire/plasma}}]$. This is consistent with physical reasoning and Lenz's law.

The wire-to-plasma complex mutual partial inductance $M^{\text{wire/plasma}}$ therefore completely describes the effect of plasma inductive coupling on the primary circuit impedance, including the reflected inductance and the reflected dissipative coupling in a single, self-consistent manner. Separate calculations [6, 9, 51] for the mutual inductance and the effective plasma dissipation are thereby circumvented. The complex mutual partial inductance depends only on geometrical dimensions (length l, and dielectric thickness plus sheath width, h), and on the plasma complex skin depth p. From appendix D.5, the latter is a function of electron density, electron–neutral collision frequency, and the RF excitation frequency. Hence the only plasma variables necessary to completely determine the inductive plasma coupling are n_{e0}, ν, and the effective vacuum sheath width. As well as the intuitive picture of image currents for calculating inductance, the complex image method therefore also offers the possibility of a self-consistent, parametric study of plasma coupling to the source, by virtue of the simple dependence on plasma complex skin depth p.

Effect of plasma on the mutual partial inductance matrix

By analogy with the effect of a conducting screen on the mutual partial inductance matrix in section 4.7.1, it suffices to replace the mutual partial inductance matrix $\bar{M}_{\text{leg/leg}}$, used throughout section 4.7, by $(\bar{M}_{\text{leg/leg}} - \bar{M}_{\text{leg/screen}} - \bar{M}_{\text{leg/plasma}})$. Again, note that each leg is mutually coupled to all the leg image currents, not only its own image. The effect of the plasma loading, $-\bar{M}_{\text{leg/plasma}}$, has been simply 'plugged in' retrospectively in the same way as the effect of the screen image currents, $-\bar{M}_{\text{leg/screen}}$, in figure 4.17. The combined effect of the inductive coupling of the antenna elements, the baseplate, and the plasma is therefore represented as in figure 6.14. The solution is obtained using the identical matrix procedure outlined in section 4.7 and the accompanying appendix B.

Finally, figure 6.11(b) shows calculations of the corresponding input impedance mode spectra using the mutual partial inductance matrix model. The calculated impedance values have been transformed through 30 cm of 50 Ω transmission line to compare with the experimental measurements. For the vacuum impedance spectra, the calculations are in very good agreement with the measured mode frequencies, impedance magnitudes, and resonance widths, thanks to the mutual partial inductance model previously presented in section 4.8.3, notably in figures 4.20 and 4.21. Furthermore, figures 6.11(a) and (b) also show reasonably good agreement for the frequency shifts and strong reductions in input impedance and quality factor due to the presence of plasma loading of the antenna; this time, thanks to the complex image method combined with the mutual partial inductance description

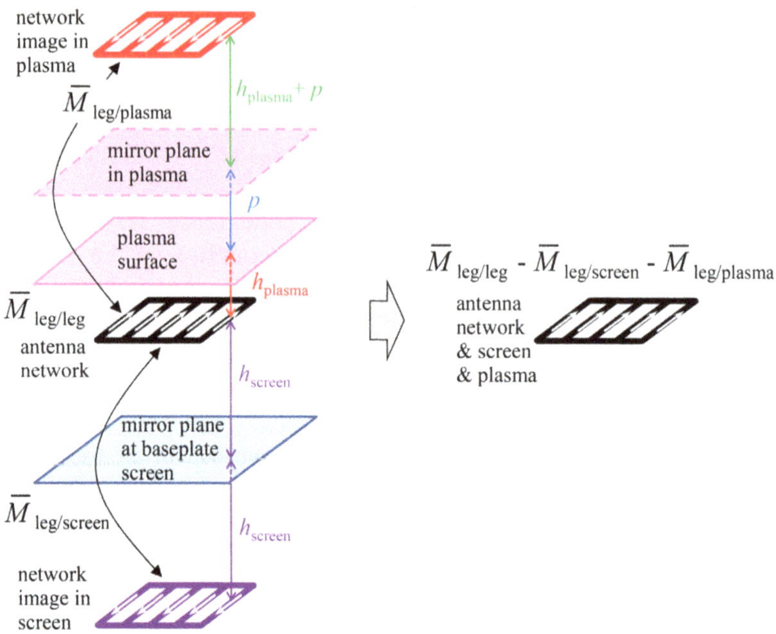

Figure 6.14. Schematic of mutual inductive coupling for the same configuration of antenna, baseplate, and plasma as in figure 6.4(b). The network is coupled to its image currents in the baseplate screen, according to the method of images, by their mutual inductance $\bar{M}_{leg/screen}$ as in figure 4.17. The network is also coupled to its image currents in the plasma, at a complex distance according to the complex image method, by their mutual inductance $\bar{M}_{leg/plasma}$. The white arrows represent the currents and image currents of the legs. The stringer couplings are not shown for clarity. Second order reflections are neglected because field strength falls with distance, i.e. images of images are ignored. (Adapted with permission from [19]. Copyright 2015 IOP Publishing.)

[19, 37]. To our knowledge, the combination of these two methods was applied here for the first time in plasma processing, as explained in the all-too-brief paper [37].

To give a quantitative example, the values of the model parameters are as follows:

1. The distance between the antenna leg axes and the plasma/sheath boundary, h_{plasma} = 1.2 cm, was chosen according to 0.9 cm from the leg axes to the glass top surface (figure 6.4(b)), plus an estimated 0.3 cm for the sheath width (corresponding to about four Debye lengths for the undriven sheath in these plasma conditions [1]).
2. The argon pressure determines the electron–neutral collision frequency ν (rad s^{-1}) $\approx 2.6 \cdot 10^7 \cdot p$ (Pa) for typical electron temperature in these plasma conditions [1]. A value of 1.3 Pa, instead of the nominal pressure of 1 Pa, was found to give the best fit and was used for all calculations in this work. Since the RF frequency and pressure were constant in figure 6.11, the collisionality $\nu/\omega \sim 0.4$.
3. The line-averaged plasma density was estimated by microwave interferometry [29] to be approximately $3 \cdot 10^{16}$ m^{-3} for 200 W delivered power. The plasma densities for the different modes were then estimated according

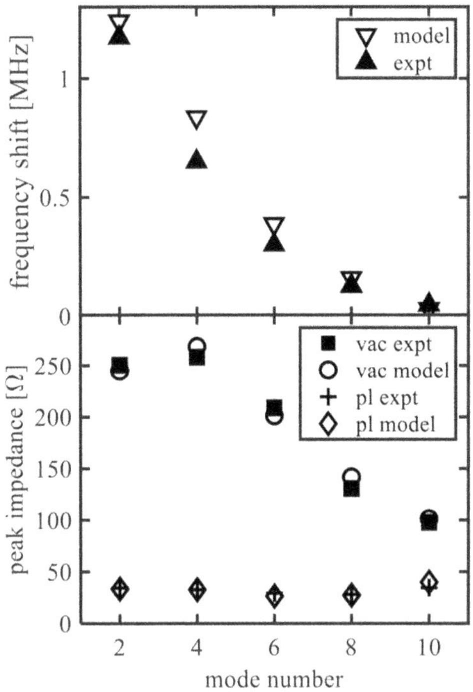

Figure 6.15. Comparison of model and experimental measurement for (a) the mode frequency shift due to plasma coupling, and (b) the impedance values, at resonance, in vacuum and with plasma. The data are taken from figures 6.11(a) and (b). (Reproduced with permission from [19]. Copyright 2015 IOP Publishing.)

to the relative ion saturated currents of the Langmuir probes, assuming that the electron temperature remained in a narrow range as could be expected by consideration of particle balance [1]. For m = {2, 4, 6, 8, 10}, $n_{e0} \approx \{2.3, 3.8, 4.6, 4.6, 3.0\} \cdot 10^{16}\ \text{m}^{-3}$. These values were used for the model calculations.

The Langmuir probes typically measured a dome profile with approximately a 50% drop in plasma density from the centre to the antenna edges [29]. Using the model, only a very small influence of different calculated profiles of plasma density was found on the frequency shifts and mode impedances. We attribute this to the spatial averaging effect of mutual inductances across all the antenna elements. Finally, the typical measured density profile was used for all model calculations shown here. The mode frequency shifts and the impedance magnitudes from figures 6.11(a) and (b) are compared in figure 6.15. The complex image method gives good and robust agreement for the effect of plasma inductive coupling for all five modes simultaneously, using only the three parameters listed above, with no need for sensitive fitting or fine tuning. These three parameters are, in any case, constrained to physically reasonable values corresponding to the experimental parameters.

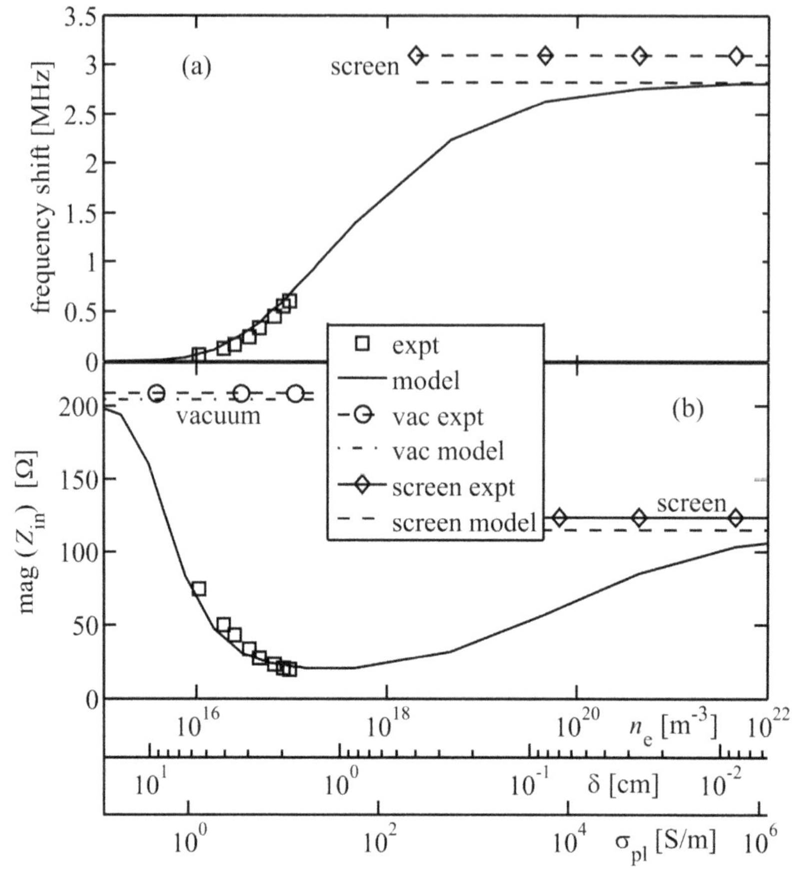

Figure 6.16. Comparison of model and measurement for (a) the mode frequency shift due to plasma coupling, and (b) the antenna input impedance at mode resonance with plasma. The x-axis is represented in terms of electron density n_{e0}, the real part δ of the plasma skin depth, and the real part of the plasma electrical conductivity $\sigma_{\rm pl}$ [37]. The measured and calculated asymptotes for the limits of low plasma density (vacuum) and high plasma density (perfectly conducting screen) are shown by horizontal lines. The input RF powers for the eight experiment points were {50, 80, 100, 150, 200, 300, 400, 500} W, in order of increasing electron density. Mode m = 6, argon pressure 1 Pa. (Reproduced with permission from [19]. Copyright 2015 IOP Publishing.)

Dependence of mode frequency and impedance amplitude on the plasma density
Figure 6.16 compares the measurements of (a) mode frequency shift and (b) impedance magnitude, with calculations using the complex image mutual partial inductance model, similarly to figure 6.15, except that now the dependent variable is the plasma density for a single mode m = 6. Again, the agreement between experiment and model is good for both the frequency shift and the impedance magnitude, using the same parameters as for figure 6.11, namely the pressure, electron density, and plasma distance.

The model results in figure 6.16 are extrapolated to the limit of low plasma density which corresponds to vacuum, and extrapolated to the limit of high plasma density to represent a metal screen at the plasma interface. This limit of high plasma density for the calculated values of frequency shift and antenna impedance were compared with experimental measurements in figure 6.16 by replacing the plasma by a metal plate at the estimated plasma interface distance of 1.2 cm from the antenna. As shown by the low-density and high-density asymptotes in figures 6.16(a) and (b), the values measured, respectively, in vacuum, or with a metal plate, are reasonably consistent with the complex image model. The measured plasma density increased by one order of magnitude, from $8 \cdot 10^{15}$ m^{-3} to $9 \cdot 10^{16}$ m^{-3} for the RF power range 50 to 500 W, but, as can be seen in figure 6.16(a), the measured frequency shift never approached that for a perfect screen, because the electron density would have to be unreasonably high for the plasma conductivity to behave as a metal conductor ($\sigma \sim 10^6$ S m^{-1} for stainless steel).

The calculated range of real skin depth in figure 6.16 was from 6 cm (at 50 W) to 2 cm (at 500 W). 50 W was the lowest power for which the plasma made the transition to the *H*-mode, and the associated skin depth was similar to the electrode gap of 8 cm. For 500 W, the small skin depth means that the power deposition now occurs close to the dielectric surface. Interestingly, there is a minimum in the impedance which corresponds to an optimum power transfer efficiency [22]. This minimum occurs because the vacuum limit (no plasma) and the infinite conductivity limit (perfect screen) both imply no dissipation, i.e. no power transfer to a plasma.

Assumptions, error evaluation, and range of validity

In order to apply the complex image method, the plasma is assumed to be laterally uniform over the length l of the excitation current element, and this length is assumed to be much greater than the wire-to-plasma distance. The plasma is also considered to be uniform in depth, for at least a few skin depths from the source conductor where the induced currents are non-negligible, although averaged homogeneous plasma properties can be supposed [42]. If necessary, the complex image method can be extended to treat multiple layers with different properties [42, 47]. The complex image methods above also assume that the plasma fills an infinite half-space, or at least a depth much greater than the plasma skin depth, so that the integrated current profile induced in the plasma is equal and opposite to the source excitation current. Otherwise, for a thin plasma slab or for low plasma density, where the skin depth becomes comparable to, or longer than, the plasma thickness, the complex image method can be adapted using appropriate boundary conditions for the magnetic field at the far side of the plasma [46, 48] as will be shown in chapter 7.

Déri *et al* [42] have compared the exact infinite series results of Carson [52] with the complex image expressions throughout the whole frequency range. It was found that the complex image method is accurate for the limits of large and small values of h/δ_c, with the largest errors no greater than 10% over an intermediate range $10^{-2} < h/\delta_c < 10^0$.

Finally, all transverse dimensions are assumed small compared with the free space wavelength so that a lumped circuit, magneto-quasistatic approximation is valid, and capacitively coupled displacement currents can be neglected compared to inductively coupled conduction currents. It should also be noted that the complex image model is only valid for RF frequencies much less than the electron plasma frequency ($\omega \ll \omega_{pe}$), because otherwise the displacement current becomes significant [1]. The complex image method is therefore only applicable when the plasma can be described as a resistive conductor (with collisional skin depth) rather than as a dielectric medium (with collisionless skin depth).

The magnitude of mutual partial inductance decreases with distance, therefore higher order reflections (such as the image in the screen of the antenna image in the plasma) have been considered negligibly weak. Experimentally, the influence of the baseplate screen was small but clearly observable, whereas the effect of the top plate (see figure 6.4(b)) was insignificant because the slots inhibited circulating image currents. Hence the top plate was neglected in the calculations of impedance matrices. The complex image method thus provides a practical, self-consistent way to calculate inductive plasma coupling of even very complicated antenna networks which would be prohibitively difficult using the transformer model [6, 9, 30, 37, 51]. In the final analysis, the ultimate justification of the complex image method lies in the good agreement with experiment in figure 6.11.

6.5 Application: RF biasing for plasma deposition

Plasma applications are first mentioned here in part II, since part I was concerned only with vacuum conditions. Plasma deposition has been employed for large area deposition of hydrogenated micro-crystalline silicon (μc-Si:H) thin films by using capacitively coupled plasma (CCP) reactors in various configurations, as reviewed in [53]. However, the deposition rate of high quality films using 13.56 MHz can be limited due to defects caused by energetic ion bombardment across sheath voltages of the order of several hundred volts [54]. On the other hand, inductively coupled plasma (ICP) can offer fast deposition because of its high plasma density [53], and the minimum ion energy—fixed by the floating sheath voltage of ~20 V—is much less than for conventional CCP [55, 56]. However, the films can be porous and exhibit post-oxidation, which is also incompatible with electronic grade material [57–59]. One conclusion is that some intermediate ion energy, between 20 eV and several hundred eV, is required to deposit high quality films [53].

The large area resonant ladder antenna in figure 6.17 is a novel ICP design, providing a spatially distributed mode resonance structure over the entire substrate. Thin films deposited on silicon wafer samples placed on the glass substrate are characterized by Fourier transform infrared (FTIR) spectroscopy [53]. The ion bombardment energy is controlled using the RF bias plate shown in figure 6.17(b); for example, the 5 MHz RF supply applies 200 V_{pp} to this plate, relative to the reactor housing. For these strongly asymmetric-area electrodes, the ion energy is $eV_{pp}/2 \sim 100$eV across the RF-biased sheath [60]. The 5 MHz voltage thus causes a time-averaged negative polarization of the bias electrode, whilst the ICP antenna at 13.56 MHz separately powers the plasma dissociation for deposition. Simultaneous comparison of thin films deposited with and without the RF bias is made by placing the silicon wafer samples inside and outside the RF bias plate region. In conclusion, the FTIR measurements show that ICP deposition

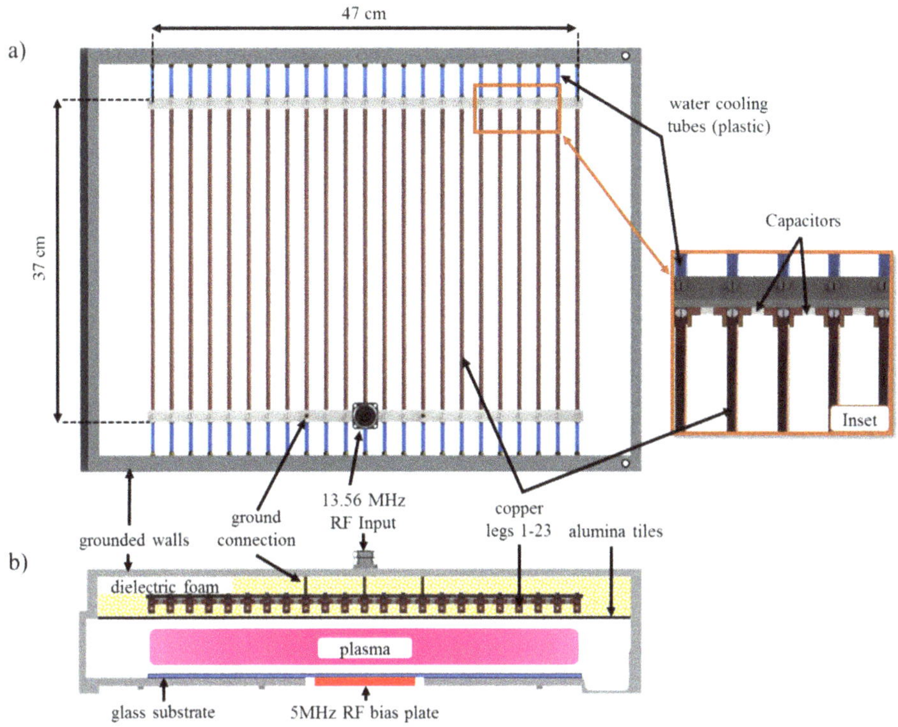

Figure 6.17. Schematic of the planar antenna RF plasma source, area 47 × 37 cm^2, drawn to scale. (a) Plan view with an inset showing the location of the capacitor assemblies, antenna legs, and the plastic water-cooling tubes. (b) Side view with the RF (13.56 MHz, 1 kW) central connection at node A12, and symmetrically placed ground stub connections at A9 and A15. A 10 × 10 cm^2 RF (5 MHz, 50 W) bias plate is incorporated at the centre of the grounded baseplate. Polyimide dielectric foam covered by 12 alumina tiles insulates the antenna from the hydrogen-diluted silane plasma. The whole assembly is placed inside a vacuum chamber. (Reproduced with permission from [53]. Copyright 2018 Elsevier.)

of dense μc-Si:H films can be obtained using a planar resonant antenna, provided that RF substrate bias is superposed for independent control of the ion energy [53].

6.6 Chapter summary for inductive, plane plasma sources

Previous chapters 3 and 4 fully characterized the ladder resonant antenna in air, without plasma. This chapter presented the plasma operation of the same antenna with emphasis on the plasma uniformity, the *E*- to *H*-mode transition typical of ICPs, and the impedance spectrum of the vacuum and plasma loading curves.

- In contrast to the fixed symmetry of a flat spiral coil, the desired plasma uniformity of a ladder resonant antenna can be chosen by design of its internal resonant mode structure, convoluted by radical diffusion in the working gas.
- The *E*- to *H*-mode transition was evidenced not only by a sharp increase in power at a given threshold, but also by two-dimensional measurements of the

plasma density distribution which correspond to the electric fields calculated according to the electro-quasistatic (E-mode) or magneto-quasistatic (H-mode) approximations.

- The complex image method of appendix E, combined with the matrix solution for the mutual partial inductances of a resonant network, provided good agreement with the measured antenna impedance spectrum, with and without plasma loading, for a range of mode numbers and plasma densities. The complex image method and mutual partial inductance approach can therefore be combined to design even complicated structures for planar inductively coupled plasma sources by accounting for the effect of plasma coupling on the source network.

Finally, various technological solutions and a plasma application were described, demonstrating the versatility of the ladder resonant antenna for planar RF plasma sources.

References

[1] Lieberman M A and Lichtenberg A J 2005 *Principles of Plasma Discharges and Materials Processing* 2nd edn (Hoboken, NJ: Wiley)

[2] Chabert P and Braithwaite N 2011 *Physics of Radio-Frequency Plasmas* (Cambridge: Cambridge University Press)

[3] Hopwood J 1992 Review of inductively coupled plasmas for plasma processing *Plasma Sources Sci. Technol.* **1** 109

[4] Okumara T 2010 Inductively coupled plasma sources and applications *Phys. Res. Int.* **2010** 164249

[5] Lee H-C 2018 Review of inductively coupled plasmas: nano-applications and bistable hysteresis physics *Appl. Phys. Rev.* **5** 011108

[6] Gudmundsson J T and Lieberman M A 1998 Magnetic induction and plasma impedance in a planar inductive discharge *Plasma Sources Sci. Technol.* **7** 83

[7] M-Fayoumi El I and Jones I R 1998 Theoretical and experimental investigations of the electromagnetic field within a planar coil, inductively coupled RF plasma source *Plasma Sources Sci. Technol.* **7** 162

[8] Forgotson N, Khemka V and Hopwood J 1996 Inductively coupled plasma for polymer etching of 200 mm wafers *J. Vac Sci. Technol.* B*14 732*

[9] Meyer J A, Mau R and Wendt A E 1996 Plasma properties determined with induction loop probes in a planar inductively coupled plasma source *J. Appl. Phys.* **79** 1298

[10] Keller J H 1996 Inductive plasmas for plasma processing *Plasma Sources Sci. Technol.* **5** 166

[11] Lee Y-J, Han H-R and Yeom G-Y 2000 Characteristics of magnetized inductively coupled plasma source for flat panel display applications *Surf. Coat. Technol.* **133–4** 612

[12] Keller J H, Forster J C and Barnes M S 1993 Novel radio-frequency induction plasma processing techniques *J. Vac Sci. Technol.* A*11 2487*

[13] Li L-S, Xu X, Liu F, Zhou Q-H, Nie Z-F, Liang Y-Z and Liang R-Q 2008 Improvement of uniformity of inductively coupled plasma with a cone spiral antenna *Chin. Phys. Lett.* **25** 2144

[14] Wendt A E and Mahoney L J 1996 Radio frequency inductive discharge source design for large area processing *Pure Appl. Chem.* **68** 1055
[15] Tsakadze E L, Ostrikov K, Tsakadze Z L and Xu S 2005 Generation of uniform plasmas by crossed internal oscillating current sheets: key concepts and experimental verification *J. Appl. Phys.* **97** 013301
[16] Lim J H, Kim K N, Park J K and Yeom G Y 2008 Effect of antenna capacitance on the plasma characteristics of an internal linear inductively coupled plasma system *Phys. Plasmas* **15** 083501
[17] Lecoultre S, Guittienne Ph, Howling A A, Fayet P and Hollenstein Ch 2012 Plasma generation by inductive coupling with a planar resonant RF network antenna *J. Phys. D: Appl. Phys.* **45** 082001
[18] Lecoultre S, Guittienne Ph, Howling A A, Fayet P and Hollenstein Ch 2012 Corrigendum: plasma generation by inductive coupling with a planar resonant RF network antenna *J. Phys. D: Appl. Phys.* **45** 409502
[19] Guittienne Ph, Jacquier R, Howling A A and Furno I 2015 Complex image method for RF antenna-plasma inductive coupling calculation in planar geometry. Part II: measurements on a resonant network *Plasma Sources Sci. Technol.* **24** 065015
[20] Kee R J, Coltrin M E and Glarborg P 2003 *Chemically Reacting Flow* (Hoboken, NJ: Wiley)
[21] Howling A A, Legradic B, Chesaux M and Hollenstein Ch 2012 Plasma deposition in an ideal showerhead reactor: a two-dimensional analytical solution *Plasma Sources Sci. Technol.* **21** 015005
[22] Hollenstein Ch, Guittienne Ch and Howling A A 2013 Resonant RF network antennas for large-area and large-volume inductively coupled plasma sources *Plasma Sources Sci. Technol.* **22** 055021
[23] Howling A A, Strahm B, Colsters P, Sansonnens L and Hollenstein Ch 2007 Fast equilibration of silane/hydrogen plasmas in large area RF capacitive reactors monitored by optical emission spectroscopy *Plasma Sources Sci. Technol.* **16** 679
[24] Guittienne Ph, Howling A A and Furno I 2018 Two-fluid solutions for Langmuir probes in collisionless and isothermal plasma, over all space and bias potential *Phys. Plasmas* **25** 093519
[25] Howling A A, Guittienne Ph and Furno I 2019 Two-fluid plasma model for radial Langmuir probes as a converging nozzle with sonic choked flow, and sonic passage to supersonic flow *Phys. Plasmas* **26** 044502
[26] Agnello R 2020 Negative hydrogen ions in a helicon plasma source *PhD Thesis* no. 7817, Ecole Polytechnique Federale Lausanne (EPFL), Switzerland
[27] Advanced Energy Industries Inc http://www.advanced-energy.com Access date 2024
[28] Bleaney B I and Bleaney B 1976 *Electricity and Magnetism* (London: Oxford University Press) 3rd edn p 1
[29] Guittienne Ph, Lecoultre S, Fayet P, Larrieu J, Howling A A and Hollenstein Ch 2012 Resonant planar antenna as an inductive plasma source *J. Appl. Phys.* **111** 083305
[30] Piejak R B, Godyak V A and Alexandrovich B M 1992 A simple analysis of an inductive RF discharge *Plasma Sources Sci. Technol.* **1** 179
[31] Kortshagen U, Gibson N D and Lawler J E 1996 On the E–H mode transition in RF inductive discharges *J. Phys. D: Appl. Phys.* **29** 1224
[32] Lieberman M A and Boswell R W 1998 Modeling the transitions from capacitive to inductive to wave-sustained rf discharges *J. Phys. IV France* **8** Pr7–145–Pr7-164

[33] Cunge G, Crowley B, Vender D and Turner M M 1999 Characterization of the E to H transition in a pulsed inductively coupled plasma discharge with internal coil geometry: bi-stability and hysteresis *Plasma Sources Sci. Technol.* **8** 576

[34] Zhao S-X, Xu X, Li X-C and Wang Y-N 2009 Fluid simulation of the E–H mode transition in inductively coupled plasma *J. Appl. Phys.* **105** 083306

[35] COMSOL Inc https://www.comsol.com Access date 2024

[36] White F M 1991 *Viscous Fluid Flow* 2nd edn (New York: McGraw-Hill)

[37] Howling A A, Guittienne Ph, Jacquier R and Furno I 2015 Complex image method for RF antenna-plasma inductive coupling calculation in planar geometry. Part I: basic concepts *Plasma Sources Sci. Technol.* **24** 065014

[38] Briefi S, Zielke D, Rauner D and Fantz U 2022 Diagnostics of RF coupling in H^- ion sources as a tool for optimizing source design and operational parameters *Rev. Sci. Instrum.* **93** 023501

[39] Wait J R 1961 On the impedance of long wire suspended over the ground *Proc. IRE* **49** 1576

[40] Wait J R and Spies K P 1969 On the image representation of the quasi-static fields of a line current source above the ground *Can. J. Phys.* **47** 2731

[41] Bannister P R 1970 Utilization of image theory techniques in determining the mutual coupling between elevated long horizontal line sources *Radio Sci.* **5** 1375

[42] Déri A, Tevan G, Semlyen A and Castanheira A 1981 The complex ground return plane: a simplified model for homogeneous and multi-layer earth return *IEEE Trans. Power Appar. Syst.* **PAS-100** 3686

[43] Déri A and Tevan G 1981 Mathematical verification of Dubanton's simplified calculation of overhead transmission line parameters and its physical interpretation *Arch. Elektrotech.* **63** 191

[44] Bannister P R 1986 Applications of complex image theory *Radio Sci.* **21** 605

[45] Boteler D H and Pirjola R J 1998 The complex-image method for calculating the magnetic and electric fields produced at the surface of the Earth by the auroral electrojet *Geophys. J. Int.* **132** 31

[46] Weisshaar A, Lan H and Luoh A 2002 Accurate closed-form expressions for the frequency-dependent line parameters of on-chip interconnects on lossy silicon substrate *IEEE Trans. Adv. Packag.* **25** 288

[47] Jiang R, Fu W and Chen C C-P 2005 EPEEC: comprehensive SPICE-compatible reluctance extraction for high-speed interconnects above lossy multilayer substrates *IEEE Trans. Comput.-Aided Des. Integr. Circuits Syst.* **24** 1562

[48] Kang K, Shi J, Yin W-Y, Li L-W, Zouhdi S, Rustagi S C and Mouthaan K 2007 Analysis of frequency- and temperature-dependent substrate eddy currents in on-chip spiral inductors using the complex image method *IEEE Trans. Magn.* **43** 3243

[49] Park D 1973 Magnetic field of a horizontal current above a conducting earth *J. Geophys. Res.* **78** 3040

[50] Thomson D J and Weaver J T 1975 The complex image approximation for induction in a multilayered earth *J. Geophys. Res.* **80** 123

[51] M-Fayoumi El I and Jones I R 1998 The electromagnetic basis of the transformer model for an inductively coupled RF plasma source *Plasma Sources Sci. Technol.* **7** 179

[52] Carson J R 1926 Wave propagation in overhead wires with ground return *Bell Syst. Tech. J.* **5** 539

[53] Demolon P, Guittienne Ph, Howling A A, Jost S, Jacquier R and Furno I 2018 RF bias to suppress post-oxidation of μc-Si:H films deposited by inductively-coupled plasma using a planar RF resonant antenna *Vacuum* **147** 58

[54] Kroll U, Meier J, Torres P, Pohl J and Shah A 1998 From amorphous to microcrystalline silicon films prepared by hydrogen dilution using the VHF (70 MHz) GD technique *J. Non-Cryst. Solids* **227–30** 68

[55] Xiao S Q, Xu S and Ostrikov K 2014 Low-temperature plasma processing for Si photovoltaics *Mater. Sci. Eng. R. Rep.* **78** 1

[56] Setsuhara Y, Takenaka K, Ebe A and Nishisaka K 2007 Development of large area plasma reactor using multiple low-inductance antenna modules for flat panel display processing *Solid State Phenom.* **127** 239

[57] Nogay G, Özkol E, Ilday S and Turan R 2014 Structural peculiarities and aging effect in hydrogenated a-Si prepared by inductively coupled plasma assisted chemical vapor deposition technique *Vacuum* **110** 114

[58] Nogay G, Saleh Z M, Özkol E and Turan R 2015 Vertically aligned Si nanocrystals embedded in amorphous Si matrix prepared by inductively coupled plasma chemical vapor deposition (ICP-CVD) *Mater. Sci. Eng.* B***196*** *28*

[59] Bugnon G, Parascandolo G, Söderström T, Cuony P, Despeisse M, Hänni S, Holovsky J, Meillaud F and Ballif C 2012 A new view of microcrystalline silicon: the role of plasma processing in achieving a dense and stable absorber material for photovoltaic applications *Adv. Funct. Mater.* **22** 3665

[60] Köhler K, Coburn J W, Horne D E, Kay E and Keller J H 1985 Plasma potentials of 13.56-MHz RF argon glow discharges in a planar system *J. Appl. Phys.* **57** 59

Chapter 7

Electromagnetic coupling to plasma in large antennas

This chapter is a watershed in the book (figure 7.1) because it is the only chapter situated at the pinnacle of figure 1.8, where electromagnetic (EM) effects are fully accounted for, as well as mutual partial inductance and complex image plasma dissipation. In contrast, the previous chapters described lumped-element ladder resonant antennas, and the following chapters will be concerned principally with lumped-element birdcage antennas. Their 'lumped-element' label indicates a quasi-static approximation which is appropriate for reactors much smaller than the RF wavelength, but incompatible with an EM standing wave solution (see appendix D.1.1).

To motivate the interest in large area resonant network antennas, we can summarize three general limitations of conventional plasma processing reactors:

- Single-coil inductively coupled reactors are usually limited to a single wafer (section 6.1), but parallel plate capacitively coupled reactors have been developed successfully for large area substrates such as for multiple flat panel displays. Currently, they can deposit thin films on areas of about 2–3 m^2 with a uniformity of ±5% [1]. One drawback is the low plasma density of capacitively coupled discharges, typically 10^{15} – 10^{16} electrons/m^3 [2]. For example, deposition rates are limited because the high ion bombardment energy, associated with high voltage sheaths, can damage the growing film.
- The input impedance of large conventional RF sources is almost purely reactive, and tends either towards zero (capacitive discharges) or towards infinity (inductive discharges) with increasing substrate area. The impedance matching between the RF generator and the source then drives very high currents and/or voltages in the matching network and power feed lines. These

Figure 7.1. 'The watershed of the streams Aganippe and Hippocrene, which are sources of poetic inspiration, is at the pinnacle of Mount Helicon, in Greek mythology.' (Illustration by Alex Howling. Copyright 2023 Alex Howling.)

limitations could be overcome by using RF resonant network antennas, as shown previously in section 5.5.

- A direct consequence of electromagnetic effects in RF systems is the standing wave effect. Standing wave non-uniformity can arise in ICP antenna sources [3–10], as well as in capacitively coupled (CCP) reactors [2, 11–13]. In the context of resonant network antennas, this occurs in two cases: at standard RF frequencies (e.g. 13.56 MHz) for large dimension (>1 m) plasma source antennas, or in smaller plasma probe antennas at very high frequencies of 370 MHz or more, where the antenna size in either case begins to become comparable to the RF wavelength. As mentioned in section 2.3, the antenna elements can no longer be considered as lumped-element components, and instead, electromagnetic effects must be taken into account. The large antenna, in fact, sustains a plasma by both inductive (H-mode) and capacitive (E-mode) coupling; it is an *electromagnetically coupled* (EM-mode) plasma source, as mentioned in section 5.2.3. Figure 7.2 shows photographs of these two cases, and each will be considered in turn, in sections 7.1 and 7.7.2, respectively.

Complete electromagnetic solutions of Maxwell's equations are therefore required for large area RF plasma reactors. Wavefield equations, using no quasistatic approximation, are often used for large area, parallel plate capacitive reactors because they resemble large cavities containing the plasma effective dielectric [2, 11–13]. In contrast, the long, parallel legs of a large area resonant network antenna are more evocative of a multi-conductor transmission line (MTL), which is inductively and capacitively coupled to an adjacent plasma dielectric slab.

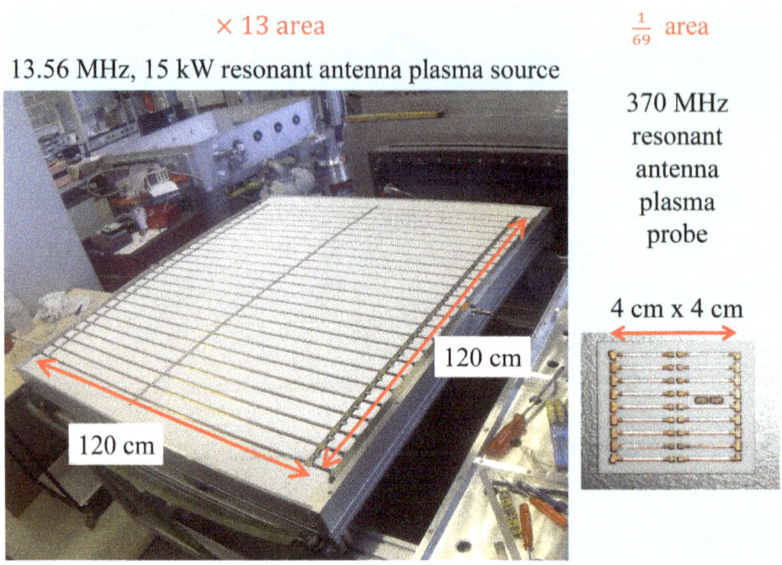

Figure 7.2. Photographs of a large plasma source antenna at 13.56 MHz with its vacuum chamber in the background; and a small plasma probe antenna operating at 370 MHz. Their areas are, respectively, 13 times larger, and 69 times smaller, than the 55×20 cm^2 resonant network antenna described in the previous chapters 3, 4, and 6.

The transmission line aspect of an MTL assures the required electromagnetic, standing wave treatment [14]. Coupling of the MTL to the plasma is facilitated by the complex image method, using a mutual partial inductance matrix, as well as a capacitance matrix, between all elements [15]. Analytic expressions are given below, and results are obtained by computation of the matrix solution. It will be seen that the model reproduces input impedance measurements on the 1.2×1.2 m^2 antenna in figure 7.2—whether grounded or electrically DC floating—with or without plasma for both electrical configurations. We explain how this method could be used to design large area EM-coupled sources in general, by applying the MTL termination impedances specific to each type of antenna.

Finally, various technological hints and industrial applications are described, including the small plasma probe antenna also shown in figure 7.2.

7.1 Electromagnetic effects in large area antennas

Plasma processing over large areas (>1 m^2) is of fundamental importance for the industrial production of solar cells, flat panel displays, packaging, surface treatment, large area electronics, multiple wafer substrates, etc. Calculations of the antenna characteristics are necessary for source design and prediction of plasma properties, as well as for comprehension and interpretation of measurements. For example, the influence of plasma coupling on the antenna input impedance determines the power transfer efficiency (6.6) [16] and the RF power matching conditions.

No significant standing wave effect is to be expected in the direction perpendicular to the antenna legs because the relevant length scale for the standing waves in this direction is not the full antenna width, but rather the size of one stringer segment containing a capacitor, which is always much smaller than the voltage/current wavelength at RF operating frequencies. Consequently, the model will be seen to be one-dimensional, along the MTL of the legs. Note that the self partial inductance of a leg, given by (4.29), increases supralinearly because the mutual partial inductance between its segments are also summed. Consequently, the EM effects of a large area ICP appear more strongly than would be expected from a simple proportional-length estimation.

7.1.1 Existing models of ICP sources

For large area ICP sources, electromagnetic (EM) standing wave effects have previously been described by treating long coils as a single, lossy transmission line [3–6], where the transformer model was used to calculate the per-unit-length coupling to the plasma. ICP antennas were also treated as segments of an immersed transmission line [7]. These models do not consider capacitance and mutual inductance matrices between all the antenna elements and the plasma [4]. However, inclusion of parasitic capacitive coupling does improve the models [2, 17] which suggests that capacitive effects are important. Mashima *et al* [18] used the moment method but plasma coupling was not taken into account. Numerical simulation can include EM effects via the wavefield equations [8–10] although the solution requires longer computation.

For the large area source in figure 7.2, experiments show that plasma can be maintained even for an antenna disconnected from ground (electrically DC floating), which means that significant capacitive currents must flow from the antenna via the plasma to the grounded backplate. Modeling and ICP source design therefore also require the calculation of the associated capacitance matrix. The self-consistent calculation of antenna currents and voltages coupled by the mutual inductance matrix and the capacitance matrix is performed for the MTL [14] coupled with the plasma. These considerations could also be relevant to other large area antenna reactor designs [19–21] such as the ladder (non-resonant) antenna [18, 22–33], the serpentine antenna [7, 9, 34], the U-type antenna [35–37], and the double comb-type antenna [38–40]. The appropriate stringer impedances enter the equations as termination impedances in appendix F.1.

7.1.2 Comparison of large area ICP sources

External coil ICP reactors have the scale-up problem that the thick dielectric window, necessary to withstand atmospheric pressure over a large area, reduces the mutual inductance with the plasma [3, 4, 8, 30, 41–43], although possible solutions are given in section 13.4.1. Internal coil ICP reactors maintain efficient inductive coupling with the plasma and avoid the necessity of a thick dielectric window [44–47] by placing the antenna inside the vacuum chamber. Large area ICP sources are often an array of parallel linear legs [30]. The legs can be *immersed* in the plasma, with the legs individually isolated by a dielectric sleeve (serpentine [7, 9, 34], U-shape [35–37], comb-type [38, 39]), or directly exposed to the plasma (for example, versions of the (non-resonant) ladder type [20, 26, 30]). Alternatively, the whole ICP antenna can be *embedded* in a dielectric, protected from the plasma by a thin window [17, 48–51]. In this context, the large area resonant antenna in this work is an internal, embedded ICP source—see section 13.4.2 for a technological description. Capacitive coupling can be mitigated by means of a slotted screen [2] or a metal film [19].

7.2 Experimental set-up for large area antennas

A schematic of the experimental set-up is shown in figure 7.3. The antenna network is made up of 25 water-cooled tubular copper legs, length 120 cm, external radius 0.4 cm, with parallel axes 5 cm apart. The legs are connected at both ends by stringers consisting of 375 pF ceramic capacitor assemblies with copper strip connectors 0.6 cm wide. The antenna assembly is set in a closed aluminium housing of internal dimensions $135 \times 135 \times 13$ cm^3. The leg axes are positioned 5 cm above the aluminium baseplate, sufficiently distant to limit eddy current power loss from inductive coupling with this plate. To avoid spurious plasma below and within the antenna network, the lower part of the reactor is filled with silica-alumina 85%-porosity foam dielectric (hence 'embedded antenna'). This low-permittivity dielectric also minimizes capacitive coupling—and hence RF capacitive current—to the grounded baseplate. The antenna surface is protected from the plasma by square glass tiles of 15 cm side length and 0.3 cm thickness with overlapping chamfered edges (section 13.4.2); this thin window promotes strong inductive coupling between the antenna and the plasma, although the close proximity to the plasma means that capacitive coupling must also be taken into account. Any remaining interstices

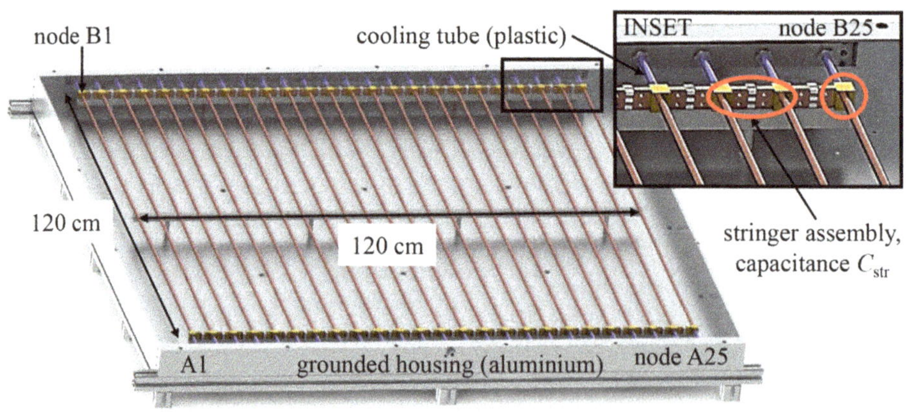

Figure 7.3. Scale drawing of the planar resonant antenna assembly inside its grounded housing, showing an open top view of the antenna emptied of dielectric, with 25 water-cooled legs considered as a multi-conductor transmission line, terminated at both ends by stringer capacitor assemblies C_{str} shown in the inset. The antenna is 120 cm long and 120 cm wide; other dimensions are given in the text. Figure 7.2 shows a photograph of the antenna filled with silica/alumina 85%-porosity dielectric. (Adapted with permission from [15]. Copyright 2017 IOP Publishing.)

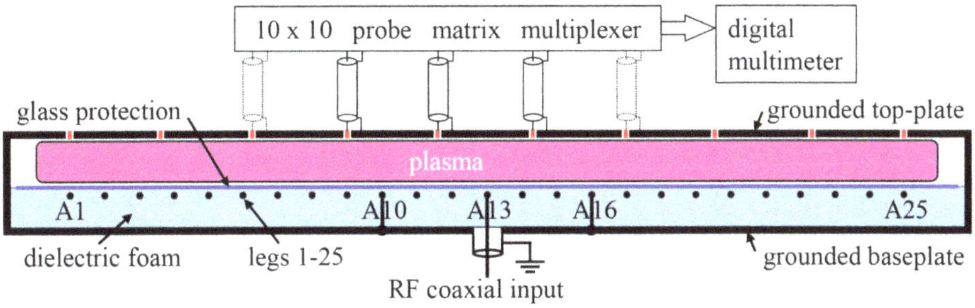

Figure 7.4. Schematic end view of the nodes A1 to A25 (see figure 7.3) in the closed reactor showing the antenna RF connections for the grounded configuration, namely, RF coaxial input at node A13 with ground connection stubs at A10 and A16. The 10 × 10 matrix of surface electrostatic probes (shown in red, with the acquisition block diagram for DC current measurements) are flush with the internal surface of the grounded top-plate. (Adapted with permission from [15]. Copyright 2017 IOP Publishing.)

between the foam, glass and legs can be filled with micro-sphere glass beads to prevent parasitic plasma below the glass. In figure 7.4, the plasma boundaries are seen to be the glass surface, the grounded metal top-plate 7 cm above the glass which defines the plasma gap, and the grounded metal sidewalls. The whole reactor was placed inside a vacuum vessel. Hence, using the terminology of section 7.1.2, the antenna is *internal* (i.e. inside the low pressure vacuum vessel), and *embedded* within a dielectric, protected from the plasma by a thin window.

7.2.1 RF electrical configurations—antenna grounded or DC floating

The RF input power is centrally connected along one side via a coaxial cable at node A13 in figure 7.3(b) whose ground shield, connected to the baseplate, defines the ground reference point. The RF current is constrained to circulate only on the inside walls of the closed reactor which therefore acts as a Faraday screen to the exterior; this is important to suppress electric field leakage and parasitic plasma ignition in the vacuum space between the reactor and the vacuum vessel walls [20]. Copper grounding straps between the aluminium housing and the vacuum chamber are also used to prevent any spurious plasma or arcing outside the reactor housing.

With reference to figure 7.4, inside the reactor, the antenna can be connected in two distinct electrical configurations:

1. The antenna can be grounded to the baseplate, for example, via stub connectors A10 and A16 to preferentially select for the chosen mode m = 8 as in figure 7.4. By current continuity (neglecting capacitive coupling), all of the RF input current returns via these ground stubs. This type of grounded configuration is used for all of the measurements involving the small antenna in chapters 3, 4, and 6 because the current flow is suited to inductive coupling (*H*-mode); capacitive currents are assumed to be negligible because of the lumped-element approximation for short legs (0.2 m) as explained in section 2.3 point 1. The good agreement between the lumped-element model and the impedance measurements in figure 6.11, and

especially figure 4.20, suggest that this assumption is justifiable. However, the same point 1 in section 2.3 implies that the long antenna (1.2 m) requires an electromagnetic treatment including the leg capacitance per unit length.

2. The antenna can be disconnected from ground, by omitting the stub connectors to A10 and A16 in figure 7.4, so that the antenna is DC floating within the grounded reactor walls, physically connected only to the RF power supply via its internal DC-blocking capacitor (not shown in the figure). The antenna RF return current to ground can then occur only by capacitive coupling (*E*-mode), principally due to the 25 legs, via the dielectric to the grounded baseplate, and via the plasma sheaths to the grounded top-plate. The observation that a plasma can be maintained in this way proves that the capacitance of long legs must be accounted for, by analogy with the transmission line consideration in section 2.3. In this DC floating configuration, the antenna behaves in some ways like a large rectangular, parallel plate, capacitively coupled reactor, except that the conventional solid parallel plate [11, 12] is replaced by a rectangular array of 25 parallel legs connected to the RF power supply.

It can be imagined that inductive (*H*-mode) and capacitive (*E*-mode) currents will flow, to some extent, in both the grounded and the DC floating cases. If the antenna impedances were perfectly balanced, and the normal mode m = 8 was perfectly symmetric and uniform with negligible perturbation from plasma currents (see chapter 3), then there would, in any case, be zero voltage at the nodes A10 and A16. The grounded and DC floating configurations are thus seen to be similar although not identical. The experimental measurements will indeed show that the grounded and floating antenna input impedances are comparable, which suggests that it is the internal resonance current circulation that dominates the antenna network impedance, as could be expected for such a high *Q* antenna.

In both the grounded and DC floating cases, therefore, there are conduction currents circulating within the antenna (*H*-mode) *and* capacitive currents from the antenna to ground (*E*-mode). This justifies an electromagnetic coupling (*EM*-mode) treatment as stated in section 5.2.3. Consideration of both inductance and capacitance is equivalent to removing the quasistatic approximations in appendix D.1.1, thus permitting an electromagnetic wave description on a transmission line.

Capacitive currents to ground do, in fact, exist also for the small antenna, but the antenna RF voltage would have to be extremely high before any significant RF current and power could be observed due to this capacitive coupling. In practice, the conduction currents circulating in the antenna (*H*-mode) totally dominate the capacitive currents to ground (*E*-mode) for small antennas. It is important to understand, therefore, that there is no dichotomy in the physical treatment between small and large antennas, only a gradual transition from a good quasistatic approximation for lumped-element antennas, to an increasingly poor approximation for larger antennas. The electromagnetic coupling (*EM*-mode) treatment is exact in every circumstance, but it is unnecessarily complicated for describing small antennas.

7.2.2 Diagnostics and plasma parameters

Similarly to the smaller ladder resonant antenna in chapter 6, the antenna impedance spectrum without plasma is measured with a network analyser as usual. The network node voltages A_n and B_n are measured with high impedance probes on the bench in air with the top-plate removed. The plasma-coupled network is electrically characterized by its input impedance and RF power using the current-voltage-phase probe (Z-Scan™ [52]) at the vacuum feedthrough between the matching box and the antenna. Similarly to the small reactor experiment in section 6.2.1, the calculated impedance values are transformed through 30 cm of 50 Ω transmission line (if mounted in the vacuum vessel) to compare with the experimental measurements, including 5 cm long stubs in parallel to both ground connections A10 and A16, as well as the A13 RF power connection.

Two-dimensional plasma density mappings (assuming density approximately proportional to the measured ion saturated current (6.1) in section 6.2.1) are performed in section 7.6.5 by means of a regular 10 × 10 square matrix array of negatively biased (–30 V), multiplexed, surface electrostatic probes 1 cm diameter, with centres spaced 13.3 cm apart, flush with the top-plate surface. These non-invasive surface probes present negligible mechanical perturbation in the plasma. As shown in figure 7.4, each probe's coaxial cable screen is connected to the ground top-plate next to the probe itself. Effective RF grounding of each probe is assured by the combined capacitance of the probe head via its alumina sleeve, ~13 pF, and its equal-length coaxial cable, ~50 pF. The total capacitance to ground of each probe is therefore much larger than its sheath capacitance to the plasma which is less than 1 pF. This RF grounding ensures continuity of any RF current from plasma to ground via the probes, so that the probe surfaces form an integral RF-grounded part of the ground top-plate; this could be important particularly in the case of the capacitive coupling configuration [53]. This capacitance is judged to be sufficient because a supplementary 1 nF capacitor across the probe connections made no difference to the DC measurements. The probe surfaces must be conducting since insulating layers would invalidate DC current measurements. The electrodes and probes were mechanically polished and only exposed to plasmas of argon. No deposition or probe discoloration was observed after four months of plasma experiments. As mentioned in section 6.2.1, we recall that the very stable steady-state plasma conditions are well suited to long-duration measurements such as the data acquisition from 100 multiplexed probes.

The plasma density profile in the 13.3 cm spaces between the probes is interpolated along one central axis by means of a single, laterally scanning Langmuir probe via a port in a sidewall. This combination of a static matrix array and a single traversing probe is mechanically better adapted to this large reactor than the 15-probe scanning array for the small reactor in section 6.2.1. Note that these density measurements cause negligible perturbation to the plasma, and are carried out with the plasma reactor completely closed.

Scans in RF plasma power and frequency of 500 to 2000 W delivered RF power over a frequency range from 12 to 14 MHz were made in 1.5–2 Pa argon pressure with an arbitrary flow rate of 200 sccm to ensure negligible accumulated contamination of the argon from de-gassing, or micro-flows from real leaks or trapped volumes. A separate high RF power experiment was performed at 15 kW with a fixed frequency of 13.56 MHz.

7.3 Single conductor lossy transmission line

As an introduction to the multi-conductor model, we first revise the case of a single conductor and ground return transmission line as introduced briefly in section 2.3 and described in standard textbooks on electromagnetism [54, 55]. By extension of figure 2.7 to the lossy transmission line in figure 7.5, the generalized phasor expressions for the one-dimensional variations in voltage and current, with implicit harmonic time dependence $e^{-j\omega t}$, along a uniform transmission line are [3, 7]

$$dV(z)/dz = (\hat{R} + j\omega\hat{L})I(z), \tag{7.1}$$

$$dI(z)/dz = (\hat{G} + j\omega\hat{C})V(z), \tag{7.2}$$

where $\hat{R}$, $\hat{L}$, $\hat{G}$, and $\hat{C}$ are the per-unit-length (p.u.l.) series resistance, series inductance, parallel conductance, and parallel capacitance, respectively. The 'hat' symbol distinguishes the p.u.l. values from lumped-element values. Dielectric conductance losses, represented by $\hat{G}$, are usually negligibly small compared series ohmic losses $\hat{R}$ when using dielectrics with a low loss tangent, but we will see that the plasma introduces significant dissipation when acting as an effective lossy dielectric. The wave equation for the line voltage $V(z)$ propagating along z, by differentiation and substitution, is therefore

$$d^2V(z)/dz^2 = -\Gamma^2 V(z), \tag{7.3}$$

and identically for $I(z)$, where $\Gamma^2 = (\hat{R} + j\omega\hat{L})(\hat{G} + j\omega\hat{C})$ is the wave propagation constant squared. It is assumed that the line is uniform along z, therefore $\hat{R}$, $\hat{L}$, $\hat{G}$, and $\hat{C}$ are independent of z although they may be functions of ω. The solutions are [55]

$$V(z) = e^{-\Gamma z}V^{+} + e^{\Gamma z}V^{-}, \tag{7.4}$$

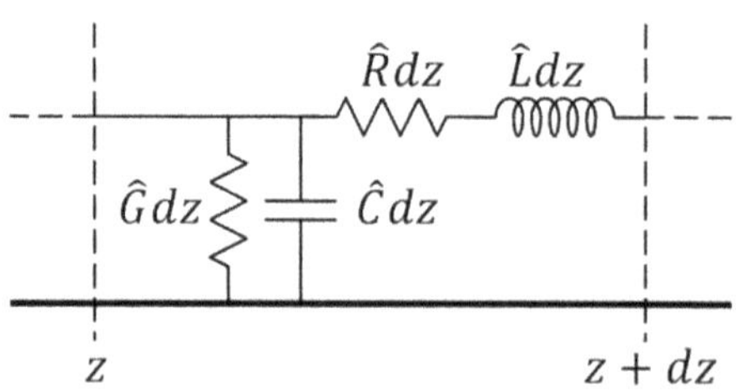

Figure 7.5. An infinitesimal section, length dz, of a general, lossy transmission line, where $\hat{R}$, $\hat{L}$, $\hat{G}$, and $\hat{C}$ are, respectively, the per-unit-length (p.u.l.) series resistance, series inductance, parallel conductance, and parallel capacitance.

$$I(z) = Y_c(e^{-\Gamma z}V^+ - e^{\Gamma z}V^-), \tag{7.5}$$

where V^+ and V^- are phasor amplitudes of forward and backward waves, to be determined by boundary conditions, and Y_c is the characteristic admittance of the line which depends on plasma coupling. The transmission line equations naturally have a one-dimensional solution, along z. All of the above parameters for the single transmission line are scalars; these parameters become vectors and matrices for the multi-conductor transmission line (MTL) in the next section 7.4.

7.4 Multi-conductor transmission line (MTL)

For the MTL model, referring to figure 7.6, the line voltages on the N conductors, with respect to the ground reference conductor, can be written as a column vector:

$$\mathbf{V}(z) = \begin{pmatrix} V_1(z) \\ V_2(z) \\ \vdots \\ V_N(z) \end{pmatrix}, \tag{7.6}$$

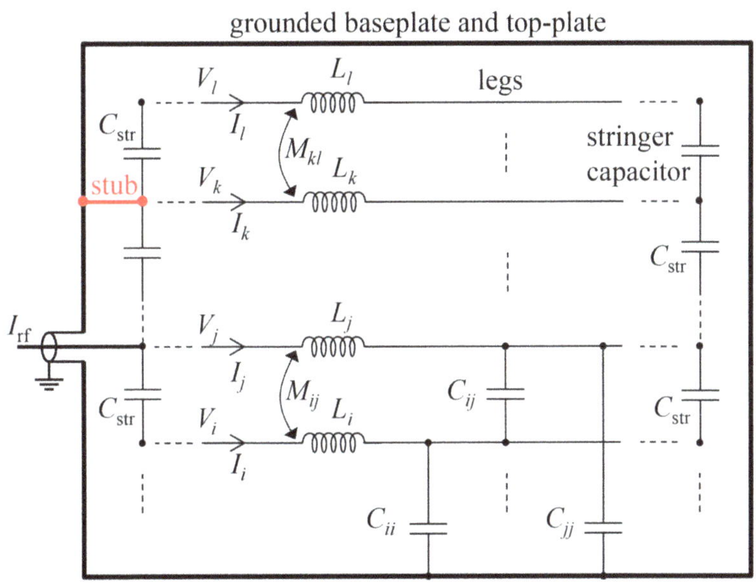

Figure 7.6. The antenna equivalent circuit comprising a multi-leg transmission line terminated at both ends by stringer capacitors C_{str}. L_i are the leg per-unit-length (p.u.l.) self partial inductances. The C_{ij} represent the p.u.l. capacitive coupling, and M_{ij} the p.u.l. mutual partial inductance, between all legs, including their images induced in the plasma and in the reactor housing in the sense of figure 6.14. The reactor housing includes the baseplate and the top-plate, and carries the total RF return current into the RF coaxial cable screen which is the ground reference point. The p.u.l. series resistances R_i and R_{ij} are implicitly included in the complex values of L_i and M_{ij}, respectively, and the p.u.l. parallel conductances G_{ij} are implicitly included in the complex value of C_{ij}. Only a few example components are shown for clarity. The red connector, labelled 'stub', represents a grounded antenna configuration (section 7.2.1). For the DC floating configuration, all stubs are omitted, and the RF current returns to ground purely by capacitive coupling via the leg capacitance C_{ii}.

and similarly for the currents. The theoretical basis for the analysis of multi-conductor transmission lines is given by Paul [14]. By analogy with single transmission line equations [3, 7, 55] in the preceding section 7.3, the generalized expressions for voltage and current are

$$d\mathbf{V}(z)/dz = j\omega\hat{M}\ \mathbf{I}(z), \tag{7.7}$$

$$d\mathbf{I}(z)/dz = j\omega\hat{C}\ \mathbf{V}(z), \tag{7.8}$$

where $\hat{M}$ is the p.u.l. mutual partial inductance $N \times N$ matrix for the antenna legs [56], and $\hat{C}$ is the p.u.l. capacitance $N \times N$ matrix. In figure 7.6, M_{ij} is the mutual partial inductance between lines i and j, and $L_i \equiv M_{ii}$ is the self partial inductance of the ith leg which is a special case of mutual partial inductance, as usual [57, 58]. The elements M_{ij} and C_{ij} of these matrices are generally complex, to account for the line series resistance and the dielectric parallel conductance, respectively. The inductance elements are partial p.u.l. inductances (section 7.6.1) and the corresponding p.u.l. capacitance elements (section 7.6.2) are discussed by Paul [14]. As a reminder, mutual partial inductance exists between conductors which are parallel, but not between orthogonal conductors [58], so one matrix can be defined for the legs, and another independently for the stringers; analytical expressions are given in chapter 4. The p.u.l. mutual partial inductance matrix for the legs, $\hat{M}$, is obtained by dividing by the leg length.

The transmission line is assumed to be uniform along z so that $\hat{M}$ and $\hat{C}$ are constants. These two matrices are considered for the antenna, respectively, without plasma and with plasma, in the next two sections. The uniform multi-conductor transmission line wave equation (see also (D.33) for the EM field wave equation), analogous to the single transmission line [3, 7, 55], is

$$d^2\mathbf{V}(z)/dz^2 = -\omega^2(\hat{M}\ \hat{C})\ \mathbf{V}(z), \tag{7.9}$$

for bi-symmetric matrices [14]. Also by analogy with single transmission line theory, the mode m voltages have the solution [14]

$$\mathbf{V}_{\mathrm{m}}(z) = \overline{e^{-\Gamma z}}\mathbf{V}_{\mathrm{m}}^{+} + \overline{e^{\Gamma z}}\mathbf{V}_{\mathrm{m}}^{-}, \tag{7.10}$$

where $\bar{\Gamma}$ represents the complex propagation constants matrix of the modes $\mathrm{m} = 1 \rightarrow N$. The worked solution of the transmission line equations is given in appendix F; note that the presence or absence of plasma is determined by the choices of $\hat{M}$ and $\hat{C}$. In view of the mathematical complexity, programs for the MTL solutions are available via a link in appendix K.

7.5 Experiment and MTL model for the vacuum case

The MTL model was first verified by comparison with antenna measurements in air, without plasma. To access the antenna with voltage probes, these measurements were made with the top-plate removed as shown in figure 7.3. The $\hat{M}$ and $\hat{C}$ p.u.l. matrices then describe coupling between antenna legs and their images in the baseplate only, for which wire-to-plane (microstrip) expressions are well known

[14, 55, 56, 58, 59]. Either $\hat{M}$ or $\hat{C}$ can be calculated directly, and the other derived using $\hat{M}\ \hat{C} = \mu_0\varepsilon_0\bar{\text{I}}$, where $\bar{\text{I}}$ is the identity matrix for this homogeneous dielectric (vacuum) [14]. It is more convenient to use $\hat{M}$ because this was already calculated in air, with a conducting baseplate, in chapter 4.

7.5.1 Measurements in a vacuum with a grounded antenna

Comparison of the MTL calculation of antenna input impedance spectrum with the measurement shows very good agreement in figure 7.7, even for the high-m modes and the overall baseline resistance. The extraneous measured peak at 17.73 MHz is attributed to a parasitic resonance, possibly due to the stub connections. The minor modeling error in m = 2 frequency and the major overestimation in the adjacent baseline resistance could be due to higher order stray impedances which are not captured in the equivalent circuit of figure 7.6.

It is revealing to compare this electromagnetic (*EM*-mode) MTL model with the previous inductive coupling model (*H*-mode) developed in chapters 4 and 6. Figure 7.8 shows that the lumped-element mutual inductance solution is not as accurate because the frequencies are over-estimated by about 1.5% for m = 8 when capacitive coupling is neglected, and the error worsens for higher frequencies, presumably because capacitive impedance decreases as $1/f$, allowing more capacitive current to flow. This is consistent with figure 6.11: inductive coupling raises the peak frequency, while capacitive coupling lowers it. The influence of capacitive coupling is nevertheless relatively weak for this case without plasma because the antenna is comparatively distant from the baseplate. Therefore, although the lumped-element

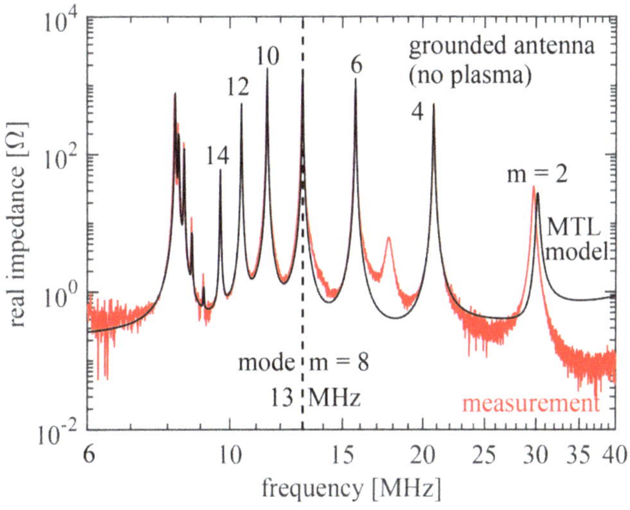

Figure 7.7. The real input impedance spectrum calculated using the electromagnetic (*EM*-mode) MTL model compared with the network analyser measurement for a grounded antenna without plasma. RF input at A13; ground connections at A10 and A16, whose even symmetry suppresses the odd-m modes. Both axes use a logarithmic scale. The m = 8 mode is used below in figures 7.9 and 7.11. (Adapted with permission from [15]. Copyright 2017 IOP Publishing.)

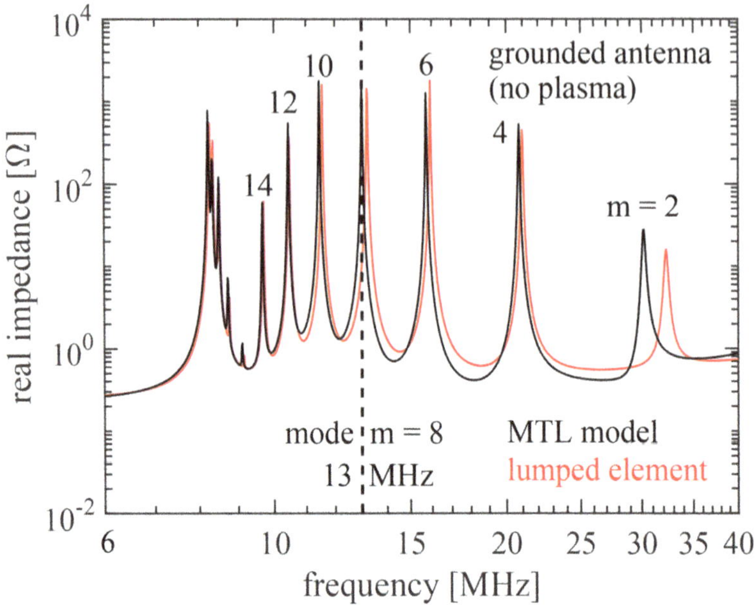

Figure 7.8. The vacuum real input impedance spectrum calculated using the electromagnetic coupling (*EM*-mode, black line) MTL model, compared with the inductive coupling, lumped-element model (*H*-mode, red line). The grounding configuration is the same as for figure 7.7. Both axes use a logarithmic scale.

model may be sufficiently accurate for these vacuum calculations, it will be shown to be less adequate for plasma operation in section 7.6.

To be clear, the improvement in modeling accuracy of the input impedance spectrum, between chapter 4 and this chapter 7, is because the electromagnetic *EM*-mode calculation is exact, and does not use the magneto-quasistatic approximation of the *H*-mode. The difference between the analogous transmission line lengths in section 2.3, point 1, shows that a 1.2 m length incurs a 4% error in effective inductance, compared to only 0.1% for 0.2 m. On the other hand, the improvement in modeling accuracy of the input impedance spectrum, between chapter 3 and chapter 4, was because the mutual partial inductance was accounted for in section 4.8.2—this is a qualitatively different phenomenon.

Comparison of the node voltage measurements A_n and B_n in figure 7.9(a) with the MTL solution in figure 7.9(b) also shows very good agreement, where the distortion of the voltage distribution due to deliberate off-resonance conditions is reproduced in detail. The amplitude envelope given by the lumped-element inductive approach in figure 7.9(c), again, is clearly not as accurate, because the effects of capacitive coupling are neglected. These effects are much stronger with plasma because the proximity of the plasma strengthens the capacitive coupling contribution, as shown later in section 7.6. The antenna parameters are now fixed and no case-to-case adjustment is applied to the reactor equivalent circuit. A link to the programs and the antenna/plasma parameters which reproduce the *EM*-mode calculations in all of the figures of this chapter is given in appendix K.

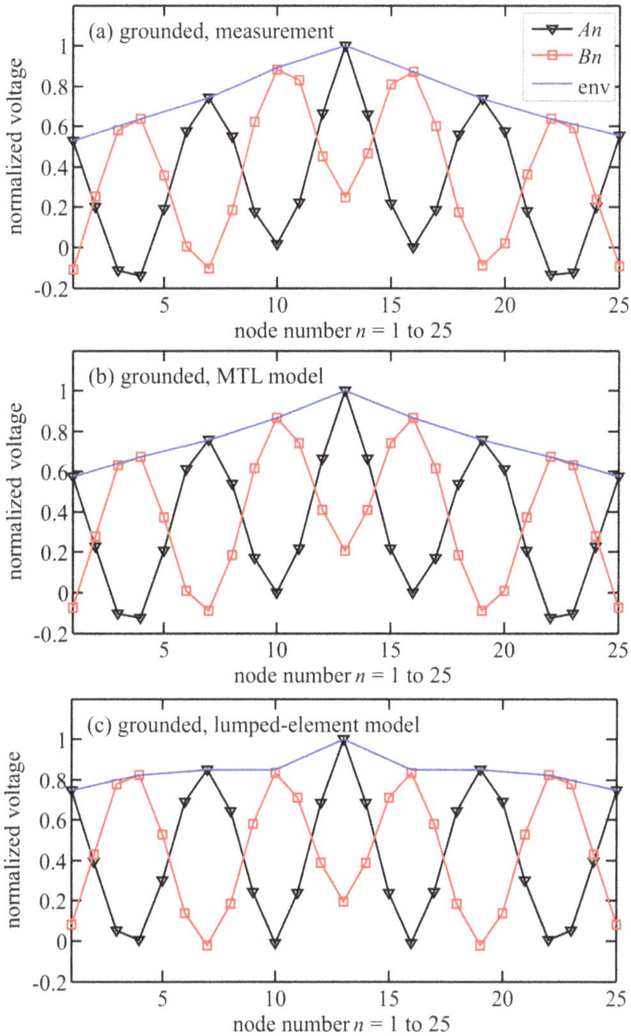

Figure 7.9. Comparison of the normalized node voltages *An* and *Bn* measured deliberately 0.35 MHz above the 13.0 MHz resonance frequency of mode m = 8 for the grounded antenna. (a) Measurements with a voltage probe and (b) calculation using the electromagnetic (*EM*-mode) MTL model. (c) Calculation using the lumped-element mutual inductance model (*H*-mode), shows poor agreement with measurement. The lines 'env' show the amplitude envelope. The grounding configuration is the same as for figure 7.7. (Adapted with permission from [15]. Copyright 2017 IOP Publishing.)

7.5.2 Measurements in a vacuum with a DC floating antenna

The MTL solution also provides good agreement with the DC floating antenna input impedance spectrum measurement in figure 7.10. In this case, the ground return connections A10 and A16 in figure 6.4(b) were removed and measurements were performed with the antenna electrically floating (DC de-coupled) with respect to ground. The symmetry of the RF central connection, with no ground connections,

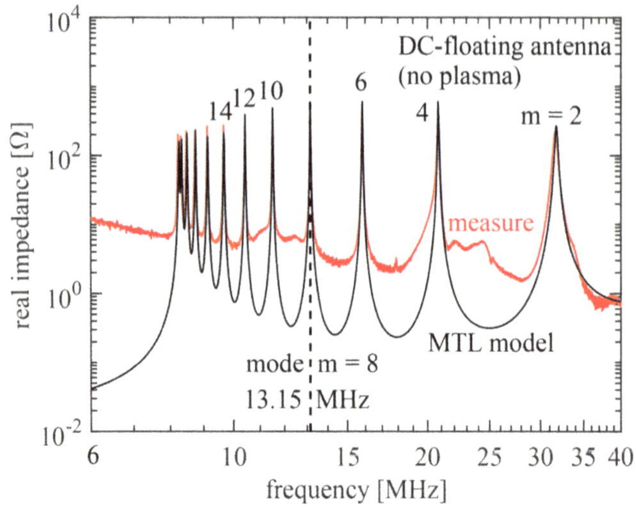

Figure 7.10. Real input impedance spectrum calculated using the MTL model compared with network analyser measurements for an electrically DC-floating antenna without plasma (RF central feeding at A13, with no ground connections). Both axes use a logarithmic scale. The mode m = 8 is used in figure 7.12 for measurements with plasma coupling. (Adapted with permission from [15]. Copyright 2017 IOP Publishing.)

means that odd-m modes are still suppressed in figure 7.10, as they were for figure 7.7 with symmetric ground connections. For an asymmetric example, placing the RF input at, say, node A1 in the floating case, would allow all the modes to appear in the spectrum.

In the MTL model, the impedances of the A10 and A16 connectors were set to a very high value to simulate an open circuit. The previous lumped-element inductive model [56] has no corresponding solution because capacitive coupling was not considered there. The calculated mode frequencies show remarkable correspondence with the measurements, although the measured baseline resistance is a few ohms higher than the calculated resistance here and in the calculations and the measurements in figure 7.7. We can speculate that this could be due to the RF current returning to ground via the skin depth resistance of the housing interior surface, whereas the stub-grounded current returns via the low-impedance stubs A10 and A16 (see also section 7.2.1). This is reminiscent of the observation by Briefi *et al* [60] that as much as 76% of the ohmic losses can occur in a Faraday screen, which is represented by the housing interior surface in our present case. Any design to diminish ohmic losses in the housing plates' skin depth depends on the particular current being considered—induced, closed-loop eddy current dissipation could be attenuated by slits; whereas return currents to ground take the path of least resistance, and their ohmic loss would be increased by imposing longer chicane paths with slits.

The measured and calculated node voltages (not shown) also agree in fine detail, similarly to figure 7.9. In one way, the antenna is acting in this case as a CCP because all of the return current to ground occurs only via capacitive coupling, although the internal resonance currents in the antenna continue to act as a powerful ICP. The total dissipated power is the sum of the CCP and the ICP powers [2, 5, 6, 61].

All of the reactor equivalent circuit parameters for the grounded case are the same as for this floating case, except for the impedance of the ground connectors A10 and A16. Inspection of figures 7.7 and 7.10 for experimental measurements and model calculations shows that the grounded and floating antenna impedances are closely comparable although some differences are visible, especially for the high m modes. This similarity again suggests that it is the internal resonance current circulation that dominates the antenna network impedance, as could be expected for such a high Q antenna (see also section 7.2.1). Capacitive coupling to ground is stronger via a plasma, and the impedance spectra will be seen to be more different comparing the grounded (section 7.6.3) and floating (section 7.6.4) cases.

The good agreement with experimental results lends confidence to the MTL model for capacitive and inductive coupling between the antenna elements without plasma. In the following sections, this approach is extended to include antenna–plasma coupling, where the antenna input impedance effects are more clearly manifested because of the close proximity of the dissipative plasma to the antenna.

7.6 Experiment and MTL model with plasma loading

For plasma experiments, the dielectric foam and glass window were installed (section 13.4.2), and the reactor was closed with the top-plate as shown in figure 7.4. The vacuum impedance measurements now correspond to a stripline because the antenna conductors are enclosed by both the baseplate and top-plate ground planes. Homogeneous stripline expressions for $\hat{M}$ and $\hat{C}$ are well known [14, 62, 63], for example, for calculations in air. In the presence of plasma above the antenna, however, we can no longer use the homogeneous dielectric equation $\hat{M}\ \hat{C} = \mu_0\varepsilon_0\bar{1}$ of section 7.5. Since we do not know the appropriate matrix for the relative permittivity of the dielectric–plasma system, we now find ourselves obliged to calculate $\hat{C}$, as well as $\hat{M}$, independently of each other. The next two sections discuss how $\hat{M}$ and $\hat{C}$ can be estimated.

7.6.1 Partial inductance matrix for the MTL plasma-coupled antenna

The mutual partial inductance matrix can be calculated in the same way as in chapter 6, except for one difference, which is that the plasma is now enclosed by a fully conducting top-plate, figure 7.4. For the slotted top-plate used previously in the small antenna, figure 6.4(b), induced currents were taken to be small, as confirmed by a much smaller frequency shift observed in the vacuum mode resonances compared with a full metal plate. However, if the plasma skin depth δ is similar to, or longer than, the electrode gap, then induced image current will now flow partly in the plasma, and partly also in the opposing, fully conducting top-plate.

Fortunately, the complex image method of chapter 6 can be easily modified to account for coupling with a resistive slab whose opposite boundary is limited by a conducting plate, as shown by Weisshaar *et al* [64]. In the expression for mutual partial inductance, the complex distance from source to mirror plane, $h + p$, is replaced by an effective complex distance $h + p\tanh(D/p)$, where D is the distance across the plasma to the top-plate, and p is the usual complex skin depth of the plasma (see figures 7.11 and 7.12 for the definitions of h and D). Hence,

$$M^{\text{leg/plasma \& top - plate}} \approx \frac{\mu_0}{2\pi} l \left(\ln\left[\frac{l}{h + p\tanh(D/p)} \right] - 1 \right), \tag{7.11}$$

where l is the leg length. The physical reasoning behind the complex distance $h + p\tanh(D/p)$ can be seen from the limiting cases:

- In the limit of large $|z|$, $\tanh z \rightarrow 1$. Therefore, in the limit $|p| \ll D$, the tanh function tends to one, so the effective complex distance from source to mirror plane tends to $h + p$. This is the same as for a semi-infinite plasma

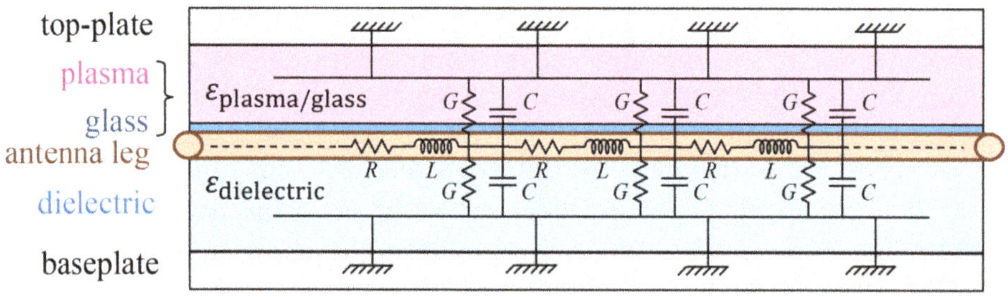

Figure 7.11. Schematic of the complete system, showing an antenna leg transmission line within the top and bottom grounded plates. The lower volume is filled with foam dielectric of relative permittivity $\varepsilon_{\text{dielectric}} \sim 1$. The upper volume comprises the glass window and the plasma, with a combined effective relative permittivity $\varepsilon_{\text{plasma/glass}}$.

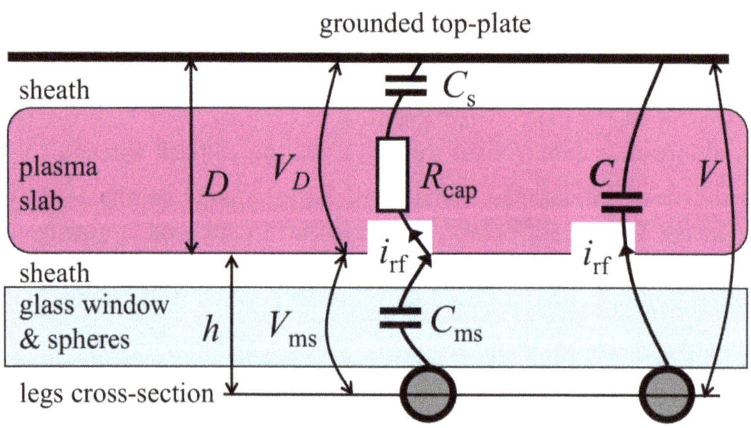

Figure 7.12. Schematic of capacitive coupling for the antenna legs to the plasma and top-plate, showing two equivalent circuits of the leg capacitances to ground. Capacitively coupled plasma dissipation is represented by the resistance R_{cap}. (Adapted with permission from [15]. Copyright 2017 IOP Publishing.)

slab in chapter 6, because the skin depth is so short that the image currents are induced entirely within the plasma thickness, and not at all in the top-plate.
- In the opposite limit of small $|z|$, $\tanh z \to z$. Hence, for $|p| \gg D$, the complex skin depth cancels out, and the mirror plane distance simplifies to the real distance $h + D$. The top-plate thus becomes the classical mirror image plane, at its distance $h + D$, because the skin depth is so long that the induced currents in the plasma do not attenuate the magnetic field of the antenna source.

We see that the expression (7.11) self-consistently accounts for skin depths much shorter than D in dense plasma, up to skin depths much longer than D for low (or no) plasma density, which is the vacuum loading case. The validity of the complex image method is thereby extended to plasma confined by a conducting boundary [64]. In the matrix description, each leg has mutual partial inductance with every leg image, not only its own image.

7.6.2 Capacitance matrix for the MTL plasma-coupled antenna

To repeat, the importance of capacitive coupling is demonstrated by the observation that plasma is maintained even when the antenna ground return connections, A10 and A16 in figure 7.4, are removed. The antenna is then electrically DC floating and it behaves as a CCP, but with an array of high voltage legs instead of an RF electrode plate. In this case, the transmission line current returns uniquely by capacitive coupling as shown in figure 7.11, principally via the plasma to the grounded top-plate. This is different from other ICP antennas which have ground connections, or at least a capacitor termination [2, 44, 47]. It is therefore necessary to estimate the p.u.l. capacitance matrix $\hat{C}$ between the antenna legs and the top-plate via the series combination of the glass window and the plasma. This series combination is detailed in figure 7.12.

The tangential magnetic field due to the leg currents decays with the characteristic skin depth [55] as it penetrates the plasma, inducing a skin depth current. In contrast, the normal electric field in the sheath does not penetrate the plasma because of Debye screening [61], therefore the antenna–plasma system can be treated as a microstrip transmission line with conducting boundaries defined by the antenna legs and the adjacent plasma surface across an effective vacuum sheath. The microstrip is assumed uniform along the legs, but it is inhomogeneous in the transverse plane because of the various dielectric regions shown in figures 7.11 and 7.12, and G.3.

The per-unit-length capacitance matrix of this inhomogeneous microstrip, $\hat{C}_{\mathrm{ms}}$, was calculated from the inverse of Maxwell's potential coefficient matrix $\hat{P}$ [14, 65] using the partial image method [14, 63], as explained in appendix G. $\hat{P}$ is the matrix of induced potentials at every leg position for unit charge separately on each leg. It is a symmetric Toeplitz matrix (symmetric Toeplitz matrices are both centro-symmetric and bi-symmetric[1]) and so $\hat{C}_{ms} = \hat{P}^{-1}$ is bi-symmetric (but not Toeplitz); it contains only real elements because the glass in figure 7.12 and figure G.3 is effectively lossless.

A particularity of the microstrip here is that, although the plasma surface is the mirror plane for induced charge, it is not the ground plane; the ground plane is the top-plate on the far side of the plasma. With reference to figure 7.12, in a simplified scalar treatment, the transverse RF currents are given by $i_{rf} = j\omega C_{ms} V_{ms}$, where V_{ms} is the microstrip voltage between the antenna legs and the plasma surface. Respecting current continuity, as shown schematically in figure 7.12, this can also be written as $i_{rf} = j\omega C\ V$, where $V = V_{ms} + V_D$ is the voltage between the antenna and ground, and C is the plasma-coupled capacitance. Hence $C = C_{ms}/(1 + V_D/V_{ms})$.

Finally, using the simplified plasma equivalent circuit in figure 7.12, $V_D/V_{ms} \sim j\omega C_{ms} R_{cap}$, for $C_{ms} \ll C_s$, where C_s is the sheath capacitance, and R_{cap} is the capacitive coupling resistance accounting for ohmic and stochastic heating of electrons [2]. This means that the MTL capacitance matrix $\hat{C}$ has complex elements, which represent an effective conductivity and capacitively coupled power loss. It is difficult to estimate R_{cap} [2] although it is expected that the plasma resistance is small compared to the capacitive impedance. In practice, $V_D/V_{ms} \sim 0.1j$, hence $\hat{C} = \hat{C}_{ms}/(1 + 0.1j)$, gives a reasonable fit to the experimental data shown below. In view of these approximations, it would be superfluous and cumbersome to calculate the capacitance matrix for the complete experimental configuration, including multiple reflections of the antenna between the plasma and the distant baseplate in presence of the inhomogeneous dielectric. Instead, the capacitance matrix described above serves as a sufficient approximate example.

7.6.3 Measurements in plasma with a grounded antenna

Figure 7.13(a) shows the grounded antenna input impedance spectra measured, with plasma, at the vacuum feedthrough for various RF powers, using a variable frequency RF generator from 12 to 14 MHz for mode resonance m = 8. The observed shifts to frequencies below the vacuum resonance frequency are a clear indication of capacitive coupling of the antenna to the plasma. Contrary to this, the plasma resonances were all shifted to higher frequencies in figure 6.11 for the small inductive antenna.

Figure 7.13(b) shows calculations of the corresponding input impedance *EM*-mode spectra using the MTL model, which show reasonably good agreement for the frequency shifts (downshifts as well as upshift) and the strong reductions in input impedance in the presence of plasma loading of the antenna. Conversely, for the purely inductive *H*-mode coupling in figure 6.11, the plasma resonance frequencies can only be above the vacuum resonance frequency because the only

[1] In a symmetric Toeplitz matrix $\hat{P}$, each descending diagonal from left to right is constant, with symmetry about both of its main diagonals (bi-symmetric and centro-symmetric). Consequently, $\hat{P}$ can be computed from a single row (or column) $\mathbf{p}$, saving computation time.

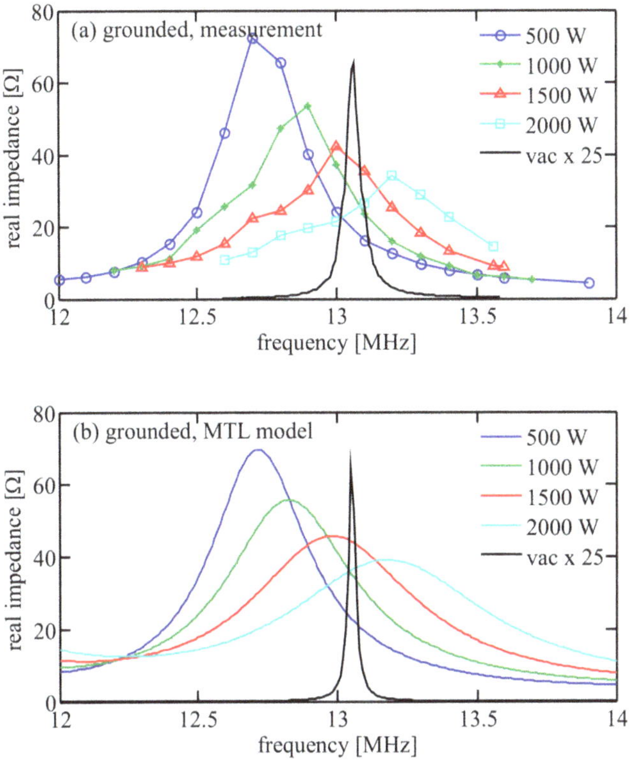

Figure 7.13. (a) Measurements of the mode m = 8 impedance spectra for a grounded antenna in vacuum and with plasma for 500, 1000, 1500, and 2000 W RF power. (b) Calculations of the same spectra using the multi-conductor transmission line model. The vacuum loading curves are 25 times higher than shown. Argon pressure 2 Pa, same grounding configuration as for figure 7.7. (Reproduced with permission from [15]. Copyright 2017 IOP Publishing.)

effect of the plasma image currents is to reduce the antenna net inductance (chapter 6 and Lenz's law). This shows the importance of accounting for capacitive coupling in large area ICP antennas.

Model predictions of ICP impedance spectra are only rarely given in the literature, and these are generally without plasma coupling [4]. This is probably because mutual inductance between antenna elements is often omitted, whereas chapter 4 showed that they are necessary to obtain good agreement with resonance frequency measurements [56, 66], and furthermore, resistive elements must be accounted for to obtain the impedance spectrum.

The mutual inductive coupling of the plasma with the antenna in figure 7.13(b) was calculated using the complex image method [56, 59] according to section 7.6.1. The mutual inductance depends on geometrical dimensions and the plasma complex skin depth, given by the electron density and the electron–neutral collision frequency [59], as follows:

1. The distance between the antenna leg axes and the plasma/sheath boundary, $h = 1.05$ cm in figure 7.12, was chosen according to 1 cm from the leg axes to the glass top surface, plus an estimated 0.3–0.5 cm for the sheath width (corresponding to about four Debye lengths for the undriven sheath in these plasma conditions [61]).
2. The argon pressure determines the electron–neutral collision frequency ν_m (s^{-1}) $\approx p$ (Pa) $\cdot 2.6 \cdot 10^7$ for typical electron temperature in these plasma conditions [61].
3. The plasma density was measured by Langmuir probe to be $n_{e0} \approx \{0.8, 1.2, 1.7, 2.4\} \cdot 10^{16}$ m^{-3} for delivered RF power $P_{rf} = \{500, 1000, 1500, 2000\}$ W.

These three parameters suffice for the complex image calculations [56] in figure 7.13(b).

The calculated range of real skin depth [56] goes from 7 cm (at 500 W) to 4 cm (at 2000 W). For the lowest power, the associated skin depth is approximately equal to the electrode gap, and capacitive coupling could be expected to dominate because significant current is induced in the top-plate instead of in the plasma. On the other hand, for the highest power, the skin depth is approximately half of the electrode gap and inductive coupling could be expected to dominate [2, 56, 61]. This is consistent with the frequency shift from negative at 500 W (dominant capacitive coupling) to positive at 2000 W (dominant inductive coupling) in figure 7.13.

Conventionally, the initial, low RF power coupled to the plasma by capacitive coupling is called the *E*-mode of power transfer, and the final, high RF power inductively coupled to the plasma from the antenna via their mutual inductance is called the *H*-mode of power transfer [2, 5, 6]; see section 5.2.3. For the small antenna, the transition from *E*- to *H*-mode is shown in section 6.3.1 and figure 6.7. The *E*-to-*H* transition might be considered to occur when the skin depth for inductive power dissipation becomes comparable to, and then smaller than, the electrode gap D in figure 7.12, as the plasma density rises with RF power. However, for the electromagnetically coupled plasma (EMCP) of the large antennas, the transition from *E*-mode to *H*-mode is indistinct, because both *E*- and *H*-modes co-exist as the *EM*-mode, and both contribute to the plasma power to some degree [2, 5, 6, 67]. In an analogous way, this is also described by Ahr *et al* [67] for the special case of phase-locked RF capacitively coupled substrate bias in a small-scale ICP. The capacitive and inductive coupling of a large area ICP antenna are inherently phase-locked because both couplings are generated simultaneously by the same, single RF source.

7.6.4 Measurements in plasma with a DC-floating antenna

For the DC floating antenna, the MTL model results in figure 7.14(b) are in fair agreement with the measurements in figure 7.14(a). Note that the mode m = 8 is determined uniquely by the choice of frequency for the floating case, since the A10 and A16 grounding stubs are omitted. The measured plasma densities were $n_{e0} \approx \{0.7, 1.7, 2.1\} \cdot 10^{16}$ m^{-3} for delivered RF power $P_{rf} = \{500, 1500, 2000\}$ W.

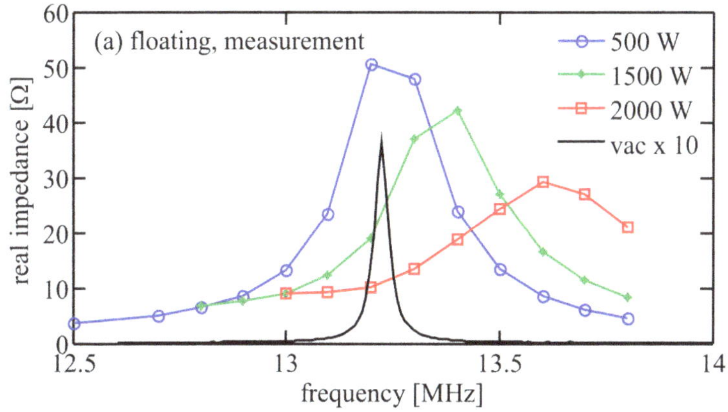

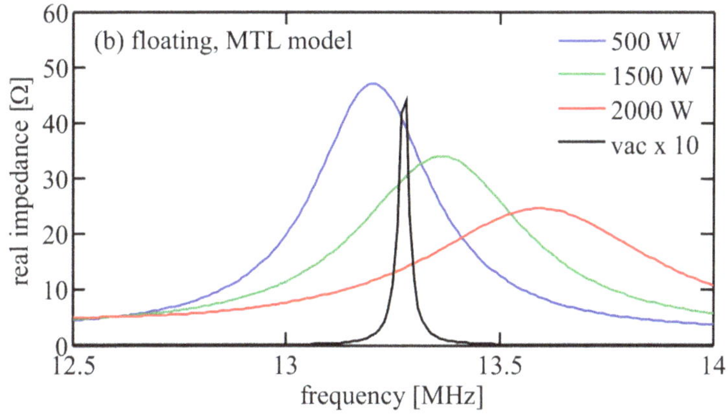

Figure 7.14. (a) Measurements of the mode m = 8 impedance spectra for a DC floating antenna in vacuum and with plasma for 500, 1000, and 2000 W RF power. (b) Calculations of the same spectra using the multi-conductor transmission line model. The vacuum loading curves are ten times higher than shown. Argon pressure 1.5 Pa, same DC floating configuration as for figure 7.10. (Reproduced with permission from [15]. Copyright 2017 IOP Publishing.)

The purely inductive coupling (lumped-element) model [56] over-estimates the mode frequencies by ~0.48 MHz (~3.6%) and the mode impedance by less than 20%. This is a smaller error compared to the grounded case, possibly because capacitive currents to the plasma are weaker for the DC floating antenna.

Comparison of the plasma-loaded impedance for grounded (figure 7.13) and floating antenna (figure 7.14) show considerably less similarity than for the vacuum impedance spectra. This can be attributed to the lower Q-factor of the plasma-loaded antenna, where the circulating internal resonance current is now not so dominant compared to the inductive and capacitive currents coupled externally via the plasma.

7.6.5 Plasma uniformity

Uniformity is a key criterion for large area plasma processing. For this antenna, Langmuir probe density measurements in figure 7.15(a) typically show a flattened dome profile with approximately 50% drop in plasma density from the centre to the

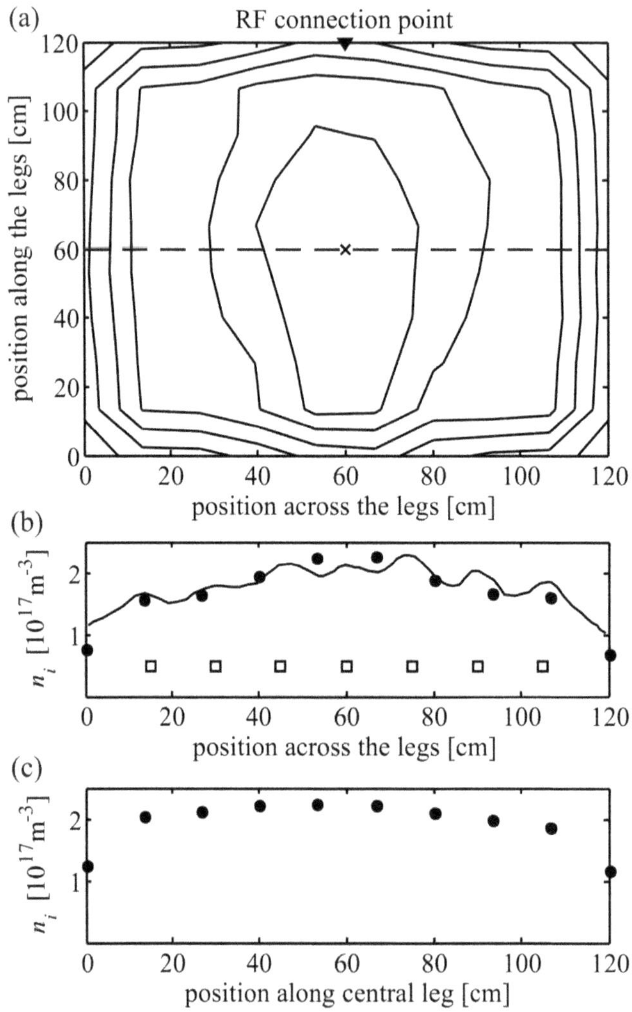

Figure 7.15. (a) Contour plot of the ion density n_i over the antenna area interpolated from the 10 × 10 regular array of surface probes (not shown). Contour levels descend successively by $\frac{1}{8}$th of the central maximum of $2.3 \cdot 10^{17}$ m^{-3}. (b) Measured density profile along the dashed line in (a), using a traversing Langmuir probe. The circles mark the surface probe measurements; the square symbols indicate the position of m = 8 mode maxima which also correspond to figure 7.9. (c) Measured density profile along the central leg. Argon pressure 1.5 Pa, 15 kW input power at 13.56 MHz. Same DC floating configuration as for figure 7.10. (Reproduced with permission from [15]. Copyright 2017 IOP Publishing.)

antenna edges[2]. This plasma non-uniformity does not fulfil the condition of transmission line uniformity in section 7.4, although averaging over the whole antenna nevertheless gives reasonable agreement with the measured plasma-loaded antenna input impedance in figure 7.14.

In CCP reactors, [11, 68, 69], the RF current circulates as a spatially distributed surface current in the skin depth of metal plate electrodes, whereas the current in ICP reactors circulates in discrete coils or antennas. For ICPs, the antenna pattern of these currents may appear on the substrate if the radical diffusion path is not long enough, requiring low pressure and wide electrode gap [9, 70]. However, the peaks in figure 7.15(b) are due to the wider m = 8 mode structure of the resonant antenna, not the individual legs. Note that low pressure ICP operation brings the advantage of no powder formation in deposition plasmas [71].

Figure 7.15(c) shows the measured variation in ion density along the length of the central leg. Using the MTL model, the calculated standing wave variation of current and voltage along the legs is between 9% and 21%, depending on whether the node position is near the middle or the end of a leg. This corresponds to a wavelength reduction, compared to vacuum, due to an effective relative permittivity $\varepsilon_r \approx 2.25$ of the antenna–plasma system. The effective RF wavelength is therefore shorter by a factor $\sim\sqrt{2.25} = 1.5$ (a 'worsening factor' [1, 11, 72]) because the speed of light in the plasma reactor is $\sim c/\sqrt{\varepsilon_r}$, thus aggravating the electromagnetic non-uniformity compared to a free space estimation, for a given antenna size. The relative permittivity is, in fact, a centro-symmetric matrix given by $\bar{\varepsilon}_r = \hat{M}\hat{C}/(\mu_0\varepsilon_0)$; it is not a diagonal matrix because of the inhomogeneous dielectric–plasma system [14] and the different plasma penetration depths of the tangential magnetic and normal electric fields. However, the size and frequency of this reactor are not high enough to unambiguously demonstrate a standing wave non-uniformity in figure 7.15(c) because of the convolution with the diffusion profile [73–76].

7.6.6 Summary for large area RF plasma sources

The large antenna generates a plasma which is simultaneously a CCP and an ICP; it would therefore be better described as an electromagnetically coupled plasma (EMCP) source. A multi-conductor transmission line (MTL) model describes electromagnetic coupling in the large area antenna. The inductive antenna–plasma coupling was calculated in terms of the per-unit-length mutual partial inductance matrix, using the complex image method [56, 59]. The per-unit-length capacitance matrix was estimated by analogy of capacitive coupling of the antenna to a plasma slab. The MTL matrix equations were solved by implementing the termination impedances of the resonant network.

The MTL model reproduces the measurements of antenna input impedance spectra, with and without plasma coupling, for grounded and electrically DC

[2] For these high power (15 kW) tests, 100 pF was added to each stringer capacitor so that the mode resonance with plasma corresponded approximately to the industrial frequency 13.56 MHz; otherwise plasma uniformity was compromised. This capacitor correction was necessary because the initial design did not account for capacitive coupling.

floating antennas. The distinguishing feature of the model is that the mutual inductances and capacitive coupling between all elements of the antenna, plasma, and reactor are taken into account. The model is valid for the whole range of plasma skin depths from much shorter to much longer than the plasma slab thickness. The MTL method is also valid for reactor lengths from shorter to longer than the effective wavelength of RF excitation in the plasma-coupled system. The model is analytic, with computation required only for the matrix solution. The MTL approach could be used to model and design other planar ICP antenna sources (ladder, U-type, serpentine, double comb, etc) by applying the termination impedances appropriate to each antenna type (appendix F.1).

7.7 Applications of EM-coupled antennas

7.7.1 Application: deposition of barrier coatings on polymers for the packaging industry

Plasma processes can be used in the packaging industries; one example is Tetra Pak [77] for the fabrication of drink cartons. These must be light-weight but resistant, watertight, and oxygen-impermeable. A 6 μm thickness of aluminium foil is the usual barrier material in the conventional packaging structure—it blocks the in-diffusion of gases and odours, and inhibits the out-diffusion of water and aromas from the food; it also prevents light penetration. However, aluminium is expensive and difficult to dispose of ecologically.

One concept is to replace the aluminium foil by a barrier coating deposited by plasma vacuum technology as a 10–20 nm thin coating on a polymer film substrate. This can be directly integrated into the packaging material within roll-to-roll machines scaled up for industrial production. Silica-like SiOx, diamond-like carbon (DLC), or alternate layers of these for optimized performance, are candidates for PECVD barrier coatings. Besides the oxygen and water vapour barrier properties, PECVD barrier films also have very good adhesion to the other materials in the packaging stack. For production costs, the key issue is the line speed of PECVD coating machines which is limited to around 200 m min^{-1}; this is the main factor that slows the whole supply chain. Doubling the line speed to 400 m min^{-1} would reduce the production cost by 5%–10% compared to the aluminium foil solution.

The previous plasma system used by Tetra Pak is schematically represented in figure 7.16(a). It is composed of three, 40 kHz magnetrons and a grounded, cooled, rotating drum around which the PET foil passes at 10 m min^{-1}. The first magnetron pre-treats the PET in an argon discharge to activate its surface. The other two magnetrons are used for the plasma deposition process with an acetylene precursor.

The application in this section was to develop an alternative RF resonant network antenna for evaluation as a pilot industrial process. The antenna was placed in a vacuum process chamber with a roll-to-roll mechanism shown in figure 7.16(b). The tensioned PET foil passes above the prototype planar source at a typical distance of 5 cm at the maximum available line speed of 40 m min^{-1}. The untreated PET foil

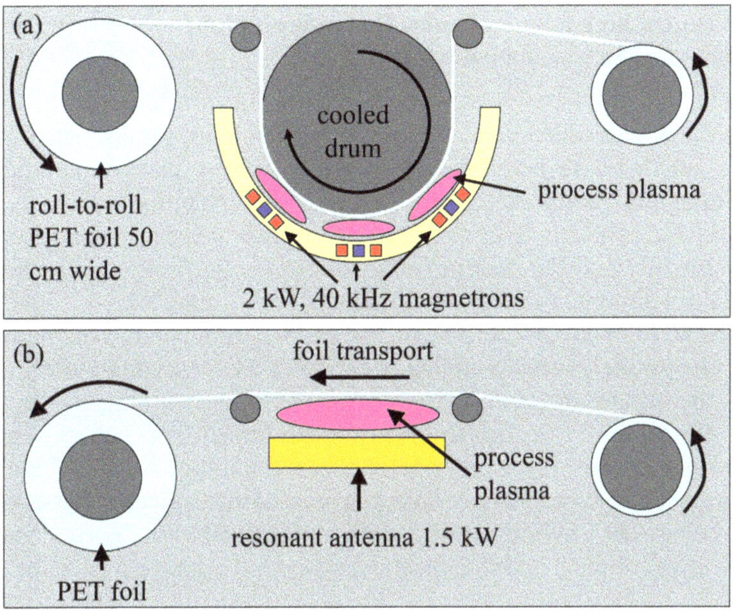

Figure 7.16. (a) Conventional plasma system for barrier coating, using three, 40 kHz magnetrons. The PET foil is exposed to the plasma during transport around the grounded, cooled drum. (b) The resonant network antenna source at 13.56 MHz. The tensioned foil is transported 5 cm above the antenna surface, defining the plasma height.

used for this project has an oxygen transmission rate (OTR) of about 120 c.c. m^{-2} atm^{-1} day^{-1}, and the aim was to obtain an OTR value below 3 c.c. m^{-2} atm^{-1} day^{-1} after plasma treatment.

Results for antenna plasma-deposited barrier coating

A prototype resonant antenna source 0.55 m wide produced good barriers at the maximum available speed of the pilot line, 100 m min^{-1}, ten times faster than the magnetron system, for 3 kW RF power. On the basis of the 65 nm s^{-1} observed deposition rate, the line speed could be raised to 60 m min^{-1}, a potential tripling of plasma processing rates. Rolls 2000 m long, 0.5 m wide, coated with DLC showed very good barrier layer performance with OTR better than 3 c.c. m^{-2} day^{-1} atm^{-1} at 100 m min^{-1} effective speed, thus achieving the required specifications.

An important advantage is that this new antenna process yields good OTR for a wide range of operation parameters, thereby eliminating the narrow constraints of the previous plasma technology. An RF bias study concluded that it can improve the quality of poor barrier layers (see also section 6.5), and for plasma parameters which already give good quality layers, RF biasing permits a higher line speed. The cost of production is expected to be competitive with other barrier film solutions on the market, such as sputtered or evaporated metallised films.

7.7.2 Application: high-frequency inductive antenna probe to measure the plasma complex conductivity

The previous sections have shown how the antenna input impedance spectrum is sensitive to the plasma–antenna coupling via their complex mutual partial inductance. This effect can therefore be exploited to measure those plasma parameters which influence this mutual partial inductance. In this section, a method for measuring plasma complex electrical conductivity is described [78], from which plasma parameters such as the electron density and the electron–neutral collision frequency can be estimated, in principle. The method relies on the measurement of the impedance of a small inductive antenna probe coupled to the plasma by mutual induction. The mutual partial inductance due to the plasma coupling is interpreted by applying the complex image method to the plasma medium; it is determined by the plasma skin depth and the distance to the plasma. For high-frequency measurements, capacitive coupling must also be accounted for as a first order correction for standing wave (transmission line) effects, hence, the electromagnetic coupling calculated in this chapter must be employed. It is shown below that a hybrid network configuration can be designed to maximize the desired inductive coupling and minimize the unwanted capacitive coupling.

Brief review of plasma density diagnostics

Control of the parameters in plasma processing is of prime importance for successful industrial manufacture of semiconductor devices. This is particularly critical for advanced process control of etching and deposition reactors. Reliable plasma diagnostics are then required for plasma process monitors and/or plasma characterization. In the domain of low temperature, non-equilibrium plasmas, the most popular devices for plasma characterization are Langmuir probes, which can lead to contamination in some processes. For deposition processes, Langmuir probes cannot normally be used because the probe surface can become coated very rapidly with insulating layers. The ion flux probe technique uses voltage pulsing to estimate the ion saturation current even when the surface is coated [79].

Microwave interferometry [61] with Lecher wires [80], hairpin probes [81], microwave cavities [82], the wave cut-off method [83], active plasma resonance spectroscopy [84], and other variations, measure the plasma frequency directly or indirectly via the plasma relative permittivity, in order to determine the electron plasma frequency and hence the electron density.

Magnetic induction probes and inductive sensors, such as flux loops, diamagnetic loops, Rogowski coils, and Bdot probes, are generally used to passively measure magnetic fields and magnetic fluxes. None of the plasma probe systems above relies on the measurement and interpretation of the mutual inductance with the plasma. This mutual inductance is derived from currents actively induced in the plasma by the probe, which are inductively coupled to the probe current. The inductive coupling affects the measured impedance of the inductive probe element in the proximity of the plasma. The reason that mutual inductance measurements were not previously made in plasma is that no convenient method for such an interpretation was given until now [56, 59].

Experimental method

A general probe device can be defined as any AC current-carrying inductive element (wire, coil, loop, network, of any shape) with self inductance L^{probe}, placed close to the surface of the plasma to be measured. Since the plasma is electrically conducting, mutual inductance with the probe elements occurs and induced currents flow in the plasma. This mutual inductance is characterized by the complex image method in terms of the mutual partial inductance $M^{\text{probe/plasma}}$ (chapter 6) which can be used to determine the complex skin depth p. Both the real and the imaginary parts of the plasma conductivity can be deduced from this measurement which could potentially be used, according to the expression for the plasma complex conductivity in appendix D.5, to determine the electron density n_{e0} as well as the electron–neutral collision frequency ν_{m}. An advantage of the complex image method, combined with partial inductance, is that it allows complex probe structures to be analysed [64, 85] by decomposing them into elementary elements for which analytical mutual partial inductance expressions are known. With reference to figure 7.17, the diagnostic method can be resumed as follows:

1. Measure the change in probe resonance frequency, Δf, and resistance, ΔR, due to coupling to the plasma.
2. Use the mutual inductance matrix model to compute the relation between the plasma complex conductivity and Δf, ΔR.
3. Deduce the electron density and collision frequency by comparison of the measurement and model.

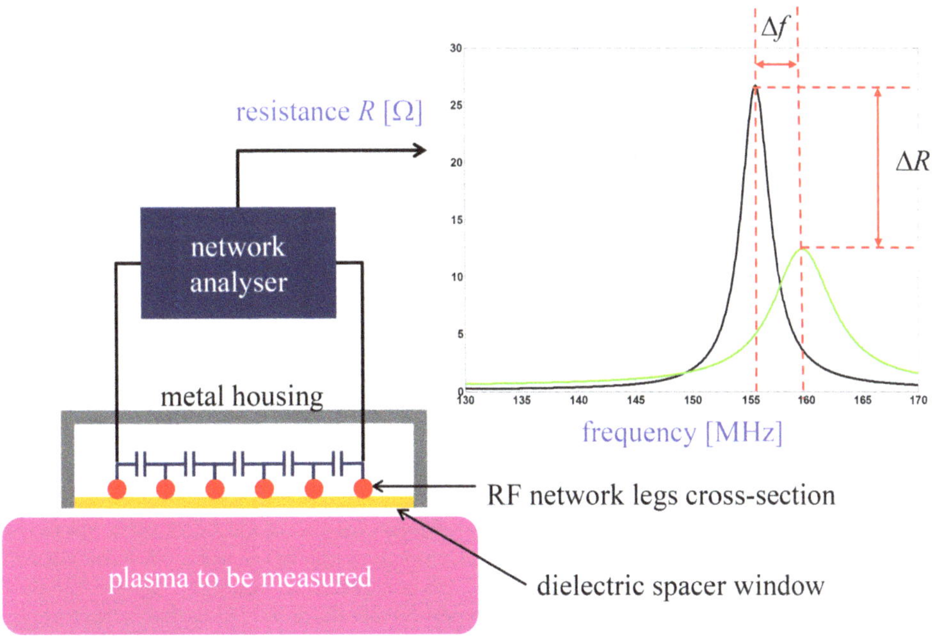

Figure 7.17. Schematic of an RF network antenna probe system, its vacuum resonance impedance (black curve), and its impedance near to a plasma (green curve), with a shift in frequency and the real part of its impedance.

Probe design and construction

Probe networks can be made up of inductive elements of any form, although resonant circuits improve sensitivity. The simplest configuration is a parallel $RL||C$ circuit. Sensitivity is further enhanced by using a resonant network antenna, whose impedance is determined by a mutual inductance matrix as shown in chapters 4, 6, and 7. However, a probe element facing a plasma is not only inductively coupled to the medium, but also capacitively, as discussed earlier in this chapter. These associated 'parasitic' capacitances are quite small (typically $C_p < 1$ pF for the 4 cm legs) but nevertheless can strongly affect the resonance frequency as a function of the plasma electron density, which becomes non-monotonic, as shown by the red curve in figure 7.18(a). This capacitive shift was described for the large antenna in sections 7.6.3 and 7.6.4. Clearly,

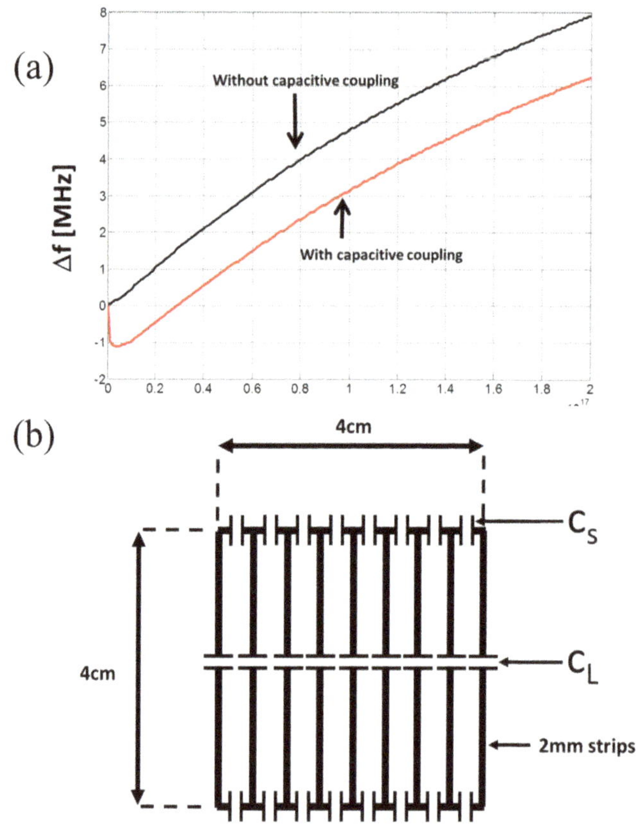

Figure 7.18. (a) The effect of capacitive coupling on the frequency response of a high-pass ladder antenna (red curve, 'With capacitive coupling'), and of a hybrid antenna design (black curve, 'Without capacitive coupling'). The hybrid design effectively eliminates the non-monotonic frequency response caused by capacitive coupling. (b) Schematic of a hybrid resonant network antenna probe. The addition of the capacitors C_L mid-way along the legs converts the high-pass ladder into a hybrid network. See also the probe photograph in figure 7.2. The RF signal was connected at opposite ends to drive the m = 1 mode which offers the highest frequency shift sensitivity to plasma density.

this non-monotonic frequency dependence is unacceptable for a diagnostic of plasma density—the capacitive coupling has to be suppressed.

For this purpose, a *hybrid* resonant network antenna probe was chosen, as distinguished from the high-pass resonant network antennas used throughout the book until now (see also section 14.1). It is shown schematically in figure 7.18(b), and in the photograph of figure 7.2, where the probe was built on a printed circuit board with surface-mounted capacitors on the reverse side of the board shown in the picture. The hybrid configuration is recognized by the supplementary capacitors C_L inserted mid-way along the leg inductors. It can be designed for high sensitivity to inductive coupling and simultaneously reduced parasitic capacitive coupling to the plasma, as explained in the following section. The probe construction can preferably be a non-invasive design, such as wall-mounted behind a dielectric window or a wafer, or otherwise mounted on a probe head in the plasma. It can also include a conducting backing plate (figure 7.17), and/or various types of segmented Faraday screen between the probe and the plasma to reduce parasitic capacitive coupling whilst maintaining effective inductive coupling.

Minimizing capacitive coupling

In the present case of 367 MHz, even the short 4 cm lengths become non-negligible (~5%) compared with the free space wavelength of 82 cm, and a transmission line treatment becomes necessary for a rigorous calculation of the antenna impedance. Nevertheless, a single π-section line impedance approximation [54] of capacitors C_p coupled to ground at both ends of the leg inductance suffices here, in figure 7.19.

Capacitive coupling can be strongly attenuated by cancelling the impedance of the legs: In the hybrid network used for the experiment, a 4 cm leg represents a self inductance L_{leg} of about 33.5 nH. In series with a capacitor $C_L \sim 5.6$ pF, the imaginary impedance for each leg becomes $Z_{leg} = j\omega L_{leg} - j/(\omega C_L)$ which vanishes

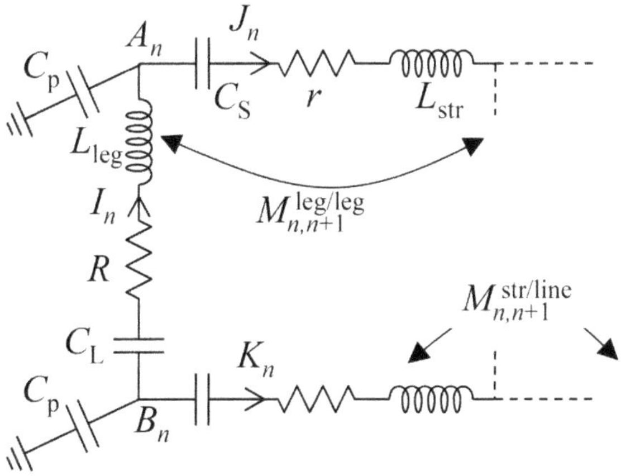

Figure 7.19. Hybrid antenna equivalent circuit, including node capacitance C_p to ground via the plasma, capacitors C_L within the legs, and the stringer capacitors C_S. The remaining notation is based on figure 4.13.

for a frequency of about 367 MHz. The value for the stringer capacitors $C_S = 47$ pF was then chosen to make the network resonate in mode m = 1, close to this value of 367 MHz. By doing so, the impact of capacitive coupling is minimized, while the sensitivity to mutual induction is maintained (see the black curve in figure 7.18(a)). The self partial inductances are eliminated, thereby replacing the leading diagonal of the leg impedance matrix (B.7) by zeroes, leaving only the off-diagonal mutual partial inductances which are sensitive to the plasma coupling. Matrix calculations for the hybrid antenna equivalent circuit in figure 7.19 are shown in appendix H.

Example results

As a practical example, we show here the results obtained with the 4 cm-by-4 cm resonant hybrid network represented in the equivalent circuit of figure 7.18(b). The network is placed as shown in figure 7.17 at a distance $h = 0.8$ cm above the plasma surface (generated by a separate RF inductive plasma source) and its input impedance spectrum is measured by means of a network analyser. In absence of plasma, the network presents a resonance (m = 1) at 370 MHz. Figure 7.20(a) shows the variation Δf of this resonance frequency as a function of the RF power injected into the plasma source. By computing the response of the network using the complex image method (Figure 7.20(b)), the measured variation of Δf is interpreted in terms of plasma density in figure 7.20(c). The estimated plasma density is in satisfactory

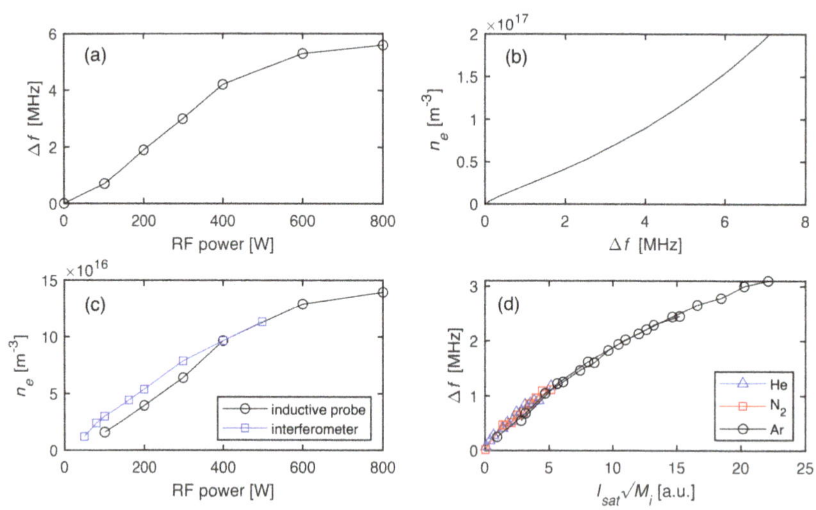

Figure 7.20. Example results using a hybrid resonant network antenna probe. (a) Measurement of the variation Δf of the probe resonance frequency as a function of the RF power injected into the plasma source. (b) The computed relation between the electron density and the change in network resonance frequency calculated using the complex image method and a mutual inductance matrix formulation. (c) The electron density, deduced from (a) and (b), as a function of the source injected power, compared with the plasma density measured by a microwave interferometer. (d) The frequency shift of the hybrid antenna probe is consistent with the plasma electron density estimated by a Langmuir probe in various gases.

agreement with a microwave interferometer, and a Langmuir probe for different gases, in figures 7.20(c) and (d), respectively.

Assessment of the inductive probe diagnostic
The hybrid inductive probe is one example of the versatile design of resonant network antennas. The dimensions, frequency, electrical configuration, and inductor and capacitor values, can all be widely varied. Standing wave effects can be accounted for, and stray capacitive coupling can be minimized. Temperature drift of the surface-mounted capacitor values can be compensated by monitoring the temperature with an integrated resistive probe, visible on the probe photograph of figure 7.2. Nevertheless, in spite of all these precautionary measures, the measurements can be confused by parasitic resonances whose frequencies drift unpredictably through the spectrum of the network analyser. Unfortunately, this renders automatic analysis unreliable in practice, and requires expert attention for interpretation of results.

7.8 Chapter summary for *EM* -coupled antennas

A large antenna generates a plasma which is simultaneously a CCP and an ICP; it would therefore be better described as an electromagnetically coupled plasma (EMCP) source. A multi-conductor transmission line (MTL) model is demonstrated to describe electromagnetic coupling in a large area planar antenna.

The MTL model reproduces the measurements of antenna input impedance spectra, with and without plasma coupling, for grounded and electrically DC floating antennas. The distinguishing feature of the model is that the mutual partial inductances and capacitive coupling between elements of the antenna, plasma, and reactor are all taken into account. The MTL approach could be used to model and design other types of planar ICP antenna sources by applying the termination impedances appropriate to each antenna type as indicated in appendix F. Two applications were described: a plasma source for roll-to-roll barrier layer deposition on industrial packaging; and a prototype inductive antenna probe diagnostic for plasma density.

References

[1] Schmidt H, Sansonnens L, Howling A A, Hollenstein Ch, Elyaakoubi M and Schmitt J P M 2004 Improving plasma uniformity using lens-shaped electrodes in a large area very high frequency reactor *J. Appl. Phys.* **95** 4559

[2] Chabert P and Braithwaite N 2011 *Physics of Radio-Frequency Plasmas* (Cambridge: Cambridge University Press)

[3] Lamm A J 1997 Observations of standing waves on an inductive plasma coil modeled as a uniform transmission line *J. Vac Sci. Technol.* A **15** 2615

[4] Khater M H and Overzet L J 2000 A new inductively coupled plasma source design with improved azimuthal symmetry control *Plasma Sources Sci. Technol.* **9** 545

[5] Sugai H, Nakamura K and Suzuki K 1994 Electrostatic coupling of antenna and the shielding effect in inductive RF plasmas *Jpn. J. Appl. Phys.* **33** 2189

[6] Lee S-H, Cho J-H, Huh S-R and Kim G-H 2014 Standing wave effect on plasma distribution in an inductively coupled plasma source with a short antenna *J. Phys. D: Appl. Phys.* **47** 015205
[7] Wu Y and Lieberman M A 2000 The influence of antenna configuration and standing wave effects on density profile in a large-area inductive plasma source *Plasma Sources Sci. Technol.* **9** 210
[8] Kushner M J, Collison W Z, Grapperhaus M J, Holland J P and Barnes M S 1996 A three-dimensional model for inductively coupled plasma etching reactors: azimuthal symmetry, coil properties, and comparison to experiments *J. Appl. Phys.* **80** 1337
[9] Park S E, Cho B U, Lee J K, Lee Y J and Yeom G Y 2003 The characteristics of large area processing plasmas *IEEE Trans. Plasma Sci.* **31** 628
[10] Kawamura E, Graves D B and Lieberman M A 2011 Fast 2D hybrid fluid-analytical simulation of inductive/capacitive discharges *Plasma Sources Sci. Technol.* **20** 035009
[11] Lieberman M A, Booth J P, Chabert P, Rax J M and Turner M M 2002 Standing wave and skin effects in large-area, high-frequency capacitive discharges *Plasma Sources Sci. Technol.* **11** 283
[12] Sansonnens L, Howling A A and Hollenstein Ch 2006 Electromagnetic field nonuniformities in large area, high-frequency capacitive plasma reactors, including electrode asymmetry effects *Plasma Sources Sci. Technol.* **15** 302
[13] Strobel C, Zimmermann T, Albert M, Bartha J W and Kuske J 2009 Productivity potential of an inline deposition system for amorphous and microcrystalline silicon solar cells *Sol. Energy Mater. Sol. Cells* **93** 1598
[14] Paul C R 2008 *Analysis of Multiconductor Transmission Lines* 2nd edn (Hoboken, NJ: Wiley)
[15] Guittienne Ph, Jacquier R, Howling A A and Furno I 2017 Electromagnetic, complex image model of a large area RF resonant antenna as inductive plasma source *Plasma Sources Sci. Technol.* **26** 035010
[16] Hollenstein Ch, Guittienne Ch and Howling A A 2013 Resonant RF network antennas for large-area and large-volume inductively coupled plasma sources *Plasma Sources Sci. Technol.* **22** 055021
[17] Colpo P, Ernst R and Rossi F 1999 Determination of the equivalent circuit of inductively coupled plasma sources *J. Appl. Phys.* **85** 1366
[18] Mashima H, Takeuchi Y, Noda M, Murata M, Naitou H, Kawasaki I and Kawai Y 2003 Uniformity of VHF plasma produced with ladder shaped electrode *Surf. Coat. Technol.* **171** 167
[19] d'Agostino R, Favia P, Kawai Y, Ikegami H, Sato N and Arefi-Khonsari F (ed) 2008 *Advanced Plasma Technology* (Weinheim: Wiley)
[20] Kawai Y, Ikegami H, Sato N, Matsuda A, Uchino K, Kuzuya M and Mizuno A (ed) 2008 *Industrial Plasma Technology* (Weinheim: Wiley)
[21] Pizzini S 2012 *Advanced Silicon Materials for Photovoltaic Applications* (New York: Wiley)
[22] Takeuchi Y, Nawata Y, Ogawa K, Serizawa A, Yamauchi Y and Murata M 2001 Preparation of large uniform amorphous silicon films by VHF-PECVD using a ladder-shaped antenna *Thin Solid Films* **386** 133
[23] Takeuchi Y, Kawasaki I, Mashima H, Murata M and Kawai Y 2001 Characteristics of VHF excited hydrogen plasmas using a ladder-shaped electrode *Thin Solid Films* **390** 217

[24] Takatsuka H, Noda M, Yonekura Y, Takeuchi Y and Yamauchi Y 2004 Development of high efficiency large area silicon thin film modules using VHF-PECVD *Sol. Energy* **77** 951
[25] Takatsuka H, Yamauchi Y, Kawamura K, Mashima H and Takeuchi Y 2006 World's largest amorphous silicon photovoltaic module *Thin Solid Films* **506–7** 13
[26] Nishimiya T, Takeuchi Y, Yamauchi Y, Takatsuka H, Kai Y, Muta H and Kawai Y 2007 Large area SiH_4/H_2 VHF plasma produced with multi-rod electrode *Plasma Process. Polym.* **4** S991
[27] Nishimiya T, Takeuchi Y, Yamauchi Y, Takatsuka H, Shioya T, Muta H and Kawai Y 2008 Large area VHF plasma production by a balanced power feeding method *Thin Solid Films* **516** 4430
[28] Yamauchi Y, Takeuchi Y, Takatsuka H, Yamashita H, Muta H and Kawai Y 2008 Characteristics of VHF H_2 plasma produced at high pressure *Contrib. Plasma Phys.* **48** 326
[29] Murata M, Takeuchi Y, Sasagawa E and Hamamoto K 1996 Inductively coupled radio frequency plasma chemical vapor deposition using a ladder-shaped antenna *Rev. Sci. Instrum.* **67** 1542
[30] Wendt A E and Mahoney L J 1996 Radio frequency inductive discharge source design for large area processing *Pure Appl. Chem.* **68** 1055
[31] Murata M, Mashima H, Yoshioka M, Nishida S, Morita S and Kawai Y 1997 Production of inductively coupled RF plasma using a ladder-shaped antenna *Jpn. J. Appl. Phys.* **36** 4563
[32] Kawai Y, Yoshioka M, Yamane T, Takeuchi Y and Murata M 1999 Radio-frequency plasma production using a ladder-shaped antenna *Surf. Coat. Technol.* **116–9** 662
[33] Mashima H, Murata M, Takeuchi Y, Yamakoshi H, Horioka T, Yamane T and Kawai Y 1999 Characteristics of very high frequency plasma produced using a ladder-shaped electrode *Jpn. J. Appl. Phys.* **38** 4305
[34] Kim K N, Lee Y J, Kyong S J and Yeom G Y 2004 Effects of multipolar magnetic fields on the characteristics of plasma and photoresist etching in an internal linear inductively coupled plasma system *Surf. Coat. Technol.* **177–8** 752
[35] Lim J H, Kim K N and Yeom G Y 2007 Inductively coupled plasma source using internal multiple U-type antenna for ultra large-area plasma processing *Plasma Process. Polym.* **4** S999
[36] Takagi T, Ueda M, Ito N, Watabe Y and Kondo M 2006 Microcrystalline silicon solar cells fabricated using array-antenna-type very high frequency plasma-enhanced chemical vapor deposition system *Jpn. J. Appl. Phys.* **45** 4003
[37] Takagi T, Ueda M, Ito N, Watabe Y, Sato H and Sawaya K 2006 Large area VHF plasma sources *Thin Solid Films* **502** 50
[38] Kim K N, Lee Y J, Jung S J and Yeom G Y 2004 Characteristics of parallel internal-type inductively coupled plasmas for large area flat panel display processing *Jpn. J. Appl. Phys.* **43** 4373
[39] Lim J H, Kim K N, Park J K, Lim J T and Yeom G Y 2008 Uniformity of internal linear-type inductively coupled plasma source for flat panel display processing *Appl. Phys. Lett.* **92** 051504
[40] Lim J H, Kim K N, Gweon G H, Park J B and Yeom G Y 2009 Study of internal linear inductively coupled plasma source for ultra large-scale flat panel display processing *Plasma Chem. Plasma Process.* **29** 251
[41] Kim S S, Chang H Y, Chang C S and Yoon N S 2000 Antenna configuration for uniform large-area inductively coupled plasma production *Appl. Phys. Lett.* **77** 492

[42] Jun H-S and Chang H-Y 2008 Development of 40 MHz inductively coupled plasma source and frequency effects on plasma parameters *Appl. Phys. Lett.* **92** 041501
[43] Forgotson N, Khemka V and Hopwood J 1996 Inductively coupled plasma for polymer etching of 200 mm wafers *J. Vac Sci. Technol.* B **14** 732
[44] Suzuki K, Konishi K, Nakamura K and Sugai H 2000 Effects of capacitance termination of the internal antenna in inductively coupled plasma *Plasma Sources Sci. Technol.* **9** 199
[45] Intrator T and Menard J 1996 Modelling and optimization of inductively coupled loop antenna plasma sources *Plasma Sources Sci. Technol.* **5** 371
[46] Menard J and Intrator T 1996 Laboratory measurements and optimization of inductively coupled loop antenna plasma sources *Plasma Sources Sci. Technol.* **5** 363
[47] Setsuhara Y, Miyake S, Sakawa Y and Shoji T 1999 Production of inductively-coupled large-diameter plasmas with internal antenna *Jpn. J. Appl. Phys.* **38** 4263
[48] Meziani T, Colpo P and Rossi F 2001 Design of a magnetic-pole enhanced inductively coupled plasma source *Plasma Sources Sci. Technol.* **10** 276
[49] Colpo P, Meziani T and Rossi F 2005 Inductively coupled plasmas: optimizing the inductive-coupling efficiency for large-area source design *J. Vac Sci. Technol.* A **23** 270
[50] Meziani T, Colpo P and Rossi F 2006 Electrical description of a magnetic pole enhanced inductively coupled plasma source: refinement of the transformer model by reverse electromagnetic modeling *J. Appl. Phys.* **99** 033303
[51] Yu Z, Shaw D, Gonzales P and Collins G J 1995 Large area radio frequency plasma for microelectronics processing *J. Vac Sci. Technol.* A **13** 871
[52] Advanced Energy Industries Inc. http://www.advanced-energy.com Access date 2024
[53] Howling A A, Derendinger L, Sansonnens L, Schmidt H, Hollenstein Ch, Sakanaka E and Schmitt J P M 2005 Probe measurements of plasma potential nonuniformity due to edge asymmetry in large-area radio-frequency reactors: the telegraph effect *J. Appl. Phys.* **97** 123308
[54] Bleaney B I and Bleaney B 1976 *Electricity and Magnetism* (London: Oxford University Press) 3rd edn p 1
[55] Paul C R and Nasar S A 1987 *Introduction to Electromagnetic Fields* (New York: McGraw-Hill)
[56] Guittienne Ph, Jacquier R, Howling A A and Furno I 2015 Complex image method for RF antenna-plasma inductive coupling calculation in planar geometry. Part II: measurements on a resonant network *Plasma Sources Sci. Technol.* **24** 065015
[57] Grover F W 1962 *Inductance Calculations: Working Formulas and Tables* (New York: Dover)
[58] Paul C R 2010 *Inductance: Loop and Partial* (Hoboken, NJ: Wiley)
[59] Howling A A, Guittienne Ph, Jacquier R and Furno I 2015 Complex image method for RF antenna-plasma inductive coupling calculation in planar geometry. Part I: basic concepts *Plasma Sources Sci. Technol.* **24** 065014
[60] Briefi S, Zielke D, Rauner D and Fantz U 2022 Diagnostics of RF coupling in H^- ion sources as a tool for optimizing source design and operational parameters *Rev. Sci. Instrum.* **93** 023501
[61] Lieberman M A and Lichtenberg A J 2005 *Principles of Plasma Discharges and Materials Processing* 2nd edn (Hoboken, NJ: Wiley)
[62] Kammler D W 1968 Calculation of characteristic admittances and coupling coefficients for strip transmission lines *IEEE Trans. Microw. Theory Tech.* **MTT-16** 925

[63] Silvester P 1968 TEM wave properties of microstrip transmission lines *Proc. IEE* **115** 43
[64] Weisshaar A, Lan H and Luoh A 2002 Accurate closed-form expressions for the frequency-dependent line parameters of on-chip interconnects on lossy silicon substrate *IEEE Trans. Adv. Packag.* **25** 288
[65] Harrington R F 1958 *Introduction to Electromagnetic Engineering* (New York: McGraw-Hill)
[66] Jin J-M 1998 *Electromagnetic Analysis and Design in Magnetic Resonance Imaging* (Boca Raton, FL: CRC Press)
[67] Ahr P, Schüngel E, Schulze J, Tsankov T V and Czarnetzki U 2015 Influence of a phase-locked RF substrate bias on the *E*- to *H*-mode transition in an inductively coupled plasma *Plasma Sources Sci. Technol.* **24** 044006
[68] Chabert P, Raimbault J-L, Levif P, Rax J-M and Lieberman M A 2006 Inductive heating and E to H transitions in high frequency capacitive discharges *Plasma Sources Sci. Technol.* **15** S130
[69] Liu Y-X, Zhang Y-R, Bogaerts A and Wang Y-N 2016 Electromagnetic effects in high-frequency large-area capacitive discharges: a review *J. Vac Sci. Technol.* A **33** 020801
[70] Kim D W, You S J, Kim J H, Chang H Y and Oh W Y 2015 Computational characterization of a new inductively coupled plasma source for application to narrow gap plasma processes *IEEE Trans. Plasma Sci.* **43** 3876
[71] Demolon P, Guittienne Ph, Howling A A, Jost S, Jacquier R and Furno I 2018 RF bias to suppress post-oxidation of μc-Si:H films deposited by inductively-coupled plasma using a planar RF resonant antenna *Vacuum* **147** 58
[72] Schmitt J P M, Elyaakoubi M and Sansonnens L 2002 Glow discharge processing in the liquid crystal display industry *Plasma Sources Sci. Technol.* **11** A206
[73] Kee R J, Coltrin M E and Glarborg P 2003 *Chemically Reacting Flow* (Hoboken, NJ: Wiley)
[74] White F M 1991 *Viscous Fluid Flow* 2nd edn (New York: McGraw-Hill)
[75] Vahedi V, Lieberman M A, DiPeso G, Rognlien T D and Hewett D 1995 Analytic model of power deposition in inductively coupled plasma sources *J. Appl. Phys.* **78** 1446
[76] Stittsworth J A and Wendt A E 1996 Reactor geometry and plasma uniformity in a planar inductively coupled radio frequency argon discharge *Plasma Sources Sci. Technol.* **5** 429
[77] Lorenzetti C, Fayet P, Larrieu J and Denecker C 2017 Laminated packaging material comprising a barrier film and packaging containers manufactured therefrom *Patent* WO2017072120A1
[78] Furno I, Guittienne Ph and Howling A A 2019 Method, measurement probe and measurement system for determining plasma characteristics *Patent* EP16721773.6
[79] Braithwaite N St J, Booth J P and Cunge G 1996 A novel electrostatic probe method for ion flux measurements *Plasma Sources Sci. Technol.* **5** 677
[80] Paris P J, Bitter M and Hollenstein Ch 1977 Measurement of ion acoustic test waves in a magnetized plasma by means of a 30-GHz Lecher wire interferometer of high spatial resolution *Rev. Sci. Instrum.* **48** 874
[81] Stenzel R L 1976 Microwave resonator probe for localized density measurements in weakly magnetized plasmas *Rev. Sci. Instrum.* **47** 603
[82] Stoffels E, Stoffels W W, Vender D, Kando M, Kroesen G M W and de Hoog F J 1995 Negative ions in a radio-frequency oxygen plasma *Phys. Rev.* E **51** 2425
[83] Kim J-H, Seong D-J, Lim J-Y and Chung K-H 2003 Plasma frequency measurements for absolute plasma density by means of wave cutoff method *Appl. Phys. Lett.* **83** 4725

[84] Styrnoll T, Bienholz S, Lapke M and Awakowicz P 2014 Study on electrostatic and electromagnetic probes operated in ceramic and metallic depositing plasmas *Plasma Sources Sci. Technol.* **23** 025013

[85] Kang K, Shi J, Yin W-Y, Li L-W, Zouhdi S, Rustagi S C and Mouthaan K 2007 Analysis of frequency- and temperature-dependent substrate eddy currents in on-chip spiral inductors using the complex image method *IEEE Trans. Magn.* **43** 3243

IOP Publishing

Resonant Network Antennas for Radio-Frequency Plasma Sources

Theory, technology and applications

Philippe Guittienne, Alan Howling and Ivo Furno

Chapter 8

Cylindrical wave functions in birdcage antennas

The aim of this chapter is to become familiar with the basic electric and magnetic fields within a birdcage antenna. These cylindrical wave functions are more complicated than the linear arrays of voltage and current in planar network antennas, and therefore require some background analytical theory. We stipulate from the outset that the birdcage solutions will always include the currents induced in a surrounding, perfect electrical conductor (PEC) cylindrical screen; this is because we assume that any industrially relevant RF antenna system will be contained within a Faraday screen for reasons of electromagnetic compatibility (EMC) [1]. This was already the case for the large planar antenna in chapter 7. Consequently, there will be no question of electromagnetic fields radiating power to infinite radii, so we do not need to concern ourselves with Hankel functions nor retarded potentials, and the electromagnetic fields will be assumed to act instantaneously over the whole circular cross-section of the cylinders [2, 3]. Remember also that the plasma is non-magnetized (no external imposed magnetic field) in this part II of the book, so we are concerned only with inductive plasma coupling. In this analytical approach, the influence of birdcage end-rings and finite-length legs will not be considered. More complex birdcage fields are postponed until the numerical simulations of chapter 12, which is concerned with magnetized plasma.

The wave functions suitable for describing electric and magnetic fields in cylindrical geometry are introduced here, step by step, building up to an analytical account of plasma–dielectric inductive coupling by a birdcage antenna. Each solution is illustrated by figures in two or three dimensions to give an intuitive grasp of the antenna field distributions within the screen. If the reader wishes to adapt the calculations and figures to their particular situation, the corresponding programs are provided via a link in appendix K. Where expressions are not written explicitly in the text, they can be found in the program listings. In the first section of this chapter, the step sequence is developed as follows:

doi:10.1088/978-0-7503-5296-3ch8

- First, plasma inductive coupling by a birdcage antenna is described in a qualitative manner by deriving a general wavefield solution in section 8.1, in order to show the ultimate goal.
- To begin the series of complete derived solutions, the simplest case of vacuum fields generated by a m = 1 normal mode shell current inside a cylindrical screen is calculated in section 8.2. This solution is then simplified, for comparison, by using the magneto-quasistatic approximation. The particular case of the m = 1 normal mode is used to show the uniform magnetic field commonly used in magnetic resonance imaging [4].
- In section 8.3, the vacuum wavefields due to a general current distribution at the shell radius are calculated, and applied first to the case of a single wire filament. The vacuum fields due to a multi-leg birdcage are then derived by extension, and it is shown that nine legs gives a reasonable compromise for uniform fields with the m = 1 normal mode. It is concluded that a shell current is a valid proxy for a birdcage, provided that the region of interest excludes the singularities at the leg positions.
- The plasma is introduced in section 8.4 as a uniform, complex-permittivity dielectric cylinder. Skin depth is investigated as a function of frequency and plasma parameters.

The last section 8.5 presents the classical Apollonius image method for line currents within a PEC screen, and the possibility of a complex image method applicable to cylindrical geometry is discussed in the context of the uniform plasma calculations.

8.1 A general wavefield solution for birdcage antennas

To put the geometrical configuration into context, figure 8.1 shows the most appropriate antenna/plasma description which we will work towards in this chapter. To facilitate analytical development, the plasma column, radius a, is considered as a uniform dielectric with complex relative permittivity ε_p, sustained by inductive coupling via a birdcage antenna radius $b > a$. The birdcage is inside a perfect electrical conductor (PEC) cylinder, radius $c > b$, which can be a metal screen or a vacuum chamber. The plasma column extends to infinity beyond both ends of the

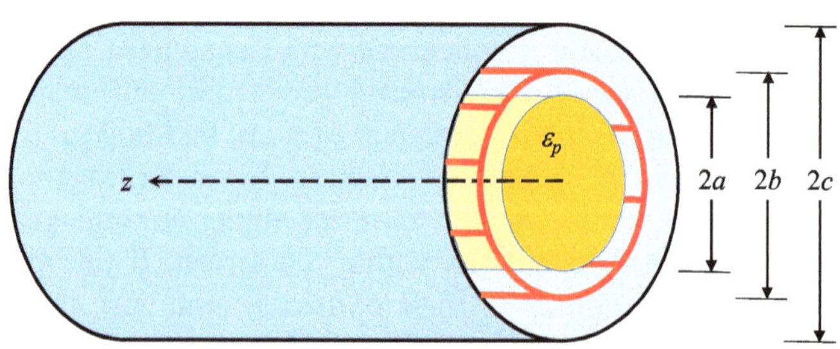

Figure 8.1. A column of plasma of uniform relative permittivity ε_p and radius a, within a birdcage antenna (red) of radius b, enclosed by a cylindrical perfectly electrically conducting (PEC) screen of radius c.

antenna and screen, i.e. the plasma is considered to be unbounded along z, and the antenna also will also be considered infinitely long to facilitate a two-dimensional solution over the cross-section for this attempt at analytical solutions.

We will solve the Helmholtz wave equation (D.34) for the axial electric field component E_z. In each dielectric, this equation is

$$\nabla^2 E_z + k_d^2 E_z = 0, \tag{8.1}$$

corresponding to electromagnetic solutions, where $k_d^2 = k_0^2 \varepsilon_r$ as derived in appendix D.4, in cylindrical coordinates. This book is concerned with planar and cylindrical antennas; spherical antennas are not considered here (but see J-M Jin [4] for applications in magnetic resonance imaging), therefore we can unambiguously define the cylindrical coordinates as $\{r, \phi, z\}$. For long birdcage legs carrying RF current along the z direction, as in figure 8.1, a reasonable starting assumption is that the magnetic field is purely in the $\{r, \phi\}$ transverse plane. These transverse magnetic (TM) modes are also called E-waves [5] because the only oscillating parameter along the z-axis is E_z (this does not necessarily mean that E_r or E_ϕ are zero). Two corollaries of course are that $B_z = 0$ everywhere in this chapter on ICP, non-magnetized plasma induced by an infinitely long, straight birdcage antenna, and that the axial component of the wavenumber, k_z, tends to zero. On the other hand, in chapter 11, we will see that the modes in magnetized plasma require $k_z \neq 0$, and B_z becomes an integral part of the wavefield solution.

Generally, the Helmholtz equation (8.1) can be written

$$\frac{\partial^2 E_z}{\partial r^2} + \frac{1}{r}\frac{\partial E_z}{\partial r} + \frac{1}{r^2}\frac{\partial^2 E_z}{\partial \phi^2} + \frac{\partial^2 E_z}{\partial z^2} + k_d^2 E_z = 0, \tag{8.2}$$

where $k_d^2 = k_0^2 \varepsilon_r$ is the square of the wavenumber for each dielectric of relative permittivity ε_r, and $k_0 = \omega/c$ is the wavenumber in free space. Note that $k_d^2 E_z$ in Helmholtz's equation includes the effective dielectric relative permittivity ε_r (appendix D.3).

The particular solutions, called cylindrical wave functions, can be solved by the standard method of separation of variables [5] by assuming a product form for $E_z(r, \phi, z) = P(r)\Phi(\phi)Z(z)$ under condition of uniform ε_r for each dielectric domain —this is an oversimplification for a typical plasma, but the analytical solution nevertheless brings useful insight. Substituting into (8.2) and dividing throughout by $P(r)\Phi(\phi)Z(z)$ gives, for each domain $0 \leqslant r \leqslant a$, $a \leqslant r \leqslant b$, and $b \leqslant r \leqslant c$ in figure 8.1,

$$\frac{1}{P}\frac{d^2 P}{dr^2} + \frac{1}{rP}\frac{dP}{dr} + \frac{1}{r^2 \Phi}\frac{d^2 \Phi}{d\phi^2} + \frac{1}{Z}\frac{d^2 Z}{dz^2} + k_d^2 = 0. \tag{8.3}$$

Following the standard method [5], we note that the fourth term is a function only of z, and all other terms are independent of z (provided that the plasma permittivity is independent of z), hence this term must be a constant, giving

$$\frac{1}{Z}\frac{d^2 Z}{dz^2} = -k_z^2; \quad Z(z) = A(k_z)e^{-jk_z z} + B(k_z)e^{+jk_z z}, \tag{8.4}$$

where we have anticipated that the solution will involve E_z waves travelling along $\pm z$ with some axial wavenumber k_z (the time dependence of the phasor E_z is implicitly $e^{-j\omega t}$, as noted at the outset in section 2.1.1). $A(k_z)$, $B(k_z)$ are constants which can be determined by boundary conditions and will depend on the value of k_z. Note that k_z is the single, common wavenumber along z for the complete solution of all domains inside $0 \leqslant r \leqslant c$, covering all the different media; the significance of this will become clear in chapter 11 and especially in appendix J.2.1. On the other hand, k_d depends on the permittivity of each separate medium. Substitute into (8.3) and multiply by r^2:

$$\frac{r^2}{P}\frac{d^2P}{dr^2} + \frac{r}{P}\frac{dP}{dr} + \frac{1}{\Phi}\frac{d^2\Phi}{d\phi^2} + (k_d^2 - k_z^2)r^2 = 0. \tag{8.5}$$

Since k_d is the magnitude of the wavenumber, and k_z is its axial component, for brevity we introduce the perpendicular component of the wavenumber $k_\perp$, where $k_\perp^2 = k_d^2 - k_z^2$ (in cylindrical geometry, $k_\perp = k_r$), which itself depends on the relative permittivity of each medium. Pursuing the separation of variables, notice that the third term depends only on ϕ, and is the only ϕ-dependent term, provided that the relative permittivity of the plasma, and hence k_d, is axisymmetric (independent of ϕ). Hence the third term equals another separation constant:

$$\frac{1}{\Phi}\frac{d^2\Phi}{d\phi^2} = -m^2; \quad \Phi(\phi) = \alpha_m e^{jm\phi} + \beta_m e^{-jm\phi}, \tag{8.6}$$

where m is an integer, since it is clear that ϕ is azimuthally periodic, and α_m, β_m are Fourier coefficients. Again, substituting into (8.5) and multiplying now by P gives

$$r^2\frac{d^2P}{dr^2} + r\frac{dP}{dr} + [k_\perp^2 r^2 - m^2]P = 0. \tag{8.7}$$

Equation (8.7) is the cylindrical Bessel equation, provided that $k_\perp$ (i.e. k_d) is a constant (independent of r) for each medium where these equations apply. Finally, therefore, for the solution by separation of variables to be valid, we require that the plasma relative permittivity ε_p be uniform over the whole plasma volume, namely, $0 \leqslant r \leqslant a$ and along all z, as stated initially for figure 8.1. Solutions to (8.7) are Bessel functions of first and second kind, $J_m(k_\perp r)$ and $Y_m(k_\perp r)$, respectively, which are linearly independent and of order m [5, 6]. The general solution to (8.7) is therefore

$$P_m(r) = a_m J_m(k_\perp r) + b_m Y_m(k_\perp r), \tag{8.8}$$

where a_m, b_m are constants which depend on the azimuthal mode number m. The first and second kind Bessel functions are appropriate for radially bounded situations; $r \leqslant c$ in this case. Other kinds of Bessel functions such as the Hankel functions are suitable for unbounded, radiating problems, but for practical purposes such as electromagnetic compatibility (EMC) and recommended engineering practice, plasma reactors should always be effectively enclosed in a Faraday screen to prevent EM radiation and perturbations to other instruments. Mathematically, the presence of a PEC screen precisely defines the outer boundary conditions, and no problems arise associated with radiated power loss, far fields, retarded potentials, magnetic flux, or return currents, at infinity. In the present case of the RAID birdcage RF source in chapter 12, the screen distance c can be adjusted for fine-tuning of the birdcage resonance frequency, and beyond the birdcage's length, screening is guaranteed by the metal vacuum vessel. The effect of the screen can be calculated by letting c tend to a large radius, without incurring mathematical or physical problems of the return current, because the current return is always assured by the screen even if its radius tends to infinity.

Finally, a *particular solution* of (8.2), using the product of individual solutions $E_z(r, \phi, z) = P(r)\Phi(\phi)Z(z)$ as assumed initially, is [5]

$$\begin{aligned} E_z(r, \phi, z)_{m,k_z} = {} & [a_m J_m(k_\perp r) + b_m Y_m(k_\perp r)] \times \left[\alpha_m e^{jm\phi} + \beta_m e^{-jm\phi}\right] \\ & \times [A(k_z)e^{-jk_z z} + B(k_z)e^{+jk_z z}]. \end{aligned} \tag{8.9}$$

As expected for three, second order differential equations, (8.4), (8.6), and (8.7), there are six unknown constants a_m, b_m, α_m, β_m, $A(k_z)$, $B(k_z)$ for each solution in each of the three domains $0 \leqslant r \leqslant a$, $a \leqslant r \leqslant b$, and $b \leqslant r \leqslant c$. Fortunately, some of these constants are redundant or related [5], and the remainder can be found by using boundary conditions, according to the properties of perfect conductors or sources of current. Although the requirement of a single uniform plasma domain is very different from experimental observation, this could be partially alleviated by using multi-layered plasma domains with different ε_p values, with the penalty of more unknowns and boundary conditions.

Until specific boundary conditions are imposed, the particular solution (8.9) is valid for any m or k_z. In the *general solution* we retain all six constants in a combination of all possible particular solutions, implying a summation over all possible m integers from $-\infty$ to $+\infty$, and an integration over all possible k_z:

$$\begin{aligned} E_z(r, \phi, z) = {} & \sum_{m=-\infty}^{\infty} [a_m J_m(k_\perp r) + b_m Y_m(k_\perp r)] \times \left[\alpha_m e^{jm\phi} + \beta_m e^{-jm\phi}\right] \\ & \times \int [A(k_z)e^{-jk_z z} + B(k_z)e^{+jk_z z}]dk_z. \end{aligned} \tag{8.10}$$

Note that for an infinitely long birdcage, or infinitely long shell current, where the current is uniquely along z, the wavenumber along z tends to 0, i.e. $k_z \to 0$ in (8.10). Furthermore, in preparation for solutions given below in terms of shell currents, we re-write the general integer summation from $m = -\infty \to +\infty$ for functions of r and ϕ in (8.10) as follows:

$$\begin{aligned} & \sum_{m=-\infty}^{\infty} P_m(r)\left[\alpha_m e^{jm\phi} + \beta_m e^{-jm\phi}\right] \\ & = F0 + \sum_{m=1}^{\infty} \left\{P_m(r)\left[\alpha_m e^{jm\phi} + \beta_m e^{-jm\phi}\right] + P_{-m}(r)\left[\alpha_{-m} e^{-jm\phi} + \beta_{-m} e^{jm\phi}\right]\right\}, \end{aligned} \tag{8.11}$$

where $F0$ is the element of the sum for $m = 0$. Note that Bessel's equation (8.7) is invariant for $\sqrt{m^2} = \pm m$, as well as for $\sqrt{k_\perp^2} = \pm k_\perp$, so the result must be independent of these choices, which are just sign conventions. Therefore, it is straightforward that we must have $P_{-m}(r) = P_m(r)$. Hence

$$\sum_{m=-\infty}^{\infty} P_m(r)\left[\alpha_m e^{jm\phi} + \beta_m e^{-jm\phi}\right] = \sum_{m=0}^{\infty} P_m(r)[\zeta_m e^{jm\phi} + \xi_m e^{-jm\phi}], \tag{8.12}$$

where the summation now runs from $m = 0, 1, 2, \ldots$ [5]. The combined Fourier coefficients for $e^{jm\phi}$, and for $e^{-jm\phi}$, in (8.11) are, respectively, defined as $\zeta_m = \alpha_m + \beta_{-m}$ and $\xi_m = \beta_m + \alpha_{-m}$.

Finally, accounting for the $e^{-j\omega t}$ time dependence multiplying every phasor quantity (see section 2.1.1), one example of a measured time-varying electric field is

$$E_z^{\text{measured}}(r, \phi, t) = \text{Re}\,[a_m J_m(k_\perp r) e^{j(m\phi\ \ \omega t)}], \tag{8.13}$$

where m is the conventional notation for the azimuthal periodicity which is not to be confused with the antenna mode number m introduced in chapter 3.

Magneto-quasistatic or electromagnetic solutions

It may be argued that all birdcage calculations are more suited to a magneto-quasistatic, lumped-element approximation (appendix D.1.1) than an electromagnetic Helmholtz wave solution, because the birdcage length and radius are much smaller than the RF wavelength. The magneto-quasistatic approximation is indeed used in cases of eddy currents induced in conducting media [7], whereupon the equations to be solved are reduced to the magnetic diffusion equation for the magnetic vector potential in resistive conductors, because the vacuum displacement current is negligible compared to the induced conduction currents. This magneto-quasistatic approximation is used to derive the complex image method for planar geometry in appendix E. Nevertheless, to maintain a consistent form of mathematical solution across all of the different examples in this chapter—from vacuum alone, to multiple zones which include conducting plasma—we choose to use the electromagnetic solution which is exact and therefore applicable to all situations. The magneto-quasistatic approximation can be recovered from the final solutions, for example, by neglecting terms associated with vacuum displacement current. Static magnetic fields (truly magnetostatic) are due to steady direct currents (DC) [4], whereas we are concerned here with induced electric fields of angular frequency ω which are non-zero, although they become very small for low frequencies.

8.1.1 Shell surface current density

In the following sections, during the development towards birdcage geometry, we shall often make use of *shell current* at the shell radius b (see figure 8.2),

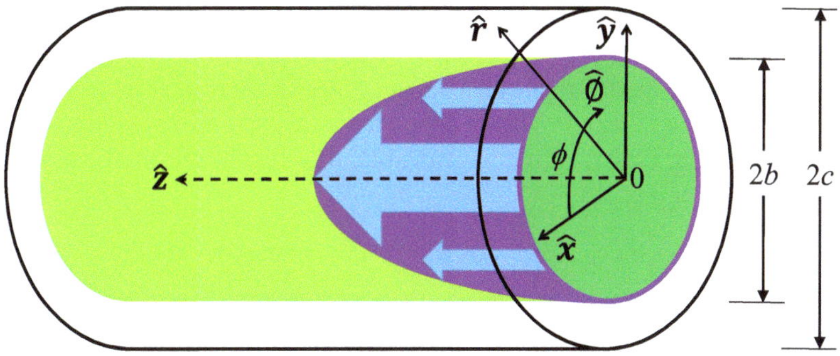

Figure 8.2. A cylindrical shell, radius b, carrying a surface current whose current density (Am^{-1}) varies as $i_s(\phi) = i_0 \cos\phi$, enclosed by a cylindrical perfectly conducting screen of radius c (transparent in this figure). The current arrows are along $+z$ in this schematic representation of $i_s(\phi)$ on the visible front face of the shell current. $\{\hat{\mathbf{r}}, \hat{\boldsymbol{\phi}}, \hat{\mathbf{z}}\}$ are the unit vectors of the right-handed cylindrical coordinate system $\{r, \phi, z\}$. Similarly, the right-handed triad of $\{\hat{\mathbf{x}}, \hat{\mathbf{y}}, \hat{\mathbf{z}}\}$ are the unit vectors of Cartesian coordinates. All of the unit vectors lie in the plane of the cross-section, except for $\hat{\mathbf{z}}$ which along the cylinder axis. The figure shows a section of the infinitely long system.

corresponding to the radius of the birdcage in figure 8.1. A *general* source current density for the wavefields, $j_z(r, \phi)$ (Am^{-2}), which can vary with r and ϕ, is defined here to be directed parallel to the z-axis, along the same direction as the (straight) birdcage legs. On the other hand, the *shell surface* current density $i_s(\phi)$ considered in this chapter varies only with angle ϕ, and is confined to the vanishingly thin shell of radius b. Therefore, the total source current I_z (A), along z, can be written as

$$I_z[\text{A}] = \int_0^{\infty}\int_0^{2\pi} j_z(r, \phi)\, r\, dr\, d\phi = \int_0^{\infty}\int_0^{2\pi} i_s(\phi)\delta(r - b)\, r\, dr\, d\phi, \tag{8.14}$$

where the Dirac delta function $\delta(r - b) = 1\,(\text{m}^{-1})$ for $r = b$, and zero otherwise; note that the physical dimensions of the delta function are the inverse of the physical dimensions of its argument. Therefore, $\int_0^{\infty} \delta(r - b)\, dr = 1$, and $\int_0^{\infty} \delta(r - b)\, r\, dr = b$. Hence

$$I_z[\text{A}] = \int_0^{2\pi} i_s(\phi)\, b\, d\phi = \int_0^{2\pi b} i_s(\phi)\, dl, \tag{8.15}$$

where $dl = b\, d\phi$ is the elemental length along the shell circumference $2\pi b$. We see that the z-directed shell surface current density $i_s(\phi)$ (Am^{-1}) is in units of current *per unit length around the shell perimeter at radius b*. Finally, then, the elemental source current dI_z (A) at angle ϕ, subtended by the elemental angle $d\phi$, is given by

$$dI_z(\phi)[\text{A}] = i_s(\phi)\, dl = i_s(\phi)\, b\, d\phi. \tag{8.16}$$

Shell currents of azimuthal harmonic $\cos(m\phi)$ are introduced in section 8.2, and specifically for the first harmonic $m = 1$, i.e. $i_s(\phi) = i_0 \cos\phi$, where i_0 (Am^{-1}) is the amplitude of the harmonic. To labour the point for the rest of this chapter, $i_0 \cos\phi$ (Am^{-1}) is the z-directed shell surface current density in units of current per unit length around the shell perimeter at radius b and angle ϕ. Note that the total net current I_z along z is zero for each shell harmonic when $\cos(m\phi)$ is integrated around 2π in (8.15), except for $m = 0$. Other special cases of shell currents such as a wire (filament) source, and birdcage current sources, are treated in section 8.3. For good measure, we recall that the $e^{-j\omega t}$ time dependence multiplies every phasor quantity (section 2.1.1), i.e. the observed, or measurable, first harmonic shell current density, similarly to (8.13) for E_z, is

$$i_s^{\text{measured}}(\phi, t) = \text{Re}[(i_0 \cos \phi)e^{-j\omega t}] = i_0 \cos(\phi)\cos(\omega t). \tag{8.17}$$

This shell current can also be seen as a standing wave made up of two counter-rotating waves (shown schematically for electric fields in figures 9.5 and 10.15), where each rotating current is a travelling wave of the type given in (8.13). This is demonstrated by the trigonometric identity

$$\cos(\phi + \omega t) + \cos(\phi - \omega t) \equiv 2\cos(\phi)\cos(\omega t). \tag{8.18}$$

Physically, what is a shell surface current?

For the physical manifestation of a shell surface current source, the first image which springs to mind is likely to be a thin metal cylinder, aligned along z, carrying an RF current on its surface. Apart from the difficulty of driving azimuthal harmonics down a continuous conductor, a moment's reflection will find that an empty metal cylinder behaves as a Faraday screen with no electromagnetic fields inside it (there are no internal sources)—this is useless for sustaining an internal plasma as depicted in figure 8.1. Therefore, the shell current within the PEC screen is *not* some sort of coaxial cable, where the internal shell carries a surface current returning via the (internal surface of the) outer screen. Instead, the shell surface current can excite fields inside and outside of itself, and the return current is shared between the plasma on its inside, and the PEC screen on its outside.

The shell surface current is therefore an idealized current source which is transparent to electromagnetic fields. Perhaps the closest physical analog would be an infinite number of separate, isolated, parallel filaments aligned along z, coincident with the cylindrical surface $r = b$, carrying a current according to their angular position ϕ. The infinitesimal separation between the wires permits the penetration of electromagnetic fields. This picture will help to understand the boundary condition of continuity of electric field across the shell surface current in (8.21) [4], and the good approximation of a $m = 1$ shell current by a multi-leg birdcage in section 8.3.3. In sum, the shell current is a birdcage with an infinite number of legs.

8.2 Vacuum wavefields for a $m = 1$ shell current inside a PEC screen

As a very first step towards the required general solution described in section 8.1 and figure 8.1, we make three simplifying conditions:

1. The plasma is removed in order to investigate vacuum solutions. It is clear that all of the shell current returns via the internal surface of the surrounding screen. Note that the external PEC screen is a permanent fixture for Faraday cage EMC reasons, as explained at the beginning of the chapter.
2. The frequency is assumed to be very low—the magneto-quasistatic limit—so that the free-space wavelength λ is much longer than the radii and the region of interest of the solutions along an axial distance $l \ll \lambda$.
3. The birdcage antenna is replaced by a cylindrical shell surface current density with a single spatial harmonic, $i_s(\phi) = i_0 \cos \phi$ (Am^{-1}), corresponding to azimuthal periodicity $m = 1$.

This elementary case is interesting in its own right to calculate the electric and magnetic field distributions due to a shell current, to understand the effects of a PEC screen, and—in section 8.5—to investigate the classical Apollonius image method for line currents or charges in conducting cylinders. Moreover, the treatment of the equations will enable us to recognize various mathematical shortcuts in more complicated models to follow.

The first condition here leaves a perfectly uniform vacuum dielectric, with only two domains, $0 \leqslant r \leqslant b$ and $b \leqslant r \leqslant c$, instead of the three zones shown in figure 8.1. The second condition means that we will ignore any z variation ($e^{\pm jk_z l} \approx 1$), although we retain the term $k_d^2 E_z = k_0^2 E_z$ in (8.2), so that the Bessel function solutions (8.8) still apply. Otherwise, in the strict limit of magnetostatics, the radial equation is a Laplace equation whose solution is a power series in r^m [4], whereas we wish to exploit section 8.1 to develop some self-consistency throughout this topic. On the other hand, we will show that the magneto-quasistatic case is correctly obtained using the small-argument expansions of Bessel functions.

To make a historical diversion, an RF resonant closed birdcage coil operating in normal mode number m = 1 is often employed in magnetic resonance imaging (MRI) to create a uniform RF magnetic field for excitation of nuclear spins in a sample (the patient); this is the so-called 'RF transmitter coil' [4]. This scheme was behind the original motivation for using resonant birdcage antennas as plasma sources by exciting electron collisions in gases [8]. The uniformity of the magnetic field will be another result demonstrated in this section. To close this MRI digression, note that other types of RF coils are used to detect z-dependent magnetic fields; these diagnostic coils are called 'gradient' or 'RF receiver' coils. In the latter case, the z-dependent term $\frac{1}{Z}\frac{d^2 Z}{dz^2}$ must obviously be properly accounted for in the wavefield equations [4]. These possibilities could be exploited in future options for birdcages with twisted legs, or other non-straight geometries, and to investigate finite-length effects of short birdcage antennas [8].

The third condition replaces the discrete leg currents by a surface current density, $i_s(\phi) = i_0 \cos \phi$, distributed over a cylindrical shell of the same diameter as the birdcage, as shown in figure 8.2. The shell surface current density has units of current per unit length around the circumference, (Am^{-1}), where the current is directed along z. The elemental current dI_z (A) along z, subtended by an elemental angle $d\phi$, is $dI_z = i_0 \cos(\phi)\, b\, d\phi$, as explained in section 8.1.1. For $\frac{\pi}{2} < \phi < \frac{3\pi}{2}$, the shell current has the opposite direction to the arrows shown in figure 8.2, and so the net total shell current I_z is zero, with current oscillating in anti-phase on opposite sides of the shell.

We will now find the solutions for the wavefields in both domains, $0 \leqslant r \leqslant b$ and $b \leqslant r \leqslant c$. It is important to note that there are no fields outside the PEC screen, $r > c$, which, of course, is the principal interest in the use of screens. Applying the first two simplifying conditions (no plasma, and $k_z \to 0$) to the general case (8.10), the wavefield expression (8.12) within the shell, $0 \leqslant r \leqslant b$, becomes

$$E_z(r, \phi) = \sum_{m=0}^{\infty} A_m J_m(k_0 r)[\zeta_m e^{jm\phi} + \xi_m e^{-jm\phi}], \tag{8.19}$$

where the second kind Bessel function $Y_m(k_0 r)$ is excluded because it diverges at the origin [4–6]. Between the shell and the screen, $b \leqslant r \leqslant c$, we must include the first and second kinds of Bessel functions:

$$E_z(r, \phi) = \sum_{m=0}^{\infty} [D_m J_m(k_0 r) + E_m Y_m(k_0 r)][\zeta_m e^{jm\phi} + \xi_m e^{-jm\phi}], \tag{8.20}$$

where the reason for using D_m, E_m instead of B_m, C_m is because the latter labels are reserved for when the plasma is introduced in section 8.4. The electric field E_z is continuous across the infinitely thin shell [4] at $r = b$ (see section 8.1.1), therefore, equating (8.19) and (8.20), a first boundary condition, for each m, is

$$A_m J_m(k_0 b) = D_m J_m(k_0 b) + E_m Y_m(k_0 b), \tag{8.21}$$

where all the summations have been multiplied by $e^{jm\phi}$ and integrated from $-\pi$ to π, using the well-known orthogonality relations for trigonometric functions of Fourier series [5], namely

$$\begin{aligned} \int_{-\pi}^{\pi} e^{j(m-m')\phi}\, d\phi &= 2\pi \quad \text{for} \quad m = m', \\ &= 0 \quad \text{for} \quad m \neq m'. \end{aligned} \tag{8.22}$$

By definition of a PEC screen at $r = c$, a second boundary condition for electric field is $E_z(c, \phi) = 0$, hence

$$D_m J_m(k_0 c) + E_m Y_m(k_0 c) = 0. \tag{8.23}$$

The third boundary condition, using Ampère's law [2] straddling the current shell at $r = b$ in figure 8.3, describes the discontinuity in azimuthal magnetic field across the shell current:

$$\begin{aligned} \oint \mathbf{H} \cdot d\mathbf{s} &= \int i_s\, ds, \\ \left(H_\phi^{\text{out}} - H_\phi^{\text{in}}\right)\left(b\, d\phi\right) &= i_s(b\, d\phi), \\ H_\phi^{\text{out}} - H_\phi^{\text{in}} &= i_s, \end{aligned} \tag{8.24}$$

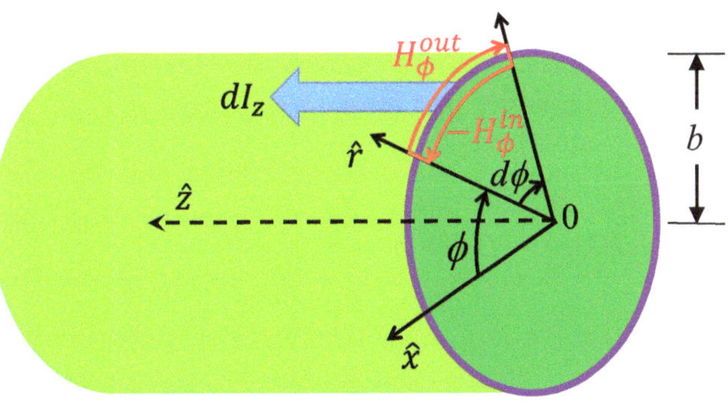

Figure 8.3. A cylindrical shell current, radius b, showing the magnetic field $H_\phi^{\rm out}$ just outside, and $-H_\phi^{\rm out}$ just inside, the shell surface current density i_s directed parallel to $\hat{\mathbf{z}}$. According to (8.16), $dI_z = i_s b\, d\phi$ is the elemental shell current flowing parallel to $\hat{\mathbf{z}}$ from the shell arc length ($b\, d\phi$) subtended by the angle $d\phi$. See also figure 8.2 and section 8.1.1.

where $i_s = i_0 \cos\phi$ is the shell surface current density directed along z as defined in section 8.1.1. From Maxwell's equation $\nabla \times \mathbf{E} = -\frac{\partial \mathbf{B}}{\partial t} = j\omega\mathbf{B}$, for the single time-Fourier component at the RF angular frequency ω (note for future reference that the factor $j = \sqrt{-1}$ in the following equations has its origin in this time derivative; it does not stem from derivatives of ϕ-dependent Fourier coefficients). Neglecting any z variation ($\frac{\partial}{\partial z} = jk_z \to 0$), E_z is related to the azimuthal magnetic field intensity H_ϕ by $\frac{\partial E_z}{\partial r} = -j\omega\mu_0 H_\phi$, which we substitute, using (8.19) and (8.20), for $H_\phi(b, \phi)$ into (8.24). To incorporate the shell current source into the boundary condition for the discontinuity of H_ϕ at $r = b$, we therefore insert $i_0 \cos\phi$ into the equation for phasors, as follows:

$$\begin{aligned}&\sum_{m=0}^{\infty}[D_m k_0 J_m'(k_0 b) + E_m k_0 Y_m'(k_0 b)][\zeta_m e^{jm\phi} + \xi_m e^{-jm\phi}] \\ &- \sum_{m=0}^{\infty} A_m k_0 J_m'(k_0 b)[\zeta_m e^{jm\phi} + \xi_m e^{-jm\phi}] = -j\omega\mu_0 i_0 \cos\phi.\end{aligned} \tag{8.25}$$

We note in passing that continuity in H_r is mathematically equivalent to continuity in E_z for these non-ferromagnetic media ($\mu_r = 1$), which is why H_r continuity is not used as a separate boundary condition. For the radial derivative of the electric field Bessel terms in (8.25), we have used $\frac{\partial}{\partial r}J_m(k_0 r) = J_m'(k_0 r) \times k_0$, from the chain rule for differentiation, where the prime denotes the Bessel function differentiated with respect to its argument, $J_m'(\alpha) = \frac{\partial}{\partial \alpha}J_m(\alpha)$, and the same for $Y_m(\alpha)$. At this point, we can again apply the orthogonality relation (8.22) to (8.25): by writing $\cos\phi = (e^{j\phi} + e^{-j\phi})/2$, by multiplying each equation by $e^{\pm jm\phi}$, and by integrating over ϕ from $-\pi$ to π, it is clear that the $\cos\phi$ dependence of the shell current density in (8.25) imposes its own first order periodicity, $m = 1$, on every J_m, Y_m Bessel function. We choose to set $\zeta_1 = \xi_1 = 1/2$, and zero for all other m, so that $[\zeta_1 e^{j\phi} + \xi_1 e^{-j\phi}] = \cos\phi$. This could have been initially suspected by inspection, and for subsequent calculations we will select the order corresponding to the shell current mode number from the outset, without explicitly implementing the general solution summation, or the orthogonality operation.

The three boundary conditions (8.21), (8.23), and (8.25) are therefore re-written for mode number $m = 1$ alone:

$$\begin{aligned} A_1 J_1(k_0 b) - D_1 J_1(k_0 b) - E_1 Y_1(k_0 b) &= 0, \\ D_1 J_1(k_0 c) + E_1 Y_1(k_0 c) &= 0, \\ A_1 J_1'(k_0 b) - D_1 J_1'(k_0 b) - E_1 Y_1'(k_0 b) &= jZ_0 i_0. \end{aligned} \tag{8.26}$$

The frequently occurring factor $\frac{\omega\mu_0}{k_0}$ has been substituted by Z_0, where $Z_0 = \sqrt{\frac{\mu_0}{\varepsilon_0}} = 376.7\ \Omega$ is the impedance of free space [2]. From these three simultaneous boundary condition equations, the three unknown constants are found to be

$$\begin{aligned} A_1 &= -jZ_0 i_0 \left[\frac{\pi k_0 b}{2}\right][Y_1(k_0 b) - \mathrm{Scrn}_1], \\ D_1 &= +jZ_0 i_0 \left[\frac{\pi k_0 b}{2}\right]\mathrm{Scrn}_1, \\ E_1 &= -jZ_0 i_0 \left[\frac{\pi k_0 b}{2}\right]J_1(k_0 b), \\ \text{where}\quad \mathrm{Scrn}_1 &= \frac{J_1(k_0 b)\, Y_1(k_0 c)}{J_1(k_0 c)}. \end{aligned} \tag{8.27}$$

The term Scrn_1 will be shown to depend on $m = 1$ currents induced in the screen. These equations have been simplified using Abel's identity for the Wronskian of these Bessel functions [4]

$$J_m(x)\, Y_m'(x) - J_m'(x)\, Y_m(x) \equiv \frac{2}{\pi x} \quad \text{for all } x, \text{ and } m > -1, \tag{8.28}$$

not forgetting that the argument x in our case includes the wavenumber, $x = k_0 b$. This identity serendipitously removes all derivatives of Bessel functions from the expressions for the constants, and therefore for E_z also. The ubiquitous factor $jZ_0 i_0$ in (8.27) arises because the field equations are solved here in terms of electric field (V m^{-1}) wave functions, whereas the boundary conditions involve a current per unit length, $i_0 \cos\phi$, which has the same (A m^{-1}) units as magnetic field intensity, H.

The solution for the E_z phasor is found by substitution into (8.19) and (8.20) for $m = 1$:

$$\begin{aligned} E_z(r, \phi) &= A_1 J_1(k_0 r)\cos\phi \quad (0 \leqslant r \leqslant b), \\ &= [D_1 J_1(k_0 r) + E_1 Y_1(k_0 r)]\cos\phi \quad (b \leqslant r \leqslant c). \end{aligned} \tag{8.29}$$

It is instructive to write out this E_z solution more fully by using (8.27):

$$
\begin{aligned}
&E_z(r, \phi) \\
&= -jZ_0 i_0\left[\frac{\pi k_0 b}{2}\right][J_1(k_0 r)(\cos\phi)Y_1(k_0 b) - J_1(k_0 r)(\cos\phi)\mathrm{Scrn}_1] \quad (0 \leqslant r \leqslant b), \\
&= -jZ_0 i_0\left[\frac{\pi k_0 b}{2}\right][Y_1(k_0 r)(\cos\phi)J_1(k_0 b) - J_1(k_0 r)(\cos\phi)\mathrm{Scrn}_1] \quad (b \leqslant r \leqslant c).
\end{aligned} \tag{8.30}
$$

The subtractive term $-J_1(k_0r)(\cos\phi)\mathrm{Scrn}_1$, both inside and outside the shell current radius b, is the only term which depends on the screen radius c, and represents the field strength reduction everywhere within the screen volume due to currents induced in the screen's inside surface. This is made more clear using the low-frequency approximation in section 8.2.1.

By way of an example of $E_z(r, \phi)$, figure 8.4 shows x, y plots of the real part of $E_z(r, \phi)$ for representative values corresponding to the RAID source (see chapter 12), namely, a birdcage radius of $b = 0.065$ m, a screen radius of $c = 0.09$ m, and an RF frequency of 13.56 MHz. Within the current shell, E_z seems to increase proportionally to y, and this behaviour persists even up to 1 GHz. In contrast, microwave cavity resonances begin to appear at higher frequencies as shown for 3 GHz in figure 8.5, which exhibits the double sombrero nature (figure 8.6) of the solution in (8.30).

From the discussion preceding (8.25), the H_ϕ phasor solution is

$$
\begin{aligned}
H_\phi(r, \phi) &= \left[\frac{j}{Z_0}\right]A_1 J_1'(k_0 r)\cos\phi \quad (0 \leqslant r \leqslant b), \\
&= \left[\frac{j}{Z_0}\right][D_1 J_1'(k_0 r) + E_1 Y_1'(k_0 r)]\cos\phi \quad (b \leqslant r \leqslant c).
\end{aligned} \tag{8.31}
$$

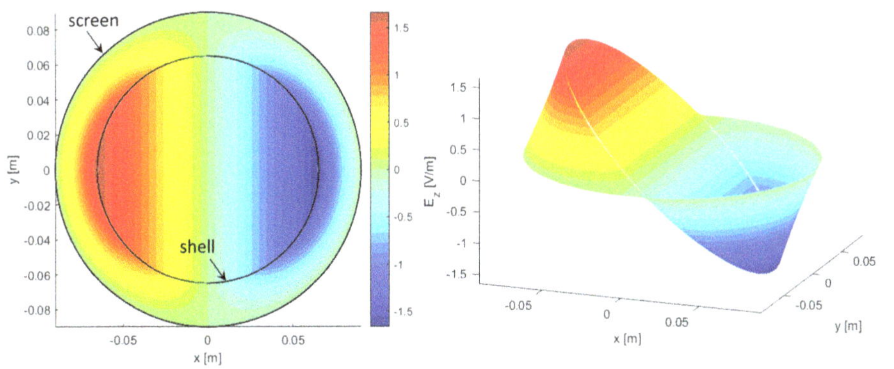

Figure 8.4. Two-dimensional plots of Re (E_z) using (8.30) for the RAID source described in chapter 12: Shell current radius $b = 0.065$ m, screen radius $c = 0.09$ m, and RF frequency is 13.56 MHz. The shell current density $i_0 \cos\phi$ is normalized to $i_0 = 1$ (Am^{-1}), with $\phi = 0$ along the x-axis. Left: surface density plot. Right: surface profile plot with axes rotated for presentation. E_z is continuous across the shell current at $r = b$, and zero at the PEC screen $r = c$. The spatial resolution of the calculation is 0.5 mm, but the colour scale is shown with only 30 gradations to aid visualization. The colorbar for E_z (V m^{-1}) applies to both graphs.

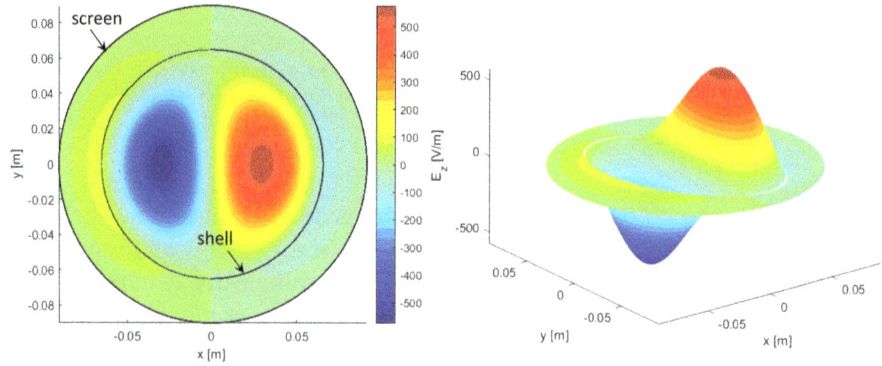

Figure 8.5. Two-dimensional plots of Re (E_z) using (8.30) for the RAID source of figure 8.4 but at a microwave frequency of 3 GHz. All other parameters are the same. Left: surface density plot. Right: surface profile plot with axes rotated for presentation. E_z is continuous across the shell current at $r = b$, and zero at the screen $r = c$.

Figure 8.6. 'The twin sombrero nature of the high-frequency solution.' (Illustration by Alex Howling. Copyright 2023 Alex Howling.)

The Bessel derivatives $J_1'(k_0r)$ and $Y_1'(k_0r)$ in (8.31) can be calculated using [4]

$$J_n'(z) = [J_{n-1}(z) - J_{n+1}(z)]/2, \tag{8.32}$$

and similarly for $Y_n'(z)$. Finally, using $\nabla \times \mathbf{E} = -\frac{\partial \mathbf{B}}{\partial t}$, with $\frac{\partial}{\partial z} = jk_z \to 0$ also for $j\omega\mu_0 H_r = \frac{1}{r}\frac{\partial E_z}{\partial \phi}$, and writing $\omega\mu_0 r = Z_0 k_0 r$, the radial magnetic field intensity is

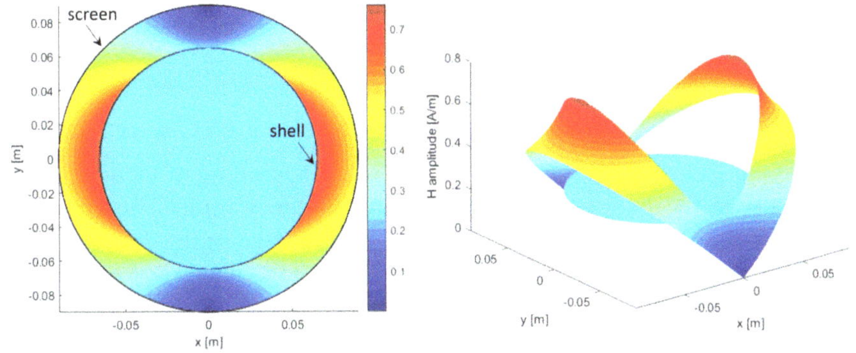

Figure 8.7. Two plots of the magnetic field amplitude $|\mathbf{H}(r,\ \phi)|$ using the expressions (8.31) and (8.33), showing a constant field strength within the shell for the parameters used in figure 8.4. The discontinuity due to the shell current harmonic $m = 1$ is clearly visible at $r = b = 0.065$ m. The same data is shown in two ways; left: surface density plot, right: surface profile plot. The spatial resolution of the calculation is 0.5 mm, but the colour scale is shown with only 30 gradations to aid visualization. The colorbar for $|\mathbf{H}(r,\ \phi)|$ (A m^{-1}) applies to both graphs.

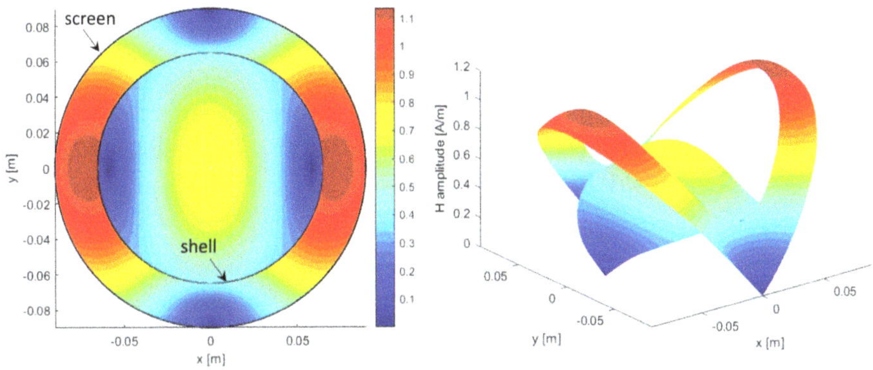

Figure 8.8. Two plots of the magnetic field amplitude $|\mathbf{H}(r,\ \phi)|$ using the same conditions as for figure 8.7, except for a much higher frequency of 1.5 GHz. The magnetic field strength is now non-uniform everywhere.

$$
\begin{aligned}
H_r(r,\ \phi) &= \left[\frac{j}{Z_0 k_0 r}\right] A_1 J_1(k_0 r)\sin\phi \quad (0 \leqslant r \leqslant b), \\
&= \left[\frac{j}{Z_0 k_0 r}\right] [D_1 J_1(k_0 r) + E_1 Y_1(k_0 r)]\sin\phi \quad (b \leqslant r \leqslant c).
\end{aligned}
\tag{8.33}
$$

There is no component of magnetic field along z, which is consistent with the TM mode where $H_z(r,\ \phi) = 0$. The amplitude of the magnetic field strength shown in figure 8.7 remains almost constant within the current shell for frequencies as high as 200 MHz; figure 8.8 shows a high-frequency example of the profile at 1.5 GHz.

The uniformity and direction of the magnetic field is more conveniently shown using the low-frequency limit in the following section.

8.2.1 Low frequency limit for $m = 1$ shell current inside a PEC screen

As stated in the second simplifying conditions at the beginning of section 8.2, we are interested here in the magneto-quasistatic, low-frequency limit. The example used above was for the RAID RF plasma source with birdcage radius $b = 0.065$ m, and screen radius $c = 0.09$ m, at RF frequency 13.56 MHz. The RF free-space wavelength of 22.1 m is much larger than the radii, so $k_0 b = 0.0185 \ll 1$ and $k_0 c = 0.0256 \ll 1$. The small-argument expansions of Bessel functions [4, 6] relevant to (8.27) are $J_1(x) \to \frac{x}{2}$ and $Y_1(x) \to -\frac{2}{\pi x}$. Applying these approximations gives

$$\begin{aligned} A_1 &\to +jZ_0 i_0\left[1 - \frac{b^2}{c^2}\right], \\ D_1 &\to -jZ_0 i_0\left[\frac{b^2}{c^2}\right], \\ E_1 &\to -jZ_0 i_0\left[\frac{\pi k_0^2 b^2}{4}\right]. \end{aligned} \tag{8.34}$$

We are finally in a position to give low-frequency, magneto-quasistatic solutions for the electric and magnetic wavefields in both domains. Inside the current shell, $0 \leqslant r \leqslant b$, taking the real part, (8.30) becomes

$$E_z(r, \phi) \to -Z_0 i_0 \frac{k_0 r \cos\phi}{2}\left[1 - \frac{b^2}{c^2}\right], \tag{8.35}$$

and between the current shell and the screen, $b \leqslant r \leqslant c$,

$$E_z(r, \phi) \to -Z_0 i_0 \frac{k_0 r \cos\phi}{2}\left[\frac{b^2}{r^2} - \frac{b^2}{c^2}\right]. \tag{8.36}$$

For the electric field within the screen, we see from (8.35) that $\mathrm{Re}(E_z)$ is proportional to $r \cos\phi$ which explains the proportionality with $x = r\cos\phi$ in figures 8.4 and 8.9. The effect of the screen is clearly seen in the dependence on c in only the last terms in (8.35) and (8.36), as mentioned previously regarding (8.30) and the term Scrn_1: Letting $c \to \infty$ diminishes these terms to zero, and the wavefields (blue dashed lines in figure 8.9) then correspond to the screen expanded to infinite radius. The presence of the screen, even at infinity, is still necessary because it carries the return current of the shell.

It is likewise instructive to calculate the magneto-quasistatic limit of the magnetic field, which can be obtained from (8.31) and (8.33), or more simply by differentiating the approximate E_z expressions (8.35) and (8.36) directly, again using the Maxwell

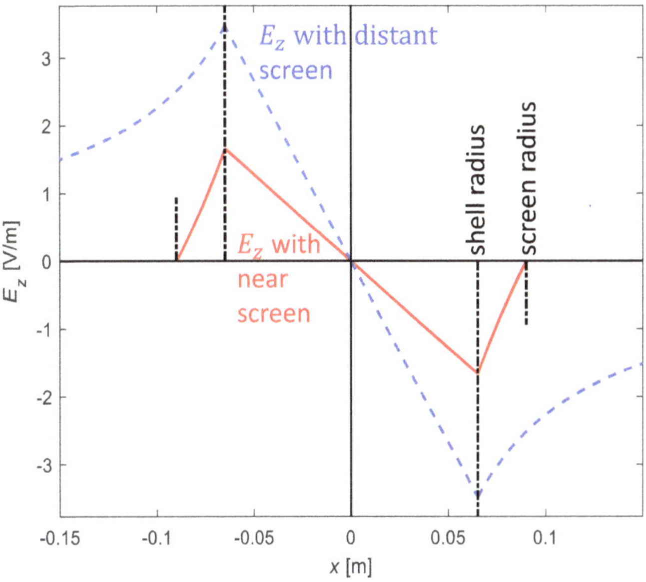

Figure 8.9. The axial vacuum electric field E_z, at $\phi = 0$, along the x-axis, in (8.35) and (8.36) due to a shell current $i_0 \cos\phi$ inside the screen ('E_z with near screen', red solid line), and when the screen is expanded to large radius ('E_z with distant screen', blue dashed line). The parameters correspond to figure 8.4. The low-frequency field is proportional to x; no significant deviation from this solution occurs until frequencies at least ten times higher than the RAID frequency of 13.56 MHz.

equation components $j\omega\mu_0 H_\phi = -\frac{\partial E_z}{\partial r}$ and $j\omega\mu_0 H_r = \frac{1}{r}\frac{\partial E_z}{\partial \phi}$. Inside the shell current the vector magnetic field intensity is (taking the real parts):

$$\mathbf{H} \rightarrow -\frac{i_0}{2}\left[1 - \frac{b^2}{c^2}\right](\hat{\phi}\cos\phi + \hat{r}\sin\phi) = -\frac{i_0}{2}\left[1 - \frac{b^2}{c^2}\right]\hat{y}. \tag{8.37}$$

We have the important result that the magnetic field intensity is constant within the $m = 1$ shell current density $i_0 \cos\phi$, and uniform along the y-axis [4], as shown in figure 8.10. As stated in the preceding section, this uniform-magnetic-field property of the m = 1 normal mode was the historical motivation for closed birdcage MRI applications. These low-frequency electric and magnetic field properties would be more difficult to discern from the exact solutions (8.30) to (8.33).

Between the current shell and the screen, $b \leqslant r \leqslant c$, the vector magnetic field intensity is non-uniform as shown in figure 8.10 (again, taking real parts):

$$\mathbf{H} \rightarrow \frac{i_0\cos\phi}{2}\left[\frac{b^2}{r^2} + \frac{b^2}{c^2}\right]\hat{\phi} - \frac{i_0\sin\phi}{2}\left[\frac{b^2}{r^2} - \frac{b^2}{c^2}\right]\hat{r}. \tag{8.38}$$

We can conveniently verify that the boundary conditions are satisfied for **H** using (8.37) and (8.38). Across the shell current, at $r = b$, the radial magnetic field is continuous, and the discontinuity in the ϕ-component equals $i_0 \cos\phi$, as specified by

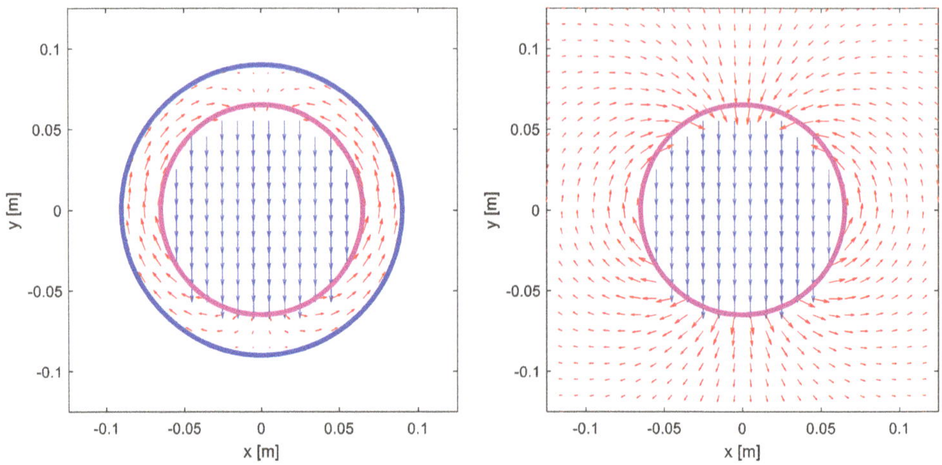

Figure 8.10. The magnetic field intensity $\mathbf{H}(r, \phi)$ inside the shell current $i_0 \cos\phi$, using (8.37) (blue arrows); and between the shell and screen, using (8.38) (red arrows). The magnetic field is uniform within the shell radius $b = 0.065$ m. The parameters correspond to figure 8.4 for the RAID birdcage. Left: for the screen radius $c = 0.09$ m. Right: when the screen is expanded to infinite radius.

the Ampère's law boundary condition (8.24) [4]. At the screen, $r = c$, the radial magnetic field is zero, just like E_z, and the azimuthal magnetic field is $H_\phi^{\text{screen}} = (\frac{b}{c})^2 i_0 \cos\phi$. Since the field outside the PEC Faraday screen must be zero, this means that an induced current density $-(\frac{b}{c})^2 i_0 \cos\phi$ (A m^{-1}) flows parallel to z along the screen's inner surface, thus cancelling the azimuthal magnetic field component.

The approximations in (8.35)–(8.38) are indistinguishable from the corresponding exact solutions (8.30)–(8.33), as shown by the low-frequency figures 8.9 and 8.10. This confirms that the RAID parameters are well within the magneto-quasistatic conditions.

Finally, the $-b^2/c^2$ terms in the magneto-quasistatic approximations show that the proximity of the PEC screen diminishes the electric and the magnetic fields both inside and outside the shell current, going from complete cancellation as $c \to b$, to negligible effect as $c \to \infty$, as expected on the physical grounds of induced currents in the Faraday screen of radius c.

8.2.2 Dynamics of the vacuum wavefields in the low-frequency limit inside an $m = 1$ shell current

This is a suitable point to summarize the relative orientation and time-dependent behaviour of the $m = 1$ shell current and vacuum wavefields in the magneto-quasistatic approximation. Remember that all parameters are represented by phasors modulated in time by the factor $e^{-j\omega t}$ as defined in section 2.1.1, and that the physical values correspond to the real part.

The cartoons in figure 8.11 summarize the space–time dependent vacuum fields within a $m = 1$ shell current $i_s(\phi, t) = \mathrm{Re}[(i_0 \cos\phi)e^{-j\omega t}]$ in (8.17). The top row shows $(i_0 \cos\phi)\cos(\omega t)$. The induced axial electric field E_z along the middle row acts to oppose the changing shell current, in the spirit of Lenz's law. The corresponding transverse magnetic field intensity H_y is shown along the bottom row, for a series of four time points in a half cycle of period τ. The time sequence in the inset cosine wave follows (a) → (b) → (c) → (d) → (c) → (b) → (a) → (b) → etc, i.e. a there-and-back oscillation. Note that all of the parameters are simple back-and-forth oscillations in time; there is no rotation.

As stated for the general solution in section 8.1, TM modes have $B_z = 0$ by definition of 'transverse magnetic'. These TM modes are also called E-waves [5] because the only oscillating parameter along the z-axis is E_z, but this does not mean that a transverse electric field cannot exist. Using Maxwell's equation $\nabla \times \mathbf{H} = -j\omega\varepsilon_0\mathbf{E}$, and simplifying to a uniform magnetic field H_y in view of figure 8.11, does in fact yield

$$E_x = \frac{1}{j\omega\varepsilon_0}\frac{\partial H_y}{\partial z} = \frac{k_z}{\omega\varepsilon_0}H_y \to 0, \tag{8.39}$$

because the RF currents are directed uniquely along z, hence $\frac{\partial}{\partial z} = jk_z \to 0$ for the infinitely long shell current. In this particular case, the electric field is directed only along z, i.e. the only component is E_z.

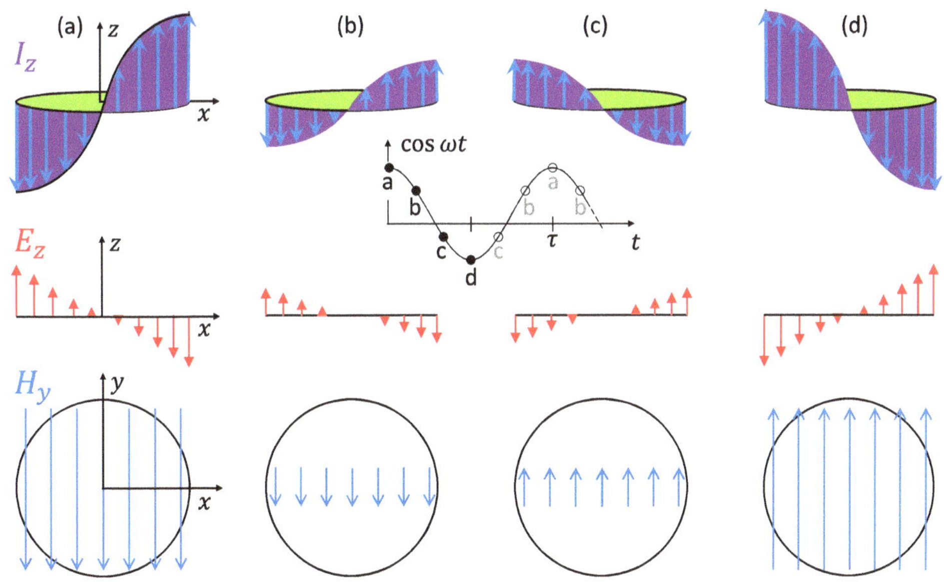

Figure 8.11. A cartoon of time- and space-dependence of the fields inside an $m = 1$ shell current at low frequency, for a half-period $\tau/2$. Top row: The shell current I_z (the colour scheme as in figure 8.2). Middle row: The see-saw of the plane of the induced E_z field (figures 8.4 and 8.9). Bottom row: Oscillation of the uniform magnetic field intensity H_y in the $\{x, y\}$ plane (figure 8.10). Column (a) $t = 0$; column (b) $t = \tau/6$; column (c) $t = \tau/3$; column (d) $t = \tau/2$, as indicated in the inset of $\cos(\omega t)$ versus t.

8.3 Vacuum wavefields of a birdcage within a PEC screen

Up to now, the seemingly artificial case of a single-harmonic shell current in section 8.2 was used to illustrate the solutions to the cylindrical wavefield equations in a convenient manner. The orthogonality relation was used to restrict the infinite number of m harmonic wavefield solutions in (8.25) to the unique shell harmonic current $m = 1$ in (8.26). This is to say that the $i_0 \cos \phi$ shell current imposed wavefields corresponding only to its $m = 1$ harmonic. By inspection, this approach applies to any order m for shell current $i_0 \cos m\phi$, and by extension, to any combination of shell current harmonics. Therefore, the wavefield due to any general current distribution on the shell radius can be constructed from a Fourier harmonic series which represents the required driving current at $r = b$.

8.3.1 Surface current density due to a wire current on the shell radius

For a specific example, we consider a single, off-axis wire parallel to the z-axis inside the screen. This wire current can also be called a filament current or a line current. We assume a current filament positioned at radius b and angle ϕ_b as shown in figure 8.12. The shell surface current density $i_s(\phi)$ was already described by a radial Dirac delta function $\delta(r - b)$ at $r = b$ in section 8.1.1. The wire is now defined by employing a second Dirac delta function at angle $\phi = \phi_b$, say, $i_s^{\text{wire}}(\phi) = K\delta(\phi - \phi_b)$. The product of the two delta functions defines the vanishingly thin current filament at radius b, angle ϕ_b. Substituting in (8.15) gives

$$I_z[\mathrm{A}] = \int_0^{2\pi} i_s^{\text{wire}}(\phi)\, bd\phi = \int_0^{2\pi} K\delta(\phi - \phi_b) b\, d\phi = bK \quad \text{at} \quad \phi = \phi_b, \tag{8.40}$$

hence $K = I/b$, where $I = I_z$ is the current carried by the wire parallel to the z-axis. Therefore, the effective shell surface current density for a wire current I at position $(r, \phi) = (b, \phi_b)$ is [5]

$$i_s^{\text{wire}}(\phi) = I\delta\left(\phi - \phi_b\right)/b. \tag{8.41}$$

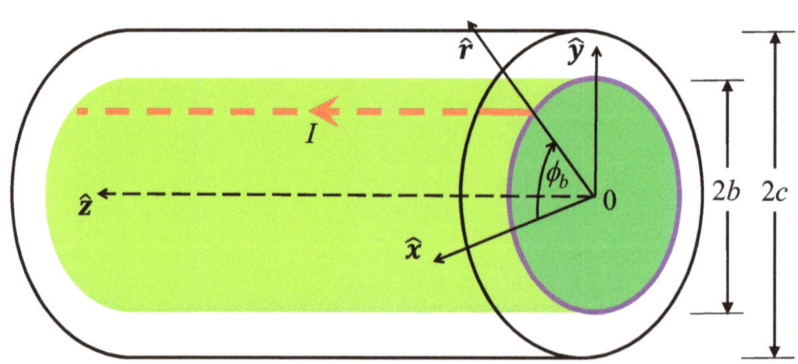

Figure 8.12. A wire filament (red), located at radius b and azimuthal angle ϕ_b in a vacuum, carrying current I, enclosed by a cylindrical PEC screen of radius c.

8.3.2 Vacuum wavefields for a wire current inside a PEC screen

The wire current expression $i_s^{\text{wire}}(\phi) = I\delta(\phi - \phi_b)/b$ now replaces the first harmonic shell current $i_0 \cos\phi$ in (8.25). In the passage from (8.25) to (8.26), the orthogonality relation (8.22) is applied by multiplying each equation by $e^{-jm\phi}$ and integrating over ϕ from $-\pi$ to π. Applying this to the wire current gives

$$\int_{-\pi}^{\pi} \frac{I}{b}\delta(\phi - \phi_b)e^{-jm\phi}\, d\phi = \frac{I}{b}e^{-jm\phi_b}. \tag{8.42}$$

Consequently, every mth harmonic now contributes to the wavefield, in contrast to the single first harmonic shell current in section 8.2. The infinite sum of the Fourier harmonics represents the Dirac delta current. Note that, although the wire is infinitely thin, this does not cause a singularity difficulty for the boundary conditions because the wire is treated as a sum of shell current harmonics, and these boundary conditions were previously described for a shell current at radius b, using $m = 1$ as an example. Replacing the first harmonic amplitude i_0 by every harmonic $\frac{I}{b}e^{-jm\phi_b}$, the boundary conditions (8.21), (8.23), and (8.25) summarized in (8.26) can be re-written by inspection as

$$\begin{aligned} A_m J_m(k_0 b) - D_m J_m(k_0 b) - E_m Y_m(k_0 b) &= 0, \\ + D_m J_m(k_0 c) + E_m Y_m(k_0 c) &= 0, \\ A_m J_m'(k_0 b) - D_m J_m'(k_0 b) - E_m Y_m'(k_0 b) &= j\frac{Z_0}{2\pi}\frac{I}{b}e^{-jm\phi_b}, \end{aligned} \tag{8.43}$$

where the factor 2π appears in the denominator because (8.42) does not include 2π in the conventional orthogonality relation (8.22) which applies to all of the other terms. From these simultaneous equations, by analogy with the solution (8.27), the coefficients are directly

$$\begin{aligned} A_m &= -jZ_0 I e^{-jm\phi_b}\left[\frac{k_0}{4}\right][Y_m(k_0 b) - \text{Scrn}_m], \\ D_m &= +jZ_0 I e^{-jm\phi_b}\left[\frac{k_0}{4}\right]\text{Scrn}_m, \\ E_m &= -jZ_0 I e^{-jm\phi_b}\left[\frac{k_0}{4}\right]J_m(k_0 b), \\ \text{where}\quad \text{Scrn}_m &= \frac{J_m(k_0 b)\, Y_m(k_0 c)}{J_m(k_0 c)}. \end{aligned} \tag{8.44}$$

The solutions for the E_z phasor are straightforwardly found by substitution into equations (8.19) and (8.20):

$$\begin{aligned} E_z(r, \phi) &= \sum_{m=-\infty}^{\infty} A_m J_m(k_0 r)e^{jm\phi} \quad (0 \leqslant r \leqslant b), \\ &= \sum_{m=-\infty}^{\infty} [D_m J_m(k_0 r) + E_m Y_m(k_0 r)]e^{jm\phi} \quad (b \leqslant r \leqslant c), \end{aligned} \tag{8.45}$$

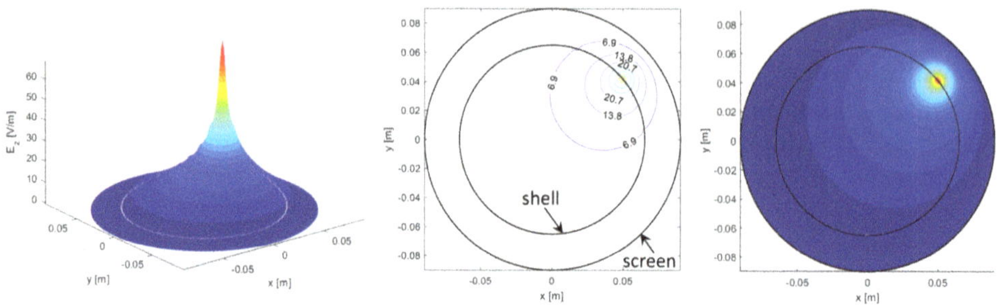

Figure 8.13. Three plots of the axial electric field E_z (real value) using (8.45) for a wire current at $(r, \phi) = (b, \phi_b)$=(0.065 m, 40°) within a screen radius $c = 0.09$ m. For display purposes, the unit current phasor is set as $I = -j = e^{-j\pi/2}$ (A). Left: Three-dimensional surface plot, middle: normalized contour plot, right: surface density plot. The parameters correspond to RAID. The spatial resolution of the calculation is 0.5 mm, but the colour scale is shown with only 30 gradations to aid visualization.

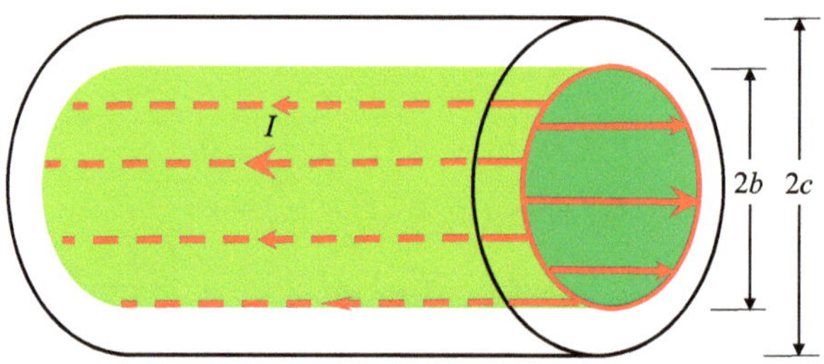

Figure 8.14. Schematic of a birdcage with N legs (red lines) positioned at regular angular intervals on radius b, in s vacuum, enclosed by a cylindrical PEC screen of radius c. Each wire carries a current $I_0 \cos\phi$ (A), according to its angle ϕ, to excite the azimuthal periodicity $m = 1$.

and the resultant E_z field is shown for RAID dimensions in figure 8.13. The effect of the screen, represented by the Scrn_m terms, can be shown explicitly by writing out (8.45) more fully in the same way as for the shell current, (8.30).

8.3.3 Vacuum wavefields for a birdcage within a screen

Figure 8.14 shows a schematic of birdcage legs carrying currents $I_0 \cos\phi$ (A) in vacuum within a screen. The electric and magnetic fields are calculated directly from the wire filament solution in section 8.3.2 by superposing the solutions of each wire according to its angle ϕ_b and corresponding current $I_0 \cos\phi_b$, to mimic the $m = 1$ distributed shell current, $i_0 \cos\phi$. Figure 8.15 shows that a birdcage with 18 legs reproduces the electric field of the $m = 1$ shell current in figure 8.4 with fair accuracy. This is consistent with the 'infinite number of separate parallel wires' concept of a shell current in section 8.1.1. On the other hand, figure 8.16 for nine legs exhibits considerable distortion in the neighbourhood of each leg singularity. Figure 8.17 confirms the partial loss of magnetic field uniformity inside the birdcage, when comparing 9 legs to 18 legs. However, given the technical economy of only nine legs, and, in any case, the expected field distortion due to the presence

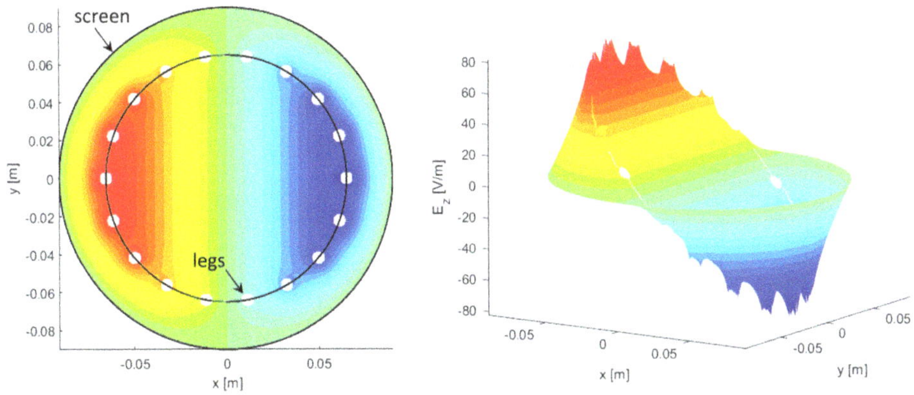

Figure 8.15. Two plots of the vacuum electric field E_z for an 18-leg birdcage of $b = 0.065$ m radius, within a PEC screen of radius $c = 0.09$ m. The leg radius is 3 mm and the leg current amplitude is $I_0 \cos \phi_b$, with $I_0 = 1$ A. The E_z linearity along x is to be compared with figure 8.4 for a $m = 1$ distributed shell current. Left: surface density plot, right: three-dimensional surface plot. The parameters correspond to RAID (chapter 12), but for twice as many legs. The spatial resolution of the calculation is 0.5 mm, but the colour scale is shown with only 30 gradations to aid visualization.

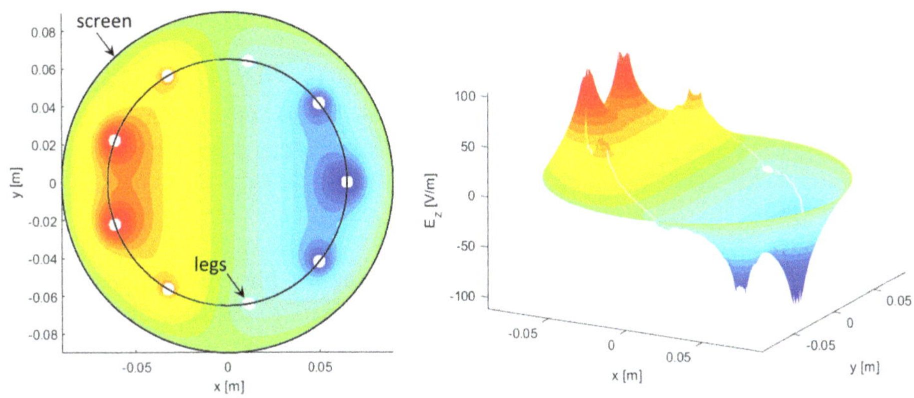

Figure 8.16. Two plots of the vacuum electric field E_z for a nine-leg birdcage within a PEC screen radius $c = 0.09$ m, where the E_z linearity is less accurate compared to figures 8.4 and 8.15. The leg current is doubled for comparison with 18 legs, because there are half as many legs. Left: surface density plot, right: three-dimensional surface plot. The parameters correspond to the RAID birdcage in chapter 12.

of plasma, it appears that a nine-leg birdcage is an acceptable compromise for exciting the azimuthal periodicity $m = 1$ in the RAID birdcage antenna (see chapter 12).

In these birdcage figures, note that the fields are calculated according to the wire filament solutions (this was already the case for mutual partial inductance expressions in chapter 4); the field plots are simply truncated at the real leg radius of 3 mm, as far as the 0.5 mm grid resolution allows, and the solution is not representative on the scale of the leg radius. The consideration, in section 8.1.1, of the transparency of a shell current to electromagnetic fields, and the concomitant continuity of E_z,

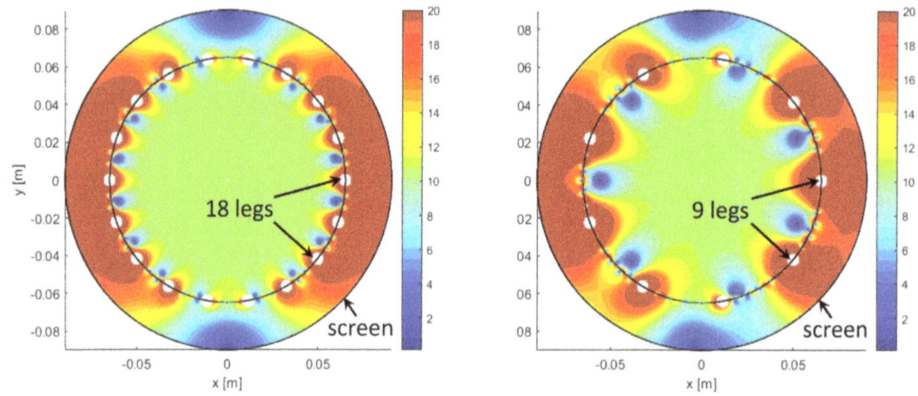

Figure 8.17. Comparison of normalized contour plots of magnetic field intensity magnitude for the same conditions as figures 8.15 and 8.16. Left: birdcage with 18 legs. Right: birdcage with nine legs and double the current to account for half as many legs. There is an arbitrary cap at $|H| = 20$ (A m^{-1}) to better show the contours of non-uniformity. Unsurprisingly, the magnetic field uniformity within the leg shell radius is less accurate for nine legs than for 18 legs, when compared to the ideal case in figure 8.7.

explains why we have not set $E_z = 0$ at the surface of each metal leg. The small number of legs is a poor approximation of the infinite legs necessary to represent a distributed shell current.

To close this section, we see that the solutions in section 8.2 have wide and general applicability, ranging from any single-harmonic shell current, through single wires, to birdcage legs. Only the first harmonic shell current $i_0 \cos\phi$ has the property of a uniform magnetic field, although a m = 1 normal mode on a closed birdcage (or normal mode m = 2 on an open birdcage—see chapter 9) gives a good approximation which improves with the number of birdcage legs [4]. Nevertheless, the $m = 1$ shell current density represents a good proxy for the wavefields in a m = 1 closed birdcage, without the time-consuming complexity of calculating the fields of N individual legs. We will therefore use an $m = 1$ shell current to evaluate the effect of plasma on the wavefield structure of a birdcage in the following section 8.4.

8.4 Plasma coupling by a shell current within a PEC screen

The magnetic field uniformity of $m = 1$ azimuthal periodicity on a birdcage antenna in vacuum was demonstrated in the previous sections. It was also shown that a single-harmonic shell current solution is a good approximation to a birdcage solution, hence we will continue with the shell current method, instead of a birdcage, for ease and speed of calculation. In the next step towards a representation of a birdcage inductive RF plasma source, a uniform dielectric is introduced inside the $i_0 \cos\phi$ shell current [4, 9], as shown in figure 8.18. A dielectric–vacuum interface now occurs at $r = a$ with a vacuum radial gap for $a < r < b$ which is intended to represent experiments where the birdcage antenna is separated from the plasma by an air gap and a dielectric wall [10]. For analytical convenience, we make no distinction here between the vacuum gap in this model, and the dielectric wall-plus-

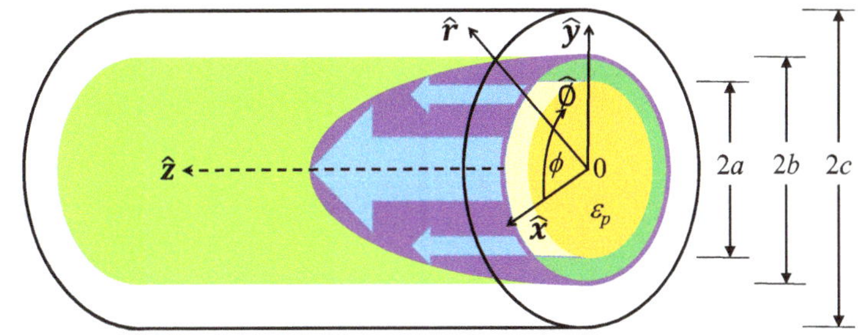

Figure 8.18. A cylindrical plasma of uniform relative permittivity ε_p, radius a, enclosed by a $i_0 \cos \phi$-shell current of radius b, within a cylindrical PEC screen of radius c [4] (transparent in this figure).

air gap in the experiments. This leaves one dielectric core domain and two concentric annular vacuum domains—one inside and one outside the shell current—as shown in the figure (there is no particular gain in algebraic simplicity by first considering the dielectric to completely fill the shell current). The oscillating fields generated by the shell current induce a current in the dielectric medium, which is the principle of an inductively coupled (H-mode) plasma source.

In the Helmholtz equation (8.2), the wavenumber k_d in the plasma effective dielectric is given by $k_d^2 = k_0^2 \varepsilon_p$, which accounts for both the vacuum displacement current and the plasma conduction current as described in appendix D.3. Continuing to assume negligible overall z variation so that $k_z \to 0$ and $k_\perp \to k_d$, the Bessel equation (8.7) and solution (8.8) in the dielectric has solutions $J_m(k_d r)$ and $Y_m(k_d r)$. Following the treatment in section 8.2, the E_z wave functions in the three domains can be directly written as $AJ_1(k_d r)$ for the dielectric, $[BJ_1(k_0 r) + CY_1(k_0 r)]$ for the vacuum between dielectric and shell current, and $[DJ_1(k_0 r) + EY_1(k_0 r)]$ for the vacuum between shell and screen [4]. For the five unknowns A, B, C, D, E, there are five boundary conditions corresponding to the continuity of E_z at $r = a$ and $r = b$; $E_z = 0$ at $r = c$; continuity of H_ϕ at $r = a$; and the discontinuity in H_ϕ equal to the shell current density at $r = b$. In this order, we have [4, 9]

$$
\begin{aligned}
AJ_1(k_d a) - BJ_1(k_0 a) - CY_1(k_0 a) &= 0, \\
BJ_1(k_0 b) + CY_1(k_0 b) - DJ_1(k_0 b) - EY_1(k_0 b) &= 0, \\
DJ_1(k_0 c) + EY_1(k_0 c) &= 0, \\
Ak_d J_1'(k_d a) - k_0 BJ_1'(k_0 a) - k_0 CY_1'(k_0 a) &= 0, \\
BJ_1'(k_0 b) + CY_1'(k_0 b) - DJ_1'(k_0 b) - EY_1'(k_0 b) &= jZ_0 i_0.
\end{aligned}
\tag{8.46}
$$

Note that the fourth boundary condition, for continuity of H_ϕ at $r = a$, makes a distinction between current flow at the edge of the plasma dielectric cylinder, and current flow on the shell at $r = b$ in the fifth boundary condition. The shell current is located on an infinitely thin surface, causing a mathematical discontinuity in an azimuthal magnetic field, whereas the current on the plasma dielectric is radially distributed due to the skin depth for induced currents in a resistive conductor. This volume distribution is then described by a gradual spatial (radial) variation in the magnetic wavefield, and not by a surface discontinuity; hence the continuity of H_ϕ at $r = a$ in the fourth boundary condition.

Following the same procedure as in section 8.2, the wavefields in the plasma dielectric, $0 \leqslant r \leqslant a$, are directly [4]

$$
\begin{aligned}
E_z(r, \phi) &= AJ_1(k_d r)\cos\phi, \\
H_\phi(r, \phi) &= A\left[\frac{jk_d}{\omega\mu_0}\right]J_1'(k_d r)\cos\phi, \\
H_r(r, \phi) &= A\left[\frac{j}{\omega\mu_0 r}\right]J_1(k_d r)\sin\phi.
\end{aligned}
\tag{8.47}
$$

To plot the wavefields over all radii up to the screen at $r = c$, all five constants are required, and these are calculated in appendix I.

For sufficiently low-frequency and small relative permittivity so that $k_d a = 2\pi f a\sqrt{\varepsilon_p}/c \ll 1$, the small-argument-expansion of (8.47) has the same spatial dependence as (8.37), hence the magneto-quasistatic limit still exhibits a uniform magnetic field in the presence of a quasi-vacuum-dielectric core as in figure 8.7 [9]. Similarly, at typical RF frequencies and low plasma density ($n_{e0} < 10^{15}$ m^{-3}), the electric field within the shell is still linear in x as in figure 8.4. However, for moderately dense plasmas, $n_{e0} > 10^{16}$ m^{-3}, the large real and/or imaginary values of $k_d = k_0\sqrt{\varepsilon_p}$ mean that uniformity is already lost even for RF frequencies as low as 13.56 MHz. In fact, the large value of wavenumber in the plasma means that $k_d r$ is no longer much less than one, so the small-argument expansions in section 8.2.1 cannot be used. To quantitatively interpret the effect of plasma, it is necessary to derive an explicit expression for $k_d^2 = k_0^2\varepsilon_p$ for which we can use the cold plasma relative permittivity approximation, namely

$$
\varepsilon_p = 1 - \frac{\omega_{\text{pe}}^2}{\omega(\omega + j\nu)}, \tag{8.48}
$$

as derived in appendix D.3. The plasma dielectric is required to be uniform in order for the separation-of-variables method to apply, as explained in section 8.1. This is admittedly a very poor approximation for the strong density gradients in a plasma, but the aim is to bring some physical insight using this analytical approach. We see from (8.48) that ε_p is complex due to the collision frequency ν, but more importantly, from section 5.1.2 we know that $\omega \ll \omega_{\text{pe}}$, so ε_p is strongly negative in these low pressure plasmas, and the effective wavenumber $k_d = k_0\sqrt{\varepsilon_p}$ is large and principally imaginary. This is mainly due to the mobile electrons, i.e. the electrical conductivity of the plasma; the RF radiating fields are cut off at the plasma boundary where their power is absorbed and reflected (see appendix D.5.4). In fact, the field confinement, by reflection within the PEC screen, means that the self-consistent steady-state solution has all of the RF power absorbed in the plasma, because absorption is impossible elsewhere in the system. The relation between plasma conductivity σ_p and the plasma complex relative permittivity ε_p is given in appendix D.3. The induced electric fields drive plasma currents which gradually screen the wavefields from penetrating the plasma; they are attenuated by a factor $1/e$ over each skin depth

δ inward across the plasma radius, degrading the field uniformity. From appendix D.5.1, the skin depth is related to the plasma wavenumber by $\delta = 1/\mathrm{Im}(k_d)$. Electron collisions with the background gas cause elastic and inelastic collisions (appendix C), dissipating the RF power. This, of course, is the basis of the ICP, H-mode reactors previously discussed in chapter 5. The skin depth is introduced for plane waves (suitable for ladder resonant antennas) in appendix D.5. Thanks to the basic wavefield model presented here, the effect of skin depth can now also be calculated for a cylindrical plasma inductively sustained by a birdcage antenna.

For the series of figures 8.19–8.22, we fix the RF frequency at 13.56 MHz and the pressure at 1 Pa, then change the order of magnitude of the (uniform) electron density

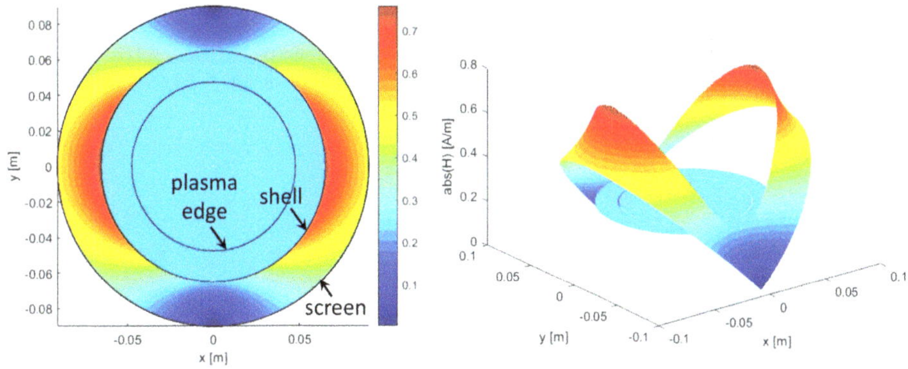

Figure 8.19. Analytical calculation of the magnetic field strength magnitude $|\mathbf{H}(r, \phi)|$ within the system of figure 8.18 for shell current amplitude $i_0 = 1$ A m^{-1}. The radii correspond to RAID, namely plasma $a = 0.0475$ m, shell $b = 0.065$ m, PEC screen $c = 0.09$ m. Argon pressure 1 Pa; electron density $n_{e0} = 10^{15}$ m^{-3}. For this low density, the field is essentially the same as in a vacuum, see figure 8.7. Left: surface density plot; right: surface profile plot. The spatial resolution of the calculation is 0.5 mm, but the colour scale is shown with only 30 gradations to aid visualization. The colorbar for $|\mathbf{H}(r, \phi)|$ (A m^{-1}) applies to both graphs. The skin depth $\delta = 17$ cm, is much longer than the plasma radius, $\delta/a = 3.7$.

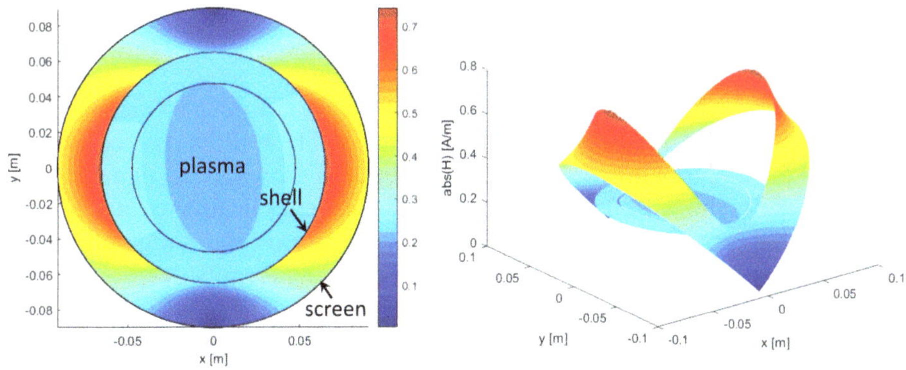

Figure 8.20. Same parameters as for figure 8.19 except for 10× higher electron density, $n_{e0} = 10^{16}$ m^{-3}. The magnetic field in the plasma has become noticeably non-uniform; The skin depth $\delta = 5.5$ cm is somewhat longer than the plasma radius, $\delta/a = 1.16$.

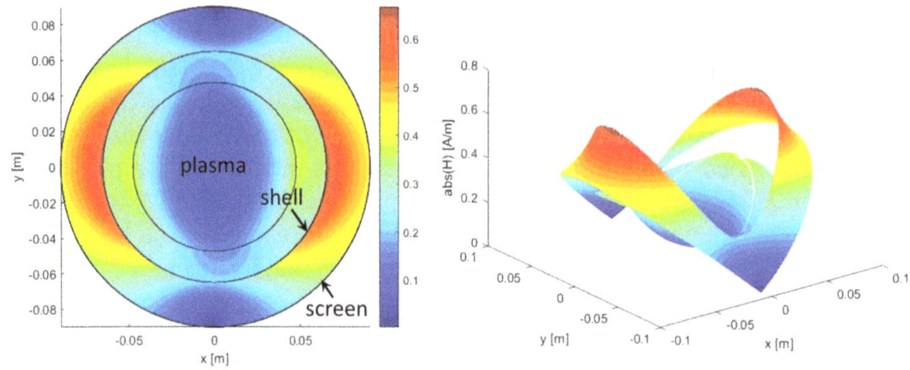

Figure 8.21. Same parameters as for figure 8.19 except for 100× higher electron density, $n_{e0} = 10^{17}$ m^{-3}. The magnetic field on axis is strongly reduced but not zero. Note the field continuity across the plasma boundary in all of these figures, as imposed by the boundary conditions (8.46). The skin depth $\delta = 1.7$ cm, is slightly more than one third of the plasma radius, $\delta/a = 0.37$.

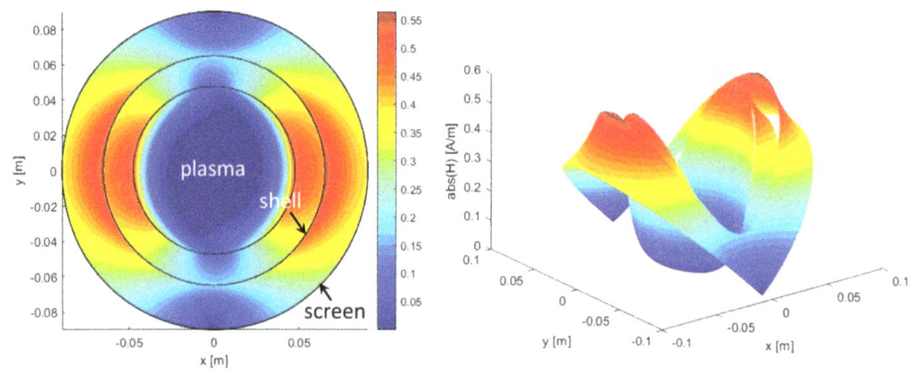

Figure 8.22. Same parameters as for figure 8.19 except for 1000× higher electron density, $n_{e0} = 10^{18}$ m^{-3}. The magnetic (and electric) field is now completely screened from a significant central zone of the plasma region. The skin depth $\delta = 0.55$ cm is just 12% of the plasma radius, $\delta/a = 0.12$.

in four logarithmic steps from $n_{e0} = 10^{15}$ m^{-3} to 10^{18} m^{-3}. Taking an electron–argon collision frequency of $\nu(\text{rad s}^{-1}) = 2.6 \cdot 10^7 \cdot p(\text{Pa})$ [11] at 1 Pa, the equations in appendix D.5.1 are used to calculate the skin depths δ shown in the figure captions. With these reactor dimensions, at 13.56 MHz RF frequency in absence of plasma, the wavefield radial profiles would correspond to the magneto-quasistatic uniformity in section 8.2.1. The figures show how the magnetic field uniformity deteriorates as the skin depth becomes a smaller and smaller fraction of the plasma radius, $\delta/a < 1$. At the highest density, $n_{e0} = 10^{18}$ m^{-3} in figure 8.22, the skin depth $\delta = 0.55$ cm, $\delta/a = 0.12$, and the magnetic field is seen to be excluded from most of the plasma volume, so ICP heating can only occur in the outer edge of the plasma.

8.4.1 Inductive power transfer in a birdcage antenna

Up to now, the field solutions have been shown to satisfy the birdcage boundary conditions, and give qualitatively reasonable pictures of the skin effect. The electric and magnetic field solutions can also be used to estimate the power transfer from the birdcage antenna to the plasma by inductive coupling; this is an electromagnetic description of the H-mode ICP mechanism. The directed power flux, $\mathbf{S}$ ($\mathrm{W\,m^{-2}}$), or the rate of energy flow per unit area, is given by Poynting's vector [2]

$$\mathbf{S} = \mathbf{E} \times \mathbf{H}, \tag{8.49}$$

which represents the instantaneous, time-dependent power density transmitted by the oscillating electromagnetic fields. Since the magnetic field is purely transverse in this small-reactor model, the radial power flux S_r from the antenna to the plasma can only come from the cross-product of the E_z and H_ϕ field components. As noted at the end of section 2.1.1, the phasor description is a useful 'shorthand' notation only for linear combinations of parameters; Poynting's vector is a non-linear combination and so must be described using the real values directly, i.e.

$$S_r(r, \phi, t) = \mathrm{Re}\,[E_z(r, \phi, t)] \cdot \mathrm{Re}\left[H_\phi(r, \phi, t)\right], \tag{8.50}$$

where it is understood from the orientation of the components in figure 8.18 that $S_r(r, \phi, t)$ is directed inwards, along $-\hat{r}$. We are interested in the time-averaged power flow $\bar{S}_r(r, \phi)$, not the back-and-forth energy oscillations in time, and it so happens that our final expression is

$$\bar{S}_r(r, \phi) = \frac{1}{2}\,\mathrm{Re}\left[E_z\, H_\phi^*\right], \tag{8.51}$$

where H_ϕ^* is the complex conjugate of H_ϕ; see appendix D.2.1 and (D.20) for explanation of the algebra.

For a preliminary physical interpretation of (8.51), we calculate the inward radial power flux for the vacuum field solutions E_z (8.29) and H_ϕ (8.32), for example, at $0 \leqslant r \leqslant b$,

$$\begin{aligned}
\bar{S}_r^{\,\mathrm{vac}}(r, \phi) &= \frac{1}{2}\,\mathrm{Re}\left[(A_1 J_1(k_0 r)\cos\phi) \cdot \left(\frac{j}{Z_0} A_1 J_1'(k_0 r)\cos\phi\right)^*\right] \\
&= \frac{1}{2Z_0}|A_1|^2 J_1(k_0 r) J_1'(k_0 r)\cos^2\phi \cdot \mathrm{Re}[-j] \\
&= 0.
\end{aligned} \tag{8.52}$$

As expected, the time-averaged radial power flux is zero in the vacuum case; this is because E_z and H_ϕ are in time phase quadrature.

For the time-averaged inward radial power flux for plasma, $0 \leqslant r \leqslant a$, the wavefields in (8.47) give

$$\begin{aligned}
\bar{S}_r^{\text{plasma}}(r,\phi) &= \frac{1}{2}\,\mathrm{Re}\left[(AJ_1(k_d r)\cos\phi)\cdot\left(A\left[\frac{jk_d}{\omega\mu_0}\right]J_1'(k_d r)\cos\phi\right)^*\right] \\
&= \frac{1}{2\omega\mu_0}|A|^2\cos^2\phi\cdot\mathrm{Re}\left[J_1(k_d r)(-jk_d^*)J_1'(k_d^* r)\right] \\
&= \bar{S}(r)\cdot 2\cos^2\phi, \\
&= \bar{S}(r)\cdot[1+\cos(2\phi)],
\end{aligned} \tag{8.53}$$

where $\bar{S}(r)$ is the average power flux over the circumference at radius r in the plasma. The inward power flux therefore has a $m = 2$ azimuthal periodicity as shown in figure 8.23(a).

Several physical observations concerning the ICP H-mode coupling in figure 8.23 are worthy of comment:

- The total power flux radiated towards the screen is zero in figure 8.23(a) because there is no power sink outside the shell current, so all power is reflected with no loss. However, there is a net power flux radiated across the vacuum gap, without loss, towards the plasma region, where the power flux diminishes as it traverses the resistive dielectric.
- The power radial flux decreases strongly within the plasma radius, and reaches zero on axis, irrespective of whether the skin depth is much longer, or much shorter, than the plasma radius. The power must fall to zero on axis by conservation of energy flux into a vanishingly small volume. This also applies to the solenoid ICP reactor ($m = 0$) in figure 5.3. Physically, when the skin depth is longer than the plasma radius, the plasma is only weakly absorbing, and the transmitted power is reflected on axis, resulting in a low power efficiency. When the skin depth is very short, the plasma

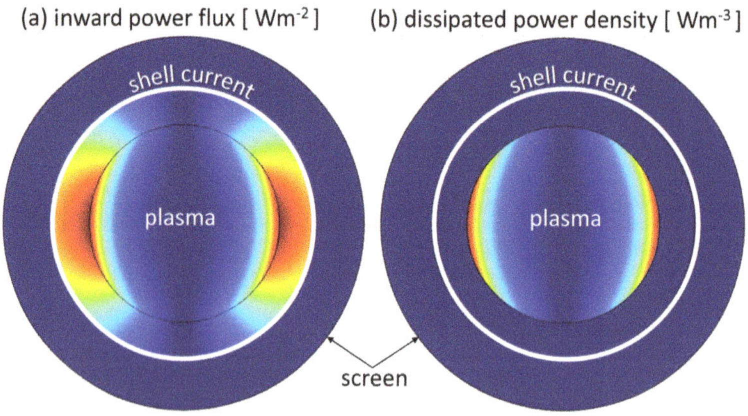

Figure 8.23. (a) The time-averaged power flux $\bar{S}_r(r,\phi) = \frac{1}{2}\,\mathrm{Re}(E_z\,H_\phi^*)$ (W m^{-2}) radiated inwards by the wavefields from the oscillating shell current (white circle). (b) The time-averaged power density $\frac{1}{2}\,\mathrm{Re}(E_z J_z^*)$ (W m^{-3}) dissipated in the plasma by the same wavefields. Parameters as for figure 8.21 with $n_{e0} = 10^{17}$ m^{-3}. The skin depth $\delta = 1.7$ cm is 37% of the plasma radius.

behaves as a conductor, reflecting power back to the source from its surface.

- Figure 8.23(b) shows the ohmic power density dissipated in the plasma region. The power dissipation is, of course, zero in the vacuum everywhere else. For reference, the radiated power flux in figure 8.23(a) and the ohmic power dissipation density in figure 8.23(b) are related by Poynting's theorem [3]. The $m = 2$ azimuthal symmetry is characteristic of the 'bean' plasma emission observed experimentally in figure 12.3 for plasma inductively sustained by a $m = 1$ shell current, ($i_0 \cos \phi$). This is because the current amplitude is maximum on both sides of the plasma, at $\phi = 0$ and $\phi = \pi$.

8.5 Image method for birdcage antennas

The interest of the image method is in the ease of calculation of the mutual partial inductance between the elements of an antenna, and between the antenna, its screen, and its inductively coupled plasma. As a reminder, the mutual partial inductance matrix is used to calculate the impedance loading curve for the design and interpretation of resonant antennas. The mutual partial inductances between wires and their filament images in plane screens and plasma are conveniently calculated using analytical formulas, rather than by performing an electromagnetic simulation for every set of parameters. The classical image theory for an ideal PEC plane screen was shown in sections 4.7.1 and 5.4.3, and the complex image method for wires parallel to a plane plasma slab was demonstrated in section 6.4 and appendix E. Those image models are appropriate for ladder resonant antennas. The aim in this section is to extend the image methods to birdcage antennas within a PEC cylindrical screen, and the holy grail is to find an analogous complex image method for birdcage legs parallel to a cylindrical plasma, which is the ultimate hurdle for determining birdcage coupling to an RF plasma. We approach this task in two steps:

- The classical Apollonius circle method is introduced to find the conjugate image of a line current, or a line charge, parallel to the axis of an ideal PEC cylindrical screen. This can be extended to a complete birdcage.
- Finally, we consider the possible existence of an analogous complex image method for cylindrical plasma sustained by inductive coupling inside a birdcage antenna. Extrapolating from the case for plane plasmas in appendix E, we anticipate that an approximation—if it exists—would incorporate a complex depth in the classical Apollonius image method.

8.5.1 Apollonius image current for a wire in a PEC cylindrical screen

To launch the discussion of the image current of a wire current inside a cylindrical screen, we will calculate the magnetic vector potential for a pair of infinitely long wires in absence of a screen, figure 8.24. The divergence difficulty of the magnetic vector potential for a single infinite wire was mentioned in

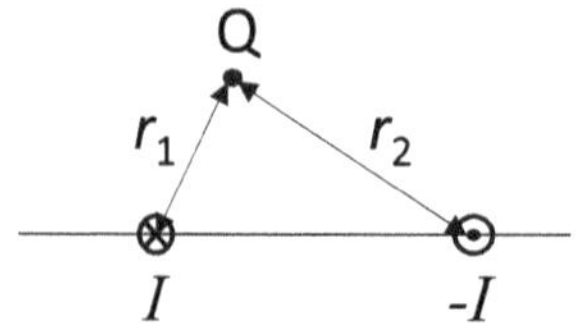

Figure 8.24. Two infinitely long wire currents with equal and opposite currents I and $-I$, which are parallel to the z-axis. The magnetic vector potential A_z is calculated at point Q.

section 4.4.1, and this question will be resolved here. First, we re-write (4.18) for A_z at a point P distance r_1 from a wire, length l, $z_p = 0$, without loss of generality:

$$\begin{aligned} A_z(r_1) &= \frac{\mu_0 I}{4\pi}\left[\sinh^{-1}\left(\frac{l/2}{r_1}\right) - \sinh^{-1}\left(\frac{-l/2}{r_1}\right)\right], \\ &= \frac{\mu_0 I}{2\pi}\left[\sinh^{-1}\left(\frac{l/2}{r_1}\right)\right], \\ &= \frac{\mu_0 I}{2\pi}\left[\ln\left(\frac{l/2}{r_1} + \sqrt{\left(\frac{l/2}{r_1}\right)^2 + 1}\right)\right], \\ &\to \frac{\mu_0 I}{2\pi}\left[\ln\left(\frac{l}{r_1}\right)\right] \quad \text{for} \quad l \gg r_1, \end{aligned} \tag{8.54}$$

using the fact that $\sinh^{-1}(x)$ is an odd function, then substituting the identity $\sinh^{-1}(x) \equiv \ln(x + \sqrt{x^2 + 1})$. This expression does indeed diverge as the wire length tends to infinity, $l \to \infty$, as noted in section 4.4.1. But if we now return to the two wires in figure 8.24 and sum A_z at point Q by superposition, we directly obtain

$$A_z(\text{Q}) = \frac{\mu_0 I}{2\pi}\left[\ln\left(\frac{l}{r_1}\right) - \ln\left(\frac{l}{r_2}\right)\right] = \frac{\mu_0 I}{2\pi}\left[\ln\left(\frac{r_2}{r_1}\right)\right], \tag{8.55}$$

which has no divergence for the pair of infinite wires[1]. The problem in (8.54) was that the single wire imposed a non-physical net source of current I to and from infinity, whereas the go-and-return currents in figure 8.24 sum to zero[2]. Currents cannot be considered in isolation; current conservation requires specification of the complete circuit.

Consider now the shape of a surface having constant magnetic vector potential A_z for the case in figure 8.24. According to (8.55), this surface is defined by a fixed ratio

---

[1] An alternative derivation [12] begins with Biot–Savart's law for an infinite wire, $\mathbf{B} = \frac{\mu_0 I}{2\pi r}\hat{\phi} = -\frac{dA_z}{dr}\hat{\phi}$, whence $A_z = -\frac{\mu_0 I}{2\pi}\ln r +$ a constant of integration.

[2] This problem of infinite wires is more usually encountered when calculating the electric potential of an infinitely long line charge [2], where zero potential at infinity is incompatible with the non-zero charge of a single line; see also appendix G.1.

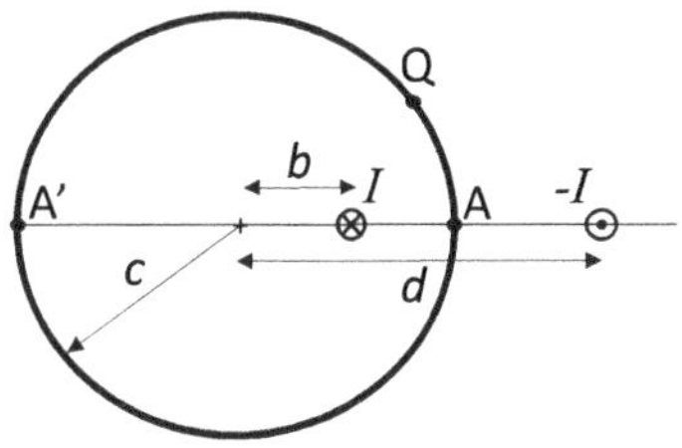

Figure 8.25. An infinitely long current I and an opposite parallel current $-I$, forming an Apollonius circle of constant magnetic vector potential through point Q of figure 8.24.

of distances from the wires, $k = r_2/r_1$ = constant. Apollonius of Perga, c. 240 BCE–c.190 BCE, discovered that a constant ratio of distances from two fixed points (known as foci) defines a circle; this is more general than the schoolboy limiting case of a set of points at equal distance from the circle's centre. Therefore, the surface having constant A_z, through point Q, is a circle as sketched in figure 8.25, because the ratio r_2/r_1 in (8.55) is the same at every point on its circumference.

For algebraic convenience, consider the value of this ratio at the points A and A′ at opposite points on a diameter through the wire in figure 8.25. The definition of a circle requires that $k = \frac{r_2}{r_1} = \frac{d-c}{c-b} = \frac{d+c}{c+b}$ for A and A′, respectively. The Apollonius relation between the circle radius c and the foci distances b and d from the circle centre is therefore

$$b\,d = c^2, \tag{8.56}$$

and also,

$$k = \frac{r_2}{r_1} = \frac{c}{b}. \tag{8.57}$$

Note that the above equations and surfaces derived for A_z apply equally to E_z, because $E_z = j\omega A_z$ for a single-harmonic $e^{-j\omega t}$ magneto-quasistatic approximation; see the development leading to (E.7).

The critical step now is to realize that E_z = constant corresponds to the boundary condition of the ideal PEC screen $E_z(c, \phi) = 0$ in (8.23), when the constant is set to zero. The equivalence principle [13] states that, if the source and image solutions satisfy the field equations and the boundary conditions, then this is the unique solution. Using the image method [13], the field due to the wire current source $+I$ at b together with its conjugate image current $-I$ at $d = c^2/b$, is identical to the field of the wire current source within the PEC screen radius c. This field equivalence applies only to the source region within the screen, namely, $r \leqslant c$. To state this in a different way, the field due to the source-induced current on the PEC screen inner surface is identical to the field produced by the conjugate image $-I$ at d, for the region $r \leqslant c$; the total field includes the current source. Therefore, calculations of vacuum fields and mutual partial inductance for a wire current source inside a cylindrical PEC screen can be performed more simply by using the wire current source and its conjugate image only, without the screen. The source-image distance relation (8.56)

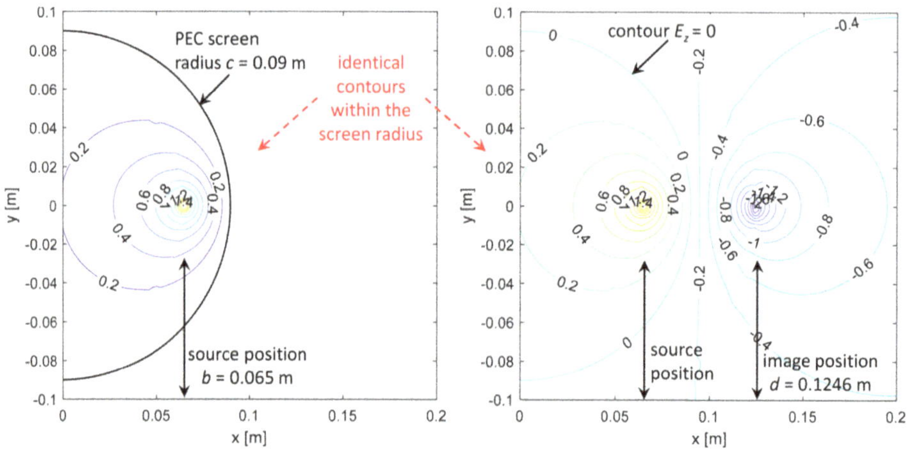

Figure 8.26. Contour plots of E_z for a wire current source in a PEC screen, or with an image current. Left: Wire current $+I$ at $b = 0.065$ m, and screen at $c = 0.09$ m (same as figure 8.13). Right: The same wire current, without a screen, but with its conjugate image current $-I$ at $d = \frac{c^2}{b} = 0.1246$ m. By arranging for the zero-E_z contour to coincide with the PEC screen, the image current generates exactly the same E_z field as the induced currents on the screen, for the region $r \leqslant c$.

for cylindrical screens replaces the equidistant source and image currents on either side of an infinite plane screen, namely, h and $-h$ in figure 4.17. The limiting cases for the cylindrical screen are easily understood: for the source $+I$ close to the inside of the screen, $b \to c$, the image current $-I$ is close to the outside of the screen, $d \to c$. This is the limiting case of equidistant images on either side of a plane screen, as used by J-M Jin [4]. For the limiting case of the source near the axis, $b \to 0$, the image moves far away, $d \to \infty$, because the surfaces tend towards concentric circles around the screen axis.

Figure 8.26 illustrates the Apollonius image method by comparing two contour plots: the left hand side shows the contour plot of E_z for a wire current within a PEC screen, as originally shown in figure 8.13, but here with the wire at $\phi = 0$, where it is now understood that all of the contours are circles. The right-hand side of figure 8.26 illustrates the contour plot of E_z for the wire current $+I$ and its conjugate image current $-I$. The image current position $d = 0.1246$ m satisfies (8.56), $bd = c^2$ for $b = 0.065$ m and the screen radius $c = 0.09$ m. Clearly, the contour shapes in both figures are the same for $r \leqslant c$, but a final adjustment is necessary to impose the required boundary condition $E_z = 0$ at the screen radius c. Note that the screen is the only reference surface in the field solutions where an absolute value of E_z is imposed, whereas (8.55) supposes that $A_z = 0$ on the mid-plane where $r_1 = r_2$ so (8.55) defines the field to within a constant. Using (8.57), it is necessary to subtract $\frac{\mu_0 I}{2\pi} \ln \frac{c}{b}$ from the A_z magnetic vector potential to satisfy $E_z = 0$ at the screen radius c. In this way, figure 8.13 of E_z in a PEC screen, originally calculated using an infinite series of shell harmonics, can more simply be mapped over all space using

$$E_z = j\omega A_z = j\omega\frac{\mu_0 I}{2\pi}\left[\ln\left(\frac{r_2 b}{r_1 c}\right)\right], \tag{8.58}$$

where different constant ratios $k_i = \left(\frac{r_2}{r_1}\right)_i$ in (8.57) trace out circles of different constant E_z values, and $k = \frac{r_2}{r_1} = \frac{c}{b}$ traces out the PEC screen at $r = c$, for which $E_z = 0$.

To summarize, the contour shapes and E_z values concur exactly for $r \leqslant c$ on both plots in figure 8.26, thus confirming the Apollonius image method.

For a complete birdcage enclosed by a screen radius c, each leg at radius b, angle ϕ, has its own image current at radius d, angle ϕ. By the superposition principle, the conjugate image of the birdcage of radius b is therefore another birdcage, with equal but opposite currents, at the larger radius d. Mutual partial inductances of the birdcage legs inside a screen (without plasma) can be conveniently calculated by replacing the screen by the birdcage current images and adding their mutual partial inductance contributions to the matrix, similarly to the ladder resonant antenna method in chapter 4.

It may be objected that the mutual partial inductance between two wires should not be calculated using image currents because the image field corresponds to the real fields only on the side of the mirror surface containing the source. However, the calculation of M from (4.15), namely

$$M_{\text{source/image}} = \frac{\int_{c_{\text{source}}} A_{\text{image}\|}\, dl}{I}, \tag{8.59}$$

involves precisely the magnetic vector potential $A_{\text{image}\|}$ due to the image calculated *at the source*, and therefore the method is valid.

To end on a practical note, the grounded screen has been assumed throughout this chapter to be a cylinder, allowing current to circulate azimuthally without hindrance. In practice, however, a screen is often split into two half-shells to enable fine-tuning of the antenna resonance frequency by moving the shell screens in or out (section 12.2.8 and 13.4.1). This breaks the symmetry of the Apollonius circles. Nevertheless, the entire assembly is enclosed in a grounded metal box Faraday screen preventing electromagnetic radiation to the environment. Therefore, the induced currents on the half-shells, combined with induced currents inside the box, might well still behave as a complete cylindrical screen. Moreover, if the split in the screen is oriented to be at $\phi = \pm\pi/2$, then the source current $I = I_0 \cos\phi$ is, in any case, zero at this position, requiring no image current.

8.5.2 Complex image method for birdcage plasma sources

The complex image method in chapters 6 and 7, and appendix E was used to calculate the mutual partial inductance between a plane plasma and a straight current filament lying parallel to its surface. This technique was able to model experimental measurements of inductive coupling of planar resonant antennas to a semi-infinite plane plasma, and also to a plasma slab. The question of birdcage loading curves in presence of plasma would require a complex image method suitable for a straight wire parallel to a resistive cylinder, within a PEC screen. To the authors' knowledge at the time of writing, such a model for a filament image in the plasma, analogous to the planar case, possibly has no equivalent in cylindrical geometry.

8.6 Chapter summary for wavefields in birdcage antennas

The electric and magnetic fields for a birdcage antenna within a Faraday screen are calculated analytically using cylindrical wave functions, for shell currents and wire currents. The magneto-quasistatic approximation is indistinguishable from the exact solution for the RF range of frequencies. A uniform magnetic field is generated by a $m = 1$ first harmonic shell current in vacuum, for which a nine-leg birdcage gives a reasonable approximation. Ideal image currents induced in the cylindrical Faraday screen are demonstrated using Apollonius circles, although a complex image theory for resistive plasma cylinders remains to be developed.

References

[1] Paul C R 2006 *Introduction to Electromagnetic Compatibility* 2nd edn (Hoboken, NJ: Wiley)

[2] Bleaney B I and Bleaney B 1976 *Electricity and Magnetism* (London: Oxford University Press) 3rd edn p 1

[3] Jackson J D 1999 *Classical Electrodynamics* 3rd edn (New York: Wiley)

[4] Jin J-M 1998 *Electromagnetic Analysis and Design in Magnetic Resonance Imaging* (Boca Raton, FL: CRC Press)

[5] Jin J-M 2010 *Theory and Computation of Electromagnetic Fields* (New York: Wiley)

[6] Abramowitz M and Stegun I A 1965 *Handbook of Mathematical Functions* (New York: Dover)

[7] Tegopoulos J A and Kriezis E E 1985 *Eddy Currents in Linear Conducting Media* (Amsterdam: Elsevier)

[8] Guittienne Ph, Chevalier E and Hollenstein Ch 2005 Towards an optimal antenna for helicon waves excitation *J. Appl. Phys.* **98** 083304

[9] Jin J-M 2010 Practical electromagnetic modeling methods *eMagRes* **2010**

[10] Furno I, Guittienne Ph and Howling A A 2019 Method, measurement probe and measurement system for determining plasma characteristics *Patent* EP16721773.6

[11] Lieberman M A and Lichtenberg A J 2005 *Principles of Plasma Discharges and Materials Processing* 2nd edn (Hoboken, NJ: Wiley)

[12] Paul C R 2010 *Inductance: Loop and Partial* (Hoboken, NJ: Wiley)

[13] Harrington R F 1958 *Introduction to Electromagnetic Engineering* (New York: McGraw-Hill)

Chapter 9

Inductive plasma generated by a birdcage antenna

The spatial structures of the oscillating electric and magnetic fields inductively generated by a birdcage resonant antenna were described in chapter 8. Now we consider experiments and applications of inductively coupled plasma using birdcage antennas. We recall that there is still no externally imposed magnetic field in this part II of the book.

9.1 Birdcage construction

Figure 9.1 shows a generic birdcage antenna which serves as a large volume ICP source [1], as mentioned in chapters 1 and 5. Various birdcage antennas were constructed in a range of sizes for various applications:

1. The 'large volume' closed birdcage reactor consists of a Pyrex tube 32 cm diameter and 55 cm long. In figure 9.1 it has 16 copper legs, 35 cm long, equally spaced by 6.7 cm, interconnected by 2.47 nF capacitor assemblies. The uncooled antenna was connected to the RF generator at one node, with the opposite node grounded via the top grounded metal end-cap, as shown in figure 9.1. These electrical connections favour the $m = 1$ mode of the birdcage antenna. Two half-cylinder screens mounted at a variable distance outside the antenna assembly (not shown) provide fine-tuning of the birdcage resonance frequency when a fixed excitation RF frequency is used [2]. The operating pressure range was 0.1 to 10 Pa in argon for a maximum RF power of 500 W at 13.56 MHz.
2. A 'high pressure' open birdcage nine-leg reactor 12 cm long and 8 cm diameter, was especially adapted for high pressure plasma up to 8300 Pa by inserting series solenoid inductors in each leg [3]. Operating at a frequency of

doi:10.1088/978-0-7503-5296-3ch9

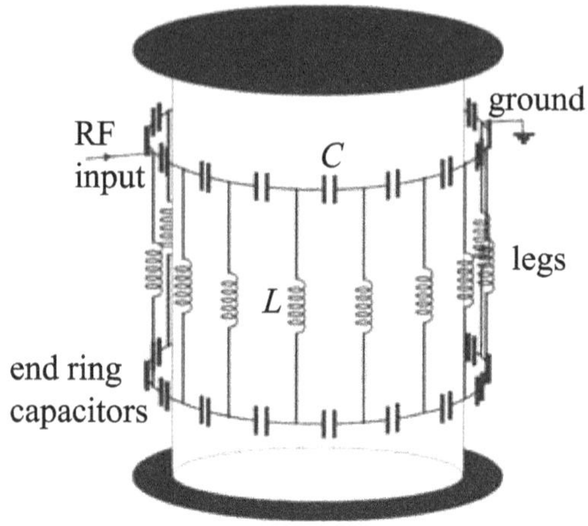

Figure 9.1. Schematic of a 16-leg birdcage antenna configured as a high-pass resonant network with straight leg conductors acting as inductances L, and end-ring capacitor assemblies C. In this example of a closed network, the RF power input and ground are connected at opposite ends of a diameter of the top end-ring, corresponding to the 'large volume' birdcage in point 1 of section 9.1. The top and bottom discs are grounded aluminium end-caps which seal the Pyrex tube. (Adapted with permission from [1]. Copyright 2013 IOP Publishing.)

40.68 MHz with 141 pF capacitors, a detailed description is given in sections 9.3.1 and 13.5.

3. 'RAID' is an open birdcage used principally for high power helicon experiments, see chapter 12 and figure 1.5. The vacuum input impedance spectrum was compared with a mutual partial inductance model which includes the Apollonius image described in section 8.5.1. The missing capacitor in the equivalent circuit of figure 1.5 acts in the same way as the fictitious stringers at the ends of an open antenna in section 3.4.1, which set the stringer current boundary conditions to $J_0 = 0$ and $J_N = 0$. An open birdcage therefore behaves as a ladder open network from the point of view of normal mode frequency and unique mode spatial distribution, without the degeneracy attributed to closed networks (section 9.2.1).

Many other examples of birdcage antennas include the MRI literature [4] and some recent plasma research [5, 6].

The advantages of resonant antennas regarding lower voltages for large ICP sources were already mentioned in section 5.5; this applies equally to large volume birdcage antennas as well as to large-area ladder resonant networks [1].

9.2 Normal modes on closed networks

The electric and magnetic fields for azimuthal periodicity m were described for a birdcage in chapter 8 with special emphasis on $m = 1$, often by replacing the birdcage currents by an effective shell current $I_0 \cos \phi$. Now we develop the normal mode model for a closed birdcage to find the mode frequencies in terms of the antenna inductances and capacitance. Compared to the open networks discussed in section 3.4, closed networks have different boundary conditions and therefore different mode frequencies, different distributions of current and voltage, and also exhibit the new property of mode degeneracy, meaning that more than one mode can have a given mode frequency. Similarly to the treatment of normal modes in chapter 3, we will again neglect the mutual inductance in this simplified analysis [4].

9.2.1 Normal mode frequencies on closed networks

Closed networks are, in fact, more widespread than open networks for applications such as MRI [4]. With reference to chapter 1, we recall that birdcage antennas with continuous end-rings are closed networks with cylindrical symmetry. Figure 3.5 is therefore modified so that the node N is now joined via another top and bottom stringer back to node 1 as in figure 9.1. There are now N stringer sections for N legs, instead of just $(N - 1)$ stringers for an open network with N legs. The normal mode equations (3.10) and (3.11) are valid also for closed networks. However, the open network boundary conditions in section 3.4.1 must now be replaced by a periodic boundary condition $J_n = J_{n+N}$, and consequently the closed network can only support integral numbers of whole wavelengths, instead of the integral numbers of half-wavelengths on open networks in section 3.4.2. By inspection, equation (3.11) for $J_n = J_{n+N}$ is now satisfied for $b_1 = +b_2$, as well as for $b_1 = -b_2$, and therefore we now have cosine solutions as well as sine solutions for the stringer currents. Indeed, any phase relation of type $b_1 = b_2 e^{j\psi}$ is a solution, meaning a sine or cosine with arbitrary phase, or, by trigonometrical expansion, a mix of sine and cosine functions. A fully defined solution would further require a phase reference, such as a choice of ground position.

The periodic boundary condition $\cos(n\beta) = \cos(n + N)\beta$, as well as $\sin(n\beta) = \sin(n + N)\beta$, both require that $N\beta = 2\mathrm{m}\pi$, with independent solutions for integer $\mathrm{m} = 0, 1, 2, \ldots, \frac{N}{2}$. In the case of odd N, the m series ends at the previous integer[1] $\frac{(N-1)}{2}$. By substituting $\beta = \frac{2\mathrm{m}\pi}{N}$ into (3.10), the normal mode frequencies for closed networks are [4]

$$\omega_{\mathrm{m}} = \frac{1}{\sqrt{C\left(L_{\mathrm{str}} + 2L_{\mathrm{leg}}\sin^2\left(\frac{\mathrm{m}\pi}{N}\right)\right)}}, \quad \mathrm{m} = 0, 1, 2, \ldots, \frac{N}{2}, \tag{9.1}$$

to be compared with the open network frequencies in (3.12). The pairs of stringer currents are

[1] The solutions $\cos(n\beta) = \cos(\frac{2n\mathrm{m}\pi}{N})$ for the final series term $\frac{(N-1)}{2}$ or $\frac{(N+1)}{2}$ in the case of odd N are in fact the same. By trigonometrical expansion, their difference, $\cos(\frac{2n\pi(N+1)}{2N}) - \cos(\frac{2n\pi(N-1)}{2N}) = -2\sin(n\pi)\sin(n\pi/N)$, is zero for integer n. In fact, the same is true for all pairs of terms symmetric about $N/2$, which is why the series of independent solutions in (9.1) ends at $\mathrm{m} = N/2$. This also complies with the Nyquist–Shannon sampling theorem, which requires at least two sample points per period: for $\mathrm{m} = N/2$, there are only two leg currents per wavelength, therefore m cannot be greater than $N/2$.

$$J_n = -K_n \propto \begin{cases} \cos\left(n\dfrac{2\mathrm{m}\pi}{N}\right), & \mathrm{m} = 0, 1, 2, \dots, \dfrac{N}{2}, \\ \sin\left(n\dfrac{2\mathrm{m}\pi}{N}\right), & \mathrm{m} = 1, 2, \dots, \dfrac{N}{2} - 1, \end{cases} \tag{9.2}$$

to compare with (3.13), and the pairs of leg currents are [4]

$$I_n \propto \begin{cases} -2\sin\left(\dfrac{\mathrm{m}\pi}{N}\right)\sin\left[\left(n - \dfrac{1}{2}\right)\dfrac{2\mathrm{m}\pi}{N}\right], & \mathrm{m} = 0, 1, 2, \dots, \dfrac{N}{2}, \\ 2\sin\left(\dfrac{\mathrm{m}\pi}{N}\right)\cos\left[\left(n - \dfrac{1}{2}\right)\dfrac{2\mathrm{m}\pi}{N}\right], & \mathrm{m} = 1, 2, \dots, \dfrac{N}{2} - 1, \end{cases} \tag{9.3}$$

to compare with (3.14). The wavelength in terms of the number of legs is $\lambda = \frac{N}{\mathrm{m}}$. Equivalently, the number of whole wavelengths around the birdcage is $\frac{N}{\lambda} = \mathrm{m}$, in comparison with the number of half-wavelengths around an open birdcage, or along a ladder resonant antenna in section 3.4.2. All half-wavelength modes are forbidden in closed networks, so the first (i.e. m = 1) current distribution in closed networks corresponds to that of m = 2 in open networks. There are $N/2$ linearly independent modes for closed networks; the $(N - 1)$ modes of open networks are almost double this number. We say 'almost' the double, because of the existence of an end-ring resonant mode for m = 0, whose current is constant in the end-rings but zero in the legs [4]. Using (9.1), this m = 0 mode also has the highest frequency, $\omega_0 = 1/\sqrt{CL_{\mathrm{str}}}$. However, for equal values of m, the modes are in degenerate pairs of equal frequency ω_{m}, which explains why there are N linearly independent modes, but only $\mathrm{m} = 0, 1, 2, \dots, \frac{N}{2}$ possible frequencies. The mode pair for m = 1 can produce uniform linearly polarized magnetic fields which are useful for MRI applications [4]. However, from the point of view of experiments to interpret wavefields generated by a closed birdcage employed as a helicon plasma source, this degeneracy is also a potential source of confusion (figure 9.2). Therefore, the RAID birdcage uses the open network configuration to lift the degeneracy of simultaneous orthogonal modes.

The closed and open birdcage resonance frequencies can be measured and compared on the same antenna, without plasma, using a network analyser as in chapters 3 and 4 for ladder resonant antennas. Figure 9.3 shows the real impedance spectra measured on a nine-leg birdcage in both closed (red) and open (black) configurations. The open network presents a total of $(N - 1) = 8$ resonance frequencies, while the closed system only shows $(N - 1)/2 = 4$ resonances (m = 1 to 4), as expected, because half of the open modes are forbidden for a closed birdcage due to the condition of azimuthal periodicity. The two m = 0 resonances shown in the figure are not included in this count, because they correspond to end-ring resonances with no current flowing in the antenna legs; clearly these end-ring currents can only circulate in the closed birdcage. The first m = 0 mode corresponds

Figure 9.2. 'Degeneracy is a potential source of confusion.' (Illustration by Alex Howling. Copyright 2023 Alex Howling.)

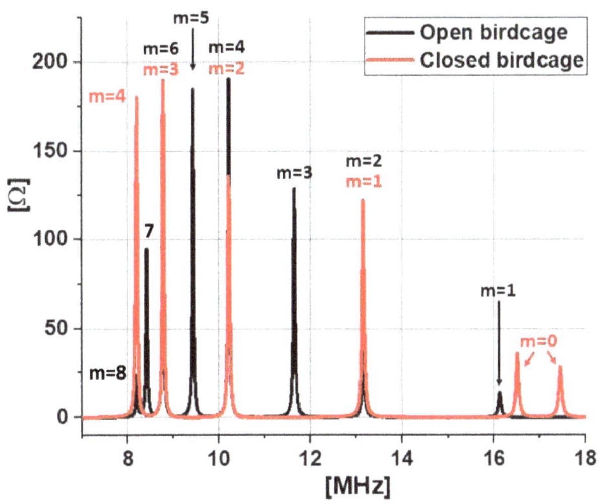

Figure 9.3. Network analyser measurements of the real impedance, as a function of frequency, for the same birdcage (N = 9 legs) in both closed (red) and open (black) configurations.

to in-phase ring currents, while the second mode corresponds to ring currents in phase opposition.

9.2.2 Mode structure in closed birdcage antennas

Similarly to the mode measurements for a ladder antenna in figure 3.7, the current distributions in this birdcage antenna were measured for the first four modes by means of an inductive B-dot coil, and are in very good agreement with the theoretical predictions in figure 9.4. Hence, with a sufficiently large number of

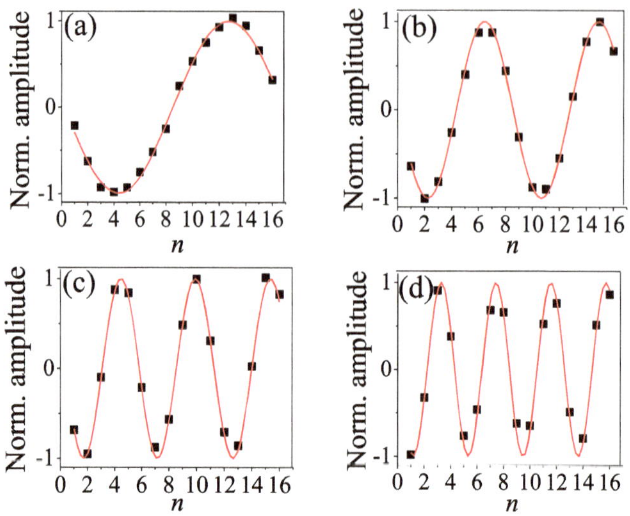

Figure 9.4. Leg current distributions for the first four modes of a 16-leg closed birdcage measured by means of an inductive B-dot coil. The points are the 16 measurements on the legs (n = 1–16), and the red line is taken from the analytical theory in section 9.2. (a) m = 1, (b) m = 2, (c) m = 3, and (d) m = 4. The end-ring mode(s) m = 0 do not appear because the leg current is zero. (Reproduced with permission from [7]. Copyright 2005 AIP Publishing.)

legs, this antenna generates a very good approximation to a sinusoidal azimuthal shell current distribution, similar to the analytical calculation shown in figure 8.15 for the case of the $m = 1$ harmonic. Conversely, conventional antennas have only two or four legs and therefore provide very poor mode selectivity.

For the cylindrical wavefields in chapter 8, there is no difference between positive or negative modes driven by shell currents $I_0 \cos(m\phi)$, because $\cos(-m\phi) = \cos(+m\phi)$. All of these modes are stationary modes with a back-and-forth oscillating amplitude as shown in figure 8.11. The circular symmetry of a closed network, with its periodic boundary conditions, means that a stationary mode such as $m = \pm 1$ can be decomposed and considered as two counter-rotating field components, as shown schematically in figure 9.5. Each component has a constant amplitude as it rotates, and summation with the opposing component results in the oscillating amplitude of the original mode. These circularly polarized fields are useful for exciting helicon waves in magnetized plasmas, see part III.

9.2.3 Inductive magnetic field measurement for mode m = 1

A B-dot probe (see also section 12.2.2) was constructed from three coils aligned along axes x, y, z to measure the RF magnetic fields in a birdcage reactor; the z-axis is the vertical axis of the large volume birdcage figure 9.1. Hybrid combiners were used to subtract any capacitively coupled component from the probe signals [8].

The B-dot magnetic pickup coil was scanned vertically along the z-axis of the large volume birdcage, point 1 of section 9.1, figure 9.1. Measurements of the field components B_x and B_y were made at intervals of 25 mm, going beyond the length of

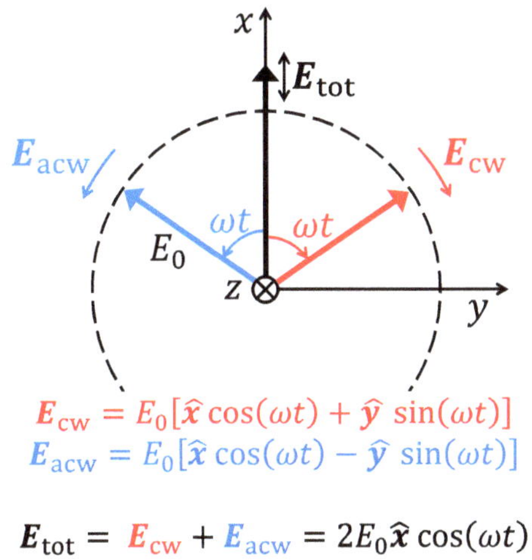

Figure 9.5. A linearly polarized field $\mathbf{E}_{tot}$, oscillating along the x-axis at angular frequency ω, is decomposed into the vector sum of a clockwise (cw) circularly polarized field $\mathbf{E}_{cw}$ and an anti-clockwise (acw) circularly polarized field $\mathbf{E}_{acw}$, rotating in opposite directions at angular frequency ω with constant amplitude E_0 (see also figure 10.15).

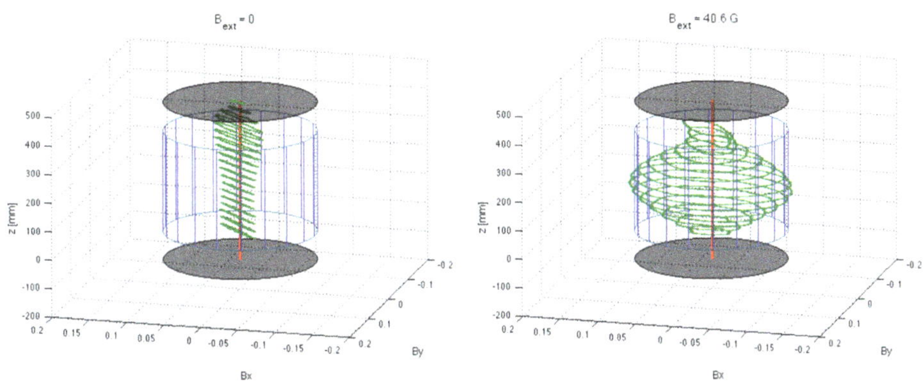

Figure 9.6. A perspective plot of the B_x and B_y components of magnetic field (arbitrary units) measured in a vacuum at 25 mm intervals along the axis of the large volume birdcage (see point 1 of section 9.1). Each green trajectory corresponds to the magnetic field over an RF period, at a given z coordinate. Left: The RF magnetic field is linearly polarized in an unmagnetized plasma ($B_{ext} = 0$), as schematized in figure 8.11. Right: Circularly polarized wavefields are measured when an external static magnetic field ($B_{ext} = 40.6$ G) is superposed along the z-axis; see part III for magnetized plasma. The birdcage antenna is shown in blue, and the metal end-plates in grey.

the birdcage legs. In figure 9.6, the z-axis labels 0 and 350 mm refer, respectively, to the bottom and top of the antenna, while −100 and 450 mm correspond to the ground end-cap positions in figure 9.1. These measurements shown on the left are in agreement with the $m = 1$ uniform linearly polarized magnetic field calculated in chapter 8, and despite the finite length of the birdcage legs, the magnetic field is uniform over the whole height of the birdcage until close to the grounded end-cap. On plasma ignition, the measured magnetic fields on the axis become much smaller, due to skin depth screening in inductively coupled plasma, as modelled in section 8.4. When an external static axial magnetic field is added, much stronger, circularly polarized wavefields are detected, as shown on the right of figure 9.6, and these are the subject of helicon modes in part III 'Resonant network antennas in magnetized plasma'.

9.2.4 RF power connection configuration

Early examples of birdcage antennas connect the RF power across a single capacitor [4, 7]. However, it is preferable to maximize the input impedance by strategically placing the RF and ground connection at the nodes of the required mode as explained in section 3.5.7 and figure 3.16. Figure 3.12 shows that the RF input current is thereby minimized, which also favours current uniformity across the antenna because the injected current minimally perturbs the resonance currents circulating within the antenna.

The input impedance is generally not a maximum across a single capacitor, and the resulting high current can overheat and damage the capacitor whose terminals are chosen to be the RF input and ground connections. This situation may not be so serious for low-loss MRI operation at the resonance of a high-Q birdcage [4], but plasma generation [5] causes much higher RF current to traverse the input capacitor, perhaps destroying it. Plasma dissipation strongly reduces the Q-factor, and the RF power is inversely proportional to Q for a given voltage (see section 2.2.2).

The ladder model developed in chapter 3 applies directly to open birdcages, and figure 3.16 can be used in this context. For an RF input across a single capacitor, we have $N_f - N_g = 1$. In the figure, this straight line crosses only the two lowest contours, whose maximum vacuum impedance is 51 Ω, compared to 301 Ω for the optimal impedance such as $(N_g, N_f) = (8, 12)$ in mode m = 6. During plasma, the input impedance example in section 6.4.1 falls by a factor 7.1, which, hypothetically, would result in only 51/7.1 = 7.2 Ω effective impedance across a single capacitor. For a nominal power of 10 kW, the input capacitor would suffer an RF current up to 50 A amplitude.

Closed birdcage antennas, because of the mode degeneracy, are susceptible to imperfections in the construction symmetry, and to external coupling influences from neighbouring grounded surfaces—even the operator's hands. Closed antennas may possibly be made more stable with additional ground connections, but open antennas have the advantage of mode stability. The dissipative model for open networks (i.e. ladder and open birdcage antennas), previously described in section 3.5, could be re-calculated for the structure and boundary conditions of closed birdcage networks. In contrast to figure 3.11, the closed birdcage has two interconnected antenna parts defined by the RF input and the ground nodes, instead of three aligned parts; the RF input current would flow in parallel around both parts I and II. However, since closed antennas have no particular advantage, they are less developed than open antennas in this book.

9.2.5 Voltage and current measurements on birdcage antennas

The impedance and RF power can be measured with reasonable accuracy directly at the input of resonant antennas by means of commercial voltage and current oscilloscope probes. This is in contrast to conventional capacitive or inductive reactors, because their dominantly reactive impedance means that the voltage–current phase difference is very close to +90° or −90° at the antenna input. In that case, the phase difference would have to be measured with extreme precision to avoid large errors in the estimation of RF power $VI\cos\phi$ because the power factor [9], $\cos\phi$, is close to zero and is critically sensitive to small errors in the relative phase between the voltage and current probes.

The amplitude and relative phase between the high frequency voltage and current probes were calibrated with respect to an RF network analyser in the absence of plasma. Figure 9.7 shows an example of electrical probe and network analyser measurements of the impedance magnitude and phase for the m = 3 mode of a birdcage antenna [1]. The calibrated measurements are reasonably accurate for the resonant antenna, but it can be imagined that phase errors of much less than 0.1° near to 90° would be unrealistic for conventional ICP or CCP sources when using voltage and current probes.

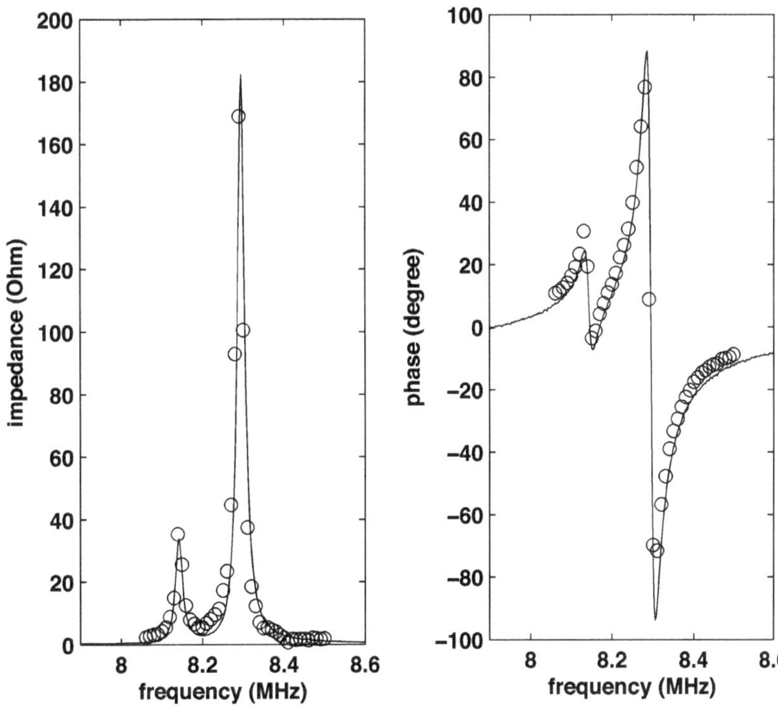

Figure 9.7. Comparison of network analyser impedance measurements (lines) with results obtained from RF voltage and RF current probe measurements (circles) in vacuum for the m = 3 mode of the birdcage RF antenna. Left: Magnitude; right: phase. (Reproduced with permission from [1]. Copyright 2013 IOP Publishing.)

9.2.6 Power measurement in a resonant network antenna

The power transfer efficiency of a resonant antenna was also estimated in section 6.4.1, and the method is explained here. This approximate method accounts for both the fall in impedance and the shift in resonance frequency; it shows the usefulness of the $RL\|C$ equivalent circuit in figure 3.14 for antenna normal modes in chapter 3. Measurement of the real impedance of the antenna at a normal mode resonance frequency with and without plasma is interpreted in terms of an equivalent circuit of a capacitance C in parallel with an inductance L with resistance R, considered previously in section 2.2.2. The values of R, L, C could be found by curve fitting to the experimental impedance measurements as a function of frequency (although this is not necessary), and they have different fitted values for each resonance. The input impedance of a $RL\|C$ parallel resonant circuit is purely resistive at resonance and is given by (2.16), which can also be written as $Z_{\text{in}}^{\text{res}} = (\omega_0^2 C^2 R)^{-1}$. The vacuum input impedance, measured without plasma, is

$$Z_{\text{vac}}^{\text{res}} \approx (\omega_{\text{vac}}^2 C^2 R_{\text{vac}})^{-1}, \tag{9.4}$$

where ω_{vac} is the angular frequency at the vacuum mode resonance. On plasma ignition, the power dissipation in the plasma due to coupling with the antenna manifests itself as an increase in R_{vac} which is the dissipative element in the equivalent parallel resonant circuit. This is responsible for the fall in the antenna resonance input resistance as shown by the inverse dependence between $Z_{\text{vac}}^{\text{res}}$ and R_{vac} in (9.4). More specifically, the transformer coupling to induced current in the plasma results in a reflected impedance in the antenna equivalent circuit as described in [10] and figure 5.4. According to (5.12), the reflected impedance increases the equivalent-circuit resistance, and reduces its inductance. The latter causes the resonance frequency to rise by $\delta\omega$ in the presence of plasma, consistent with the inductive coupling of small antennas in figures 6.11 and 9.8.

Since the equivalent-circuit capacitance is dominated by the antenna capacitors whose capacitance is much larger than any antenna–plasma sheath capacitive coupling, we will assume here that the mode frequency increase, $\delta\omega$, is due solely to the decrease in inductance. In presence of plasma, (9.4) therefore becomes

$$Z_{\text{plasma}}^{\text{res}} \approx (\omega_{\text{plasma}}^2 C^2 R_{\text{plasma}})^{-1}, \tag{9.5}$$

where $Z_{\text{plasma}}^{\text{res}}$ is the reduced input impedance when the resonant antenna is coupled to the plasma, and $\omega_{\text{plasma}} = \omega_{\text{vac}} + \delta\omega$. R_{plasma} is the increased equivalent-circuit resistance due to plasma coupling. Because the vacuum circuit resistance and the reflected impedance are in series, the power transfer efficiency η from the antenna to the plasma is given by the fraction of the equivalent resistance due to the plasma coupling, i.e.

$$\eta = \frac{R_{\text{plasma}} - R_{\text{vac}}}{R_{\text{plasma}}} = 1 - \frac{\omega_{\text{plasma}}^2 Z_{\text{plasma}}^{\text{res}}}{\omega_{\text{vac}}^2 Z_{\text{vac}}^{\text{res}}} \approx 1 - \frac{Z_{\text{plasma}}^{\text{res}}}{Z_{\text{vac}}^{\text{res}}}\left(1 + 2\frac{\delta\omega}{\omega_{\text{vac}}}\right), \tag{9.6}$$

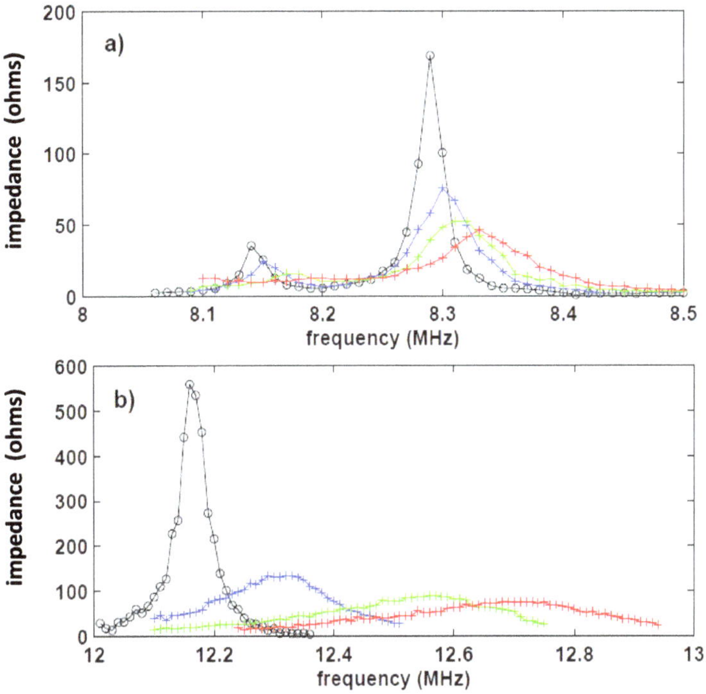

Figure 9.8. Magnitude of the impedance for a birdcage RF network antenna at gas pressure 10 Pa as a function of applied RF powers 0 W (vacuum, black), 30 W (blue), 50 W (green), and 70 W (red). (a) m = 3; (b) m = 1. (Adapted with permission from [1]. Copyright 2013 IOP Publishing.)

where $\delta\omega$ is the frequency shift due to the plasma, and $Z_{\text{vac}}^{\text{res}}$ and $Z_{\text{plasma}}^{\text{res}}$ are the measured antenna real input impedances at resonance in a vacuum, and with plasma, respectively.

Figure 9.8 shows the impedance measurements for the birdcage antenna, which essentially show similar behaviour to the small planar antenna in figure 6.11. However, in this case the observed frequency shift due to dominantly inductive coupling in the presence of the plasma is larger than for the planar antenna. The frequency shift for the birdcage m = 1 mode is also more pronounced than for the m = 3 mode which is generally the case for these high-pass resonant antennas. Such frequency shifts could be a disadvantage when using a fixed RF frequency generator in plasma processing, since the antenna impedance could shift out of the resonance condition when the plasma ignites, leading to impedance mismatch and high reflected power if the shift is large. The frequency shift can be compensated by adjusting the distance of metal screens mounted around the birdcage antenna [2, 4], because this alters the antenna–screen Apollonius image spacing and hence the mutual partial inductance with the birdcage legs as discussed in chapter 8. An alternative strategy is considered in section 13.5.

Finally, figure 9.9 shows the estimated power transfer efficiencies of the planar and cylindrical RF antennas operated at different normal modes. Both RF network

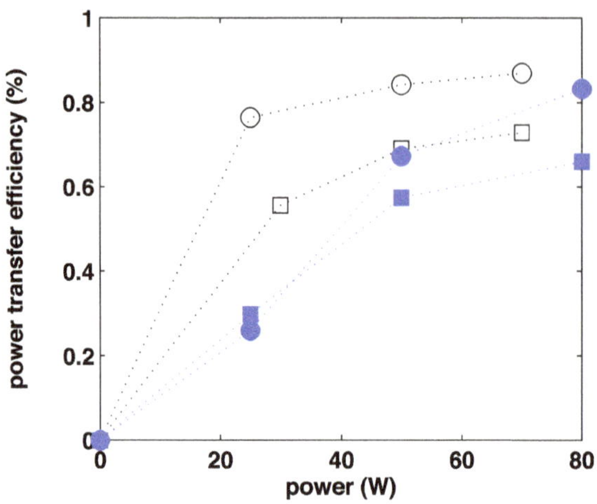

Figure 9.9. Measured power transfer efficiency for a cylindrical RF antenna (open circles m = 1, open squares m = 3, argon pressure 1 Pa); and for a planar RF network antenna (blue circles m = 2, blue squares m = 4, argon pressure 0.2 Pa). (Reproduced with permission from [1]. Copyright 2013 IOP Publishing.)

antennas show a power transfer efficiency of 70% to 90% for the inductively driven discharge obtained at higher RF power (see also section 6.4.1). These values for RF resonant network antennas are comparable to power transfer efficiencies of conventional ICP devices [10].

9.3 Applications of birdcage inductive antennas

9.3.1 Application: plasma treatment of powder in a fluidized bed

There is strong commercial interest in dry, non-vacuum treatment of powders for mass production markets such as the food, pharmaceutical, and other industries [11]. Applications include particle coatings, food sterilization, seed disinfection [12], and so on, but present plasma sources are either too weak (dielectric barrier discharges [13, 14]) or too powerful (thermal plasmas) for high throughput of heat-sensitive products. To fill this niche, the high pressure birdcage antenna in section 13.5 can be compared with a conventional coil-ICP in a pilot industrial application for powder treatment using a fluidized bed reactor [3].

To paraphrase section 5.1, in low pressure (<1 mbar), non-thermal plasmas, the electron temperature (typically ~30 000 °C is much higher than the ion, gas, and substrate ambient temperatures. This is exploited in many industrial processes to perform very high temperature, exotic plasma chemistry whilst maintaining a low temperature of the gases and substrates. When the operating pressure is increased, however, due to higher gas collision frequencies, the ion and gas temperatures rise to converge at the common temperature of all three species in thermal equilibrium which can be at several hundred degrees celsius or more. Consequently, thermal management of high gas pressure plasmas is necessary to prevent pyrolysis of temperature-sensitive gaseous precursors and melting of powders. Therefore, for optimal plasma treatment of powders, a working range must be found where the pressure is low enough to maintain non-thermal plasma operation, yet with the gas pressure high enough for fluid dynamic forces to act on particles in a circulating fluidized bed or in a downer reactor.

In usual fluidized bed technologies, the diameter of the tubular reactor is only a few millimetres because, for wider tubes, fine powder particles stick together and clog the reactor. With the high pressure birdcage reactor, the tube diameter is enlarged to 4 cm which could process up to 20 kg of powder per hour, assuming similar power density to section 13.5. For higher power, where enhanced monomer dissociation increases the flowability treatment rate, the powder throughput could possibly reach 50 kg h^{-1}.

Operation and results

A downer reactor [11] with a screw feeder was installed with both a birdcage and a conventional four-turn coil-ICP source mounted side-by-side on the tube in figure 9.10 for comparison in identical conditions during treatment of lactose powder [3, 14]. The plasma intensity was comparable, if rather lower for the birdcage, but the coil-ICP plasma exhibited strong instabilities; this is probably linked to the high coil voltage which couples capacitively to the plasma column, striking intermittent arcs to the grounded chassis. The birdcage has a lower operating voltage for a given RF power, thanks to its real input impedance, and is thus a more stable source of plasma. It can also be scaled up to longer lengths, whereas a longer coil would suffer from even

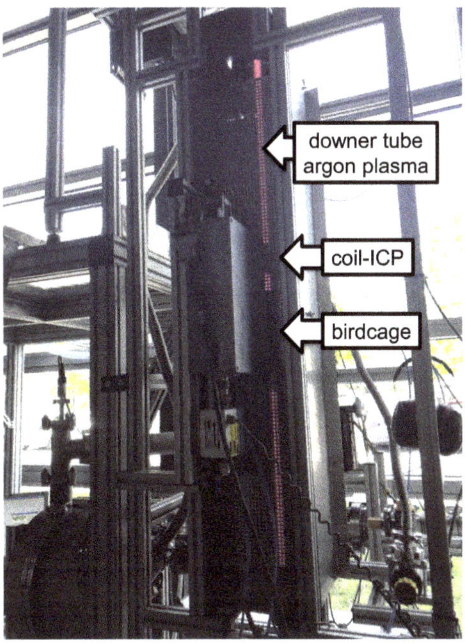

Figure 9.10. The downer reactor, ~1.5 m long, with the screw feeder at the top, showing the high pressure birdcage and the coil-ICP mounted side-by-side along the tube. The argon plasma is almost obscured by the tube protection grid.

Table 9.1. Flowability of plasma-treated powder using the birdcage and coil-ICP sources.

Powder treatment	Flowability
Untreated powder	2.63 ± 0.06
Birdcage plasma treated	5.55 ± 0.26
Coil-ICP plasma treated	6.00 ± 0.26

higher voltage. The birdcage is thus suited to long downer tubes with longer transit times for particle treatment in a lower power density plasma, without risk of melting the powder in a short, high power coil-ICP plasma. The birdcage power capability was limited only by the failure risk of the capacitors, as discussed in section 13.2. The flowability of lactose powder [14], after plasma treatment with a 200 W, 150 Pa, gas mixture of HMDSO/oxygen/argon at 50/250/1450 sccm, was measured using a ring shear tester to compare the powder processing capabilities of the plasma sources as shown in table 9.1. The birdcage and coil-ICP sources are found to be equally efficient for powder treatment. RF power pulsing, or distribution of a lower-intensity plasma along the whole length of the tube using a scaled-up birdcage, would help to reduce local heating if higher powers were to be used.

In the final analysis, the physics principle of inductive coupling is the same, whether for resonant network antennas or for conventional ICP coils. Therefore, the optimal reactor tube radius for a given plasma skin depth is similar for both sources. However, the advantage of the birdcage antenna lies in its impedance properties; for example, it can easily be scaled up in length for industrial use without incurring the high voltage of the coil-ICP (see section 5.5).

9.3.2 Application: plasma deposition in a large volume birdcage

The large volume birdcage reactor in figure 9.1, described in section 9.1 point 1, is well suited for plasma batch processing of multiple arrays of heat-sensitive work-pieces [1]. A low density plasma of $n_{e0} \sim 10^{16}$ m^{-3}, pictured in figure 9.11, has a skin depth of several centimeters, as calculated in section 8.4, which is appropriate for the reactor radius of 16 cm. Such conditions can be used for thin film technology with low temperature substrates. Figure 9.12 shows the broad profiles of plasma density vertically and radially, which are typical of diffusion-dominated profiles [15].

One specific application is for parylene-based hermetic multilayer coating of bio-sensors. Parylene stands for a polymer group of poly(para-xylylene) derivatives; it is used for high conformity coatings with excellent electrical insulation, hydrophobicity, and biocompatibility. Deposited without plasma by means of chemical vapour deposition (CVD) at low pressure and low temperature, parylene forms a transparent, pinhole-free coating from 50 nm to 50 μm thick. However, it is not optimal for barrier layer encapsulation and adhesion compatibility for long-life sensors in bio-implants, which are exposed to a particularly aggressive chemical environment. In these circumstances, a multilayer film would be more able to compensate for micro-defects in individual layers. Since the parylene CVD process takes place in a large

Figure 9.11. Photograph of a diffuse argon plasma (13.56 MHz RF frequency at 70 W delivered power, 5 Pa pressure) in the large volume birdcage reactor of figure 9.1. Several of the 16 birdcage legs are clearly visible in the foreground; the shadow within the plasma is a scanning Langmuir probe for the vertical and radial profiles in figure 9.12.

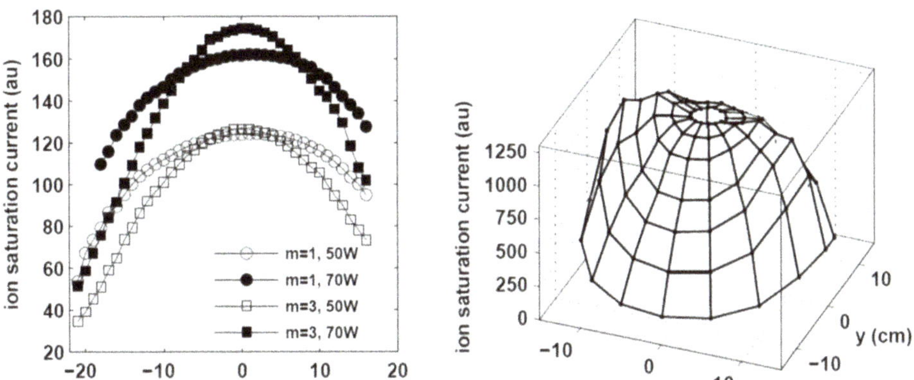

Figure 9.12. Profiles of the ion saturation current measured by a scanning Langmuir probe in the large volume argon plasma of figure 9.11. Left: axial profiles for 10 Pa pressure. Right: radial profile measured at the mid-plane, for m = 1, RF power 65 W, 4 Pa pressure. (Reproduced with permission from [1]. Copyright 2013 IOP Publishing.)

volume vacuum vessel, its deposition could be alternated conveniently with plasma deposition of a thin film such as silicon oxide (SiOx) without breaking the vacuum, thus avoiding surface contamination between the layers [16], and enhancing the barrier properties by the addition of inorganic layers. Plasma-enhanced chemical

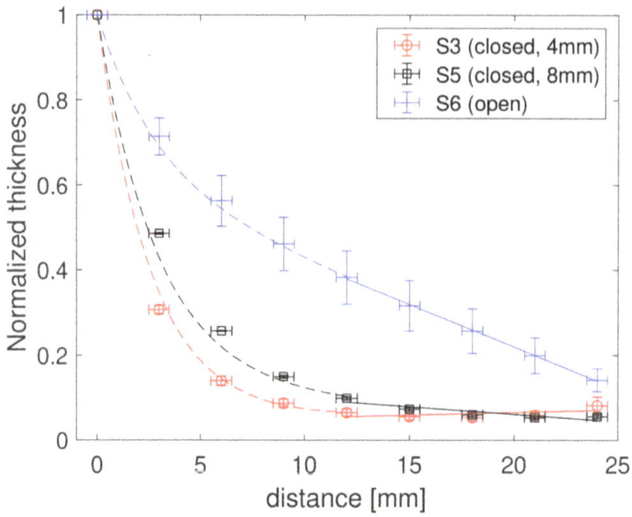

Figure 9.13. Thickness profiles for SiOx plasma deposition, measured inwards from the edge of two parallel plates. Red points: on the inside of the plates spaced 4 mm apart. Black points: on the inside of the plates spaced 8 mm apart. Blue points: on the outside of the plates.

vapour deposition (PECVD) of SiOx using a hexamethyldisiloxane (HMDSO) vapour precursor is faster than other processes such as atomic layer deposition (ALD), and the large volume ICP birdcage antenna is a promising plasma source candidate, provided that the multi-layers satisfy the stringent barrier criteria.

One obvious criterion is the uniformity of the individual layers in the stack. The birdcage ICP provided better global coverage than a CCP arrangement of internal, stacked parallel-plate electrodes, because the plates act as large-area sinks of the active radicals, draining the source species and causing density gradients between the plates. The birdcage ICP is superior in this regard because it is an external antenna, sustaining a diffuse source of radicals throughout the reactor volume with its batch arrays of suspended workpieces. However, PECVD conformity over all the convoluted surfaces of the individual workpieces depends on their scale and shape, as well as on the diffusivity, volume reactivity, sticking factor, surface mobility and deposition rate of the plasma-generated radicals [17–19]. Measurements in figure 9.13 of the SiOx film thickness profile deposited between parallel plates 4 mm or 8 mm apart showed an exponential decrease inwards from the leading edge, indicative of diffusion-limited deposition. Even the thickness profile on the outside of the plates was strongly non-uniform—clearly unsatisfactory for a process requiring conformity of a multilayer. The profile shape was due to a diffusive combination of low and high sticking factor radicals, formed by plasma dissociation of the HMDSO precursor with uncontrolled branching ratios. Different dilutions of various buffer gases, and different precursors (TEOS) were also tested, finishing with a gas mixture of 54 sccm oxygen, 9 sccm of HMDSO, and 18 sccm of argon, at a total pressure of 5 Pa. The observed strongly limited coating penetration and large edge effects meant that the PECVD process was not capable of conformal coating on small-scale, complex workpieces.

Another critical observation is that the film porosity, more than the SiOx stoichiometry, controls the SiOx properties, including the permeability to water vapour and oxygen [16]. To densify the films, a second electrode was installed, but this time inside the birdcage reactor and powered at a different frequency (6 MHz), to control the ion bombardment on the samples. This approach resembles the microcrystalline silicon deposition application [20] in section 6.5. By optimizing the ion bombardment at different bias voltages, the barrier properties of the SiOx thin films were improved by a factor 100, but the reactor configuration was no longer compatible with the original requirement of a free, open volume, whose plasma is sustained by an external antenna.

Comparison of PECVD and ALD methods

At this point, we consider some general implications of the obstacles to satisfactory PECVD coatings based on the experience with this particular application. The advantages of plasma processing were briefly extolled in section 5.1.1—nevertheless, with a reference to this section to flag challenges up ahead. The plasma's energetic electrons create all manner of highly reactive radicals and ions by indiscriminate inelastic collisions with the working gases. This can be (somewhat disparagingly) referred to as a plasma 'soup', in which almost any reaction is permitted by the high energies available. The plasma properties [15] entail the electron density and the electron energy distribution function (determined by the electric field and collision cross-sections), and the radical composition which depends on dissociation branching ratios, including all possible combinations of secondary reactions between molecules, radicals, and ions occurring in the gas (homogeneous reactions) and with the surfaces (heterogeneous reactions). Gas-phase polymerization creates clusters and particles on the timescale of the gas residence time in the reactor, and negative ions can be electrostatically confined indefinitely in the electric potential well between the plasma sheaths, possibly growing to large size [21]. The quality of the growing film is influenced by the types of clusters, radicals, positive ions, and their flux and energy transport to the surface, their sticking coefficients and surface mobilities, which in turn depend on the film topology, dangling bonds, and its surface temperature, including any hyperthermal contribution from ion bombardment.

In face of this complexity, the plasma physicist disposes only of the RF power level and frequency, the input gas composition, flowrate, and pressure, the substrate temperature (limited by the specific application), and the reactor design. Hence, the dilemma for PECVD is the lack of selectivity compared with the inherent versatility of plasma chemistry. In contrast, atomic layer deposition (ALD) is a thin film deposition method based on chemical vapour deposition, generally without plasma enhancement [22]. The majority of ALD processes use two gaseous precursors which react alternately with the surface one at a time in a sequential, self-limiting manner. A thin film is gradually built up, one atomic step at a time, through iterative exposure to the individual precursors. Conformity and selectivity are therefore ideal in principle, although the choice of layers is limited by the known precursors, and by the substrate temperature permitted by the device tolerance. The slow deposition rate and low throughput restricts industrial application to high-added-value devices such as medical implants. Plasma-assisted ALD [23] widens the available scope of precursors and materials, and has the potential to combine the best (fast deposition of compact films) or worst (rate-limited, porous layers) of both of the PECVD and ALD worlds.

9.4 Chapter summary for inductive birdcages

Birdcage resonant antennas are characterized in terms of their input impedance and mode structure, which resemble ladder resonant antennas except for the two-fold mode degeneracy of closed birdcage networks. Two pilot industrial applications were also described: high pressure operation for powder treatment in a fluidized bed, and PECVD deposition for multilayer batch processing on large volume arrays of small workpieces. The high pressure was limited by the rise in gas temperature, and the PECVD film uniformity was perturbed by the unfavourable radical composition of plasma-dissociated HMDSO. It must be recognized that some processes are limited by physical phenomena which cannot be resolved only by choice of the plasma source. Non-magnetized resonant network antennas are of interest for several reasons (for example, their impedance in section 5.5, as well as for upscaling), but they are nevertheless constrained by the laws of plasma physics in the same way as for any other inductively coupled source.

Another interest of birdcage resonant antennas lies in the generation of helicon waves [7] for which the cylindrical geometry of the birdcage is well suited. This is the subject of part III 'Resonant network antennas in magnetized plasma', beginning with an introduction to whistler waves in chapter 10.

References

[1] Hollenstein Ch, Guittienne Ch and Howling A A 2013 Resonant RF network antennas for large-area and large-volume inductively coupled plasma sources *Plasma Sources Sci. Technol.* **22** 055021

[2] Guittienne Ph, Fayet P, Larrieu J, Howling A A and Hollenstein Ch 2012 Plasma generation with a resonant planar antenna *55th Annual SVC (Surface Vacuum Coaters) Technical Conf. (Santa Clara, CA, 28 April–3 May)* p 3

[3] Furno I, Guittienne P, Howling A A, Plyushchev G and Rudolf von Rohr P 2018 High pressure resonant network antennas for powder treatment *5th workshop Plasma Science and Interfaces (St Gallen, 18–19 October)*

[4] Jin J-M 1998 *Electromagnetic Analysis and Design in Magnetic Resonance Imaging* (Boca Raton, FL: CRC Press)

[5] Romano F *et al* 2020 RF helicon-based inductive plasma thruster (IPT) design for an atmosphere-breathing electric propulsion system (ABEP) *Acta Astronaut.* **176** 476

[6] Drexler P, Jurik K and Stary J 2023 Design and fabrication of birdcage resonators for low-pressure plasma excitation *Radioengineering* **32** 44

[7] Guittienne Ph, Chevalier E and Hollenstein Ch 2005 Towards an optimal antenna for helicon waves excitation *J. Appl. Phys.* **98** 083304

[8] Borg G G and Jahreis T 1994 Radio-frequency power combiner for CW and pulsed applications *Rev. Sci. Instrum.* **65** 449

[9] Bleaney B I and Bleaney B 1976 *Electricity and Magnetism* (London: Oxford University Press) 3rd edn p 1
[10] Piejak R B, Godyak V A and Alexandrovich B M 1992 A simple analysis of an inductive RF discharge *Plasma Sources Sci. Technol.* **1** 179
[11] Arpagaus C, Oberbosse G and Rudolf von Rohr P 2018 Plasma treatment of polymer powders–from laboratory research to industrial application *Plasma Process. Polym.* **15** e1800133
[12] Butscher D, Zimmermann D, Schuppler M and Rudolf von Rohr P 2015 Inactivation of microorganisms on granular materials: reduction of *Bacillus amyloliquefaciens* endospores on wheat grains in a low pressure plasma circulating fluidized bed reactor *J. Food Eng.* **159** 48
[13] Oberbossel G, Güntner A T, Kündig L, Roth C and Rudolf von Rohr P 2015 Polymer powder treatment in atmospheric pressure plasma circulating fluidized bed reactor *Plasma Process. Polym.* **12** 285
[14] Wallimann R, Roth C and Rudolf von Rohr P 2018 Lactose powder flowability enhancement by atmospheric pressure plasma treatment *Plasma Process. Polym.* **15** e1800088
[15] Lieberman M A and Lichtenberg A J 2005 *Principles of Plasma Discharges and Materials Processing* 2nd edn (Hoboken, NJ: Wiley)
[16] Framil D, Van Gompel M, Bourgeois F, Furno I and Leterrier Y 2019 The influence of microstructure on nanomechanical and diffusion barrier properties of thin PECVD SiO_x films deposited on parylene C substrates *Front. Mater.* **6** 319
[17] Smith D L 1995 *Thin Film Deposition: Principles and Practice* (New York: McGraw-Hill)
[18] Ohring M 2001 *Materials Science of Thin Films* (New York: Academic)
[19] Theil J A 1995 Sticking probability and step coverage studies of SiO_2 and polymerized siloxane thin films deposited by plasma enhanced chemical vapor deposition *MRS Online Proc. Libr.* **385** 97
[20] Demolon P, Guittienne Ph, Howling A A, Jost S, Jacquier R and Furno I 2018 RF bias to suppress post-oxidation of μc-Si:H films deposited by inductively-coupled plasma using a planar RF resonant antenna *Vacuum* **147** 58
[21] Howling A A, Descoeudres A and Hollenstein Ch 2020 Multiple dehydrogenation reactions of negative ions in low pressure silane plasma chemistry *Plasma Sources Sci. Technol.* **29** 105015
[22] George S M 2010 Atomic layer deposition: an overview *Chem. Rev.* **110** 111
[23] Profijt H B, Potts S E, van de Sanden M C M and Kessels W M M 2011 Plasma-assisted atomic layer deposition: basics, opportunities, and challenges *J. Vac Sci. Technol.* A **29** 050801

Part III

Resonant network antennas in magnetized plasma

Chapter 10

Whistler waves in an infinite uniform magnetized plasma

It is well known that electromagnetic (EM) waves cannot propagate freely into unmagnetized plasma for frequencies lower than the electron plasma frequency, which is typically in the GHz range for most applications. When an EM wave arrives on the plasma surface it is attenuated over a distance of typically a few centimeters, called the skin depth. If the electron–neutral collision frequency ν is greater than the wave frequency ω, the wave energy is absorbed over the collisional skin depth, see appendix D.5.2. Consequently, for the inductively coupled RF plasma sources of chapter 5, all the energy deposition (i.e. the ionization, excitation, and gas collisional heating) occurs in this collisional skin depth region and the plasma is concentrated in the vicinity of the driving currents. If, on the other hand, the collision frequency ν is much less than the wave frequency ω, the wave energy is reflected by the freely oscillating electrons and the wavefield is evanescent over the collisionless skin depth, see appendices D.5.3 and D.5.4. Other possibilities such as the anomalous skin depth [1] are not considered here.

The situation is very different for a plasma magnetized by an externally imposed static magnetic field, whereupon the electrons are constrained to follow an orbital or spiral motion around the magnetic field lines [2–10]. The plasma conductivity is then described by an anisotropic tensor, and several lower-frequency circularly polarized EM modes can propagate into the plasma. One example is whistler waves, which were first observed in ionospheric plasma studies [11–16]. In that case, the plasma is a pre-existing medium wherein very low frequency whistlers are generated, notably by lightning. Later, in the second part of the twentieth century, the successful excitation of whistlers in a cylindrical plasma led to so-called helicon plasma sources [17]. Helicon waves are essentially whistler waves in a bounded geometry. The wave heated regimes are obtained using antennas [18–20] to produce an RF field that, ideally, should match, more or less, the electromagnetic structure of the cylindrically bounded whistler modes.

Helicon discharges are characterized by high electron densities, typically one order of magnitude higher than conventional inductively coupled discharges, although the underlying power deposition mechanisms are not yet well understood [5, 21–25]. These could include collisional (classical, stochastic, anomalous) and/or collisionless (Landau damping, resonance and kinetic effects) heating, and/or—simply—the access of propagating waves to a larger volume of magnetized plasma than for the skin depth of non-magnetized, inductively coupled plasma. The contrast between non-magnetized plasma in part II, and magnetized plasma in this part III, is more evident when electron–neutral collisions do *not* disrupt the electron orbital motion causing collisional damping of the plasma waves. Therefore, we will mostly be concerned with *collisionless* conditions, $\nu \ll \omega$, in this chapter.

The helicon source geometry is intrinsically cylindrical as the discharge is generated in a tube with an axial static magnetic field. Specially shaped RF antennas are used to excite the whistler wave, the more popular being the Nagoya III [18], the double saddle coil [19], and the helical antenna [20]. Despite their ability to generate high density plasmas, these conventional helicon sources have not yet met with a very wide range of industrial applications because of certain constraints, notably in terms of operating pressure (typically limited to less than 1 Pa) and higher powers required with electronegative[1] gases [27]. To overcome these limitations, and in the context of this book, a cylindrical resonant network, called a birdcage, has been shown to be a very efficient alternative antenna design for helicon wave excitation [28], with extended operation parameters. This improved performance is, in large part, because the birdcage coil is not only efficient for wave excitation, but also for inductive coupling, which produces the critical electron densities necessary to make the transition toward wave heated regimes.

In this chapter, we will introduce the basic theory of plane waves in an infinite, uniform magnetized plasma, and show why we place a special emphasis on whistlers. To cut a long story short, whistlers are a branch of right-hand circularly polarized waves (so-called 'R-waves') which are, in fact, the only waves to propagate in low temperature magnetized plasma at RF frequencies. The transformation of whistler waves into helicon modes will be treated in the following chapter, chapter 11, and helicon plasma sources using birdcage antennas and novel planar antennas are presented in chapter 12 to complete this part III of the book.

10.1 Introduction and classification of plasma waves

In plasma physics, an 'electromagnetic electron wave' is a wave in a plasma which has oscillating magnetic and electric field components and in which primarily the electrons oscillate. There may or may not be a static magnetic field. In an unmagnetized plasma (see appendix D), an electromagnetic electron wave is simply

[1] Atoms or molecules with positive affinity for electron attachment are said to be electronegative, and can form stable negative ions [26].

a light wave (electric and magnetic fields perpendicular to each other, and to the direction of propagation) modified by the plasma relative permittivity.

In this part III of the book, we now consider an externally imposed, uniform, static magnetic field for the first time. The electrons undergo driven oscillations at the RF frequency, and their fluid motion is determined by the RF electric field force, the Lorentz force in the static magnetic field B_0, and electron–neutral collisions (see sections 10.3.1 and 10.9). A plasma in which the electrons perform several oscillations between collisions is called a *magnetized plasma*, or a *magnetoplasma*. A plasma magnetized in this way must have a neutral density that is not too high and must be located in a strong enough magnetic field. Clearly, the frequency ω of the exciting wave and the driven electron oscillation must be at least several times higher than the electron–neutral collision frequency ν, i.e. $\omega/\nu \gg 1$; otherwise, the oscillatory motion is disrupted by frequent collisions, the magnetic field has no influence, and waves are collisionally damped.

10.1.1 Waves considered here: electromagnetic electron waves in unbounded, magnetized plasma

Waves are classified according to their propagation direction along $\mathbf{k}$, and their electric field $\mathbf{E}$ orientation, with respect to the static magnetic field $\mathbf{B}_0$ [1, 8, 29, 30].

1. $\mathbf{k} \perp \mathbf{B}_0$ and $\mathbf{E} \| \mathbf{B}_0$, linearly polarized: 'ordinary wave', O-wave.
2. $\mathbf{k} \perp \mathbf{B}_0$ and $\mathbf{E} \perp \mathbf{B}_0$, elliptically polarized: 'extraordinary wave', X-wave.
3. $\mathbf{k} \| \mathbf{B}_0$ and $\mathbf{E} \perp \mathbf{B}_0$, circularly polarized: R-wave (includes whistlers) and L-wave.

The O, X, R, and L-waves are called the 'principal waves' in magnetized plasma. As will be shown below, in the part of the frequency spectrum between the ion and electron cyclotron frequencies, the R-wave is also called a 'whistler wave' or simply, a 'whistler'[2]. Our aim in this chapter is to demonstrate why RF plasma sources are principally concerned with whistlers, and not with the alternative panoply of plasma waves. To complicate matters, *whistler* refers to the R-wave in *unbounded* magnetized plasma, but in RF plasma sources—which, naturally, are bounded in space—the terminology changes to *helicon* mode, which will be the subject of chapter 11.

10.1.2 Waves not considered here

Within the taxonomy of plasma waves [29], the following are not electromagnetic electron waves in infinite, uniform plasma, and therefore do not concern us further:

1. Electrostatic electron waves (mentioned briefly in section 10.6.1), which have no magnetic component. This excludes plasma oscillations (Langmuir waves) and upper hybrid oscillations. However, dominantly electrostatic

[2] To nip any confusion in the bud, we note from the outset that R- and L-waves correspond, respectively, to right-hand polarization (RHP) and left-hand polarization (LHP). The R and L labels do *not* refer to the right- or left-handed pitch of the wavefields.

Trivelpiece–Gould modes will be necessary for the different case of boundary condition matching in confined plasmas; see chapter 11.

2. Electrostatic ion waves also have no magnetic component, and their frequency is far below the RF spectrum of interest in this book. This, therefore, excludes acoustic waves, electrostatic ion cyclotron waves, and lower hybrid oscillations (see section 10.7 for typical frequencies).
3. Electromagnetic ion waves also have frequencies far below the RF spectrum. This excludes magnetosonic waves and Alfven waves.

10.2 Revision of polarization in magnetized plasma

In appendix D.4, it was shown that electric and magnetic fields are perpendicular to each other and the direction of propagation. There are three possible states of polarization: linearly (or plane) polarized; elliptically (including circularly) polarized; and unpolarized. The latter means that there is a random direction of **E** as the wave propagates. Polarization becomes an important issue for magnetized plasma in general, and for our discussion of R- and L-waves in particular.

10.2.1 Linear polarization

Cartesian convention often takes the wave electric field to be $\mathbf{E} = E(t)\hat{\mathbf{x}}$ (where $\hat{\mathbf{x}}$ is the unit vector along x), and the wave magnetic field strength to be $\mathbf{H} = H(t)\hat{\mathbf{y}}$, where $E(t)$ and $H(t)$ are both real, so that the wave propagates along the Poynting vector direction $\mathbf{S} = \mathbf{E} \times \mathbf{H}$ which is along the $+z$-axis. The wave is said to be plane polarized in the x direction, i.e. it is the electric field which is chosen to define the polarization plane. If the electric field also has an in-phase y component, then the wave would be plane polarized in the **E**-vector-resultant-direction $\hat{\mathbf{n}}$ according to the relative strength of the x and y components. For the general plane polarization at angle ϕ to the x-axis, $\mathbf{E} = E(t)\hat{\mathbf{n}} = E(t)(\hat{\mathbf{x}}\cos\phi + \hat{\mathbf{y}}\sin\phi)$, and a travelling sinusoidal wave electric field is

$$\mathrm{Re}(\mathbf{E}e^{j(kz-\omega t)}) = E_0(\hat{\mathbf{x}}\cos\phi + \hat{\mathbf{y}}\sin\phi)\cos(kz - \omega t); \tag{10.1}$$

where E_0 is the electric field amplitude.

10.2.2 Circular polarization, right and left

Circular polarization, right- or left-handed, refers uniquely to the *temporal* rotation of a field, not to its helical spatial structure (the spatial handedness, or thread). By definition of polarization, the electric field of a right-hand polarized wave rotates clockwise (cw) in time when viewed along the direction of the static, externally imposed magnetic field $\mathbf{B}_0$, *irrespective of the wave's propagation direction*, parallel, or anti-parallel, to $\mathbf{B}_0$. Both polarizations are explained schematically in figure 10.1.

Right-handed polarization, RHP

The polarization can be deduced from a movie of the electric field at any fixed point in space. The thread handedness can be deduced from a snapshot of the electric field structure at any given time. Figure 10.2 is intended to show the temporal and spatial

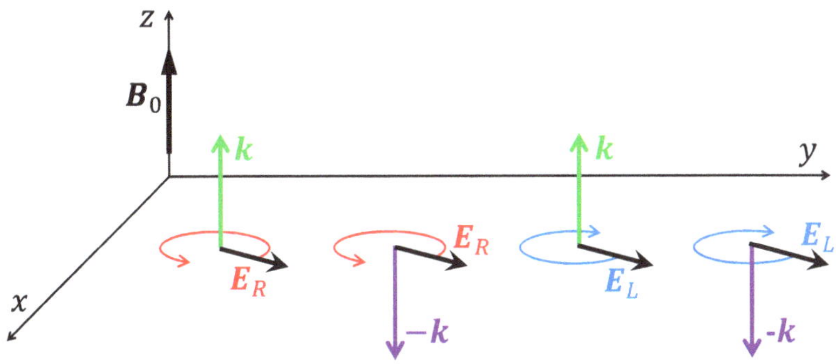

Figure 10.1. Four combinations of wavenumber direction and electric wavefield rotation are shown. The right-hand polarization (E_R, red circles) and left-hand polarization (E_L, blue circles) are defined with respect to the direction of $\mathbf{B}_0$ (along $+z$ in this example). The black arrows indicate planar electric wavefields which are rotating in the $\{x, y\}$ plane. The wavenumbers $+\mathbf{k}$ (green) and $-\mathbf{k}$ (purple) signify that the polarizations are defined irrespective of the direction of wave propagation, whether parallel or anti-parallel to $\mathbf{B}_0$.

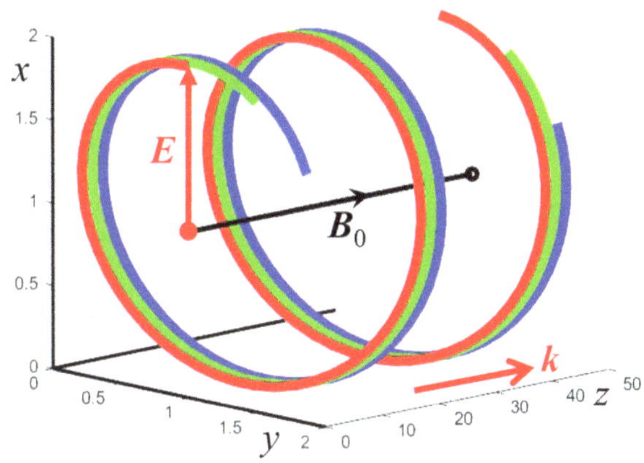

Figure 10.2. Red, green, and blue snapshots at three successive times (0, $\tau/12$, $\tau/6$, where τ is the wave period) showing two wavelengths of a wavefield as it propagates along $+z$. The wavefield exhibits a left-handed thread. Looking along $\mathbf{B}_0$, the $R \rightarrow G \rightarrow B$ colour sequence shows that the rotation is clockwise in time, so the wave is right-hand polarized (RHP).

variations in the same figure, using an example of left-handed thread and right-handed polarization. We note here that the handedness of polarization of a circularly polarized wave depends on the direction of view—this is why it is necessary to specify a view direction which is defined to be along $\mathbf{B}_0$. On the other hand, the thread of the wave is the same independent of the direction of view along a snapshot of the wave. This can be verified in figure 10.2 (as well as in the later figures of section 10.6.2).

In contrast to the general linear polarization in (10.1), elliptical polarization is caused by a dephasing in time between the x and y directions, which causes the electric field to rotate. Dephasing occurs for complex electric field phasors, which incorporate a phase shift. For circular polarization, the x and y amplitudes are

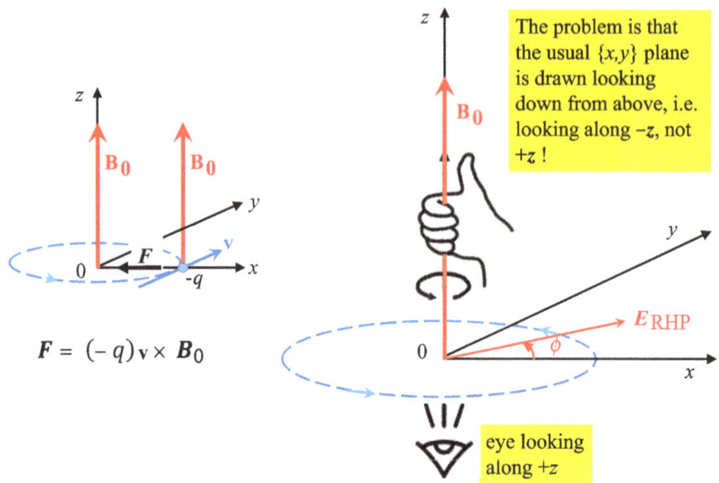

Figure 10.3. Left: The right-hand rule for negative charge gyration in a uniform magnetic field $\mathbf{B}_0$, where the charge $-q$ experiences a force $\mathbf{F} = (-q)\mathbf{v} \times \mathbf{B}_0$. This also gives the same direction of rotation in time of the right-hand polarized (RHP) electric wavefield. Right: When looking from *below* the $\{x, y\}$ plane, along the $+z$-axis, the electric field $\mathbf{E}_{\mathrm{RHP}}$ is turning *clockwise*; this is right-hand polarization (RHP) if $\mathbf{B}_0$ is along $+z$. This can be confusing until it is realized that the conventional right-handed coordinate system is drawn with the x–abscissa and y–ordinate axes looking from above, i.e. along $-z$, not $+z$.

equal, but the x, y components differ by a factor $\pm j = e^{\pm j\pi/2}$ (quadrature) according to the polarization. For RHP, we have

$$\begin{aligned} \text{RHP:} \quad \mathrm{Re}(\mathbf{E}) &= \mathrm{Re}\left[\frac{E_0}{\sqrt{2}}(\hat{\mathbf{x}} + j\hat{\mathbf{y}})e^{j(kz-\omega t)}\right] \\ &= \frac{E_0}{\sqrt{2}}[\hat{\mathbf{x}}\cos(kz - \omega t) - \hat{\mathbf{y}}\sin(kz - \omega t)], \end{aligned} \tag{10.2}$$

where E_0 is the electric field amplitude[3]. The electric field is *uniform* over the infinite transverse plane at any one time and position; it is not radial, nor is around a circumference. Circular polarization just means that the direction of $\mathbf{E}$ is rotating in time.

For a fixed *position* (say, $z = 0$), from (10.2) we have directly

$$\begin{aligned} \text{RHP:} \quad \mathrm{Re}(\mathbf{E}) &= \frac{E_0}{\sqrt{2}}[\hat{\mathbf{x}}\cos(-\omega t) - \hat{\mathbf{y}}\sin(-\omega t)] \\ &= \frac{E_0}{\sqrt{2}}[\hat{\mathbf{x}}\cos(\omega t) + \hat{\mathbf{y}}\sin(\omega t)], \quad \text{at} \quad z = 0. \end{aligned} \tag{10.3}$$

At time $t = 0$, $\mathbf{E}_0$ is uniform and lies along the $+x$-axis. As time starts to increase, the y component starts to rise (grow positively) along $+y$ in (10.2). In figure 10.3, when

[3] The amplitude squared is $\mathbf{E}\cdot\mathbf{E}^* = \left[\frac{E_0}{\sqrt{2}}(\hat{\mathbf{x}} + j\hat{\mathbf{y}})e^{j(kz-\omega t)}\right]\cdot\left[\frac{E_0}{\sqrt{2}}(\hat{\mathbf{x}} - j\hat{\mathbf{y}})e^{-j(kz-\omega t)}\right] = \frac{E_0^2}{2}\times 2 = E_0^2$.

looking from *below* the x, y plane, i.e. along the $+z$-axis, the electric field is turning *clockwise*; this is *right-hand polarization* (RHP) if $\mathbf{B}_0$ is along $+z$. If the right-hand thumb is aligned with $\mathbf{B}_0$, the fingers indicate the direction of rotation of $\mathbf{E}$. While we are on the subject of RHP, we also note that negative charges gyrate according to the right-hand rule [1]: the magnetic component of the Lorentz force for a negative charge $-q$ is $(-q)\mathbf{v} \times \mathbf{B}_0$, which provides the centripetal force $\mathbf{F}$ as shown in figure 10.3.

For a fixed *time*, (say $t = 0$), at the origin $z = 0$, $\mathbf{E}_0$ lies along the $+x$-axis, as before. However, if the electric field at further distance along the wave, along $+z$, has its y component decreased along the $+y$ direction as in (10.2), the field has an anti-clockwise (acw) thread progressing along the $+z$-axis, i.e. a left-hand thread, see figure 10.2. To summarize, (10.2) represents an RHP wave with right-hand polarization and a left-hand thread.

Left-handed polarization (LHP)

Clearly, the opposite sign for the y component of electric field in (10.2) will give the opposite case:

$$\begin{aligned}\text{LHP:}\quad \mathrm{Re}(\mathbf{E}) &= \mathrm{Re}\left[\frac{E_0}{\sqrt{2}}(\hat{\mathbf{x}} - j\hat{\mathbf{y}})e^{j(kz-\omega t)}\right] \\ &= \frac{E_0}{\sqrt{2}}[\hat{\mathbf{x}}\cos(kz - \omega t) + \hat{\mathbf{y}}\sin(kz - \omega t)],\end{aligned} \tag{10.4}$$

which corresponds to anti-clockwise rotation in time, left-hand polarization (LHP); with a clockwise helical thread (right-handed thread), looking along $+z$. This LHP wave structure is like a regular right-handed corkscrew being turned anti-clockwise so as to twist it off a cork.

To labour the point, for a fixed *position* (say, $z = 0$), from (10.4) we have directly

$$\begin{aligned}\text{LHP:}\quad \mathrm{Re}(\mathbf{E}) &= \frac{E_0}{\sqrt{2}}[\hat{\mathbf{x}}\cos(-\omega t) + \hat{\mathbf{y}}\sin(-\omega t)] \\ &= \frac{E_0}{\sqrt{2}}[\hat{\mathbf{x}}\cos(\omega t) - \hat{\mathbf{y}}\sin(\omega t)], \quad \text{at} \quad z = 0.\end{aligned} \tag{10.5}$$

At time $t = 0$, $\mathbf{E}_0$ is uniform and lies along the $+x$-axis. As time starts to increase, the y component starts to grow negatively, along $-y$, in (10.4). In figure 10.3, when looking from *below* the x, y plane, i.e. along the $+z$-axis, the electric field is turning *anti-clockwise*; this is *left-hand polarization* (LHP) if $\mathbf{B}_0$ is along $+z$.

Elliptical polarization

The RHP and LHP circular polarizations are special cases—the most general polarization is elliptical, where the complex amplitudes of the x and y directions are described by the Jones vector $(E_{0x}e^{j\phi_x}\hat{\mathbf{x}} + E_{0y}e^{j\phi_y}\hat{\mathbf{y}})$. This encompasses linear, elliptical, and circular polarizations generally.

10.3 Conductivity and permittivity tensors in uniform magnetoplasma

Plasma properties for unmagnetized plasma were the object of study in part II and its accompanying appendix D. The present chapter is the natural sequel to appendix D by adding the Lorentz force into the electron fluid force equation in the presence of an RF electric field and electron–neutral collisions. If the static magnetic field $\mathbf{B}_0$ is set to zero, the unmagnetized results are recovered.

10.3.1 Solution using electron fluid velocity components

A uniform, static magnetic field $\mathbf{B}_0 = B_0\hat{\mathbf{z}}$ is imposed along the z-axis by means of external coils. We assume a sinusoidal time variation for the small-amplitude perturbations in electric field, the electron fluid velocity, and the electron current [1]. The cold plasma is therefore defined to be a linear, anisotropic medium, in first order terms. The electron fluid force balance equation neglecting pressure and fluid convection, (D.13), is modified by the static magnetic field [1]; apart from this, the following treatment is analogous to the method used for unmagnetized plasma in (D.13) to (D.27). Electron–neutral collisions are accounted for by writing $\hat{\omega} = \omega + j\nu$ in the following linearized equations:

$$m_e n_{e0}\frac{d\mathbf{u}}{dt} = -qn_{e0}(\mathbf{E} + \mathbf{u} \times \mathbf{B}_0) - m_e\nu n_{e0}\mathbf{u}, \tag{10.6}$$

$$-j\omega m_e\mathbf{u} = -q(\mathbf{E} + \mathbf{u} \times \mathbf{B}_0) - m_e\nu\mathbf{u}, \tag{10.7}$$

$$-j\hat{\omega} m_e\mathbf{u} = -q(\mathbf{E} + \mathbf{u} \times \mathbf{B}_0), \quad \text{where} \tag{10.8}$$

$$\hat{\omega} = \omega + j\nu; \quad \text{whence} \tag{10.9}$$

$$\mathbf{E} = \frac{m_e}{nq^2}\left[j\hat{\omega}nq\mathbf{u} - \frac{qB_0}{m_e}nq\mathbf{u} \times \hat{\mathbf{z}}\right], \tag{10.10}$$

$$\mathbf{E} = \frac{1}{\varepsilon_0\omega_{\text{pe}}^2}[-j\hat{\omega}\mathbf{J} + \omega_{\text{ce}}\mathbf{J} \times \hat{\mathbf{z}}], \tag{10.11}$$

where the electron charge and mass are $(-q)$ and m_e, respectively, the time-constant number density is n_{e0}, the electron fluid velocity is $\mathbf{u}$, and the electric current $\mathbf{J} = -qn_{e0}\mathbf{u}$ is assumed to be carried only by the electrons as in (D.15). The electron plasma frequency, $\omega_{\text{pe}} = \sqrt{\frac{n_{e0}e^2}{\varepsilon_0 m_e}}$, and the electron cyclotron frequency, $\omega_{\text{ce}} = \frac{qB_0}{m_e}$ (defined positive), are both angular frequencies in these expressions, with units (rad s^{-1}). The pressure and magnetic viscosity terms are neglected (cold plasma theory [8]). The linearized fluid force equation (10.6) superficially resembles the force equation for single particle motion multiplied by the electron density as some sort of statistical ensemble of electrons, i.e. a fluid element [8]; however, this is a consequence of the several approximations used in its derivation [1, 8], see also appendix D.2. The resulting equation (10.11), in conjunction with Maxwell's equations, has spawned a large number of theories of waves in magnetized plasma.

In accordance with the $e^{-j\omega t}$ choice in section 2.1.1, the time derivatives d/dt have been written as $-j\omega$ because a Fourier component, at the RF frequency, has been implicitly assumed [30]. In fact, this is only valid for steady-state *finite* motion [29], i.e. where the electron motion does not grow indefinitely. For exact, collisionless resonance conditions, or certain configurations in static electric and magnetic fields which cause drift, an explicit time solution must instead be calculated as in section 10.8. Moreover, the unstated assumption of just one single frequency component is not trivial—simultaneous motion at the electron cyclotron frequency can also occur, which is generally different from the RF frequency. This is not a question of harmonics; the RF frequency and electron cyclotron frequency are unrelated, independent values. However, section 10.9 shows that the only steady-state motion is indeed at the RF frequency because it is continuously driven by the RF electric field, whereas other frequencies must eventually be damped, even in nominally collisionless conditions. RF fields therefore can drive waves at the RF frequency, but do not excite other frequencies such as the electron cyclotron frequency. Consequently, for steady-state, linear theory in cold plasma, the RF frequency ω is the only Fourier component to be considered.

Directly writing the linearized electron velocity Fourier components in (10.8) gives [1, 8]:

$$\begin{aligned} j\hat{\omega}u_x &= \frac{q}{m_e}E_x + \omega_{\text{ce}}u_y, \\ j\hat{\omega}u_y &= \frac{q}{m_e}E_y - \omega_{\text{ce}}u_x, \\ j\hat{\omega}u_z &= \frac{q}{m_e}E_z. \end{aligned} \tag{10.12}$$

Solving the simultaneous equations for u_x and u_y gives

$$\begin{aligned} u_x &= -\frac{q}{m_e}\frac{j\hat{\omega}E_x + \omega_{\text{ce}}E_y}{\hat{\omega}^2 - \omega_{\text{ce}}^2}, \\ u_y &= -\frac{q}{m_e}\frac{j\hat{\omega}E_y - \omega_{\text{ce}}E_x}{\hat{\omega}^2 - \omega_{\text{ce}}^2}, \\ u_z &= \frac{q}{m_e}\frac{E_z}{j\hat{\omega}}. \end{aligned} \tag{10.13}$$

The plasma conductivity tensor, $\bar{\sigma}_p$, is defined by the relation between the current density $\mathbf{J}$ and the electric field $\mathbf{E}$, namely, $\mathbf{J} = -qn_{e0}\mathbf{u} = \bar{\sigma}_p \cdot \mathbf{E}$. The terms of $\bar{\sigma}_p$ are generally complex because of electron inertia, as also in (D.16); the magnetized plasma conductivity tensor is

$$\bar{\sigma}_p = \frac{\varepsilon_0\omega_{\rm pe}^2}{(\hat{\omega}^2 - \omega_{\rm ce}^2)}\begin{pmatrix} j\hat{\omega} & \omega_{\rm ce} & 0 \\ -\omega_{\rm ce} & j\hat{\omega} & 0 \\ 0 & 0 & -\dfrac{\hat{\omega}^2 - \omega_{\rm ce}^2}{j\hat{\omega}} \end{pmatrix}, \tag{10.14}$$

where $(q^2 n_{e0}/m_e)$ has been written as $\varepsilon_0\omega_{\rm pe}^2$. To find the plasma relative permittivity tensor, we can repeat the analogous procedure used for the unmagnetized plasma in appendix D.3, by adding the vacuum displacement current $\varepsilon_0\frac{\partial \mathbf{E}}{\partial t}$ to $\mathbf{J}$, to obtain one of Maxwell's equations [31]:

$$\begin{aligned} \nabla \times \mathbf{H} &= \mathbf{J} + \varepsilon_0\frac{\partial \mathbf{E}}{\partial t} \\ &= \bar{\sigma}_p \cdot \mathbf{E} - j\omega\varepsilon_0\bar{1} \cdot \mathbf{E} \quad \text{(for moving charges in vacuum)} \\ &= -j\omega\varepsilon_0\bar{\varepsilon}_p \cdot \mathbf{E}, \quad \text{(for effective dielectric medium)} \end{aligned} \tag{10.15}$$

$$\text{where} \quad \bar{\varepsilon}_p = \left(\frac{j\bar{\sigma}_p}{\omega\varepsilon_0} + \bar{1}\right), \tag{10.16}$$

where $\bar{1}$ is the identity matrix, and $\bar{\varepsilon}_p$ is the plasma relative permittivity tensor. Note that ω, in (10.15) and (10.16) and the following equations, is due to the vacuum displacement current density, $-j\omega\varepsilon_0\bar{1} \cdot \mathbf{E}$, and is distinct from $\hat{\omega}$ derived from the force equations (10.6) to (10.11). Substituting (10.14) into (10.16) gives

$$\bar{\varepsilon}_p = \frac{j}{\omega}\frac{\omega_{\rm pe}^2}{(\hat{\omega}^2 - \omega_{\rm ce}^2)}\begin{pmatrix} j\hat{\omega} & \omega_{\rm ce} & 0 \\ -\omega_{\rm ce} & j\hat{\omega} & 0 \\ 0 & 0 & -\dfrac{\hat{\omega}^2 - \omega_{\rm ce}^2}{j\hat{\omega}} \end{pmatrix} + \bar{1}, \quad \text{so finally:} \tag{10.17}$$

$$\bar{\varepsilon}_p = \begin{pmatrix} \varepsilon_\perp & j\varepsilon_\times & 0 \\ -j\varepsilon_\times & \varepsilon_\perp & 0 \\ 0 & 0 & \varepsilon_\parallel \end{pmatrix}, \quad \text{where} \tag{10.18}$$

$$\varepsilon_\perp = 1 - \frac{\hat{\omega}}{\omega}\left(\frac{\omega_{\rm pe}^2}{\hat{\omega}^2 - \omega_{\rm ce}^2}\right), \tag{10.19}$$

$$\varepsilon_\times = \frac{\omega_{\rm ce}}{\omega}\left(\frac{\omega_{\rm pe}^2}{\hat{\omega}^2 - \omega_{\rm ce}^2}\right), \tag{10.20}$$

$$\varepsilon_\parallel = 1 - \frac{\omega_{\rm pe}^2}{\omega\hat{\omega}}. \tag{10.21}$$

$\varepsilon_{\parallel}$ is the same expression as ε_p in unmagnetized plasma, equation (D.27), because the external magnetic field has no effect on electrons moving parallel to it. To permit real solutions for the following work on dispersion relations, it is necessary to assume collisionless conditions ($\hat{\omega} \rightarrow \omega$) so that $\varepsilon_{\perp}$, $\varepsilon_{\times}$, and $\varepsilon_{\parallel}$ are real. The off-diagonal elements of $\bar{\varepsilon}_p$ are then still complex, but complex conjugates, hence the tensor is Hermitian, with real eigenvalues.

10.4 Plane wave dispersion relations in collisionless magnetoplasma

We return to the wave equation which is the combination of Maxwell's equations describing electromagnetic waves (D.35), but where the plasma relative permittivity $\bar{\varepsilon}_p$ is now a tensor appropriate for magnetized plasmas:

$$\mathbf{k} \times (\mathbf{k} \times \mathbf{E}) = -k_0^2 \bar{\varepsilon}_p \cdot \mathbf{E}, \tag{10.22}$$

where $k_0 = \frac{\omega}{c}$ is the wavenumber of light in vacuum. This wave equation is more complicated than for unmagnetized plasmas, (D.36), because of the coupled components of $\mathbf{E}$. The terms $\varepsilon_{\perp}$, $\varepsilon_{\times}$, and $\varepsilon_{\parallel}$ in $\bar{\varepsilon}_p$, (10.18), are taken to be real (collisionless approximation) by setting $\hat{\omega} \rightarrow \omega$. We already have $\mathbf{B}_0 = B_0\hat{\mathbf{z}}$, and without loss of generality, we take $\mathbf{k}$ to lie in the (x, z) plane as in figure 10.4(a), i.e. we set $k_y = 0$ so $\mathbf{k} = [k_x, 0, k_z]$. Using (D.28), assuming plane waves, the wavefield components vary as

$$\mathbf{E}, \mathbf{B} \propto e^{j(\mathbf{k}\cdot\mathbf{r} - \omega t)} = e^{j(k_x x + k_z z - \omega t)}, \tag{10.23}$$

in Cartesian geometry. To expand $\mathbf{k} \times \mathbf{k} \times \mathbf{E}$ we begin with a vector identity

$$\begin{aligned}
\mathbf{k} \times \mathbf{k} \times \mathbf{E} &\equiv (\mathbf{k} \cdot \mathbf{E})\mathbf{k} - k^2\mathbf{E} \\
&= (k_x E_x + k_z E_z)\mathbf{k} - k^2\mathbf{E} \\
&= [k_x^2 E_x + k_x k_z E_z,\ \ 0,\ \ k_x k_z E_x + k_z^2] - k^2\mathbf{E} \\
&= \left[-k_z^2 E_x + k_x k_z E_z,\ -(k_x^2 + k_z^2)E_y,\ \ k_x k_z E_x - k_x^2\right] \\
&= \begin{pmatrix} -k_z^2 & 0 & k_x k_z \\ 0 & -(k_x^2 + k_z^2) & 0 \\ k_x k_z & 0 & -k_x^2 \end{pmatrix} \begin{pmatrix} E_x \\ E_y \\ E_z \end{pmatrix}.
\end{aligned} \tag{10.24}$$

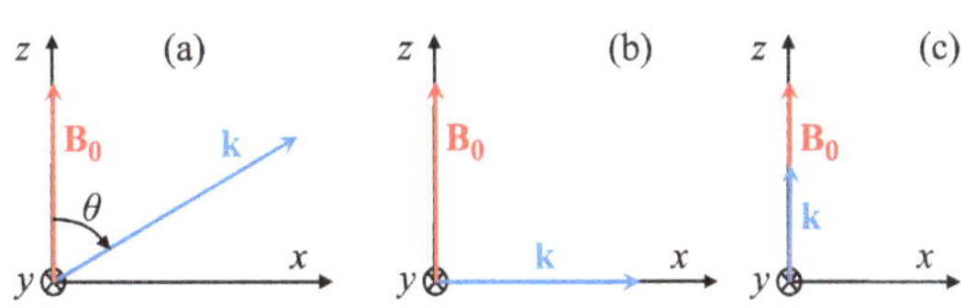

Figure 10.4. Static magnetic field $\mathbf{B}_0$ along $+z$ for all cases. (a) Wavenumber $\mathbf{k}$ at angle θ, in the (x, z) plane, for all E_x, E_y, E_z components, as used in section 10.4 [1]. (b) Wavenumber $\mathbf{k}$ along x: O-wave with E_z only; X-wave with E_x and E_y components [8]. (c) Wavenumber $\mathbf{k}$ along z: R- (whistler) and L-waves with E_x and E_y components [8].

So the tensor wave equation for magnetized plasma (10.22), in contrast to unmagnetized plasma (D.36), is

$$-\mathbf{k}\times(\mathbf{k}\times\mathbf{E}) = +k_0^2\bar{\varepsilon}_p\cdot\mathbf{E},$$

$$\begin{pmatrix} k_z^2 & 0 & -k_xk_z \\ 0 & k_x^2+k_z^2 & 0 \\ -k_xk_z & 0 & k_x^2 \end{pmatrix}\begin{pmatrix} E_x \\ E_y \\ E_z \end{pmatrix} = k_0^2\begin{pmatrix} \varepsilon_\perp & -j\varepsilon_\times & 0 \\ j\varepsilon_\times & \varepsilon_\perp & 0 \\ 0 & 0 & \varepsilon_\parallel \end{pmatrix}\begin{pmatrix} E_x \\ E_y \\ E_z \end{pmatrix}, \tag{10.25}$$

$$\text{i. e.}\begin{pmatrix} k_z^2-k_0^2\varepsilon_\perp & jk_0^2\varepsilon_\times & -k_xk_z \\ -jk_0^2\varepsilon_\times & k_x^2+k_z^2-k_0^2\varepsilon_\perp & 0 \\ -k_xk_z & 0 & k_x^2-k_0^2\varepsilon_\parallel \end{pmatrix}\begin{pmatrix} E_x \\ E_y \\ E_z \end{pmatrix} = 0.$$

We now define $k_z = k\cos\theta$, so $k_x = k\sin\theta$, and the index of refraction of the wave[4] $N = k/k_0$.

For a non-trivial solution of the magnetoplasma wave equation (10.25), its determinant must be zero and its roots give the general plane wave dispersion relations [1]

$$\det=\begin{vmatrix} N^2\cos^2\theta-\varepsilon_\perp & j\varepsilon_\times & -N^2\cos\theta\sin\theta \\ -j\varepsilon_\times & N^2-\varepsilon_\perp & 0 \\ -N^2\cos\theta\sin\theta & 0 & N^2\sin^2\theta-\varepsilon_\parallel \end{vmatrix} = 0. \tag{10.26}$$

Expanding the determinant of the general plane wave dispersion relation gives a biquadratic equation

$$aN^4 + bN^2 + c = 0, \quad \text{where} \tag{10.27}$$

$$a = \varepsilon_\perp\sin^2\theta + \varepsilon_\parallel\cos^2\theta, \tag{10.28}$$

$$b = -(\varepsilon_\perp^2-\varepsilon_\times^2)\sin^2\theta - \varepsilon_\parallel\varepsilon_\perp(1+\cos^2\theta), \tag{10.29}$$

$$c = (\varepsilon_\perp^2-\varepsilon_\times^2)\varepsilon_\parallel. \tag{10.30}$$

The terms in N^6 cancel out.

10.5 Solution of the principal wave dispersion relations

In the quadratic equation (10.27) for N^2, the discriminant is always positive and N^2 is always real in the collisionless approximation. If N^2 is positive, then N is real, and the wave propagates. If N^2 is negative, then N and k are purely imaginary, so the wave is purely evanescent. Equation (10.27) could be directly solved as $N^2 = N^2(\theta, \bar{\varepsilon}_p)$ (see section 10.11.2 for an approximate solution by neglecting the

[4] The refractive index N of unmagnetized plasma is the ratio of the speed of light in vacuum $c = 1/\sqrt{\mu_0\varepsilon_0}$ to the speed (phase velocity) in the plasma $v = 1/\sqrt{\mu_0\varepsilon_p\varepsilon_0}$. Therefore, $N = \frac{c}{v} = \frac{\omega/k_0}{\omega/k} = \frac{k}{k_0}$, and $N^2 = \varepsilon_p$.

vacuum displacement current), but an exact solution instead solves as $\theta = \theta(N^2, \bar{\varepsilon}_p)$, as follows [1].

Equation (10.27) can be re-written using $\sin^2\theta + \cos^2\theta = 1$ in b and c, so the biquadratic can be divided throughout by $\cos^2\theta$,

$$\begin{aligned}&(\varepsilon_\perp \sin^2\theta + \varepsilon_\parallel \cos^2\theta)N^4\\ &-\left[(\varepsilon_\perp^2 - \varepsilon_\times^2)\sin^2\theta + \varepsilon_\parallel\varepsilon_\perp(\sin^2\theta + 2\cos^2\theta)\right]N^2\\ &+(\varepsilon_\perp^2 - \varepsilon_\times^2)\varepsilon_\parallel(\sin^2\theta + \cos^2\theta) = 0,\end{aligned}$$

yielding

$$\tan^2\theta = \frac{-\varepsilon_\parallel(N^4 - 2\varepsilon_\perp N^2 + (\varepsilon_\perp^2 - \varepsilon_\times^2))}{\varepsilon_\perp N^4 - (\varepsilon_\perp^2 - \varepsilon_\times^2)N^2 - \varepsilon_\parallel\varepsilon_\perp N^2 + (\varepsilon_\perp^2 - \varepsilon_\times^2)\varepsilon_\parallel},$$

which can be factorized:

$$\tan^2\theta = \frac{-\varepsilon_\parallel\left[(N^2 - (\varepsilon_\perp + \varepsilon_\times)\right]\left[N^2 - (\varepsilon_\perp - \varepsilon_\times)\right]}{(N^2 - \varepsilon_\parallel)(\varepsilon_\perp N^2 - (\varepsilon_\perp^2 - \varepsilon_\times^2))},$$

where it clearly makes sense to define new terms to replace the bracket factors, such as

$$\varepsilon_R = \varepsilon_\perp + \varepsilon_\times, \tag{10.31}$$

$$\varepsilon_L = \varepsilon_\perp - \varepsilon_\times. \tag{10.32}$$

Finally, we have [1]

$$\tan^2\theta = \frac{-\varepsilon_\parallel(N^2 - \varepsilon_R)(N^2 - \varepsilon_L)}{(N^2 - \varepsilon_\parallel)(\varepsilon_\perp N^2 - \varepsilon_R\varepsilon_L)}. \tag{10.33}$$

We can immediately see that dispersion relations will correspond to each of the four bracketed factors equal to zero (and a special fifth case for $\varepsilon_\parallel = 0$), when $\tan^2\theta = 0$ or ∞. We now consider each of these four principal electron waves R, L, O, and X, which correspond to propagation parallel to $\mathbf{B}_0$ ($\theta = 0$; R- and L-waves); and propagation perpendicular to $\mathbf{B}_0$ ($\theta = \pm\frac{\pi}{2}$; O- and X-waves). Our principal interest will be in the parallel propagation because whistlers—transformed into helicons—will be seen to propagate along the axis of a cylinder aligned with $\mathbf{B}_0$. Moreover, we will see that whistlers are the only waves in the RF spectrum which propagate in magnetoplasma.

10.6 Wave number parallel to the magnetic field

An advantage of the approach in section 10.5 is that it is only necessary to choose the angle θ in figure 10.4(a) in order to cover all the waves. For $\pm\mathbf{k}\|\mathbf{B}_0$, we simply set $\theta = 0$ or π, and solutions for $\tan^2\theta = 0$ in (10.33) yield three possibilities for the numerator to be equal to zero: Langmuir 'waves' in section 10.6.1, R-waves in section 10.6.2, and L-waves in section 10.6.3. The collisionless approximation still applies, so the terms $\varepsilon_\perp$, $\varepsilon_\times$ and $\varepsilon_\parallel$ are purely real.

10.6.1 Langmuir 'wave'

A trivial solution for zero numerator in (10.33) is $\varepsilon_\parallel = 0$. Since $\varepsilon_\parallel = 1 - \frac{\omega_{pe}^2}{\omega\hat{\omega}} \to 1 - \frac{\omega_{pe}^2}{\omega^2}$, in (10.21), the obtained dispersion relation is special because it directly imposes the wave frequency. The phase velocity ω/k does not have any meaning in this context because k, the magnitude of the wave vector, is not defined. Therefore, we have an electron oscillation parallel to the confining magnetic field $\mathbf{B}_0$. The magnetic field exerts no force on electrons moving parallel to it, so the electrons respond as in an unmagnetized plasma at the electron plasma frequency ω_{pe}; see also appendix D.3. Since $\mathbf{k}$ is parallel to $\mathbf{E}$, this characterizes an electrostatic wave, also known as a plasma oscillation or a Langmuir wave. The wave does not, in fact, propagate; it is the response of the charged particles (essentially the electrons, owing to their high mobility) to any deviation of the plasma from quasi-neutrality [32]; it is not one of the four principal waves listed in section 10.1.1.

10.6.2 R-wave

For the next solution of $\tan^2\theta = 0$ in (10.33), we have

$$N^2 = \frac{k^2}{k_0^2} = \varepsilon_R = \varepsilon_\perp + \varepsilon_\times = 1 - \frac{\omega_{pe}^2}{\omega(\hat{\omega} - \omega_{ce})} \to 1 - \frac{\omega_{pe}^2}{\omega(\omega - \omega_{ce})}, \tag{10.34}$$

for collisionless plasma. Note that in the limit of zero magnetic field, $\omega_{ce} = 0$, we recover the relative permittivity for unmagnetized plasma, (D.26), as expected. The R-wave has an electron cyclotron resonance ($k \to \infty$) for $\omega = \omega_{ce}$, and a cut-off ($k = 0$) for the higher frequency

$$\omega_R = \frac{\omega_{ce} + \sqrt{\omega_{ce}^2 + 4\omega_{pe}^2}}{2}. \tag{10.35}$$

Cut-off and resonance are summarized in appendix D.5.4, and calculated for all the waves together in section 10.7. To determine the R-wave polarization, we need the ratio of its electric field components, E_x: E_y (because $\mathbf{k}$ is parallel to $\mathbf{B}_0$ along z). The easiest way to do this is to substitute the R dispersion relation solution $\frac{k^2}{k_0^2} = \varepsilon_\perp + \varepsilon_\times$ (10.34) into the tensor wave equation (10.25), where $k_z = k$ and $k_x = 0$ in our present case

$$\begin{pmatrix} k^2 - k_0^2\varepsilon_\perp & ik_0^2\varepsilon_\times & 0 \\ -ik_0^2\varepsilon_\times & k^2 - k_0^2\varepsilon_\perp & 0 \\ 0 & 0 & k_0^2\varepsilon_\parallel \end{pmatrix}\begin{pmatrix} E_x \\ E_y \\ 0 \end{pmatrix} = 0. \tag{10.36}$$

The first, or second, rows individually give

$$
\begin{aligned}
&(k^2 - k_0^2\varepsilon_\perp)E_x + jk_0^2\varepsilon_\times E_y = 0,\\
&\left(\frac{k^2}{k_0^2} - \varepsilon_\perp\right)E_x + j\varepsilon_\times E_y = 0,\\
&[(\varepsilon_\perp + \varepsilon_\times) - \varepsilon_\perp]E_x + j\varepsilon_\times E_y = 0,\\
&E_y = jE_x. \qquad (10.37)
\end{aligned}
$$

Substituting into (10.2), we recognize a right-hand circularly polarized (RHP) wave,

$$
\begin{aligned}
\mathrm{Re}(\mathbf{E}) &= \mathrm{Re}\left[\frac{E_0}{\sqrt{2}}(\hat{\mathbf{x}} + j\hat{\mathbf{y}})e^{j(kz-\omega t)}\right],\\
&= \frac{E_0}{\sqrt{2}}[\hat{\mathbf{x}}\cos(kz - \omega t) - \hat{\mathbf{y}}\sin(kz - \omega t)],
\end{aligned} \qquad (10.38)
$$

hence the name ‘R-wave’. Note from section 10.2.2 that the R-wave E-field turns in the same direction as the electron gyration (the right-hand rule for negative charges in figure 10.3). Therefore, at the electron cyclotron resonance, the R-wave rotates synchronously with the electrons which consequently experience a constant E-field leading to resonant energy absorption (see section 10.8). The determining role of the direction of the static magnetic field is fixed by the originating equation (10.6) for the force on electrons via the $\mathbf{u} \times \mathbf{B}_0$ term, and this is why polarization is defined with respect to the magnetic field, rather than the direction of propagation. It fixes the RHP temporal lag for both directions of propagation.

For the $+k$ solution, (propagation along $+z$ and $\mathbf{B}_0$), for a given time (e.g. $t = 0$), E_y *decreases* (goes negative) with *distance* z in (10.38). Therefore, this is *left*-hand thread and *right*-hand polarization, see figure 10.5. For the $-k$ solution, (propagation along $-z$, counter to $\mathbf{B}_0$), for a given time (e.g. $t = 0$), E_y *decreases* with *distance*

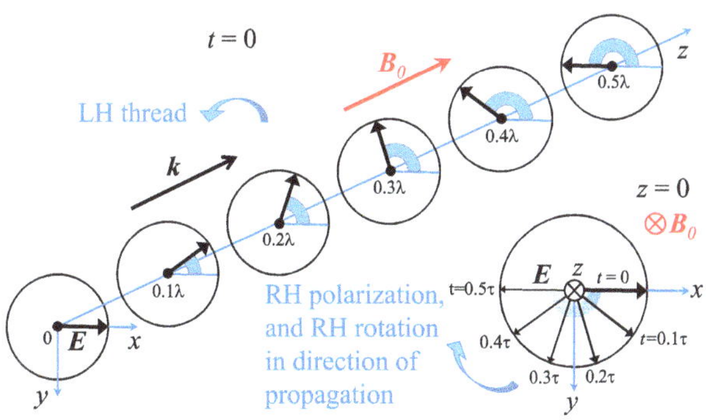

Figure 10.5. R-wave, with $\mathbf{B}_0$ along $+z$, and propagation in the direction of $\mathbf{k}$ which is parallel to $\mathbf{B}_0$: RH polarization (the inset shows clockwise rotation in time when viewed along $\mathbf{B}_0$). The wave snapshot along $+z$, shown for half a wavelength, has a left-hand thread, viewed from either direction. The rotation in time viewed along the direction of propagation is also right-handed.

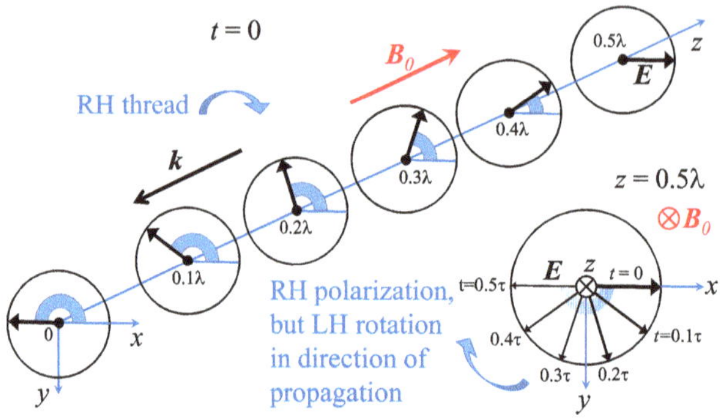

Figure 10.6. R-wave, $\mathbf{B}_0$ still along $+z$, but propagation in the direction of $\mathbf{k}$ which is anti-parallel to $\mathbf{B}_0$. The polarization is still right-handed (still clockwise in time when viewed along $\mathbf{B}_0$ in the inset). A snapshot of the wave is shown along the $-z$ direction for half a wavelength, exhibiting a right-hand thread, viewed from either direction. The $\mathbf{E}$ vector rotation in time viewed along the direction of propagation is now left-handed, opposite to the polarization.

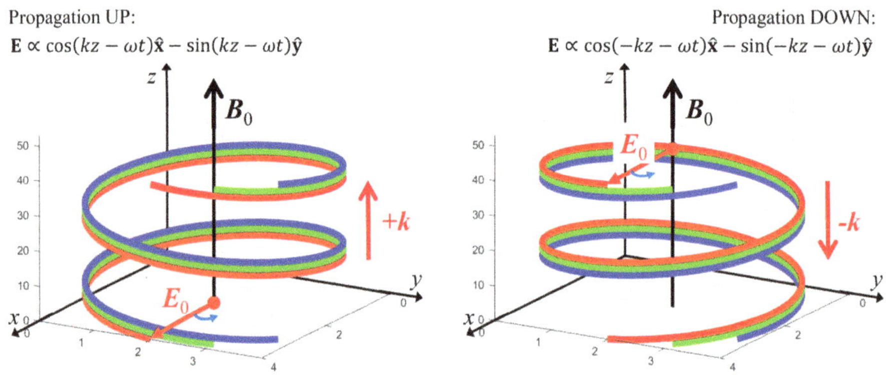

Figure 10.7. Summary of two R-waves, with $\mathbf{B}_0$ along $+z$; see equation (10.38). Left: $\mathbf{k}$ parallel to $\mathbf{B}_0$, propagation upwards. Right: $\mathbf{k}$ anti-parallel to $\mathbf{B}_0$, propagation downwards. With period τ, the time sequence, starting from $\mathbf{E}_0$, is $t = 0 \rightarrow \tau/12 \rightarrow \tau/6$, following the colour scheme $R \rightarrow G \rightarrow B$. The right-hand polarization (clockwise rotation in time when viewed along $\mathbf{B}_0$) is confirmed for both the left and right figures, as in figures 10.5 and 10.6. However, the threads (viewed from either direction), and the rotation in time viewed along the direction of propagation, have opposite handedness between the left and right figures.

along $-z$ in (10.38). This is *right*-hand thread and *right*-hand polarization in figure 10.6. Note that the thread handedness is the same when viewed in either direction. Both figures 10.5 and 10.6 are resumed in figure 10.7. Note that the dispersion relation (10.34) is a function of k^2, so the R-wave solution is the same for $\pm k$, i.e. for both signs of k. This was made clear in figure 10.1 [8]. Finally, all four configurations are resumed in table 10.1.

Table 10.1. For R-waves only: A summary of circular polarization; thread handedness; rotation in time viewed along the direction of propagation; and rotation in time viewed along $+z$, for the wavefields. 'cw' indicates clockwise, and 'acw' means anti-clockwise. All four possible combinations of $\pm\mathbf{B}_0$ and $\pm k$ with respect to the $+z$ direction are shown. The first row corresponds to figure 10.5, the second row to figure 10.6, and both of these rows are summarized in figure 10.7. In chapter 12 for RAID, where the z-axis goes from source to target, only $k > 0$ is possible, if reflections are ignored.

$\mathbf{B}_0$	**k**	Polarization	Thread	Rotation along k	Rotation along $+z$
along $+z$	$k > 0$	RHP	LH thread	RH (cw)	RH (cw)
along $+z$	$k < 0$	RHP	RH thread	LH (acw)	RH (cw)
along $-z$	$k < 0$	RHP	LH thread	RH (cw)	LH (acw)
along $-z$	$k > 0$	RHP	RH thread	LH (acw)	LH (acw)

10.6.3 L-wave

We can repeat section 10.6.2 for the second numerator bracket of $\tan^2\theta = 0$ in (10.33), namely

$$N^2 = \frac{k^2}{k_0^2} = \varepsilon_L = \varepsilon_\perp - \varepsilon_\times = 1 - \frac{\omega_{pe}^2}{\omega(\hat{\omega} + \omega_{ce})} \to 1 - \frac{\omega_{pe}^2}{\omega(\omega + \omega_{ce})}, \tag{10.39}$$

for collisionless plasma. The L-wave has no electron resonance because it turns in the opposite direction to electrons (although it does resonate at the ion cyclotron resonance at much lower frequency when ion motion is accounted for) but it still has a cut-off ($k = 0$) for

$$\omega_L = \frac{-\omega_{ce} + \sqrt{\omega_{ce}^2 + 4\omega_{pe}^2}}{2}. \tag{10.40}$$

To determine the L-wave polarization, we need the ratio of its electric field components, E_x: E_y (because $\mathbf{k}$ is parallel to $\mathbf{B}_0$). Again, the easiest way to do this is to substitute the L dispersion relation solution $\frac{k^2}{k_0^2} = \varepsilon_\perp - \varepsilon_\times$ (10.39) into the tensor wave equation (10.25), where we still have $k_z = k$ and $k_x = 0$ in (10.36). The first, or second, rows individually give

$$\begin{aligned}
&(k^2 - k_0^2\varepsilon_\perp)E_x + jk_0^2\varepsilon_\times E_y = 0,\\
&\left(\frac{k^2}{k_0^2} - \varepsilon_\perp\right)E_x + j\varepsilon_\times E_y = 0,\\
&[(\varepsilon_\perp - \varepsilon_\times) - \varepsilon_\perp]E_x + j\varepsilon_\times E_y = 0,\\
&E_y = -jE_x.
\end{aligned} \tag{10.41}$$

Substituting in (10.4), we recognize a left-hand polarized (LHP) wave,

$$\begin{aligned}\mathrm{Re}(\mathbf{E}) &= \mathrm{Re}\left[\frac{E_x}{\sqrt{2}}(\hat{\mathbf{x}} - j\hat{\mathbf{y}})e^{j(kz-\omega t)}\right],\\ &= \frac{E_0}{\sqrt{2}}[\hat{\mathbf{x}}\cos(kz - \omega t) + \hat{\mathbf{y}}\sin(kz - \omega t)],\end{aligned} \tag{10.42}$$

hence the name 'L-wave'. The E-field of a propagating L-wave rotates anti-clockwise when viewed along $\mathbf{B}_0$, and its spatial structure has a right-hand thread if it propagates along $\mathbf{B}_0$. We could reproduce the R-wave figures and table 10.1 for the L-wave, but the L-wave is not a priority in our context of whistlers, which are R-waves (see section 10.1.1).

10.6.4 Wave number perpendicular to the magnetic field

For $\mathbf{k}\perp\mathbf{B}_0$, we set $\theta = \pi/2$ in figure 10.4(a) and seek solutions for $\tan^2\theta \to \infty$ in (10.33). We intuit from section 10.1.1 that O- and X-waves are the two solutions corresponding to the denominator brackets equal to zero. To demonstrate why we are concerned exclusively with R-waves (whistlers), the O and X dispersion relations and dispersion diagram will be given in section 10.7 without investigating their properties further here.

10.7 Electromagnetic electron wave cut-offs and resonances

In order to compare the four principal waves in a magnetized plasma, we will concentrate on exact solutions for collisionless plasma. By 'exact' we mean that the ion motion contributions are included here, for the first time, to obtain the low frequency resonances [1]. Collisionless conditions continue to be used, i.e. $\hat{\omega} \to \omega$, because there is no simple way of representing complex numbers on the ω versus k dispersion diagram. In the following lists, the plasma angular frequency ω_p squared is the sum of the squares of the electron plasma frequency ω_{pe}, and the ion plasma frequency ω_{pi}, where

$$\omega_p^2 = \omega_{\mathrm{pe}}^2 + \omega_{\mathrm{pi}}^2 = \frac{n_{e0}q^2}{\varepsilon_0}\left(\frac{1}{m_e} + \frac{1}{M}\right), \tag{10.43}$$

assuming $n_{e0} = n_i$, i.e. singly charged ions, where m_e and M are the electron and ion masses, respectively. The 'exact', collisionless, cold plasma dispersion relations for the four principal waves are [1]

$$\begin{aligned}
&\text{O wave:} \quad N_O^2 = \frac{k^2}{k_0^2} = 1 - \frac{\omega_p^2}{\omega^2},\\
&\text{R wave:} \quad N_R^2 = \frac{k^2}{k_0^2} = 1 - \frac{\omega_{\mathrm{pe}}^2}{\omega(\omega - \omega_{\mathrm{ce}})} - \frac{\omega_{\mathrm{pi}}^2}{\omega(\omega + \omega_{\mathrm{ci}})} = 1 - \frac{\omega_p^2}{(\omega - \omega_{\mathrm{ce}})(\omega + \omega_{\mathrm{ci}})},\\
&\text{L wave:} \quad N_L^2 = \frac{k^2}{k_0^2} = 1 - \frac{\omega_{\mathrm{pe}}^2}{\omega(\omega + \omega_{\mathrm{ce}})} - \frac{\omega_{\mathrm{pi}}^2}{\omega(\omega - \omega_{\mathrm{ci}})} = 1 - \frac{\omega_p^2}{(\omega + \omega_{\mathrm{ce}})(\omega - \omega_{\mathrm{ci}})},\\
&\text{X wave:} \quad N_X^2 = \frac{k^2}{k_0^2} = \frac{N_R^2 N_L^2}{1 - \dfrac{\omega_{\mathrm{pe}}^2}{\omega^2 - \omega_{\mathrm{ce}}^2} - \dfrac{\omega_{\mathrm{pi}}^2}{\omega^2 - \omega_{\mathrm{ci}}^2}}.
\end{aligned} \tag{10.44}$$

Table 10.2. Approximate expressions for the cut-off and resonance angular frequencies ω [rad s^{-1}] of the principal electromagnetic plane waves (collisionless, cold uniform plasma, but including mobile ions) [1]. The plasma angular frequency is $\omega_p = (\omega_{pe}^2 + \omega_{pi}^2)^{1/2}$. ω_{UH} and ω_{LH} are the upper and lower hybrid angular frequencies, respectively. Example frequencies f (Hz) are given for H_2^+ and Ar^+ at $B_0 = 200$ G and $n_{e0} = 10^{18}$ m^{-3} for comparison with the RF frequency 13.56 MHz. (Adapted with permission from [34]. Copyright IOP Publishing.)

Wave	ω cut-off	$f^{\text{cut-off}}$	ω resonance	f^{res}
R	$\approx \frac{\omega_{ce} + \sqrt{\omega_{ce}^2 + 4\omega_p^2}}{2}$	9.27 GHz	ω_{ce}	560 MHz
L	$\approx \frac{-\omega_{ce} + \sqrt{\omega_{ce}^2 + 4\omega_p^2}}{2}$	8.71 GHz	ω_{ci}	H_2:153, Ar:7.63 kHz
X	Both as above	9.27 GHz 8.71 GHz	$\omega_{UH}^2 \approx \omega_p^2 + \omega_{ce}^2$ $\frac{1}{\omega_{LH}^2} \approx \frac{1}{\omega_{pi}^2} + \frac{1}{\omega_{ce}\omega_{ci}}$	9.00 GHz H_2:9.23, Ar:2.06 MHz
O	ω_p	8.98 GHz	None	—

The new terms are the ion contributions involving the ion cyclotron frequency $\omega_{ci} = \frac{qB_0}{M}$, where M is the ion mass. The 'exact', collisionless expressions for the cut-off frequencies, when $k \to 0$ in the dispersion relation are

$$\begin{aligned}
&\text{R, X wave cut-off:} \quad \omega_{R,X}^{\text{cut-off}} = \left(\omega_{ce} - \omega_{ci} + \sqrt{(\omega_{ce} - \omega_{ci})^2 + 4\left(\omega_{ce}\omega_{ci} + \omega_p^2\right)}\right)\Big/2,\\
&\text{L, X wave cut-off:} \quad \omega_{L,X}^{\text{cut-off}} = \left(\omega_{ci} - \omega_{ce} + \sqrt{(\omega_{ce} - \omega_{ci})^2 + 4\left(\omega_{ce}\omega_{ci} + \omega_p^2\right)}\right)\Big/2,\\
&\text{O wave cut-off:} \quad \omega_O^{\text{cut-off}} = \omega_p.
\end{aligned} \tag{10.45}$$

The 'exact', collisionless expressions for the resonance frequencies, when $k \to \infty$ in the dispersion relation, are given by:

$$\begin{aligned}
&\text{R wave resonance:} \quad \omega_R^{\text{res}} = \omega_{ce},\\
&\text{L wave resonance:} \quad \omega_L^{\text{res}} = \omega_{ci},\\
&\text{X wave resonance:} \quad \omega_X^{\text{res}} = \omega_{UH} \quad \text{and} \quad \omega_{LH},
\end{aligned} \tag{10.46}$$

where ω_{UH} and ω_{LH} are the upper and lower hybrid resonances given by the roots of the quartic $1 = \frac{\omega_{pe}^2}{\omega^2 - \omega_{ce}^2} - \frac{\omega_{pi}^2}{\omega^2 - \omega_{ci}^2}$. Approximate expressions for the cut-off frequencies and resonances [1] are summarized in table 10.2.

Wavefield behaviour near cut-off or resonance

Near a resonance (approaching from below the resonance frequency), the wavefield amplitude diminishes, the index of refraction becomes infinite, the wavelength decreases to zero ($k \to \infty$), and the wave energy is absorbed. The wave energy is lost

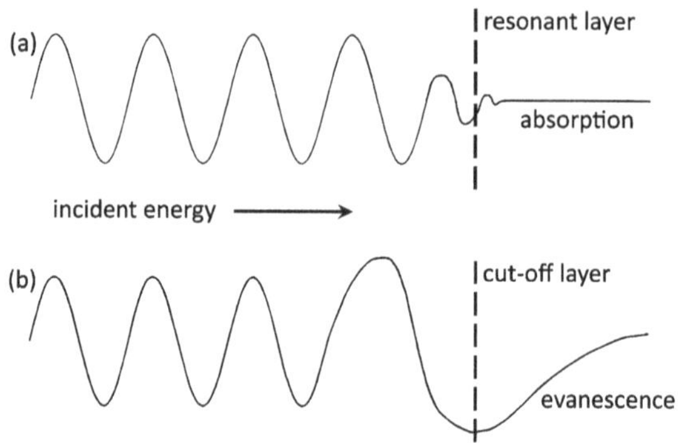

Figure 10.8. (a) Near to a resonance, the wavefield amplitude and wavelength decrease, and the wave energy is absorbed in the plasma. (b) Near to a cut-off layer in a collisionless plasma, the wavefield amplitude and wavelength increase, the wave energy is reflected and the wavefield is evanescesent into the plasma.

by transfer to the particle energy, and its amplitude falls to zero, as shown schematically in figure 10.8(a) [33]. For the R-wave, the resonance is at ω_{ce}.

Near a cut-off in a collisionless plasma (approaching from above the cut-off frequency), the wavefield amplitude increases as the reflected wave energy contributes, the index of refraction goes to zero, and the wavelength becomes infinite ($k \to 0$). The wave is reflected at the cut-off layer, and it is evanescent beyond this point into the plasma, as shown schematically in figure 10.8(b) [33]; see also section 10.10 and appendix D.5.4.

Dispersion diagrams

Figure 10.9 shows an illustrative dispersion diagram for artificial conditions of extremely low effective ion mass, so that the cut-off frequencies are well distinguished for clarity. The dispersion curves of the four principal waves R, L, O and X are all represented, often with multiple branches. The branch which will be our main interest is the R-wave branch labelled 'R whistler', because the R-wave is known as the whistler wave in the frequency range $\omega_{ci} \ll \omega \ll \omega_{ce}$ [1]. As shown in (10.45) and (10.46), there are three cut-offs, for the O, R, X and L-waves; and four resonances at the ion and electron cyclotron frequencies, and the upper and lower hybrid frequencies are so called because they involve combinations of cyclotron and plasma frequencies.

In order to situate these frequencies with respect to realistic experimental parameters, figure 10.10 shows a dispersion diagram for argon plasma in RAID (chapter 12) with $B_0 = 200$ G and $n_{e0} = 10^{18}$ m^{-3} in a semilog plot to encompass the wide range of frequencies. There are four resonances as before, but now the three cut-offs are almost superposed at the plasma frequency. Between the lower hybrid and electron cyclotron resonances, $\omega_{LH} < \omega < \omega_{ce}$, the whistler branch of the R-wave exists in isolation. The dispersion diagram is similar for hydrogen plasma in the same conditions, except that the 13.56 MHz RF frequency is just above the X-wave's lower hybrid resonance. We therefore see that the whistler branch of the

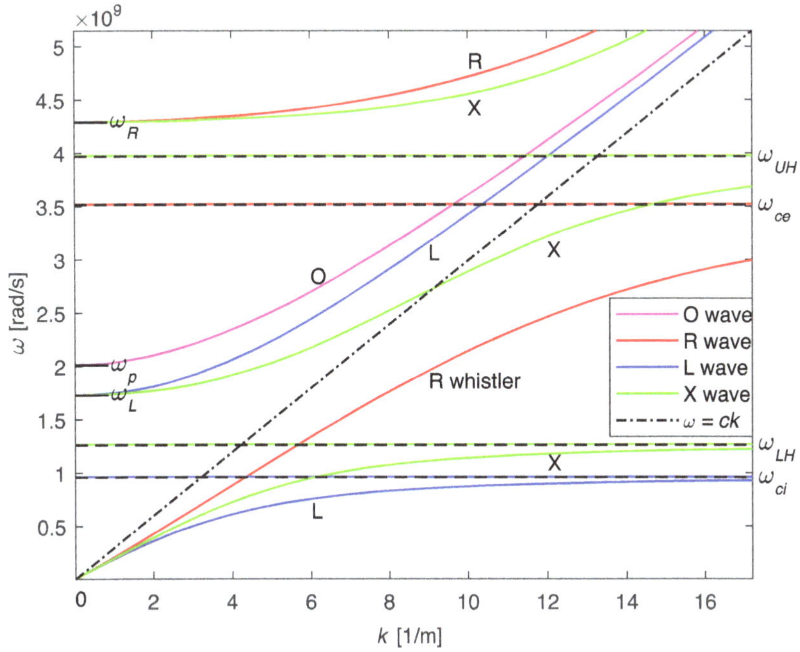

Figure 10.9. An illustrative linear/linear dispersion diagram to show the different waves, resonances and cut-offs [1]. Parameters B_0 = 200 G; $n_{e0} = 10^{15}$ m^{-3}; with an artificially low ion mass 0.002 amu. $\omega_{ce} > \omega_p$ for these conditions. The dash-dot line corresponds to light waves in vacuum, $\omega = ck$. The three cut-off frequencies are at the y-axis intersections where $k = 0$, and the four resonance frequencies correspond to the asymptotes (dashed lines) as $k \to \infty$. (Adapted with permission from [1]. Copyright 2005 John Wiley.)

R-wave is the only wave of interest for RF magnetized hydrogen plasma excited at 13.56 MHz and above. All the frequencies for hydrogen and argon plasmas in these conditions are summarized in table 10.2.

We are now in a position to state why RF sources of magnetized plasma are principally concerned with whistlers—a simple reason is that *the whistler branch of the R-wave is the only principal wave which is excited in the RF range of frequencies* for conventional low temperature plasma conditions. In contrast, the corresponding branch of the L-wave is below the ion cyclotron frequency (kHz range), the X-wave branch is below the lower hybrid frequency (below 13.56 MHz), and all other wave branches are in the region of the plasma frequency and above (GHz range).

The case of R-wave whistlers lends itself to an analytical treatment, even when the propagation angle is arbitrary, for the intermediate range of frequencies $\omega_{ci}, \omega_{pi} \ll \omega \ll \omega_{ce}, \omega_{pe}$ [32]. Moreover, the propagation of R-wave whistlers is not affected by finite temperature effects [32], therefore we are justified in limiting our analytical treatment to cold plasma theory for the next development of the wave equations in magnetized plasma in section 10.11. Before this, however, we investigate some basic properties of R-waves in sections 10.8 and 10.9 to arrive at the

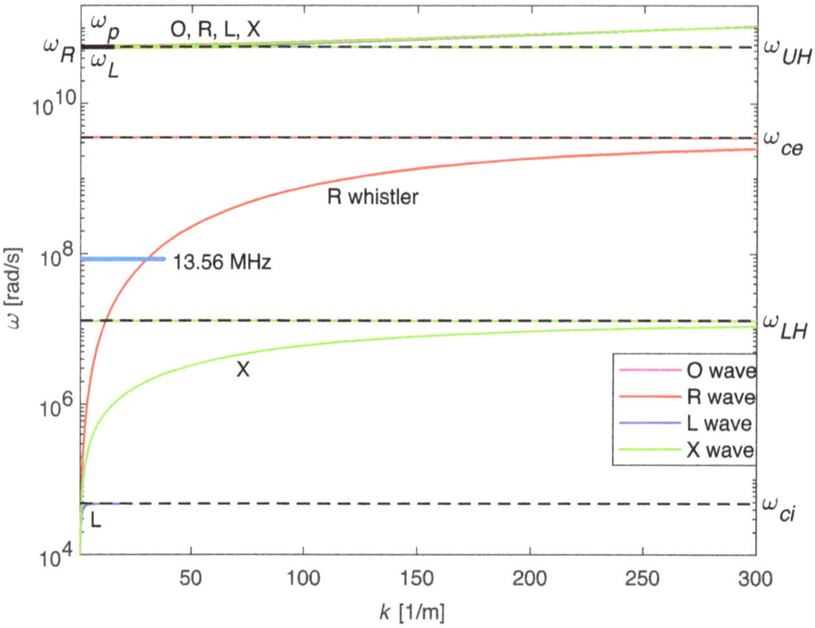

Figure 10.10. A semilog dispersion diagram for Ar (40 amu) plasma in RAID (chapter 12) for $B_0 = 200$ G and $n_{e0} = 10^{18}$ m^{-3}. The electron density is high and the magnetic field is relatively weak, so $\omega_{ce} \ll \omega_p$ for these conditions. On this scale, the O, R, L, X branches above ω_{UH} are almost superposed, and the three cut-off frequencies are almost coincident with the ω_{UH} resonance asymptote. The line and label types correspond to figure 10.9. The R whistler wave above 13.56 MHz and below the electron cyclotron frequency exists in splendid isolation.

physical mechanisms of R-wave propagation, and evanescence and reflection, across the electron cyclotron frequency, in section 10.10.

10.8 Unbounded collisionless motion at the electron cyclotron resonance: explicit time solution

As mentioned in section 10.3.1, finite oscillatory motion of the electrons can be dealt with using RF Fourier component solutions, but resonance in collisionless conditions is a special case which has to be treated by an explicit time solution [29]. Figure 10.11 shows schematically why the R-wave transverse electric field[5] at the electron cyclotron frequency continuously transfers energy to resonant electrons; in fact, the electrons spiral outwards as in figure 10.12(a). On the other hand, the L-wave causes only energy oscillation, with no net power transfer [1, 35]. The diagram again justifies why the rotation of E and the direction of B_0 determine the wave polarization, independently of the wave propagation direction. We will examine the electron cyclotron resonance for the R- and L-waves starting again from the equation of fluid motion for electrons (10.6) in collisionless conditions:

[5] The transverse electric field generated by the birdcage end-rings, coupled with the transverse magnetic field generated by the legs, is favourable for helicon excitation.

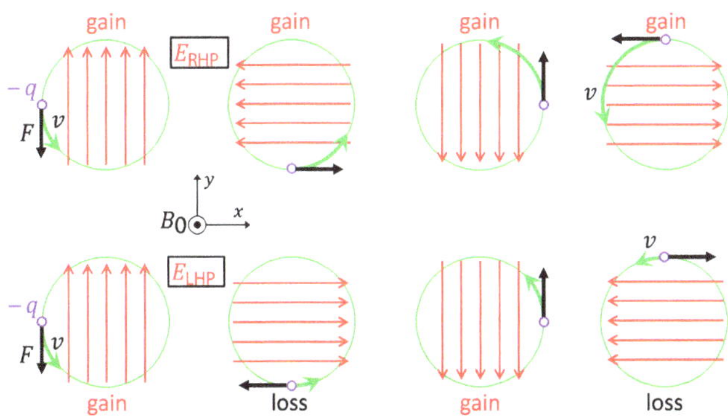

Figure 10.11. At the electron cyclotron resonance frequency in collisionless plasma, the electrons gain energy continuously from the R-wave (top row), because the electric force $F = -qE_{\mathrm{RHP}}$ is aligned with the electron velocity v at all times; the electron and wavefield rotate in phase, in accordance with the right-hand rule in figure 10.3. In contrast, the L-wave (bottom row) causes only oscillating electron energy with no net power transfer, because the electric force is alternately parallel and anti-parallel to the electron velocity, as they rotate in opposite directions, during a cycle. (Adapted with permission from [35]. Copyright 1994 Aacademic Press. Published by Elsevier.)

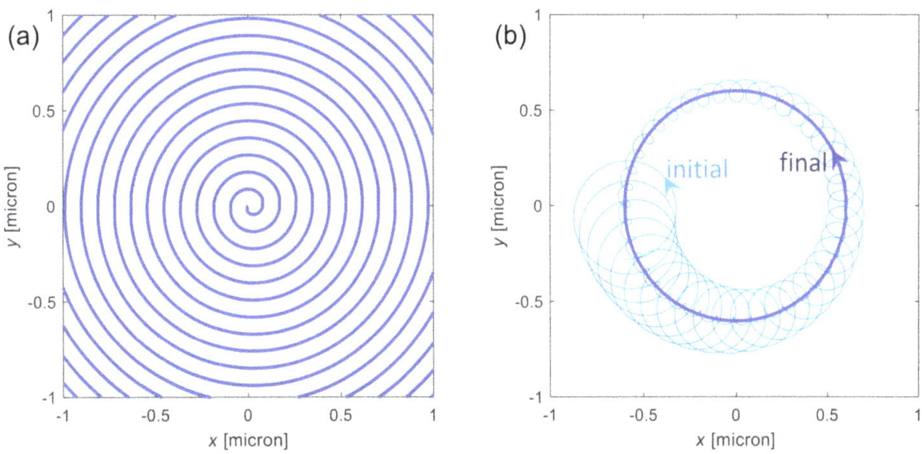

Figure 10.12. (a) The electron trajectory spiralling outwards at the electron cyclotron resonance driven by an R-wave, $\omega = \omega_{\mathrm{ce}}$, in collisionless conditions; see section 10.8. (b) For this example with $\omega < \omega_{\mathrm{ce}}$, any initial motion (pale blue arrow and line) at the electron cyclotron frequency ω_{ce}, is damped at the collisional frequency ν, leaving only the final, steady-state rotation (dark blue arrow and line) at the RF driving frequency ω, according to section 10.9. In this example, the collision frequency is 4.52 MHz, RF frequency 13.56 MHz, and electron cyclotron resonance frequency 560 MHz (corresponding to 200 G). There are $\approx$41 cyclotron orbits per RF cycle. All motions are right-hand polarized for B_0 along $+z$ (see figure 10.3).

$$m_e \frac{d\mathbf{u}}{dt} = -q(\mathbf{E} + \mathbf{u} \times \mathbf{B}_0), \tag{10.47}$$

but this time without taking the Fourier component. The Cartesian components of the equation of collisionless motion are

$$\begin{aligned} \frac{du_x}{dt} &= -\frac{q}{m_e}\left(E_x + u_y B_0\right), \\ \frac{du_y}{dt} &= -\frac{q}{m_e}\left(E_y - u_x B_0\right). \end{aligned} \tag{10.48}$$

By taking derivatives to eliminate $\frac{du_y}{dt}$, we find the second-order inhomogeneous differential equation for u_x,

$$\frac{d^2 u_x}{dt^2} + \omega_{ce}^2 u_x = -\frac{q}{m_e}\left(\frac{dE_x}{dt} - \omega_{ce} E_y\right). \tag{10.49}$$

For the R-wave polarization as in section 10.2.2, we take the electric field to be circularly polarized such that

$$\begin{aligned} E_x &= \mathrm{Re}\,(E_0 e^{-j\omega_{ce}t}) = E_0 \cos(\omega_{ce} t), \\ E_y = jE_x &= \mathrm{Re}\left(jE_0 e^{-j\omega_{ce}t}\right) = E_0 \sin(\omega_{ce} t). \end{aligned} \tag{10.50}$$

For the L-wave polarization,

$$\begin{aligned} E_x &= \mathrm{Re}\,(E_0 e^{-j\omega_{ce}t}) = E_0 \cos(\omega_{ce} t), \\ E_y = -jE_x &= \mathrm{Re}\left(-jE_0 e^{-j\omega_{ce}t}\right) = -E_0 \sin(\omega_{ce} t). \end{aligned} \tag{10.51}$$

Substituting in (10.49), we obtain

$$\text{R wave:} \quad \frac{d^2 u_x}{dt^2} + \omega_{ce}^2 u_x = \frac{2qE_0\omega_{ce}}{m_e} \sin(\omega_{ce} t), \tag{10.52}$$

$$\text{L wave:} \quad \frac{d^2 u_x}{dt^2} + \omega_{ce}^2 u_x = 0, \tag{10.53}$$

for u_x, and a similar procedure for u_y. The cancellation observed for the L-wave occurs only at resonance, when $\omega = \omega_{ce}$ which is the present case, but not generally. The R-wave case is a linear equation of second order whose general solution is the sum of the complementary function and the particular integral[6]. The homogeneous equation gives rise to the R-wave complementary functions

[6] The complementary function is the solution of the homogeneous equation [36, 37]; the particular integral [36], or particular solution [37], is the solution of the inhomogeneous equation; and the general solution is the sum of the complementary function and the particular integral.

$$u_x = u_0 \cos(\omega_{ce}t + \phi),$$
$$u_y = u_0 \sin(\omega_{ce}t + \phi), \tag{10.54}$$

where the two constants of integration u_0 and ϕ are, respectively, a velocity amplitude and phase constant depending on initial conditions. The R-wave also has a particular integral which can be found by testing with $u_x = \omega_{ce}t[A \sin(\omega_{ce}t) + B \cos(\omega_{ce}t)]$. Setting $u_0 = 0$ for negligible initial velocity, the solution of the inhomogeneous equation (10.52) yields [29]

$$u_x = -\frac{qE_0}{m_e}t \cos(\omega_{ce}t),$$
$$u_y = -\frac{qE_0}{m_e}t \sin(\omega_{ce}t). \tag{10.55}$$

The electron fluid speed $u = \sqrt{u_x^2 + u_y^2} = (\frac{qE_0}{m_e})t$ increases linearly with time, and the kinetic energy is given by

$$\frac{1}{2}m_e u^2 = \frac{(qE_0 t)^2}{2m_e}. \tag{10.56}$$

The R-wave energy increases without bound in time, with no steady-state solution, which is precisely why we avoided using the Fourier component, since that method assumes a steady-state, finite amplitude. The velocities u_x, u_y in (10.55) can be time-integrated to obtain the electron trajectory during cyclotron resonance, using integration by parts. The electron co-ordinates for the R-wave with zero initial velocity are therefore

$$x = -\frac{E_0}{\omega_{ce}B_0}[\omega_{ce}t \sin(\omega_{ce}t) + \cos(\omega_{ce}t) - 1], \tag{10.57}$$

$$y = -\frac{E_0}{\omega_{ce}B_0}[-\omega_{ce}t \cos(\omega_{ce}t) + \sin(\omega_{ce}t)]. \tag{10.58}$$

Finally, the electron trajectory at the electron cyclotron resonance, driven by an R-wave, is shown in figure 10.12(a). At resonance, the radius increases uniformly in time and no steady-state solution exists [29, 31, 38]. In this case, we expect the cold plasma approximation to fail and either collisions, thermal motion, gradients, or nonlinear effects to dominate the dynamics [29]. In contrast, the L-wave homogeneous equation, from (10.53), contains only oscillatory terms, hence the R- and L-wave solutions correspond to the pictures in figure 10.11.

To summarize, it is clear that the R-wave transfers energy to electrons exactly at the electron cyclotron resonance, whereas the L-wave does not. This makes a distinction between the interaction of R- and L-waves with plasma, but only for this particular frequency, and not generally. It is the assumption of strictly collisionless conditions which allows the resonant orbits to grow indefinitely, and this non-physical situation is rectified in section 10.9.

10.9 Bounded collisional motion: explicit time solution

We can repeat the collisional solution in section 10.3.1, but with an explicit time solution instead of the assumed single Fourier component. Our starting point is again the fluid equation of motion for electrons (10.6):

$$m_e \frac{d\mathbf{u}}{dt} = -q(\mathbf{E} + \mathbf{u} \times \mathbf{B}_0) - m_e \nu \mathbf{u}, \tag{10.59}$$

without taking the Fourier component, as in section 10.8, in order to obtain an explicit time solution, but now persevering with the electron–neutral collision term ($-m\nu\mathbf{u}$) still included. For magnetized plasma, $\nu \ll \omega$, $\omega_{\rm ce}$, as throughout this chapter. The transverse components of the equation of collisional electron motion are

$$\frac{du_x}{dt} = -\frac{q}{m_e}\left(E_x + u_y B_0\right) - \nu u_x, \tag{10.60}$$

$$\frac{du_y}{dt} = -\frac{q}{m_e}\left(E_y - u_x B_0\right) - \nu u_y. \tag{10.61}$$

By taking a time derivative of (10.60) to eliminate $\frac{du_y}{dt}$ using (10.61), we find

$$\frac{d^2u_x}{dt^2} + \omega_{\rm ce}^2 u_x = -\frac{q}{m_e}\left(\frac{dE_x}{dt} - \omega_{\rm ce}E_y\right) - \nu\frac{du_x}{dt} + \nu\omega_{\rm ce}u_y. \tag{10.62}$$

The last term includes u_y which can be substituted using (10.60) once more. We obtain a second-order inhomogeneous differential equation for an explicitly time-dependent $u_x(t)$:

$$\frac{d^2u_x}{dt^2} + 2\nu\frac{du_x}{dt} + (\omega_{\rm ce}^2 + \nu^2)u_x = -\frac{q}{m_e}\left(\frac{dE_x}{dt} - \omega_{\rm ce}E_y + \nu E_x\right). \tag{10.63}$$

This is the same as (10.49) except for the three terms in ν, which, mathematically and intuitively, will prevent the electron motion becoming infinite at resonance when $\omega = \omega_{\rm ce}$. For this reason, according to the discussion in section 10.3.1, we are permitted to solve for the particular integral by trying a single Fourier component solution for $u_x(t)$. Knowing that the electric field components E_x and E_y vary as $e^{-j\omega t}$, we substitute $u_x(t) = u_x e^{-j\omega t}$ to obtain a solution for the phasors:

$$\text{particular integral:} \quad u_x = -\frac{q}{m_e}\left[\frac{j(\omega + j\nu)E_x + \omega_{\rm ce}E_y}{(\omega + j\nu)^2 - \omega_{\rm ce}^2}\right]. \tag{10.64}$$

Having re-arranged (10.64) judiciously, we notice that substitution of $\hat{\omega} = \omega + j\nu$ immediately gives the previous solution for u_x in (10.13), as expected because the procedure is effectively identical. The same goes without saying for u_y. However, the new and interesting point here is that the time-explicit equation (10.63) also possesses a complementary function which is the solution of the homogeneous equation of motion:

$$\frac{d^2u_x}{dt^2} + 2\nu\frac{du_x}{dt} + (\omega_{\mathrm{ce}}^2 + \nu^2)u_x = 0. \tag{10.65}$$

By factorizing, the two terms of the complementary function are remarkably simple:

$$\text{complementary function:} \quad u_x = C_1 e^{-(\nu + j\omega_{\mathrm{ce}})t} + C_2 e^{-(\nu - j\omega_{\mathrm{ce}})t}, \tag{10.66}$$

which can be verified by back-substitution into (10.65). C_1 and C_2 represent the two constants of integration of the second-order equation, to be determined by initial conditions. The general solution, then, is the sum of the particular integral (10.64) and the complementary function (10.66); an example trajectory is shown in figure 10.12(b) for $\nu < \omega < \omega_{\mathrm{ce}}$. Two important intermediate conclusions can be drawn:

- The complete solution for the electron oscillation perpendicular to the magnetic field, excited by an arbitrary wave electric field now involves *two* frequencies in contrast to the single-Fourier-component assumption in (10.7), namely, the RF driving frequency ω, and the electron cyclotron resonance frequency ω_{ce}. The ω_{ce} term is tacitly ignored when considering only the Fourier component at the RF frequency. The amplitude of the electron cyclotron oscillation depends on the initial conditions which could include some transient impulse upon plasma ignition, for example.
- However, the cyclotron oscillation terms in (10.66) are damped exponentially with a $1/e$ time equal to the collision interval, $1/\nu$. Therefore, the steady-state solution, which will always be attained eventually due to some effective collisional damping in real conditions, is indeed the same as the single RF Fourier component, justifying the implicit assumption in section 10.3.1 for a steady state.

This clears up any confusion where readers could expect the electrons to orbit the magnetic field at the electron cyclotron frequency ω_{ce}, instead of (or, as well as) at the RF frequency ω.

Collisional bounded motion and resonance for R- and L-waves

The particular integral (10.64) can be studied for the specific case of the R-wave by setting $E_y = jE_x$, see (10.37) and (10.38), and the L-wave by setting $E_y = -jE_x$, see (10.41) and (10.42). We find

$$\text{R wave:} \quad u_x = -j\frac{q}{m_e}\left[\frac{E_x}{\hat{\omega} - \omega_{\mathrm{ce}}}\right], \quad u_y = ju_x, \tag{10.67}$$

$$\text{L wave:} \quad u_x = -j\frac{q}{m_e}\left[\frac{E_x}{\hat{\omega} + \omega_{\mathrm{ce}}}\right], \quad u_y = -ju_x, \tag{10.68}$$

with reference to (10.2) and (10.4) for the electric field amplitude $E_x = E_0/\sqrt{2}$. These velocities are the same as in the simultaneous equation solutions (10.13), for the special case of R- and L-waves. When the RF frequency enters into resonance at the electron cyclotron frequency ω_{ce}, we have

$$\text{R wave:} \quad u_x = -\frac{qE_x}{m_e \nu}, \tag{10.69}$$

$$\text{L wave:} \quad u_x = -\frac{q}{m_e}\left[\frac{E_x}{\nu - 2j\omega_{ce})}\right] \approx -j\frac{q}{m_e}\left[\frac{E_x}{2\omega_{ce}}\right], \quad \text{for} \quad \nu \ll \omega_{ce}. \tag{10.70}$$

Inclusion of the collision frequency therefore means that the cyclotron resonance electron trajectory for the R-wave is not the infinite spiral obtained in section 10.8, although the R-wave steady-state electron kinetic energy will reach high values for low collision frequency. Note that u_x is in phase with E_x for the R-wave resonance, whereas u_x and E_x are approximately in quadrature for the L-wave. This contrast motivates further investigation into the power transfer from the wavefields to the driven electrons.

Power transfer for R- and L-waves

Using the power density expressions in appendix D.2.1, and the electron velocities for R- and L-waves in (10.67) and (10.68), the time-averaged power density transfer from wave to electrons is the same as in (D.21), except that ω^2 is replaced by $(\omega - \omega_{ce})^2$ for R-waves, or by $(\omega + \omega_{ce})^2$ for L-waves:

$$\bar{P} = \frac{1}{2}\sigma_{dc}\left[\frac{\nu^2}{\nu^2 + (\omega \mp \omega_{ce})^2}\right]E_0^2. \tag{10.71}$$

Apart from the singular resonance case for collisionless R-waves when $\omega = \omega_{ce}$, note that continuous power transfer occurs only because the collision frequency ν is not zero, similarly to the unmagnetized case explained in appendix D.2.1. The R-wave power dissipation at resonance is then much higher than for the L-wave, especially for low collisionality, but for RF frequencies, $\omega \ll \omega_{ce}$, there is almost no difference between the power for R- and L-waves. Finally, there appears to be no qualitative difference between R- and L-wave power in (10.71), nor for the passage through the electron cyclotron resonance, to explain their very different behaviours in the dispersion figure 10.10.

General power transfer for waves in magnetized plasma

This lack of distinction of wave-to-electron power transfer mechanism between the different waves is shown even more starkly by returning to the originating electron fluid force equation, (10.11). Substituting (10.11) into the time-averaged power density expression (D.20) gives

$$
\begin{aligned}
\bar{P} &= \frac{1}{2}\,\mathrm{Re}(\mathbf{E}_0 \cdot \mathbf{J}_0^*) \\
&= \frac{1}{2\varepsilon_0\omega_{\mathrm{pe}}^2}\,\mathrm{Re}\left[(-j\hat{\omega}\mathbf{J}_0 + \omega_{\mathrm{ce}}\mathbf{J}_0 \times \hat{\mathbf{z}}) \cdot \mathbf{J}_0^*\right] \\
&= \frac{1}{2\varepsilon_0\omega_{\mathrm{pe}}^2}\,\mathrm{Re}\,[-j\hat{\omega}J_0^2 + (\omega_{\mathrm{ce}}\mathbf{J}_0 \times \hat{\mathbf{z}}) \cdot \mathbf{J}_0^*] \\
&= \frac{1}{2\varepsilon_0\omega_{\mathrm{pe}}^2}\left\{\nu J_0^2 + \mathrm{Re}\,[(\omega_{\mathrm{ce}}\mathbf{J}_0 \times \hat{\mathbf{z}}) \cdot \mathbf{J}_0^*]\right\} \\
&= \frac{1}{2}\frac{m_e}{n_{e0}q^2}\left\{\nu J_0^2 + 0\right\} \\
&= \frac{1}{2}\rho_{\mathrm{dc}}J_0^2,
\end{aligned}
\tag{10.72}
$$

recalling that $\hat{\omega} = \omega + j\nu$, by using (D.18) for ρ_{dc}, and noting that the vector cross product terms are purely imaginary, as shown in the footnote[7]. This time-averaged power density expression applies to all the plasma fluid waves, and is, in fact, the same as for the *un*magnetized plasma of (D.23), because ohmic dissipation is the only power transfer mechanism in the fluid model. It should not be surprising that the magnetic force makes no contribution to the power in (10.72), because it is perpendicular to the electron velocity, and so does no work. The magnetic field does influence the electron fluid velocity itself, via the ω_{ce} term in (10.67) and (10.68), and hence influences the power density in (10.72) via the electron current amplitude J_0.

Ohmic dissipation by electron–neutral collisions in fluid models may seem to be a rather simplistic view of power transfer in plasmas, but this approach is successful for the non-magnetized, inductive plasmas in part II. Indeed, the complex image method is entirely based on ohmic dissipation in a resistive medium (see appendix E). This part III, however, is concerned with wave propagation and damping in magnetized plasma, where kinetic effects might play a role and the fluid model may have its limitations. Nevertheless, the RF whistler waves have frequencies so far below the electron plasma frequency that the resistive fluid model will be assumed to remain valid.

In view of these last two sub-sections, the contrasting physical mechanisms for R-wave propagation and L-wave evanescence are not to be found in the wave-to-electron power transfer. Instead, the distinction must arise in the linear interactions which determine the plasma wave's electron velocity (10.67), permittivity and refractive index (10.44), because these linear relations do change sign on passage through the resonance, in contrast to the power (10.71). These linear relations are discussed in section 10.10.

[7] $(\mathbf{J}_0 \times \hat{\mathbf{z}}) \cdot \mathbf{J}_0^* = (J_y\hat{\mathbf{x}} - J_x\hat{\mathbf{y}}) \cdot (J_x^*\hat{\mathbf{x}} + J_y^*\hat{\mathbf{y}}) = J_x^*J_y - J_xJ_y^* = J_x^*J_y - (J_x^*J_y)^*$, and since $z - z^* = 2j\,\mathrm{Im}(z)$, this is purely imaginary.

10.10 Whistler propagation, or evanescence and reflection in collisionless plasma

Figure 10.13 plots the plasma relative permittivity for R-waves using (10.34), and for L-waves using (10.39), considering only the electrons (the frequency range is assumed much higher than the ion cyclotron frequency), and in collisionless conditions. Below the electron cyclotron frequency, we see that k^2 is positive for the whistler R-waves, therefore k has a real value in $e^{j(kz-\omega t)}$, and the R-wave propagates as if in a dielectric. Conversely, where k^2 is negative for R- or L-waves, the wavenumber k is imaginary in $e^{j(kz-\omega t)}$, and the waves are evanescent in the plasma. The passage through the electron cyclotron resonance frequency for R-waves corresponds to propagation, absorption at resonance, then evanescence and reflection. Mathematically, this interpretation is straightforward, but we would like to know the physical mechanism for why R-waves can propagate, or evanesce, in uniform, collisionless RF magnetoplasma.

An explanation is sought in the physics behind the dispersion relation (10.34), which describes steady-state, cold, collisionless plasma for a single, small-amplitude

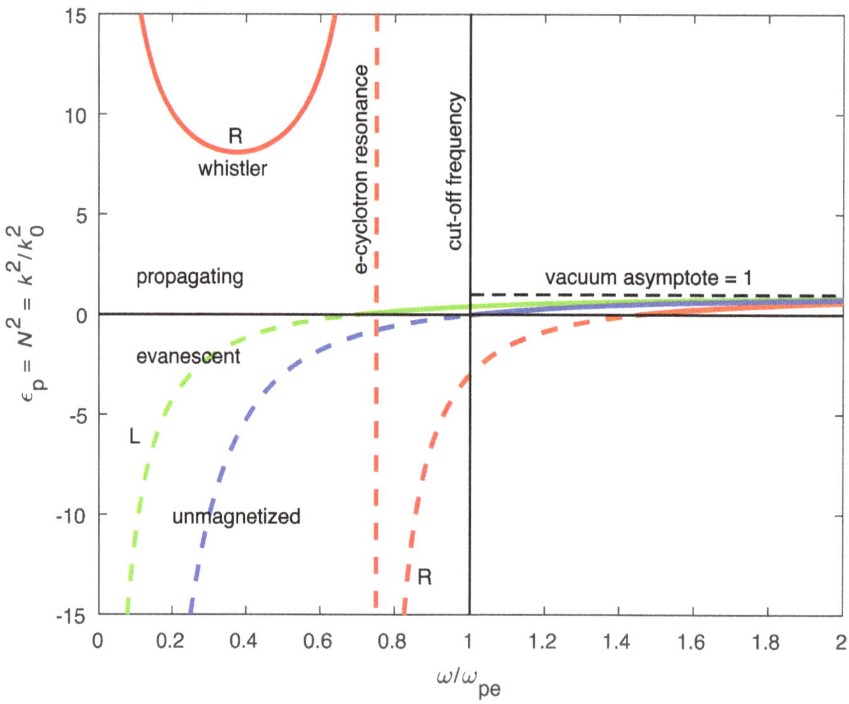

Figure 10.13. The plasma relative permittivity ε_p in a collisionless magnetized plasma, as a function of ω/ω_{pe}. For the R-wave (red curve), ε_p is given by (10.34); ε_p for the L-wave (green curve) is given by (10.39), with $\hat{\omega} = \omega$ for both. For comparison with figure D.1, the blue curve represents the unmagnetized case. For illustrative purposes, the electron cyclotron frequency has been chosen to be $\omega_{ce} = 0.75\omega_{pe}$. The salient effect of the magnetic field is that ε_p is now large and positive for the R-wave below the electron cyclotron frequency (whistlers), permitting R-wave propagation through the plasma at low frequencies.

Fourier component at the RF frequency, without invoking thermal, collisional, Landau damping or other kinetic effects, nor minimum constraints on energy or entropy such as the Taylor states [39, 40]. Moreover, from the preceding section 10.9, the explanation is not to be found in the wave–electron power transfer. The answer must therefore be contained within the initial equation of electron fluid motion (10.6) in the limit $\nu \to 0$, i.e. $\hat{\omega} \to \omega$.

A starting point is given in appendix D.5.4 by considering the wave electric field vacuum displacement current in competition with the opposing electron conduction current which arises in response to the wave electric field. The rotating conduction current density for the R-wave and L-wave is calculated from the electron velocity components in (10.67) and (10.68), respectively,

$$\mathbf{J}_e = -n_{e0}q\mathbf{u} = +j\left(\frac{n_{e0}q^2}{m_e(\omega \mp \omega_{\text{ce}})}\right)\mathbf{E}, \tag{10.73}$$

which is nothing more than than the solution of the initial simultaneous equations (10.13) in the special case of circular polarization (R- and L-waves). The rotating vacuum displacement current density for the same electric field is

$$\mathbf{J}_d = \varepsilon_0\frac{\partial \mathbf{E}}{\partial t} = -j(\omega\varepsilon_0)\mathbf{E}. \tag{10.74}$$

The unmagnetized plasma conduction current (D.44) is recovered for $B_0 \to 0$, i.e. $\omega_{\text{ce}} \to 0$ in (10.73), where the conduction current opposes the vacuum displacement current and reflects the wavefield below the electron plasma frequency cut-off as in appendix D.5.4. But now there is a fundamental difference—when the R-wave RF frequency ω is less than the electron cyclotron frequency ω_{ce}, the conduction current $\mathbf{J}_e$ in (10.73) *changes sign*, becomes negative, and now effectively *adds to the vacuum displacement current* $\mathbf{J}_d$ instead of opposing it. Therefore, according to the point of view in appendix D.5.4, the magnetoplasma for the R-wave becomes a 'super dielectric' with an effective relative permittivity (10.34) which is large and positive:

$$\varepsilon_p = \left(1 + \frac{\omega_{\text{pe}}^2}{\omega(\omega_{\text{ce}} - \omega)}\right), \quad \text{for} \quad \omega < \omega_{\text{ce}} < \omega_{\text{pe}}, \tag{10.75}$$

whence $\varepsilon_p \gg 1$, as indicated by the continuous red curve labelled 'R whistler' in figure 10.13. This 'super dielectric' property is why whistler waves can propagate in magnetized plasma at RF frequencies far below the electron plasma frequency which otherwise cuts off lower frequencies in unmagnetized plasma. From (10.75), for whistlers, $\varepsilon_p \to \infty$ as the RF angular frequency $\omega \to \omega_{\text{ce}}$, or to low values. The minimum $\varepsilon_p^{\min} \approx 4\omega_{\text{pe}}^2/\omega_{\text{ce}}^2$ occurs at $\omega = \omega_{\text{ce}}/2$.

Conversely, on the higher frequency side of the electron cyclotron resonance, $\omega > \omega_{\text{ce}}$, the conventional positive-sign conduction current applies, dominating the vacuum displacement current, and the R-wave is evanescent in the magnetoplasma, i.e. the fields are reflected from the plasma surface as for the unmagnetized plasma (see appendix D.5.4). It is straightforward to see that the L-wave conduction current

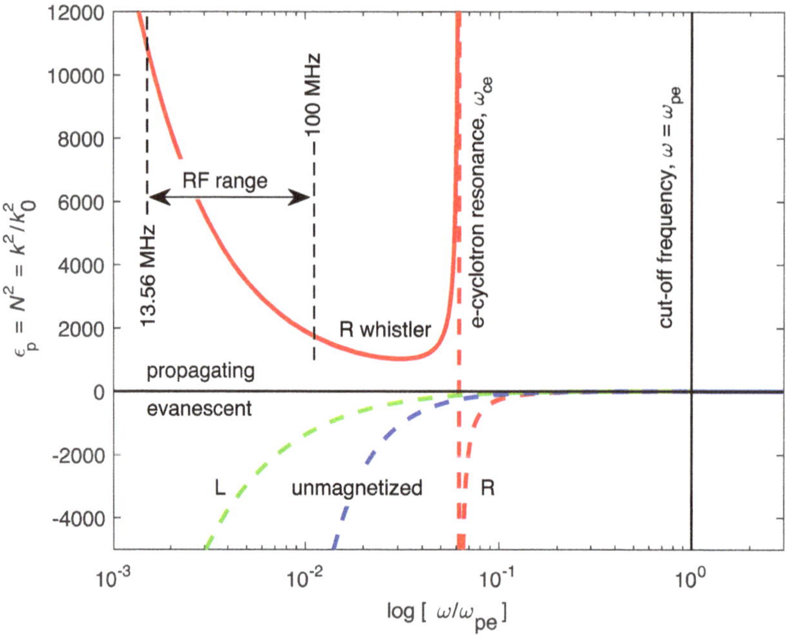

Figure 10.14. The same graph as figure 10.13, but for RAID parameters B_0 = 200 G, $n_{e0} = 10^{18}$ m^{-3}, and with ω/ω_{pe} on a log scale. For RAID, the electron cyclotron frequency is 560 MHz and the electron plasma frequency is 8.98 GHz (table 10.2). In the RF range from 13.56 MHz to 100 MHz, the effective plasma relative permittivity ε_p for whistlers varies from approximately 11 000 down to 1760.

due to the electron velocity (10.68) does not change sign across the electron cyclotron frequency in figure 10.13, and the L-wave remains evanescent unless the excitation frequency falls below the ion cyclotron frequency in (10.44). In this way, the propagation and evanescence of the R- and L-waves can be understood in figure 10.10. The special case of R-wave resonance at $\omega = \omega_{ce}$ was considered in section 10.8.

Figure 10.14 re-plots figure 10.13 for typical low temperature plasma in RAID, table 10.2. The effective plasma relative permittivity for whistlers, (10.34), at 13.56 MHz is $\varepsilon_p = N^2 = k^2/k_0^2 \approx 11\,000$. The relative permittivity is so very large that the RF free-space wavelength of 22 m is shrunk by a factor $\sqrt{\varepsilon_p} \approx 100$ down to around 22 cm in the magnetized plasma, as confirmed by measurements in RAID reported in section 12.2.3. The minimum relative permittivity, $\varepsilon_p^{\min} \approx 4\omega_{pe}^2/\omega_{ce}^2 = 1030$, occurs at $\omega = \omega_{ce}/2$, at frequency 280 MHz.

Alternative solution by transforming to a rotating frame of reference

It may be argued that the expression for the R-wave conduction current, (10.73), is no more enlightening than the relative permittivity (10.34) obtained via a chain of algebra from the simultaneous equations (10.13). Another approach is to transform to a frame precessing clockwise (when looking along $\mathbf{B}_0$, see figure 10.3) at the electron cyclotron frequency in which the static magnetic field is eliminated [41]. This is a mathematical sleight of hand to solve the original equation of motion (10.6) in a different way, but perhaps affording more physical insight[8]. Using the textbook general expression for vector differentiation [31], $\frac{d\mathbf{u}}{dt} = \frac{D\mathbf{u}}{Dt} + \omega_B \times \mathbf{u}$, where $\frac{D}{Dt}$ denotes the time derivative in the frame rotating at frequency ω_B with respect to the laboratory, the equation of motion (10.6) in collisionless conditions becomes

$$\begin{aligned}\frac{d\mathbf{u}}{dt} = \frac{D\mathbf{u}}{Dt} + \omega_B \times \mathbf{u} &= -\frac{q}{m_e}(\mathbf{E} + \mathbf{u} \times \mathbf{B}_0)\\ &= -\frac{q}{m_e}\mathbf{E} + \frac{q}{m}\mathbf{B}_0 \times \mathbf{u}\end{aligned} \tag{10.76}$$

$$= -\frac{q}{m_e}\mathbf{E} + \frac{qB_0}{m}\hat{\mathbf{z}} \times \mathbf{u}. \tag{10.77}$$

By choosing $\omega_B = \frac{qB_0}{m_e}\hat{\mathbf{z}} = \omega_{ce}\hat{\mathbf{z}}$, the vector cross products cancel, leaving

$$\frac{D\mathbf{u}}{Dt} = -\frac{q}{m_e}\mathbf{E}, \tag{10.78}$$

so that the equation of motion in the frame rotating clockwise at the electron cyclotron frequency is as if the collisionless plasma were unmagnetized, namely, equation (D.13) and appendix D.5.4. To abbreviate the following discussion, 'cw' means clockwise, and 'acw' signifies anti-clockwise, when looking along $\mathbf{B}_0$.

In the stationary reference frame of the laboratory, the relative quadratures of the vector components u_x and $u_y = \pm ju_x$, oscillating as $e^{-j\omega t}$, give rise to the right (cw) and left (acw) circular polarization respectively, rotating at RF frequency ω. For this laboratory reference frame, $\frac{d\mathbf{u}}{dt} = -j\omega\mathbf{u}$, as used everywhere previously in this chapter.

In the rotating frame, self-consistent right- and left-polarization solutions are now obtained for oscillations $e^{-j(\omega \mp \omega_{ce})t}$ respectively, because the cw frame precession at ω_{ce} is relative to the R-wave cw rotation at ω (frequency subtraction), but adds to the L-wave acw counter-rotation (frequency addition). Hence, $\frac{D\mathbf{u}}{Dt} = -j(\omega \mp \omega_{ce})\mathbf{u}$ for R- and L-waves, respectively. Substitution into (10.78) gives

$$-j(\omega \mp \omega_{ce})\mathbf{u} = -\frac{q}{m_e}\mathbf{E}, \tag{10.79}$$

in the rotating frame. Therefore, the electron conduction current is

$$\mathbf{J}_e = -n_{e0}q\mathbf{u} = +j\left(\frac{n_{e0}q^2}{m_e(\omega \mp \omega_{ce})}\right)\mathbf{E}, \tag{10.80}$$

[8] This bears a passing resemblance to Larmor's theorem [31, 41, 42], where the Coriolis force apparently cancels the Lorentz force, but the physical analogy goes beyond the scope of the purely mathematical operation here.

for R- and L-waves, respectively, which recovers the earlier solution for R-waves, (10.73).

Finally, the proposed physical picture to explain propagation or evanescence of R-waves in plasma, below ω_{pe}, as shown in figures 10.13 and 10.14, is as follows. The precessing frame, in which the static magnetic field $\mathbf{B}_0$ is eliminated, is rotating cw at the electron cyclotron frequency ω_{ce} relative to the laboratory, so the relative rotation frequency of the R-wave (cw by definition) in the spinning frame is $(\omega - \omega_{ce})$.

- For $\omega < \omega_{ce}$, where the R-wave is a whistler, the cw-rotating wave is overtaken by the frame's faster cw rotation, and the electron fluid appears to rotate backwards, i.e. acw at frequency $(\omega_{ce} - \omega)$. This inverses the effective electron current which now has the effect of reinforcing the vacuum displacement current; the plasma behaves as a 'super dielectric' for the propagating whistler, with extremely high relative permittivity, as per appendix D.5.4 and the discussion following (10.73).
- For $\omega = \omega_{ce}$, electron cyclotron resonance occurs and the wave energy is absorbed as described in sections 10.7 and 10.8.
- For $\omega > \omega_{ce}$, the R-wave spins cw faster than the frame rotates cw, so the wave effective electron conduction current remains in the conventional cw direction for R-waves, at frequency $(\omega - \omega_{ce})$. The conduction current dominates the vacuum displacement current, so R-wave radiation is reflected from the plasma by rotating dipole currents [41] similarly to light reflection from a metal surface [43]. The R-wave field exhibits only evanescent penetration into the plasma, as per the discussion following (10.73) and appendix D.5.4, until its cut-off frequency (10.45).

The L-wave electron current rotates acw, so the cw frame rotation frequency always adds to the L-wave conduction current with net frequency $(\omega + \omega_{ce})$, hence the L-wave is always reflected, and evanescent in the plasma, until its cut-off frequency in (10.45).

Wave heating in magnetoplasma

The physical mechanisms of wave heating in magnetoplasma are complex and still not fully understood [1]. In the literature, when linearly polarized wavefields are decomposed into a sum of right- and left-handed circularly polarized wavefields, as shown schematically in figure 10.15, it is often simply stated that the energy associated with the left-handed wave component is ineffective [31, 38]—see section 14.3. In the present case, the power associated with the L-wave is perhaps converted to R-waves owing to multiple reflections between the exciting antenna, the plasma resonance layer, and the vacuum vessel [1]. An investigation into RF heating by whistlers could be relevant for understanding the heating by helicon modes. However, this mechanism is most likely linked to their particular bounded geometry and boundary conditions, whereas the whistler waves propagating in an infinite, uniform medium (figure 10.16) in this chapter do not possess boundaries by

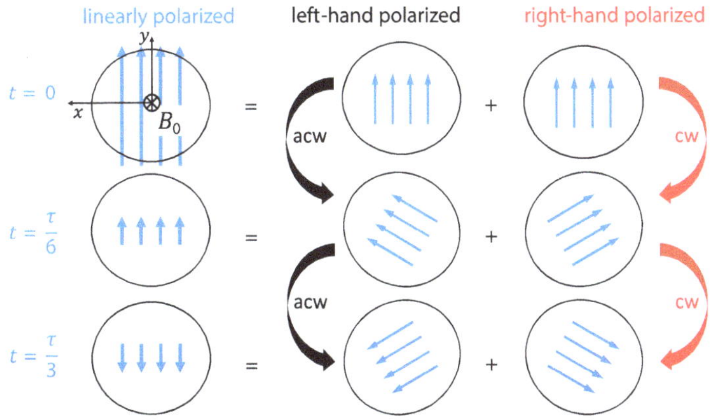

Figure 10.15. The left column shows an oscillating, linearly polarized field for three successive times (see also figure 9.5). This uniform field can be decomposed into the vector sum of two oppositely rotating uniform fields, left-hand circularly polarized and right-hand circularly polarized, both with constant amplitude. The polarizations are defined with respect to a static imposed magnetic field B_0 oriented along $+z$.

Figure 10.16. '#1: Infinite whistler waves.' (Illustration courtesy of Alex Howling. Copyright 2023 Alex Howling.)

definition. Obviously, the reflection of waves from surfaces or gradients of density and/or temperature cannot be modelled in a uniform plasma, so the details of wave energy absorption are postponed to the more relevant chapter 11.

10.11 Two approximate methods for an arbitrary angle of plane waves in magnetized plasma

In the preceding sections 10.3 to 10.7, the plane wave dispersion relation was solved using the cold plasma dielectric tensor by setting the determinant of the Helmholtz wave equation to zero [1]. The four principal waves were studied because of the simple analytical expressions afforded by setting $\theta = 0$ or $\frac{\pi}{2}$. To finish this chapter on plane waves in infinite uniform magnetized plasma, we will calculate a more general analytical dispersion relation for plane waves in magnetized plasma with *arbitrary* angles of propagation, for which some approximation will be necessary, namely, the vacuum displacement current will be neglected.

10.11.1 Neglecting the vacuum displacement current

The necessary conditions for neglecting the vacuum displacement current can be obtained from the general expression (10.16) relating the plasma relative permittivity tensor $\bar{\varepsilon}_p$ to the plasma conductivity tensor $\bar{\sigma}_p$, arising from the time harmonic ansatz and the equivalence between the dielectric and free charge descriptions of the plasma. As a reminder, (10.16) is

$$\bar{\varepsilon}_p = \left(\frac{j\bar{\sigma}_p}{\omega\varepsilon_0} + \bar{\mathrm{I}}\right).$$

The vacuum displacement current contribution arises from the unity tensor $\bar{\mathrm{I}}$ in (10.16), so it can be generally neglected, compared to the conductivity terms, provided that

$$\begin{aligned} \left|\frac{j\sigma_{11}}{\omega\varepsilon_0}\right| &\gg 1, \\ \left|\frac{j\sigma_{22}}{\omega\varepsilon_0}\right| &\gg 1, \\ \left|\frac{j\sigma_{33}}{\omega\varepsilon_0}\right| &\gg 1, \end{aligned} \tag{10.81}$$

where σ_{nn} are the diagonal elements of $\bar{\sigma}_p$. Applied to the magnetized plasma conductivity tensor (10.14) (collisionless for simplicity, $\hat{\omega} \to \omega$), this leads to the pair of conditions:

$$\begin{aligned} \omega^2 &\ll \omega_{\mathrm{pe}}^2, \\ \omega_{\mathrm{ce}}^2 &\ll \omega_{\mathrm{pe}}^2. \end{aligned} \tag{10.82}$$

The first condition is comfortably satisfied for $f_{\mathrm{RF}} = 13.56$ MHz and f_{pe} typically about 10 GHz (table 10.2). The second condition, however, is much more stringent, because the electron cyclotron frequency can be of the order of GHz (0.56 GHz for 200 G, table 10.2). Nevertheless, for most cases, the vacuum displacement current can still be neglected, even if this second condition in (10.82) is not satisfied, by considering the vacuum displacement current in the full context of the wave equation, provided that the wave's axial wavenumber k_z (along the magnetic field) is sufficiently large compared to the vacuum wavenumber k_0.

To demonstrate this point, we consider the general wave equation (D.33) which can be written as

$$\nabla \times \nabla \times \mathbf{E} = k_0^2\, \bar{\varepsilon}_p \cdot \mathbf{E} = (j\omega\mu_0\bar{\sigma}_p + k_0^2\, \bar{\mathrm{I}}) \cdot \mathbf{E}, \tag{10.83}$$

for magnetized plasma. The $k_0^2\,\bar{\text{I}}$ term in (10.83) corresponds to the vacuum displacement current. Under the $e^{jk_z z}$ ansatz ($\mathbf{E} \propto e^{jk_z z}$), (10.83) can be re-written as

$$\begin{pmatrix} -\partial_y^2 & \partial_{xy} & jk_z\partial_x \\ \partial_{xy} & -\partial_x^2 & jk_z\partial_y \\ jk_z\partial_x & jk_z\partial_y & -(\partial_x^2+\partial_y^2) \end{pmatrix}\mathbf{E} = j\omega\mu_0\bar{\sigma}_p\cdot\mathbf{E} + \begin{pmatrix} k_0^2-k_z^2 & 0 & 0 \\ 0 & k_0^2-k_z^2 & 0 \\ 0 & 0 & k_0^2 \end{pmatrix}\mathbf{E}, \quad (10.84)$$

where the partial derivatives in x, y remain general so that the result depends explicitly only on the $e^{jk_z z}$ ansatz. It can be seen from the right-hand side of (10.84) that the perpendicular components (the two first lines of the tensor) of the vacuum displacement current can be neglected, for any $\bar{\sigma}_p$, when $k_z^2 \gg k_0^2$. To estimate the orders of magnitude, at 13.56 MHz the free-space wavelength is about 22 m, while in the framework of helicon/whistler propagation we typically consider axial wavelengths shorter than 1 m (see the 'super dielectric' discussion of figure 10.14 in section 10.10, and section 12.2.3), which makes the condition $k_z^2 \gg k_0^2$ generally well satisfied.

If $k_z^2 \gg k_0^2$, then the sole condition related to $\bar{\sigma}_p$ in order to totally neglect the vacuum displacement current is given by the third line of (10.84), which corresponds, in the magnetized plasma context, to the first condition in (10.82), namely $\omega^2 \ll \omega_{\text{pe}}^2$, which is easily satisfied.

Finally, therefore, using the magneto-quasistatic approximation (appendix D.1.1) corresponding to negligible vacuum displacement current, two different approaches are used here to calculate the dispersion relation for arbitrary angles of plane wave propagation:

- In section 10.11.2, we continue directly onwards from the cold plasma dielectric tensor (10.18) derived above, by neglecting the vacuum displacement current, following a treatment by Swanson [29].
- In section 10.11.3, we instead recommence from the initial electron force equation (10.11) and Maxwell's equations, by neglecting the vacuum displacement current from the very start, to obtain a quadratic equation in $\nabla\times\mathbf{B}$, following Klozenberg [2]. The expressions obtained will come in useful for when whistlers are transformed into helicons in cylindrical bounded plasmas in chapter 11.

10.11.2 Tensor approximation of the plane wave dispersion relation

Taking up again the discussion at the beginning of section 10.5, it was noted there that N^2 can be found as a function $N^2(\theta, \bar{\varepsilon}_p)$, but exact results would generally require numerical solution of the plane wave determinant. Therefore, we will make one important concession in order to obtain analytical expressions for approximate solutions—the vacuum displacement current will be neglected in this and the following section 10.11.3, as justified in the preceding section 10.11.1.

D G Swanson [29] suggests solving the biquadratic equation (10.27) for N^2 in a direct way using the schoolboy quadratic formula:

$$N^2 = (-b \pm \sqrt{b^2 - 4ac})/2a, \tag{10.85}$$

which serendipitously turns out to give a simple analytical solution provided that the vacuum displacement current is neglected. As mentioned above, it suffices to drop the unit tensor terms in (10.18), for which the tensor elements become

$$\varepsilon_\perp = \frac{\hat{\omega}}{\omega}\left(\frac{\omega_p^2}{\omega_c^2 - \hat{\omega}^2}\right), \tag{10.86}$$

$$\varepsilon_\times = \frac{\omega_c}{\omega}\left(\frac{\omega_p^2}{\omega_c^2 - \hat{\omega}^2}\right), \tag{10.87}$$

$$\varepsilon_\parallel = -\frac{\omega_p^2}{\omega\hat{\omega}}, \tag{10.88}$$

dropping the electron suffix e since ions are not considered here in any case. In section 10.11.1, we saw that the vacuum displacement current can be neglected. Note that collisions are still symbolically included (via $\hat{\omega}$) at this point, although this is more as a reminder of the different origins of $\hat{\omega}$ (via the electron fluid equation of motion) and ω (via the vacuum displacement current). Of course, inequalities and solutions for real quantities are only valid in the limit of collisionless collisions, where $\hat{\omega} \to \omega$. Substituting (10.86), (10.87), and (10.88), the coefficients of the quadratic in N^2, (10.27), are easily shown to be as follows:

$$a = \varepsilon_\perp \sin^2\theta + \varepsilon_\parallel \cos^2\theta = \frac{\omega_p^2}{\omega\hat{\omega}(\omega_c^2 - \hat{\omega}^2)}(\hat{\omega}^2 - \omega_c^2\cos^2\theta), \tag{10.89}$$

$$b = -(\varepsilon_\perp^2 - \varepsilon_\times^2)\sin^2\theta - \varepsilon_\parallel\varepsilon_\perp(1 + \cos^2\theta) = \frac{2\omega_p^4}{\omega^2(\omega_c^2 - \hat{\omega}^2)}, \tag{10.90}$$

$$c = (\varepsilon_\perp^2 - \varepsilon_\times^2)\varepsilon_\parallel = \frac{\omega_p^6}{\omega^3\hat{\omega}(\omega_c^2 - \hat{\omega}^2)}. \tag{10.91}$$

After substitution into (10.85), and remarkable cancellations, we find a general expression for plane electron waves in magnetized plasma in this approximation [44]:

$$\text{arbitrary } \theta: \quad N^2 = \frac{k^2}{k_0^2} = \frac{\omega_p^2}{\omega(\pm\omega_c\cos\theta - \hat{\omega})}, \tag{10.92}$$

$$\theta = 0: \quad N^2 = \frac{k^2}{k_0^2} = \frac{\omega_p^2}{\omega(\pm\omega_c - \hat{\omega})}, \quad \text{R, L waves}, \tag{10.93}$$

$$\theta = \frac{\pi}{2}: \quad N^2 = \frac{k^2}{k_0^2} = -\frac{\omega_p^2}{\omega\hat{\omega}}, \quad \text{O, X waves.} \tag{10.94}$$

For the principal waves there are two angles, $\theta = 0$ and $\theta = \pi/2$. The four principal waves correspond to the well known O-, X-, R-, and L-waves. However, at $\theta = \frac{\pi}{2}$, the neglect of the vacuum displacement current degenerates the O- and X-waves, which are now always evanescent. Equation (10.93) with the plus sign for R-waves corresponds to (10.34) when the unity for vacuum displacement current is neglected; propagation occurs for $\omega < \omega_c$. Equation (10.93) with the minus sign for evanescent L-waves corresponds to (10.39) in the same approximation.

Two RHP wavenumber solutions

Reconsider the positive sign in (10.92), which corresponds to the R-wave when $\theta = 0$:

$$\text{arbitrary } \theta: \quad N^2 = \frac{k^2}{k_0^2} = \frac{\omega_p^2}{\omega(\omega_c \cos\theta - \hat{\omega})}. \tag{10.95}$$

This RHP wave propagates for $\omega < \omega_c \cos\theta$, which shows that the wave is confined to a cone centred on $\mathbf{B}_0$ of angular half-width satisfying $\cos\theta > \omega/\omega_c$. This is the range of possible angles for any given RHP wavenumber k. Alternatively, the permitted RHP wavenumbers can be calculated for a given angle θ by writing $\cos\theta = k_z/k$ (see figure 10.4(a)); this results in a quadratic equation for k for a given k_z. The two RHP-wavenumber roots, which we choose to label[9] k_H and k_T, are

$$k_{\mathrm{H,T}} = \frac{\omega_c k_z}{2\hat{\omega}}\left(1 \mp \sqrt{1 - \frac{4\hat{\omega}}{\omega}\frac{\omega_p^2 k_0^2}{\omega_c^2 k_z^2}}\right), \tag{10.96}$$

which is similar[10] to [2, 46]. This is the dispersion relation for RHP plane waves at an arbitrary angle to the magnetic field in cold, uniform, magnetized plasma, when the vacuum displacement current is neglected.

10.11.3 Curl factorization approximation of the plane wave dispersion relation

All of the electromagnetic electron wave calculations above were derived using three equations: the electron fluid force equation (10.11) (also called the Boltzmann first moment, or Ohm's law), and two Maxwell equations which involve just three variables $\mathbf{E}$, $\mathbf{B}$, and $\mathbf{J}$. As a reminder, these are (dropping the subscript e for electrons):

[9] The reason for the subscript labels H (helicon mode) and T (Trivelpiece–Gould mode [45]) will become clear in chapter 11. For the Trivelpiece–Gould mode, we use the single letter subscript T to avoid any ambiguity which could be caused by a (more appropriate) double letter subscript T–G, with apologies to R W Gould [45].

[10] Klozenberg *et al* [2], followed by others [46], defined the wavenumber roots as β instead of k, and the wavenumber z component as k instead of k_z.

$$\mathbf{E} = \frac{1}{\varepsilon_0\omega_p^2}[-j\hat{\omega}\mathbf{J} + \omega_c\mathbf{J} \times \hat{\mathbf{z}}], \tag{10.97}$$

$$\nabla \times \mathbf{E} = j\omega\mathbf{B}, \tag{10.98}$$

$$\nabla \times \mathbf{B} \approx \mu_0\mathbf{J} \quad \text{(vacuum displacement current neglected)}. \tag{10.99}$$

These three equations can be solved vectorially, as shown by Klozenberg [2], without needing to derive the dielectric tensor. The results must be identical to the tensor method provided that the same approximation is used, and this can be demonstrated as follows.

In the vector method, the technique is to directly take the curl of Ohm's law (10.97), instead of first obtaining the Helmholtz wave equation in section 10.4 by using the curl of (either of) the Maxwell equations. Hence,

$$\nabla \times \mathbf{E} = \frac{1}{\varepsilon_0\omega_p^2}[-j\hat{\omega}\nabla \times \mathbf{J} + \omega_c\nabla \times (\mathbf{J} \times \hat{\mathbf{z}})]. \tag{10.100}$$

Expand the last term using a vector identity:

$$\begin{aligned}\nabla \times (\mathbf{J} \times \hat{\mathbf{z}}) &= \mathbf{J}(\nabla \cdot \hat{\mathbf{z}}) - \hat{\mathbf{z}}(\nabla \cdot \mathbf{J}) + (\hat{\mathbf{z}} \cdot \nabla)\mathbf{J} - (\mathbf{J} \cdot \nabla)\hat{\mathbf{z}},\\ &= -\hat{\mathbf{z}}(\nabla \cdot \mathbf{J}) + (\hat{\mathbf{z}} \cdot \nabla)\mathbf{J},\\ &= -\hat{\mathbf{z}}(\nabla \cdot \mathbf{J}) + jk_z\mathbf{J},\end{aligned} \tag{10.101}$$

because $\hat{\mathbf{z}}$ is constant, and $\hat{\mathbf{z}} \cdot \nabla = jk_z$ for plane waves, $e^{j(k_x x + k_z z - \omega t)}$. Substituting for $\mu_0\mathbf{J} \approx \nabla \times \mathbf{B}$ using the Maxwell equations (10.98) and (10.99), and neglecting the vacuum displacement current (hence $\nabla \cdot \mathbf{J} \approx 0$), the curl of Ohm's law (10.100) becomes

$$\nabla \times \mathbf{E} = j\omega\mathbf{B} = \frac{1}{\varepsilon_0\omega_p^2}[-j\hat{\omega}\nabla \times \mathbf{J} + jk_z\omega_c\mathbf{J}],$$

$$j\omega\mathbf{B} \approx \frac{c^2}{\omega_p^2}[-j\hat{\omega}\nabla \times \nabla \times \mathbf{B} + jk_z\omega_c\nabla \times \mathbf{B}],$$

$$\nabla \times \nabla \times \mathbf{B} - \frac{k_z\omega_c}{\hat{\omega}}\nabla \times \mathbf{B} + \frac{\omega\omega_p^2}{c^2\hat{\omega}}\mathbf{B} \approx 0 \quad \text{(no vacuum displacement current)}. \tag{10.102}$$

We can rapidly relate this vector equation in $\mathbf{B}$ to the well known Helmholtz equation (D.34) for unmagnetized plasma by noticing that the unfamiliar term in $\nabla \times \mathbf{B}$ becomes zero because $\omega_c = 0$ when the static magnetic field $B_0 = 0$. This leaves

$$\begin{aligned}\nabla \times \nabla \times \mathbf{B} &= -\frac{\omega\omega_p^2}{c^2\hat{\omega}}\mathbf{B}\\ \nabla^2\mathbf{B} &= -\frac{\omega^2}{c^2}\varepsilon_p\mathbf{B} = -k_0^2\varepsilon_p\mathbf{B} \quad \text{(unmagnetized)},\end{aligned} \tag{10.103}$$

by using the vector identity $\nabla \times (\nabla \times \mathbf{a}) \equiv \nabla(\nabla \cdot \mathbf{a}) - \nabla^2\mathbf{a}$, and recognizing that $\varepsilon_p = -\frac{\omega_p^2}{\omega\hat{\omega}}$ from (D.27) when vacuum displacement current is neglected ($\bar{\varepsilon}_p \to \varepsilon_p$ as $B_0 \to 0$). Thus the magnetoplasma equation (10.102) reduces directly to the unmagnetized plasma wave equation (D.34) for zero applied magnetic field, provided that the vacuum displacement current is negligible. This clearly identifies the term $-\frac{k_z\omega_c}{\hat{\omega}}\nabla \times \mathbf{B}$ in (10.102) as being solely responsible for the magnetoplasma effects introduced by a static magnetic field $\mathbf{B}_0$.

Returning to the magnetoplasma equation (10.102), and following Klozenberg [2], we factorize (10.102) as a vector quadratic in curl with wavenumber roots[11] k_H and k_T:

$$[(\nabla \times) - k_\mathrm{H}][(\nabla \times) - k_\mathrm{T}]\mathbf{B} = 0, \tag{10.104}$$

$$\nabla \times \nabla \times \mathbf{B} - (k_\mathrm{H} + k_\mathrm{T})\nabla \times \mathbf{B} + k_\mathrm{H}k_\mathrm{T} = 0. \tag{10.105}$$

By comparison of these coefficients with (10.102),

$$k_\mathrm{H} + k_\mathrm{T} = \frac{k_z\omega_c}{\hat{\omega}}, \quad \text{and} \quad k_\mathrm{H}k_\mathrm{T} = \frac{\omega\omega_p^2}{c^2\hat{\omega}}, \quad \text{whence} \tag{10.106}$$

$$k_\mathrm{H,T} = \frac{1}{2}\left(\frac{k_z\omega_c}{\hat{\omega}} \mp \sqrt{\frac{k_z^2\omega_c^2}{\hat{\omega}^2} - 4\frac{\omega\omega_p^2}{c^2\hat{\omega}}}\right) = \frac{\omega_c k_z}{2\hat{\omega}}\left(1 \mp \sqrt{1 - \frac{4\hat{\omega}}{\omega}\frac{\omega_p^2k_0^2}{\omega_c^2k_z^2}}\right), \tag{10.107}$$

using $\omega = ck_0$, which is identical to the tensor solution (10.96), as required. In the absence of the external magnetic field, $\omega_c = 0$, we find that $k_\mathrm{H} = -k_\mathrm{T}$ in (10.106); the wavenumber magnitudes become unequal only due to $\mathbf{B}_0$. The general solutions, $\mathbf{B} = \mathbf{B}_1 e^{j(\mathbf{k}_\mathrm{H}\cdot\mathbf{r}-\omega t)}$ or $\mathbf{B}_2 e^{j(\mathbf{k}_\mathrm{T}\cdot\mathbf{r}-\omega t)}$ can be shown to be pure plane waves which can only exist independently in the infinite uniform plasma, without superposition. The interest in these expressions will be developed in the helicon theory of chapter 11, specifically in section 11.1.3 on uniform plasma density.

10.12 Chapter summary for whistler waves in uniform, magnetized plasma

Whistlers, which are a branch of R-waves between the ion and electron cyclotron frequencies, are the only waves in the RF spectrum to propagate in typical low temperature plasma conditions ($\omega_\mathrm{ci} \ll \omega_\mathrm{rf} \ll \omega_\mathrm{ce} \ll \omega_p$) as shown in figure 10.10. This is the obvious reason why we concentrate on whistlers amongst the panoply of other plasma waves. The polarization, thread handedness, conductivity, permittivity, and dispersion relation of these plasma waves were studied, along with the associated electron motion in collisionless and collisional magnetoplasma. The physical mechanism of propagation, or evanescence and reflection, of R-waves

[11] The curl factorization method introduced by [2], and followed by others [46], defined the wavenumber roots as β instead of k, and the wavenumber z component as k instead of k_z.

below or above the electron cyclotron frequency was interpreted in terms of the sign of effective conduction current in a rotating reference frame. It was also demonstrated that the RF excitation frequency is the only wave frequency driven in steady-state plasma, because any electron cyclotron motion is eventually damped.

Having selected the relevant RF wave type for magnetized plasma, the ultimate aim is to elaborate the role of whistlers in plasma heating excited by resonant network antennas. The birdcage antenna is the most common example of this, for which whistler waves must now be studied in their bounded cylindrical manifestation in chapter 11, where they are known as helicon modes.

References

[1] Lieberman M A and Lichtenberg A J 2005 *Principles of Plasma Discharges and Materials Processing* 2nd edn (Hoboken, NJ: Wiley)

[2] Klozenberg J P, McNamara B and Thonemann P C 1965 The dispersion and attenuation of helicon waves in a uniform cylindrical plasma *J. Fluid Mech.* **21** 545

[3] Lehane J A and Thonemann P C 1965 An experimental study of helicon wave propagation in a gaseous plasma *Proc. Phys. Soc.* **85** 301

[4] Boswell R W 1975 Measurements of the far-field resonance cone for whistler mode waves in a magnetoplasma *Nature* **258** 58

[5] Chen F F 1991 Plasma ionization by helicon waves *Plasma Phys. Control. Fusion* **33** 339

[6] Boswell R W and Chen F F 1997 Helicons—the early years *IEEE Trans. Plasma Sci.* **25** 1229

[7] Chen F F 2015 Helicon discharges and sources: a review *Plasma Sources Sci. Technol.* **24** 014001

[8] Chen F F 2016 *Introduction to Plasma Physics and Controlled Fusion* 3rd edn (Cham: Springer)

[9] Stenzel R L 2019 Whistler modes excited by magnetic antennas: a review *Phys. Plasmas* **26** 080501

[10] Tsankov T V, Chabert P and Czarnetzki U 2022 Foundations of magnetized radio-frequency discharges *Plasma Sources Sci. Technol.* **31** 084007

[11] Helliwell R A 1965 *Whistlers and Related Ionospheric Phenomena* (Stanford, CA: Stanford University Press)

[12] Storey L R O 1953 An investigation of whistling atmospherics *Phil. Trans. R. Soc.* A **246** 113

[13] Kurth W S and Gurnett D A 1991 Plasma waves in planetary magnetospheres *J. Geophys. Res.* **96** 18977

[14] Al'pert Y 1980 40 years of whistlers *J. Atmos. Terr. Phys.* **42** 1

[15] Stenzel R L 1999 Whistler waves in space and laboratory plasmas *J. Geophys. Res.* **104** 14379

[16] Amatucci W E 2006 A review of laboratory investigations of space plasma waves *Radio Sci. Bull.* **319** 32

[17] Boswell R W 1970 Plasma production using a standing helicon wave *Phys. Lett.* A **33** 457

[18] Chen F F 1996 Physics of helicon discharges *Phys. Plasmas* **3** 1783

[19] Boswell R W 1984 Very efficient plasma generation by whistler waves near the lower hybrid frequency *Plasma Phys. Control. Fusion* **26** 1147

[20] Miljak D G and Chen F F 1998 Density limit in helicon discharges *Plasma Sources Sci. Technol.* **7** 537

[21] Tarey R D, Sahu B B and Ganguli A 2012 Understanding helicon plasmas *Phys. Plasmas* **19** 073520

[22] Chen F F and Blackwell D D 1999 Upper limit to Landau damping in helicon discharges *Phys. Rev. Lett.* **82** 2677
[23] Shamrai K P, Pavlenko V P and Taranov V B 1997 Excitation, conversion and damping of waves in a helicon plasma source driven by an $m = 0$ antenna *Plasma Phys. Controlled Fusion* **39** 505
[24] Shamrai K P and Taranov V B 1996 Volume and surface RF power absorption in a helicon plasma source *Plasma Sources Sci. Technol.* **5** 474
[25] Krämer M, Aliev Y M, Altukhov A B, Gurchenko A D, Gusakov E Z and Niemi K 2007 Anomalous helicon wave absorption and parametric excitation of electrostatic fluctuations in a helicon-produced plasma *Plasma Phys. Control. Fusion* **49** A167
[26] Howling A A, Descoeudres A and Hollenstein Ch 2020 Multiple dehydrogenation reactions of negative ions in low pressure silane plasma chemistry *Plasma Sources Sci. Technol.* **29** 105015
[27] Chakraborty M, Sharma N, Neog N K and Bandyopadhyay M 2020 Study on negative ion production by electronegative gases in a helicon source *Jpn. J. Appl. Phys.* **59** SHHC01
[28] Guittienne Ph, Chevalier E and Hollenstein Ch 2005 Towards an optimal antenna for helicon waves excitation *J. Appl. Phys.* **98** 083304
[29] Swanson D G 2003 *Plasma Waves (Series in Plasma Physics)* 2nd edn (Bristol: Institute of Physics Publishing)
[30] Stix T H 1992 *Waves in Plasmas* (New York: AIP Press, Springer)
[31] Bleaney B I and Bleaney B 1976 *Electricity and Magnetism* (London: Oxford University Press) 3rd edn p 1
[32] Dumont R 2017 Waves in plasmas *Lecture Notes* cel-01463091 Master. France https://cea.hal.science/cel-01463091
[33] Howard J 2002 Introduction to plasma physics C17 *Lecture Notes* Plasma Research Laboratory, Australian National University
[34] Guittienne Ph, Jacquier R, Pouradier Duteil B, Howling A A, Agnello R and Furno I 2021 Helicon wave plasma generated by a resonant birdcage antenna: magnetic field measurements and analysis in the RAID linear device *Plasma Sources Sci. Technol.* **30** 075023
[35] Lieberman M A and Gottscho R A 1994 Design of high-density plasma sources for materials processing *Plasma Sources for Thin Film Deposition and Etching* ed M Francombe (New York: Academic) (Physics of Thin Films Series)
[36] Stephenson G 2003 *Mathematical Methods for Science Students* 2nd edn (London: Longman)
[37] Boas M L 2006 *Mathematical Methods in the Physical Sciences* 3rd edn (Hoboken, NJ: Wiley)
[38] Buttrill S E 1969 Measurement of ion–molecule reaction rate constants using ion cyclotron resonance *J. Chem. Phys.* **50** 4125
[39] Stenzel R L, Urrutia J and Rousculp C 1995 Nonlinear phenomena associated with large amplitude whistler pulses *J. Physique IV* **05** C6–61
[40] Bellan P M 2018 Relaxation of an isolated configuration to the Taylor state *Magnetic Helicity, Spheromaks, Solar Corona Loops, and Astrophysical Jets* (Singapore: World Scientific) ch 4
[41] Jackson J D 1999 *Classical Electrodynamics* 3rd edn (New York: Wiley)
[42] Royer A 2011 Why is the magnetic force similar to a Coriolis force? arXiv:1109.3624v1 [physics.gen-ph]

[43] Weisskopf V F 1968 How light interacts with matter *Sci. Am.* **219** 60
[44] Borg G G and Boswell R W 1998 Power coupling to helicon and Trivelpiece–Gould modes in helicon sources *Phys. Plasmas* **5** 564
[45] Trivelpiece A W and Gould R W 1959 Space charge waves in cylindrical plasma columns *J. Appl. Phys.* **30** 1784
[46] Chen F F and Arnush D 1997 Generalized theory of helicon waves. I. Normal modes *Phys. Plasmas* **4** 3411

IOP Publishing

Chapter 11

Helicon modes in a magnetized plasma column

The preceding chapter 10 identified RF electromagnetic (EM) waves in an infinite, uniform, magnetized plasma as the whistler branch of R-waves. Now we consider the influence of boundaries in a cylindrical plasma column, where propagating whistler waves are transformed into helicon modes. The theory is presented step-by-step, considering collisionality (dissipation), plasma radial non-uniformity, and boundary conditions, to arrive at a model sufficiently realistic to compare with experiment in chapter 12. In a nutshell, whistler waves are right-hand circularly polarized electromagnetic waves (R-waves) which can propagate in an infinite plasma immersed in a static magnetic field B_0. If these waves are generated in a finite plasma, such as a cylinder, the existence of boundary conditions cause a left-hand circularly polarized mode (L-wave) to exist simultaneously, together with an electrostatic contribution to the total wavefield. These bounded whistler wave solutions are known as helicon modes (figure 11.1). Helicon plasma science is reviewed in the recent book by S Shinohara [1].

It turns out that collisionless models (no dissipation) are quite unrewarding for the understanding of helicon waves. Notably, the relative roles of helicon (H) and Trivelpiece–Gould (T–G) modes [2] cannot be clarified with this simplification because there is no selective damping. Furthermore, the apparent preferential excitation of certain modes, as observed experimentally, cannot be explained by a collisionless approach; it is nevertheless an instructive initial step. The introduction of dissipation changes the situation significantly. Dominant H and dominant T–G modes can be clearly identified, and wave beating patterns can be calculated. New dominant T–G modes of small axial wavelength, forbidden in a collisionless approach, are now predicted. One important result of this collisional, uniform density model with a dielectric boundary at the plasma edge, is that the T–G component determines the power deposition radial profile, even with a dominant H

doi:10.1088/978-0-7503-5296-3ch11

Figure 11.1. '#2: Helicons are confined whistlers. Note that helicons are also large, brass musical instruments similar to sousaphones and tubas.' (Illustration by Alex Howling. Copyright 2023 Alex Howling.)

mode—damping of the T–G mode component leads to strong power deposition at the plasma edge.

However, uniform density profiles are not very relevant for comparison with experiment. For radially varying density profiles, a fourth order differential equation for the magnetic induction field is obtained, which can only be integrated numerically. The wave propagation is influenced by the shape of the radial density profile. For example, for bell-shaped plasma profiles, as in the RAID vacuum chamber [3] of chapter 12, the power deposition is predicted to be mainly peaked on the column axis, and not on the edge. Furthermore, new modes of longer axial wavelength are now predicted. These are notably characterized by an off-axis power deposition profile, which could well explain the results from COMSOL [4] numerical simulations, and some experimental observations which show that several wavelengths are excited simultaneously.

This chapter is organized as follows. Section 11.1 introduces the cylindrical configuration, the terminology, and the fundamental equations. On this basis, section 11.2 derives the analytical solutions for the case of uniform plasma density. Section 11.3 then presents numerical solutions for radially non-uniform density profiles. Some details of the derivations and methods are resumed in appendix J.

11.1 Introduction to the helicon mode equations

Figure 11.2 shows a plasma column which is magnetized by a field $\mathbf{B}_0$ along the z-axis. The geometry is naturally cylindrical, as opposed to the Cartesian system used for plane whistler waves in chapter 10. The plasma is infinitely long and axisymmetric, without any antenna source nor shell current surrounding the plasma used in the inductively coupled birdcage of chapter 8—for mathematical tractability, the helicon modes are assumed to be excited by some unspecified source, from which they propagate along the magnetized plasma. This is therefore a study of normal modes, as opposed to the source and wavefield calculations in chapter 8.

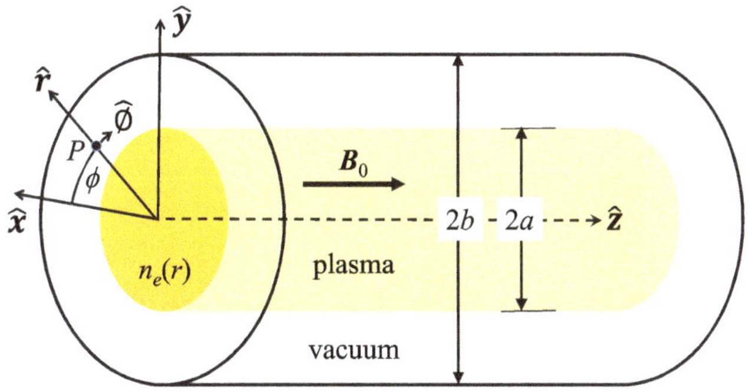

Figure 11.2. A schematic of the axially magnetized plasma column, radius a, within an annular vacuum gap bounded by a perfect electrically conducting (PEC) cylindrical screen, radius b. The plasma density axisymmetric radial profile $n_e(r)$ may be uniform or peaked on axis, depending on the model. The arbitrary point P can be written as $P(x, y, z)$ or $P(r, \phi, z)$, where $\{\hat{\mathbf{x}}, \hat{\mathbf{y}}, \hat{\mathbf{z}}\}$ are the unit vectors of the right-handed Cartesian system, and $\{\hat{\mathbf{r}}, \hat{\boldsymbol{\phi}}, \hat{\mathbf{z}}\}$ are the unit vectors of the right-handed cylindrical coordinate system.

A self-consistent numerical simulation model of a birdcage source and its helicon plasma will be described for RAID in chapter 12. The plasma density $n_e(r)$, as well as the electron–neutral effective collision frequency ν, are generally functions of the radial coordinate r, but in the following, we consider radial variations only of the density, with constant collision frequency assumed. At least, in the COMSOL models, taking a uniform temperature or a temperature radial profile does not change the result noticeably. Beyond the plasma radius a, the annular vacuum region extends up to $r = b$ where we posit a perfectly electrically conducting (PEC) vacuum vessel wall, corresponding to the surrounding Faraday screen at $r = c$ in chapter 8. The case of a PEC boundary directly at the plasma edge ($b = a$) will also be treated.

The general form of solution in cylindrical geometry, assuming time harmonic Fourier components, satisfying azimuthal periodicity, and allowing for wave propagation along the z-axis, is taken to be

$$\mathbf{E}(\mathbf{r}, t) = \mathbf{E}(\mathbf{r})e^{-j\omega t} = \mathbf{E}(r)e^{j(k_z z + m\phi - \omega t)}, \tag{11.1}$$

as derived in chapter 8, where $\mathbf{E}(r) = E_r(r)\hat{\mathbf{r}} + E_\phi(r)\hat{\boldsymbol{\phi}} + E_z(r)\hat{\mathbf{z}}$ represents the general phasor (complex amplitude) of the electric wavefield, and similarly for the magnetic wavefield $\mathbf{B}(\mathbf{r}, t)$. The general solution for a wave of angular frequency ω is a superposition of multiple (m, k_z) Fourier components, as shown in (8.10)—we recall that the azimuthal periodicity m here is not the antenna normal mode number m. The equivalent wavefield in Cartesian coordinates is given by (10.23). As a reminder of the terminology [5], with reference to figure 10.4(a):

- The wavenumber k is the magnitude of the wave-vector $\mathbf{k}$, i.e. $k = |\mathbf{k}|$.
- $k_z = k \cos\theta$ is the axial, z component of the wave-vector, in both Cartesian and cylindrical coordinate systems, along the externally imposed, constant, uniform magnetic field $\mathbf{B}_0$ which is aligned with the z-axis.

- $k_\perp$ is the transverse component of the wave-vector, which is perpendicular to $\mathbf{B}_0$. In the Cartesian coordinates for plane waves in chapter 10, $k_\perp = k \sin\theta$. In the cylindrical coordinates of this chapter, $k_\perp = k_r$ is the r-component, aligned with the r-axis.
- To summarize [5], $k = (k_\perp^2 + k_z^2)^{1/2}$.

Do not confuse the azimuthal cylindrical coordinate ϕ with the wave-vector angle θ relative to $\mathbf{B}_0$ in figure 10.4(a). We point out here that the wavefield component $B_z \neq 0$ in this chapter, in contrast to the TM modes studied in section 8.1. The magnetized plasma supports simultaneous modes which impose an axial wave-number k_z, therefore it cannot be set to zero, as opposed to the infinitely long, straight birdcage antenna in chapter 8, where $k_z \to 0$. In the present chapter, we will see that B_z becomes an integral part of the wavefield solution.

11.1.1 The constitutive relations for the magnetized plasma column

Although the conductivity and permittivity tensors will not be used in this chapter, their transformation into cylindrical geometry is of general interest. The tensors defining the plasma properties were given in chapter 10, namely, the plasma conductivity tensor $\bar{\sigma}_p$ (10.14), and the relative permittivity tensor $\bar{\varepsilon}_p$ (10.18). Note that these tensors were derived using a Cartesian system and it is legitimate to question their validity in the cylindrical coordinates of the present chapter. The vector equation (10.6) and Maxwell's equations could be re-derived from scratch using cylindrical coordinates, but appendix J.1 proves that these tensors are, in fact, invariant in the transformation between Cartesian and cylindrical coordinate systems, thanks to the particular symmetry of the magnetic force in (10.6). Therefore, despite the introduction of cylindrical wave functions, a column axis, and azimuthal periodicity in figure 11.2, $\bar{\sigma}_p$ (10.14) and $\bar{\varepsilon}_p$ (10.18) can be employed directly for analysis of the magnetized plasma column. These questions did not arise in chapter 8 because the conductivity and permittivity of the non-magnetized medium are isotropic and the tensors are simply unit tensors.

11.1.2 The equations to be solved for various simplifying assumptions

We have already seen in section 10.11 that considerable simplification of the wave equations is obtained by neglecting the vacuum displacement current. Section 10.11.1 showed that this magneto-quasistatic approximation (D.1.1) is justified for whistler waves at RF frequencies. Moreover, the equations and tensors which include the vacuum displacement current are very unwieldy and will only be resumed below in section 11.1.5.

As well as neglecting the vacuum displacement current, other assumptions affect the set of equations to be solved. In order of increasing accuracy, difficulty, and relevance for experimental comparison, section 11.1.3 assumes uniform plasma density, section 11.1.4 considers radially non-uniform plasma density, and section 11.1.5 briefly re-introduces the vacuum displacement current. Section 11.1.6 considers a special case, called the helicon approximation, which neglects electron inertia and collisions, although it does maintain the vacuum displacement current. A whole gamut of arbitrary approximations can easily be imagined—the literature contains almost as many authors as possible combinations.

Throughout this treatment of the initial electron fluid force equation (10.11) and Maxwell's equations, the reader (and the authors, it must be admitted) may well get a distinct impression of horse trading one approximation for another, while searching for a trade-off between mathematical tractability and experimental relevance. Approximations can bring insight and/or illusion, so careful judgement must be exercised. With hindsight, for the experimental parameters of table 10.2 and chapter 12, it appears that the vacuum displacement current can justifiably be neglected, but radial non-uniformity and collisional dissipation are both mandatory. Section 11.1.4 is therefore the most relevant for our purposes, whereas the other sections are interesting for comparison, also for the vast literature on this subject.

11.1.3 Uniform plasma density and collisionality, but no vacuum displacement current

To recall the basic equations for reference, the curl factorization method in section 10.11.3 employs the electron fluid force and two of Maxwell's equations, as follows:

$$\mathbf{E} = \frac{1}{\varepsilon_0 \omega_p^2}[-j\hat{\omega}\mathbf{J} + \omega_c \mathbf{J} \times \hat{\mathbf{z}}], \tag{11.2}$$

$$\nabla \times \mathbf{E} = j\omega \mathbf{B}, \tag{11.3}$$

$$\nabla \times \mathbf{B} \approx \mu_0 \mathbf{J} \quad \text{(vacuum displacement current neglected).} \tag{11.4}$$

Taking the curl of the fluid equation gave (10.100):

$$\nabla \times \mathbf{E} = \frac{1}{\varepsilon_0 \omega_p^2}[-j\hat{\omega}\nabla \times \mathbf{J} + \omega_c \nabla \times (\mathbf{J} \times \hat{\mathbf{z}})]. \tag{11.5}$$

Regarding (10.100), note that it was implicitly assumed that the plasma density n_e is uniform, because the electron plasma frequency (5.1) ω_p is taken to be constant[1]. The last term was expanded using the vector identity

$$\begin{aligned} \nabla \times (\mathbf{J} \times \hat{\mathbf{z}}) &= \mathbf{J}(\nabla \cdot \hat{\mathbf{z}}) - \hat{\mathbf{z}}(\nabla \cdot \mathbf{J}) + (\hat{\mathbf{z}} \cdot \nabla)\mathbf{J} - (\mathbf{J} \cdot \nabla)\hat{\mathbf{z}}, \\ &= -\hat{\mathbf{z}}(\nabla \cdot \mathbf{J}) + (\hat{\mathbf{z}} \cdot \nabla)\mathbf{J}, \\ &= -\hat{\mathbf{z}}(\nabla \cdot \mathbf{J}) + jk_z \mathbf{J}, \end{aligned} \tag{11.6}$$

because $\hat{\mathbf{z}} \cdot \nabla = jk_z$ for both plane waves, $e^{j(k_x x + k_z z - \omega t)}$, and cylindrical waves, $e^{j(m\theta + k_z z - \omega t)}$. Using $\nabla \cdot \mathbf{J} = 0$ because the vacuum displacement current is neglected in (11.4), the vector wave equation (10.106) was obtained and factorized[2] as follows:

[1] Only the electron fluid is considered in this chapter, hence the subscript e is dropped throughout.

[2] The curl factorization method introduced by [6], and followed by others [7], defined the wavenumber roots as β instead of k, and the wavenumber z component as k instead of k_z.

$$\nabla \times \nabla \times \mathbf{B} - \frac{k_z \omega_c}{\hat{\omega}} \nabla \times \mathbf{B} + \frac{\omega \omega_p^2}{c^2 \hat{\omega}} \mathbf{B} = 0 \text{ (uniform; no vac. displacement current)}, \quad (11.7)$$

$$[(\nabla \times) - k_{\mathrm{H}}][(\nabla \times) - k_{\mathrm{T}}]\mathbf{B} = 0, \quad (11.8)$$

$$k_{\mathrm{H,T}} = \frac{\omega_c k_z}{2\hat{\omega}} \left(1 \mp \sqrt{1 - \frac{4\hat{\omega}}{\omega} \frac{\omega_p^2 k_0^2}{\omega_c^2 k_z^2}} \right), \quad (11.9)$$

where the subscript 'H' (associated with the '−' sign, and therefore with the smaller value of $k_{\mathrm{H,T}}$ and the longer wavelength) stands for 'helicon' and 'T' (associated with the '+' sign, and therefore with the larger value of $k_{\mathrm{H,T}}$ and the shorter wavelength) stands for Trivelpiece–Gould[3], as explained below. It can be seen that the wave equation (11.7) and wavenumber $k_{\mathrm{H,T}}$ solutions (10.96), (10.107), or (11.9), are the same for the Cartesian and cylindrical coordinate systems because they both use the ansatz $e^{jk_z z}$. A general solution of (11.8) is given by the superposition of the two roots:

$$\mathbf{B} = \mathbf{B}_{\mathrm{H}} + \mathbf{B}_{\mathrm{T}} \quad \text{with} \quad \nabla \times \mathbf{B}_{\mathrm{H,T}} = k_{\mathrm{H,T}} \mathbf{B}_{\mathrm{H,T}}. \quad (11.10)$$

In contrast to the purely independent plane whistler waves of infinite, uniform magnetoplasma in section 10.11.3, helicon modes require a superposition of both solutions in order to satisfy the boundary conditions of confined magnetoplasma, as explained in section 11.1.7 and appendix J.2.

The branch of the smaller wavenumber k_{H}, associated with the minus sign in (11.9), is conventionally called the H (helicon) branch, with the longer wavelength—hence the subscript label 'H'. For $k_z \gg k_0 = \omega/c$, the wavenumber in free space, the square root in (11.9) can be expanded binomially [7], and with the minus sign for k_{H}, the $\hat{\omega}$ terms cancel, yielding

$$k_{\mathrm{H}} = \frac{\omega_p^2 k_0^2}{\omega \omega_c k_z}, \quad (k_z^2 \gg k_0^2). \quad (11.11)$$

This is called the 'helicon approximation' [3, 5, 7], and it is the same as the dispersion relation (10.96) or (10.107) for whistler waves in infinite, uniform plasma for the same approximations. See section 11.1.6 and (11.40) for more discussion of the helicon approximation.

The larger k_{T} root, associated with the plus sign in (11.9), is designated as the Trivelpiece–Gould [2] (T–G) branch, with the shorter wavelength. By just considering these names, one could think that these $k_{\mathrm{H,T}}$ branches are specific to helicon modes. But, as shown immediately above, and in section 10.11, the separation into two branches is solely a consequence of the ansatz $e^{jk_z z}$, and it notably has nothing to do with cylindrical symmetry, and therefore neither with helicon modes specifically.

[3] For the Trivelpiece–Gould mode, we use the single letter subscript T to avoid any ambiguity which could be caused by a (more appropriate) double letter subscript T–G, with apologies to R W Gould [2].

In the vacuum region, an equation for the magnetic field wave equation is

$$\nabla \times \nabla \times \mathbf{B} - k_0^2\mathbf{B} = \mathbf{0}, \tag{11.12}$$

which can be factorized as

$$[(\nabla \times) - k_0][(\nabla \times) + k_0]\mathbf{B} = 0, \tag{11.13}$$

with general solution

$$\mathbf{B} = \mathbf{B}_+ + \mathbf{B}_- \quad \text{with} \quad \nabla \times \mathbf{B}_\pm = \pm k_0 \mathbf{B}_\pm. \tag{11.14}$$

For cylindrical geometry in figure 11.2, a general solution to (11.10) for uniform density plasma and neglecting the vacuum displacement current, and also for vacuum (11.14), can be derived in terms of Bessel functions, as follows: The unit vector coefficients of $\nabla \times \mathbf{B} = k\mathbf{B}$ are

$$\hat{\mathbf{r}}: \quad kB_r = \frac{1}{r} jmB_z - jk_z B_\phi, \tag{11.15}$$

$$\hat{\boldsymbol{\phi}}: \quad kB_\phi = jk_z B_r - B_z', \tag{11.16}$$

$$\hat{\mathbf{z}}: \quad kB_z = \frac{1}{r}\left[B_\phi + rB_\phi' - jmB_r\right]. \tag{11.17}$$

Elimination of B_r and B_ϕ from these equations leads to a Bessel equation in $B_z(r)$:

$$r^2 B_z'' + rB_z' + [(k^2 - k_z^2)r^2 - m^2]B_z = 0, \tag{11.18}$$

which is the same as (8.7) in the development of the solution for E_z, provided that the uniform-dielectric wavenumber $k_d = k_0\sqrt{\varepsilon_r}$ is now replaced by the magnetoplasma wavenumber k in (11.9). The radial solutions are therefore Bessel functions of the first and second kinds, $J_m(r\sqrt{k^2 - k_z^2})$ and $Y_m(r\sqrt{k^2 - k_z^2})$, exactly as obtained in chapter 8 to solve the Helmholtz equation for a uniform density plasma column. The relation between these two different, but equivalent, solution methods can be recognized by taking the curl of the curl factorization general solution (11.10), namely $\nabla \times \mathbf{B} = k\mathbf{B}$, and using $\nabla \cdot \mathbf{B} = 0$ in the vector expansion of $\nabla \times (\nabla \times \mathbf{B})$:

$$\begin{aligned} \nabla \times (\nabla \times \mathbf{B}) &= \nabla \times (k\mathbf{B}), \\ -\nabla^2\mathbf{B} &= k\nabla \times \mathbf{B} = k(k\mathbf{B}), \\ \nabla^2\mathbf{B} &= -k^2\mathbf{B}, \end{aligned} \tag{11.19}$$

which is the same as the Helmholtz wave equation (8.1), solved by the more familiar method of the separation of variables in chapter 8 for the isotropic plasma dielectric.

The transverse wavenumber $k_\perp$ is defined by $k_\perp^2 = k^2 - k_z^2$, which is along the r-axis in this cylindrical coordinate system, so $k_r = k_\perp$. The solution for B_z is then expressed as

$$B_z = C_1 J_m(k_\perp r) + C_2 Y_m(k_\perp r), \tag{11.20}$$

where C_1, C_2 are the two integration constants of the second order equation, to be determined by boundary conditions. This is the first and last time in this chapter that an analytical expression can be calculated for the wavefield. The radial magnetic wavefield B_r can also be cast in the form of Bessel functions by using (11.15) and (11.16) to write

$$B_r = \frac{j}{rk_\perp^2}(kmB_z + k_z r B_z'), \tag{11.21}$$

and substituting for B_z using (11.20) to obtain several possible combinations such as[4]

$$\begin{aligned} B_r = &\frac{jC_1}{2k_\perp}[(k - k_z)J_{m+1}(k_\perp r) + (k + k_z)J_{m-1}(k_\perp r)] \\ &+ \frac{jC_2}{2k_\perp}[(k - k_z)Y_{m+1}(k_\perp r) + (k + k_z)Y_{m-1}(k_\perp r)]. \end{aligned} \tag{11.22}$$

The azimuthal magnetic wavefield B_ϕ is obtained in the same way:

$$\begin{aligned} B_\phi = &\frac{C_1}{2k_\perp}[(k - k_z)J_{m+1}(k_\perp r) - (k + k_z)J_{m-1}(k_\perp r)] \\ &+ \frac{C_2}{2k_\perp}[(k - k_z)Y_{m+1}(k_\perp r) - (k + k_z)Y_{m-1}(k_\perp r)]. \end{aligned} \tag{11.23}$$

The $Y_m(r)$ Bessel function diverges at $r = 0$. As a consequence, for the plasma region, the constant C_2 will always vanish, for both roots $k = k_\mathrm{H}$ and $k = k_\mathrm{T}$, because diverging solutions at $r = 0$ are not physical. Therefore, in the plasma region we only have two unknown integration constants that need to be determined, one C_1 amplitude for each wavenumber k_H and k_T. In the surrounding vacuum region there remain four unknown amplitudes, C_1 and C_2 for each of $+k_0$ and $-k_0$. Without loss of generality, for normal modes we can always set one of these constants to be equal to 1, by normalization. Then, at the end we are left with five unknown amplitudes that will be determined by applying boundary conditions. Naturally, if a PEC boundary is stated at the plasma interface, only one constant needs to be determined. The modes are orthogonal functions, therefore each excited mode must simultaneously satisfy its own set of boundary conditions. These boundary conditions are discussed further in section 11.1.7.

11.1.4 Radially non-uniform plasma density and collisionality, but no vacuum displacement current

Chapter 10 dealt with the idealized case of plane waves in infinite, uniform plasma. The more realistic plasma column in figure 11.2 now requires the equations (10.97)–(10.99), which neglect the vacuum displacement current, to account for radially non-uniform plasma density in cylindrical geometry. Plasma density intervenes only via

---

[4] There are many mathematical properties of Bessel functions [8, 9], among which $B_n'(z) = [B_{n-1}(z) - B_{n+1}(z)]/2$, and $\frac{2m}{z}B_m(z) = [B_{n-1}(z) + B_{n+1}(z)]$, are used here.

the electron plasma frequency, hence we introduce a new variable $\alpha(r) = \frac{1}{\varepsilon_0\omega_p^2} = \frac{m_e}{q^2 n_e(r)}$ and re-write the electron fluid force equation (10.97) as

$$\mathbf{E} = \alpha[-j\hat{\omega}\mathbf{J} + \omega_c\mathbf{J} \times \hat{\mathbf{z}}]. \tag{11.24}$$

Now, when we take the curl of this equation, as for (10.100), the r-dependence of α introduces a new term, seen here on the second line:

$$\begin{aligned}\nabla \times \mathbf{E} = j\omega\mathbf{B} = \alpha[-j\hat{\omega}\nabla \times \mathbf{J} + \omega_c\nabla \times (\mathbf{J} \times \hat{\mathbf{z}})]\\ + \alpha'\hat{\mathbf{r}} \times [-j\hat{\omega}\mathbf{J} + \omega_c\mathbf{J} \times \hat{\mathbf{z}}],\end{aligned} \tag{11.25}$$

obtained by using the vector identity $\nabla \times (f\mathbf{A}) = f(\nabla \times \mathbf{A}) + (\nabla f) \times \mathbf{A}$, and recognizing that $\nabla\alpha = \alpha'\hat{\mathbf{r}}$, where $\alpha' = \frac{\partial\alpha}{\partial r}$. The first line of (11.25) is expanded as before, using the vector identity (10.101). The cross products (vector products) in the new second line can be expanded as $\hat{\mathbf{r}} \times \mathbf{J} = (-J_z\hat{\boldsymbol{\theta}} + J_\theta\hat{\mathbf{z}})$, and $\hat{\mathbf{r}} \times (\mathbf{J} \times \hat{\mathbf{z}}) = -J_r\hat{\mathbf{z}}$. Substituting these expressions into (11.25) gives

$$j\omega\mathbf{B} = \alpha\left[-j\hat{\omega}\nabla \times \mathbf{J} + jk_z\omega_c\mathbf{J}\right] + \alpha'[j\hat{\omega}(J_z\hat{\boldsymbol{\theta}} - J_\theta\hat{\mathbf{z}}) - \omega_c J_r\hat{\mathbf{z}}]. \tag{11.26}$$

The various coefficients of the $\mathbf{J}$ components can be collected into a tensor $\bar{\Pi}$, and (11.26) multiplied throughout by $-j\frac{\omega_p^2}{c^2\hat{\omega}}$ with an eye on (10.102):

$$\begin{aligned}&\frac{\omega\omega_p^2}{c^2\hat{\omega}}\mathbf{B} = \nabla \times \mu_0\mathbf{J} - \frac{k_z\omega_c}{\hat{\omega}}\bar{\Pi} \cdot \mu_0\mathbf{J},\\ &\text{where} \quad \bar{\Pi} = -\frac{k_z\omega_c}{\hat{\omega}}\begin{pmatrix} 1 & 0 & 0 \\ 0 & 1 & \dfrac{\hat{\omega}\Lambda'}{k_z\omega_c} \\ j\dfrac{\Lambda'}{k_z} & -\dfrac{\hat{\omega}\Lambda'}{k_z\omega_c} & 1 \end{pmatrix},\end{aligned} \tag{11.27}$$

and $\Lambda' = \frac{\alpha'}{\alpha} = \frac{\partial \ln \alpha}{\partial r} = -\frac{\partial \ln n_e}{\partial r}$. Using Maxwell's equation (10.99) neglecting vacuum displacement current, $\nabla \times \mathbf{B} \approx \mu_0\mathbf{J}$, to conveniently substitute for $\mathbf{J}$, our final destination is the wave equation for magnetized plasma with radially non-uniform density:

$$\begin{aligned}&\nabla \times \nabla \times \mathbf{B} + \bar{\Pi} \cdot \nabla \times \mathbf{B} + \frac{\omega\omega_p^2}{c^2\hat{\omega}}\mathbf{B} = 0\\ &\text{(non-uniform; no vacuum displacement current)},\end{aligned} \tag{11.28}$$

where $\bar{\Pi}$ (11.27) replaces the scalar wavenumber coefficient $-\frac{k_z\omega_c}{\hat{\omega}}$ of $\nabla \times \mathbf{B}$ in the wave equation (11.7) for uniform magnetoplasma, because of radial non-uniformity. For uniform plasma, $\Lambda' = 0$, and the tensor component of $\bar{\Pi}$ becomes the unit tensor. This is the second and last time in this chapter that the wave equation can easily be derived by hand.

A fourth order expression for the radial component of the magnetic wavefield can be established from (11.28); to simplify the notation, we introduce the terms $\delta = \frac{\omega_c}{\hat{\omega}}$ and $\Delta(r) = \frac{\omega\omega_p^2}{c^2\hat{\omega}}$. We first take from $\nabla \cdot \mathbf{B} = 0$ the relation

$$B_z = \frac{j}{k_z r}(B_r + rB_r' + jmB_\theta). \tag{11.29}$$

With (11.29), the first line of (11.28) can be used to express B_θ as a function of B_r and its derivatives to obtain expressions of the form

$$\begin{aligned} B_\theta &= f_{\theta 2}(r)B_r'' + f_{\theta 1}(r)B_r' + f_{\theta 0}(r)B_r, \\ B_z &= f_{z2}(r)B_r'' + f_{z1}(r)B_r' + f_{z0}(r)B_r, \end{aligned} \tag{11.30}$$

where

$$\begin{aligned} f_{\theta 0}(r) &= j(1 + k_z^2 r^2 + m(m + \delta) + r^2\Delta)/\text{den}, \\ f_{\theta 1}(r) &= jr(m\delta - 1)/\text{den}, \\ f_{\theta 2}(r) &= -jr^2/\text{den}, \\ f_{z0}(r) &= -j(m^3 + m(k_z^2 r^2 - 1) - k_z^2 r^2 \delta + mr^2\Delta)/(k_z r \ \text{den}), \\ f_{z1}(r) &= j(3m + k_z^2 r^2 \delta)/(k_z \ \text{den}), \\ f_{z2}(r) &= jmr/(k_z \ \text{den}), \\ \text{and} \quad \text{den} &= (2m + m^2\delta + k_z^2 r^2 \delta). \end{aligned} \tag{11.31}$$

It can be verified, at this point, that the last two lines of (11.28) are equivalent. From the definitions in (11.31), the fourth order equation for the B_r component can be expressed as

$$c_4(r)B_r'''' + c_3(r)B_r''' + c_2(r)B_r'' + c_1(r)B_r' + c_0(r)B_r = 0, \tag{11.32}$$

where the coefficients are

$$\begin{aligned} c_4(r) &= f_{\theta 2}, \\ c_3(r) &= \left(\frac{2}{r} + \Lambda'\right)f_{\theta 2} + 2f_{\theta 2}', \\ c_2(r) &= f_{\theta 0} + \left(k_z^2(\delta^2 - 1) + \frac{3m\delta - 2 - m^2}{r} - \frac{m\delta - 2}{r}\Lambda' - \Delta\right)f_{\theta 2} \\ &\quad + \left(\frac{3 - m\delta}{r} + \Lambda'\right)f_{\theta 2}' + f_{\theta 2}'', \\ c_1(r) &= \left(\frac{m\delta + 1}{r} + \Lambda'\right) + 2f_{\theta 0}' + \left(\frac{m\delta(m^2 - 6) - m^2}{r^3} + \frac{(m\delta - 1)(k_z^2 + \Delta)}{r}\right)f_{\theta 2} \\ &\quad + \left(k_z^2\delta^2 + \frac{4m\delta - 1}{r^2} - \frac{m\delta - 1}{r}\Lambda'\right)f_{\theta 2}' + \frac{1 - m\delta}{r}f_{\theta 2}'', \\ c_0(r) &= j\left(k_z^2\delta + \frac{2m + \delta}{r^2} - \frac{m}{r}\Lambda'\right) - \left(k_z^2 + \frac{m^2 + m\delta + 1}{r^2} - \frac{1}{r}\Lambda' + \Delta\right)f_{\theta 0} \\ &\quad + \left(\frac{m\delta + 1}{r} + \Lambda'\right)f_{\theta 0}' + f_{\theta 0}''. \end{aligned} \tag{11.33}$$

The fourth order in (11.32) is a consequence of the radial non-uniformity [10]. This equation is solved numerically for given radial density profiles in section 11.3. It is naturally the most interesting case because it gives the best fit to the experimental results in chapter 12.

11.1.5 Radially non-uniform plasma density, collisionality, and vacuum displacement current

The wave equation for non-uniform plasma density which includes the vacuum displacement current in (10.15) or (D.4), can be solved with the computer aid of mathematical software such as Mathematica [11]. Two algebraic complications, for example, are that $\nabla \times \mathbf{B} \neq \mu_0 \mathbf{J}$ in (11.4), so $\nabla \cdot \mathbf{J}$ is no longer zero in (11.6). For the sake of completeness, and with no pretence of solving by hand, the wave equation is of the form

$$\nabla \times \nabla \times \mathbf{B} + \bar{\Pi} \cdot \nabla \times \mathbf{B} + \bar{Z} \cdot \mathbf{B} = 0 \quad \text{(non-uniform; with vacuum displacement current)}, \tag{11.34}$$

where both scalar coefficients of the initial wave equation (11.7) are now tensors. The first tensor in (11.34) is

$$\bar{\Pi} = -\frac{k_z \omega_c}{\hat{\omega}} \left[\frac{\omega_p^2 F}{F^2 - \omega_D^2 \omega_c^2} \right] \times \begin{pmatrix} 1 & j\frac{\omega_D \omega_c}{F} & 0 \\ -j\frac{\omega_D \omega_c}{F} & 1 & \frac{\hat{\omega}\Lambda'}{k_z \omega_c}\left[1 - \frac{\omega_D^2 \omega_c^2}{F^2}\right] \\ j\frac{\Lambda'}{k_z}\left[\frac{\hat{\omega}(\omega_D^2 \omega_c^2 + 2\hat{\omega}\omega_D F + F^2)}{F(\hat{\omega}F + \omega_D \omega_c^2)}\right] & -\frac{\hat{\omega}\Lambda'}{k_z \omega_c}\left[\frac{\hat{\omega}F^2 + \hat{\omega}\omega_D^2 \omega_c^2 + 2\omega_D \omega_c^2 F}{F(\hat{\omega}F + \omega_D \omega_c^2)}\right] & \left[\frac{\hat{\omega}(F^2 - \omega_D^2 \omega_c^2)}{F(\hat{\omega}F + \omega_D \omega_c^2)}\right] \end{pmatrix}, \tag{11.35}$$

$$\text{where} \quad F = \omega_p^2 - \hat{\omega}\omega_D.$$

The new term ω_D labels the RF angular frequency ω which is introduced when the vacuum displacement current is accounted for; this brings considerable complication. However, every term in square brackets is equal to one when the vacuum displacement current is neglected, i.e. when ω_D is set to zero. The second tensor in (11.34) is

$$\bar{Z} = \frac{\omega \omega_p^2}{c^2 \hat{\omega}} \left[\frac{F}{\omega_p^2} \right] \begin{pmatrix} 1 & 0 & 0 \\ 0 & 1 & 0 \\ 0 & 0 & \left[\frac{\hat{\omega}(F^2 - \omega_D^2 \omega_c^2)}{F(\omega_D \omega_c^2 + \hat{\omega} F)} \right] \end{pmatrix}, \tag{11.36}$$

where, again, the terms in square brackets equal one when the vacuum displacement current is neglected, i.e. when ω_D is set to zero.

The beauty of this beast is that all possible approximations of the wave quadratic equations can be conveniently derived from the general wave equation (11.34) and its tensors (11.35) and (11.36) by deleting the terms to be neglected. To summarize, the following approximations can be applied singly or together:

1. Negligible vacuum displacement current: $\omega_D \to 0$ (square bracket terms $\to$1).
2. Uniform radial density profile: $\Lambda' \to 0$.
3. Unmagnetized plasma: $\omega_c \to 0$.
4. Collisionless plasma: $\hat{\omega} \to \omega$.
5. Negligible electron inertia: $\hat{\omega} \to j\nu$.

For example, the wave equation of section 11.1.4 is recovered by using point 1, and the wave equation of section 11.1.3 is recovered by using points 1 and 2. In practice, the vacuum displacement current makes insignificant difference to the solutions in our plasma conditions, so the vacuum displacement current can be justifiably neglected (in accordance with section 10.11.1) in what follows.

11.1.6 Helicon approximation equations

The helicon approximation is a qualitatively different attempt to solve the magneto-plasma equations [7, 12–14] by neglecting the $\hat{\omega}\mathbf{J}$ term in (11.2), which has the effect of eliminating the T–G mode. The electron fluid force equation reduces to

$$\mathbf{E} \approx \frac{\omega_c}{\varepsilon_0 \omega_p^2} \mathbf{J} \times \hat{\mathbf{z}}, \tag{11.37}$$

which neglects electron inertia and power dissipation via electron collisions, because both terms in $\hat{\omega} = \omega + j\nu$ have effectively been set to zero. The necessary condition for this approximation to hold is $\omega \ll \omega_c$. The vacuum displacement current and radially non-uniform density are maintained, but unfortunately, because the order of the wavefield equation is reduced, (11.34) cannot be used because the $\bar{\bar{\Pi}}$ tensor is not invertible in the special 'helicon approximation' condition $\hat{\omega} \to 0$. In combination with Maxwell's equations, and after much algebraic manipulation [3], the equation for the wavefield is

$$\nabla \times \mathbf{B} = \frac{k_0^2 \omega_p^2}{k_z \omega \omega_c} \begin{pmatrix} 1 & j\dfrac{\omega_D \omega_c}{\omega_p^2} & 0 \\ -j\dfrac{\omega_D \omega_c}{\omega_p^2} & 1 & 0 \\ -j\dfrac{k_z}{k_z^2 - k_0^2}\Lambda' & 0 & \dfrac{k_z^2}{k_z^2 - k_0^2} \end{pmatrix} \mathbf{B}. \tag{11.38}$$

This equation superficially resembles $\nabla \times \mathbf{B} = k\mathbf{B}$ in (11.10), except that the wave-number k is replaced here by a tensor. Similarly to (11.15)–(11.18), by equating field components in (11.38), a second order differential equation can be established [3] for B_r:

$$B_r'' + f(r)B_r' + g(r)B_r = 0, \quad \text{with:}$$
$$f(r) = \frac{k_d^2 r^2 - 3m^2}{r(k_d^2 r^2 - m^2)},$$
$$g(r) = \frac{1 - m^2}{r^2} + k_d^2\left(1 - \frac{2}{k_d^2 r^2 - m^2}\right) - \frac{k_d^2 \gamma^2}{\varphi^2} - \frac{2k_d^2 m\gamma}{\varphi(k_d^2 r^2 - m^2)} + \frac{m\gamma'}{r\varphi}, \tag{11.39}$$
$$\varphi = k_d^2 \omega \omega_c,$$
$$\gamma = k_0^2 \omega_p^2,$$
$$k_d^2 = k_0^2 - k_z^2.$$

The functions $f(r)$ and $g(r)$ are non-singular, except at $r = 0$ [3]. This helicon approximation, when applicable, can be useful for evaluating the dominant H mode wavenumber [3], especially for radially non-uniform density because (11.39) is much easier to numerically integrate than the fourth order equation (11.32). The drawback of this helicon approximation model is the total suppression of all dissipation [7], therefore there is no damping of axial modes, so no mode selectivity to determine which modes are experimentally observable.

The equation with neglected vacuum displacement current is obtained from (11.38) with $\omega_D = 0$ and $k_z^2 \gg k_0^2$, and with $\Lambda' = 0$ for uniform plasma density, which together lead to a unit matrix so that

$$\nabla \times \mathbf{B} = \frac{\omega_p^2 k_0^2}{\omega \omega_c k_z} \mathbf{B} = k_{\mathrm{H}} \mathbf{B}, \tag{11.40}$$

which is the same as the helicon wavenumber for the same approximations in (11.11) where the $\hat{\omega}$ term cancels, hence the name 'helicon approximation'.

11.1.7 Boundary conditions and mode quantization

The boundary conditions appropriate for the plasma column in figure 11.2 are in principle the same as for the inductive plasma column in chapter 8 (except that the shell current source is not considered here because we are concerned with a normal mode study), and the magnetized plasma now, generally, has two simultaneous wavenumbers k_{H} and k_{T}, for a given k_z. For the present case, we have only to consider boundary conditions at $r = a$ and $r = b$ in figure 11.2. The former is the plasma–vacuum interface, requiring continuity of the tangential components of the electric field and of all the components of the magnetic field[5]. Designating the plasma fields as $\mathbf{E}_p$ and $\mathbf{B}_p$, and the vacuum fields as $\mathbf{E}_v$ and $\mathbf{B}_v$, we then must generally have

[5] In chapter 8, $B_z = 0$ (TM mode), and the B_r boundary condition is redundant due to the E_z continuity. Therefore, only the continuity of B_ϕ needs to be considered there.

$$
\begin{aligned}
E_{v\phi}(a) &= E_{p\phi}(a), \\
E_{vz}(a) &= E_{pz}(a), \\
B_{v\phi}(a) &= B_{p\phi}(a), \\
B_{vz}(a) &= B_{pz}(a),
\end{aligned}
\tag{11.41}
$$

where the fields are the sum of the H and T–G mode fields (see appendix J.2), and which apply to each mode m. The boundary at $r = b$ is a PEC, therefore we only have to satisfy the continuity of the tangential components of the electric field:

$$
\begin{aligned}
E_{v\phi}(b) &= 0, \\
E_{vz}(b) &= 0,
\end{aligned}
\tag{11.42}
$$

because all fields are zero in the PEC screen (or the vacuum vessel metal wall). Note that, in both the plasma and the vacuum regions, Maxwell's equation $\nabla \times \mathbf{E} = j\omega\mathbf{B}$ is generally applicable, and with $\mathbf{E}, \mathbf{B} \propto e^{j(k_z z + m\phi)}$, this leads to

$$
B_r = \frac{mE_z - k_z r E_\phi}{r\omega}. \tag{11.43}
$$

Therefore, in the present context, the continuity of the tangential components E_ϕ, E_z of the electric field implies the continuity of the radial component of the magnetic field component B_r, which is why the B_r continuity has been omitted in (11.41) and (11.42). Naturally, any pair can be chosen from $E_z = 0$, $E_\phi = 0$, and $B_r = 0$, at a PEC boundary. Consequently, at the plasma/vacuum interface we have four boundary conditions to fulfil, and only two at the PEC boundary (for each orthogonal mode number m).

Constraint condition on the axial wavenumber

We have seen in section 11.1.3 that generally, five integration constants need to be determined by boundary conditions. But we showed here that six boundary conditions are to be satisfied. Therefore, once we have determined all the required amplitudes (integration constants), we are left with one equation, which is a constraint, and it leads to a quantization of the possible axial wavenumbers k_z. In the following, this remaining condition on k_z is referred to as $\mathrm{Cond}(k_z) = 0$. The significance of this condition is that the roots corresponding to $\mathrm{Cond}(k_z) = 0$ define the possible mode solutions in terms of k_z.

The expression for $\mathrm{Cond}(k_z)$ depends on the model. If we posit a PEC boundary at the plasma boundary, only one constant needs to be determined for two boundary conditions to be satisfied, which leads to a quantization of k_z. In the simple helicon approximation (section 11.2.1) it arises directly from $B_r(a) = 0$, because it is the sole boundary condition for the H mode alone. The example case of {H,T–G} modes for uniform density and PEC boundary is described in full detail in appendix J.2 as an illustration of the constraint $\mathrm{Cond}(k_z) = 0$. For the {H,T–G} model with dielectric boundary, the expression for $\mathrm{Cond}(k_z) = 0$ is much more cumbersome because six boundary conditions are involved.

11.2 Normal mode solutions for uniform plasma density

In the first three sections, we will consider the PEC/plasma boundary case. To begin with, we show what can be expected in the framework of the helicon approximation, which is intrinsically collisionless. Then we consider a collisionless {H,T–G} model, and finally a collisional {H,T–G} model. For the last two sections, we treat the case of a plasma/vacuum boundary for collisionless, then collisional, {H,T–G} models.

For many examples and illustrations in this 'uniform density' section we refer to standard RAID conditions, as follows:

$$\begin{aligned} \text{plasma density} \quad & n_e = 5 \cdot 10^{18}\ \text{m}^{-3} \\ \text{magnetic flux density} \quad & B_0 = 200\ \text{G} = 0.02\ \text{T} \\ \text{plasma radius} \quad & a = 0.05\ \text{m}. \end{aligned} \tag{11.44}$$

Note also that we almost always take the $m = 1$ case for illustration, because it appears to be the most studied mode in the literature, and indeed appears to be the strongly dominant excited mode in the RAID experiment in chapter 12.

11.2.1 PEC at the plasma boundary, helicon approximation

As a first tutorial step in the study of the possible modes that can propagate into the plasma column, we consider the helicon approximation in section 11.1.6, i.e. $\hat{\omega} = \omega + j\nu \to 0$, in other words, we set the electron collision frequency to vanish (a non-dissipative case), and electron inertia is also neglected. Another very usual simplification consists in setting a perfectly electrically conducting (PEC) boundary at the plasma edge, although this situation is not very relevant for real experimental set-ups. In fact, this over-simplified system can be solved only for a PEC boundary.

In this uniform density helicon approximation, there is only one single helicon root $k_{\rm H} = \omega_p^2 k_0^2/(\omega\omega_c k_z)$ from (11.40), and the fields are described by Bessel functions of argument $(k_\perp r)$ as in (11.20). For this collisionless framework, the transverse wavenumber $k_\perp$ is real, given by $k_\perp^2 = k^2 - k_z^2 > 0$. The maximum value of the longitudinal wavenumber k_z is when the wavenumber is aligned along the z-axis, i.e. $k_z < k_z^{\max} = \omega_p k_0/\sqrt{\omega\omega_c}$. Because there is only one k branch, there is just one integration constant C_1 to be determined, which can be arbitrarily set to $C_1 = 1$ by normalization. We are then left with a single and simple condition to be satisfied at $r = a$ from (11.42):

$$\text{Cond}(k_z) = 0 \quad \Rightarrow \quad B_r(a) = 0, \quad \text{equivalently,} \quad E_\phi(a) = 0. \tag{11.45}$$

$E_z(a) = 0$ by definition in the helicon approximation. Figure 11.3 shows the function $\text{Cond}(k_z) = B_r(a)$ for the RAID standard parameters (11.44), and for $m = 1$. It can be seen that the condition $\text{Cond}(k_z) = 0$ is satisfied for infinite values of $k_z < k_z^{\max}$. As a convention in the collisionless cases, we designate these eigenvalues as $k_{z\ m,n}$, where m refers to the azimuthal mode number (usually assumed equal to 1 here), and n designates the successive eigenvalue numbers of the possible k_z longitudinal wavenumbers, where $n = 1$ corresponds to the highest possible k_z value as shown in figure 11.3. We will see that this numbering becomes problematic when

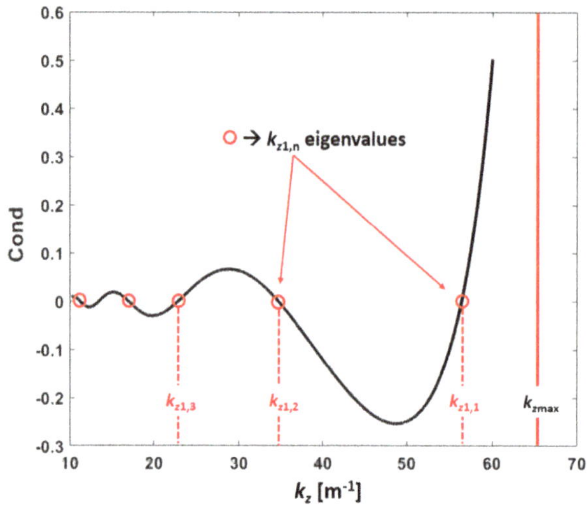

Figure 11.3. Plot of the function $\text{Cond}(k_z) = B_r(a)$, for RAID standard conditions (11.44) and taking $m = 1$, with roots $B_r(a) = 0$ at the zero crossing points.

dissipation is considered. Figure 11.4 shows the modulus of the same function, $|\text{Cond}(k_z)|$, using a semilog scale, because the possible eigenvalues appear very clearly as downward peaks. Figure 11.5 shows the radial profile of the magnetic field components for the first three n eigenvalues $k_{z1,\,1}$, $k_{z1,\,2}$, and $k_{z1,\,3}$. As a general rule, modes of lower n have smaller axial wavelength (larger wavenumber k_z) and higher radial wavelength (smaller wavenumber $k_\perp$). Finally, figure 11.6 shows the influence of the azimuthal mode number m on the $k_{z\ m,1}$ value. The smallest longitudinal wavelength excited is obtained for $m = 1$, but this is the only notable fact in this study of the azimuthal mode number m.

11.2.2 PEC at the plasma boundary, collisionless {H,T–G} model

We now consider a case more general than the helicon approximation of section 11.2.1 by re-introducing the electron inertia so that the $\omega\mathbf{J}$ term is retained in (11.2) to (11.4) of section 11.1.3. Collisions are still neglected, $\nu = 0$, however, by writing $\omega\mathbf{J}$ instead of $\hat{\omega}\mathbf{J}$. The wavefield curl equation is now quadratic so both wavenumbers k_H and k_T are now present, and the collisionless version of (11.9) is obtained directly:

$$k_{\text{H,T}} = \frac{1}{2\omega}\left(\omega_c k_z \mp \sqrt{\omega_c^2 k_z^2 - 4\omega_p^2 k_0^2}\right). \tag{11.46}$$

As for the helicon approximation case, in this collisionless approach we have to take care of the fact that $k_{\text{H,T}}$ are to be real numbers, which implies that the determinant is greater than zero, $\omega_c^2 k_z^2 - 4\omega_p^2 k_0^2 > 0$, as well as the transverse wavenumbers $k_{\perp\text{H,T}}$, which implies that $k_{\text{H, T}}^2 - k_z^2 > 0$. These two conditions constrain the range of possible k_z values to the interval:

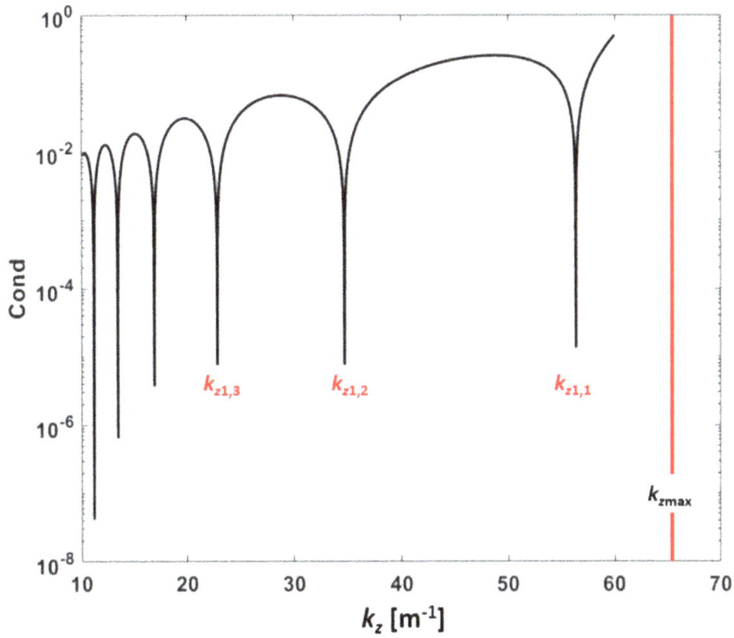

Figure 11.4. Plot of the function log(|Cond(k)|), for RAID standard conditions (11.44) and taking $m = 1$ as in figure 11.3.

$$\frac{2k_0\omega_p}{\omega_c} \leqslant k_z \leqslant \frac{k_0\omega_p}{\sqrt{\omega(\omega_c - \omega)}}. \tag{11.47}$$

To fix the order of magnitudes, figure 11.7 shows the evolution as a function of $\lambda_z = 2\pi/k_z$ of the H and T–G radial wavelengths $\lambda_{r\mathrm{H},r\mathrm{T}} = 2\pi/k_{\perp\mathrm{H},\perp\mathrm{T}}$ for standard RAID parameters (11.44) and $m = 1$. It can be seen that the H and T–G branches have opposite behaviour, meaning that the radial wavelength of H modes decreases with λ_z while the T–G radial wavelength increases. Therefore, generally speaking, in the major part of the k_z spectrum, and especially for small λ_z, H and T–G components have quite different radial wavelengths. In this example, H wavelengths are several centimetres long, while T–G radial wavelengths are typically in the millimetre range.

An illustration of the influence of plasma density and magnetic field strength on H and T–G longitudinal and radial wavelengths is shown in figure 11.8. These are typically the most basic curves to be considered when designing a helicon experiment, to check the compatibility of the desired density and magnetic field parameters with the typical dimensions of the experiment. As a matter of fact, for a given (λ_z, λ_r) mode to propagate, we could state as a rule of thumb that the length and radius (L, R) of the desired plasma column should be greater, or at least equal to, (λ_z, λ_r). For example, taking $L = 100$ cm and $R = 10$ cm for the plasma column, represented by the red dashed lines in figure 11.8, this is not compatible with the excitation of the H branch for a plasma column with $n_e = 10^{17}$ m^{-3} and $B_0 = 200$ G, and only a

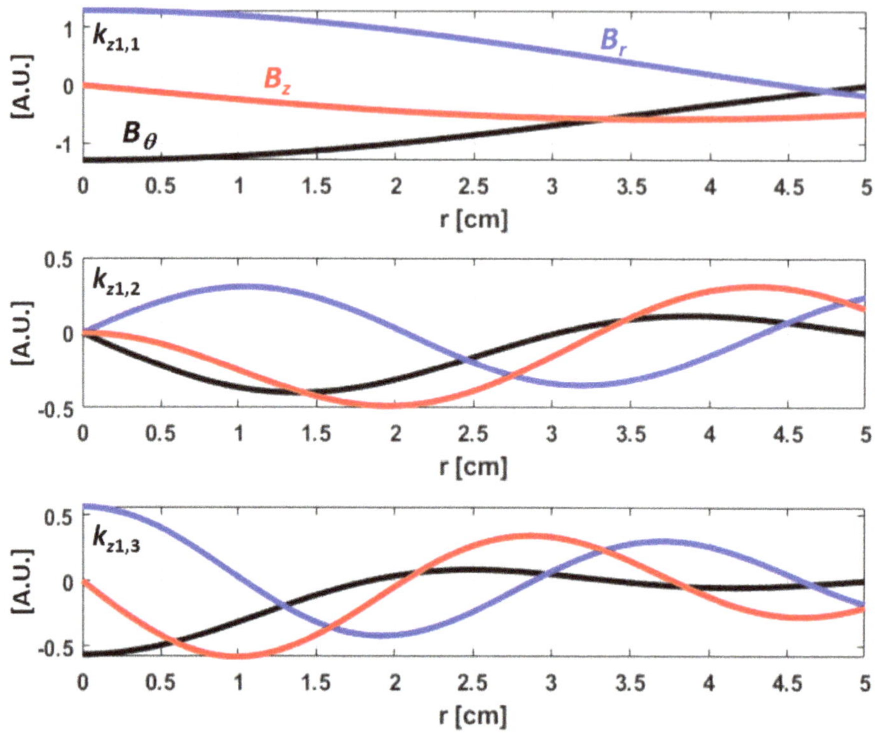

Figure 11.5. Plots of B_r, B_ϕ, B_z for the first three n eigenvalues $k_{1,n}$ of figures 11.3 and 11.4.

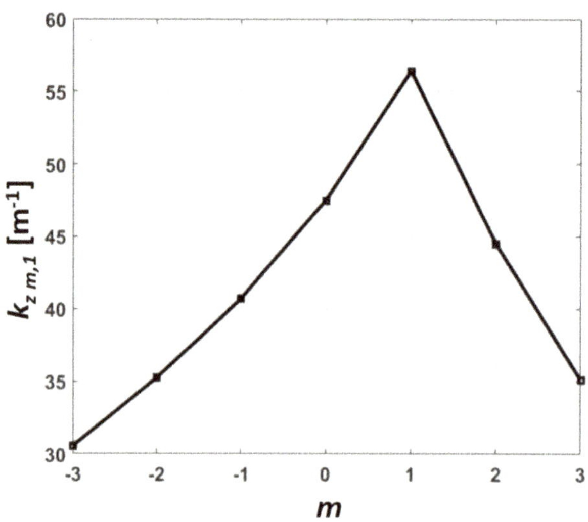

Figure 11.6. Value of the first eigenvalue (n = 1) as a function of the azimuthal mode number m.

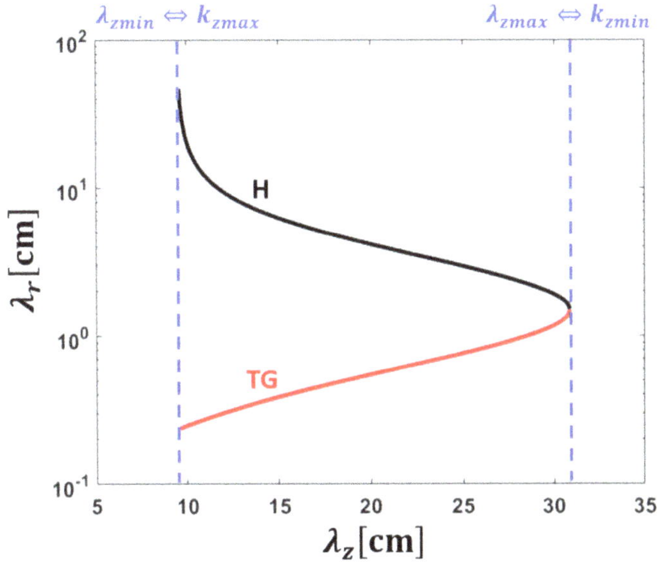

Figure 11.7. Plot of the functions $\lambda_{r1,\,2} = 2\pi/k_{\perp 1,\,2}$ as a function of λ_z, for RAID standard conditions (11.44). The blue dashed lines represent the $\lambda_z = 2\pi/k_z$ range limits in (11.47).

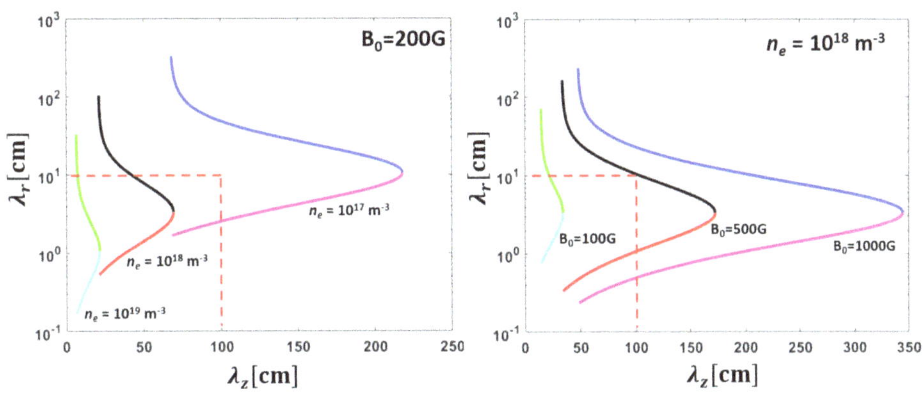

Figure 11.8. Plots of the wavelengths in figure 11.7. Left: for fixed magnetic field strength and variable plasma density. Right: for fixed density and variable B_0. The red dashed lines correspond to a plasma column length (along the z direction) and radius (along the r direction) of 100 cm and 10 cm, respectively.

couple of T–G wavelengths are accessible. For the same magnetic field, a $n_e = 10^{18}$–10^{19} m^{-3} plasma would allow for a much wider range of the (λ_z, λ_r) modes to be excited, which we could expect to be a much better condition for operating the plasma source.

Figure 11.9 shows the function log(|Cond(k_z)|) for RAID standard conditions (11.44), and $m = 1$. As mentioned in section 11.1.7, for a PEC boundary at the plasma interface, only one constant is to be determined for two boundary conditions to be satisfied, which leads to a quantization of k. This case is described in full detail in appendix J.2 as an illustration of the constraint Cond(k_z) = 0. Remember in figure 11.9 that each downward peak corresponds to a possible eigenvalue $k_{z1,n}$. It can be seen that there is now a much denser spectrum of eigenvalues than for the case of the helicon approximation in figure 11.4, meaning that the introduction of the T–G branch has considerably enhanced the number of possible radial modes. It is difficult to make a direct connection between the $k_{z1,n}$ spectrum predicted by the present collisionless {H,T–G} model in figure 11.9, and the one predicted in the framework of the helicon approximation, figure 11.4. This connection can be visualized by considering the ratio between the T–G component amplitude and the H amplitude for the different $k_{z1,n}$, as shown in figure 11.10. Clearly, for most of the eigenvalues, the T–G component amplitude is much higher than for the H amplitude (for example the green circled eigenvalue in figure 11.10). Only a few eigenvalues correspond to the opposite situation, meaning the dominant H mode (for example the blue circled eigenvalues in figure 11.10). It must be noted that we cannot, strictly speaking, have pure H modes or pure T–G modes; each eigenvalue corresponds to a superposition of a H component and a T–G component, although one of the components can strongly dominate the other.

Note that the ratio of H to T–G amplitudes characterizes the dominance of one or the other component at $r = 0$. In the framework of this collisionless solution, if one component, H or T–G, dominates over the other at $r = 0$, this remains true also for the whole plasma radius, because the H and T–G transverse wave vectors $k_{H\perp}$ and $k_{T\perp}$ are purely real. Considering a collisional case, as will be shown in the coming

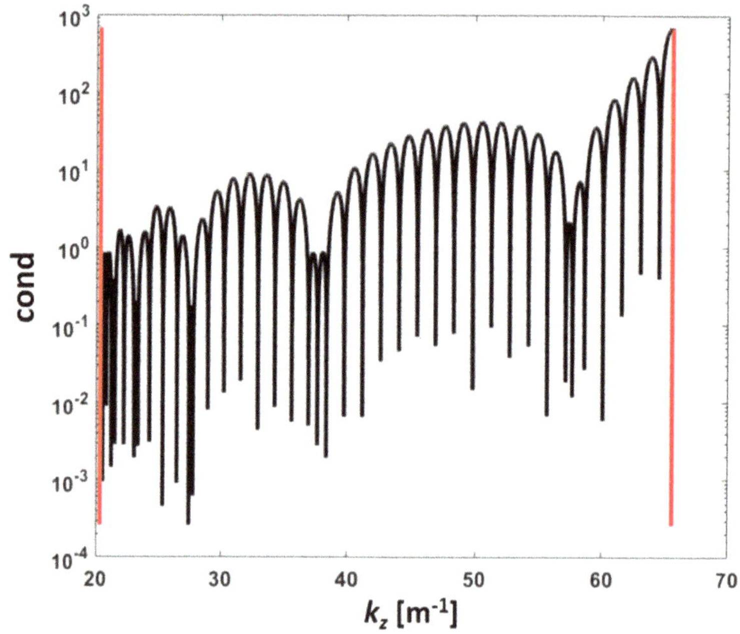

Figure 11.9. Plot of the function log(|Cond(k_z)|), in the {H,T–G} collisionless PEC model, for RAID standard conditions (11.44) and taking $m = 1$. The red lines represent the same k_z(min) and k_z(max) range limits in (11.47) as used also for figure 11.7. The function Cond(k_z) is given by (J.15) in appendix J.2.

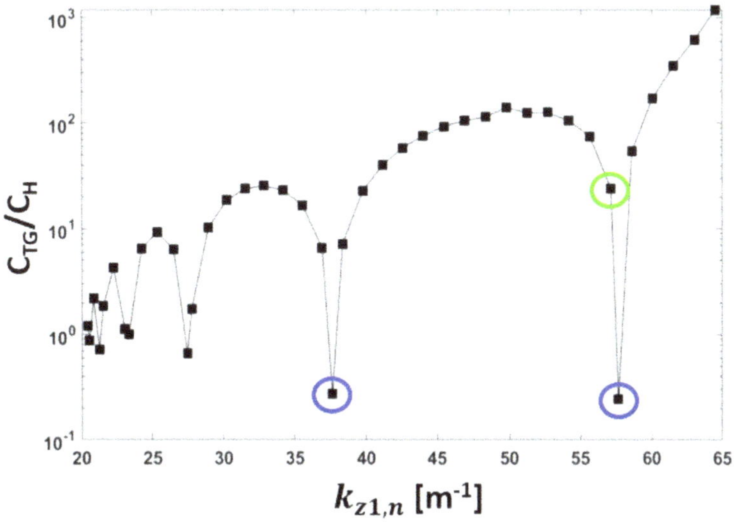

Figure 11.10. Evolution of the ratio of the T–G and H component amplitudes, as a function of $k_{z1,n}$. Blue circles mark examples of dominantly helicon modes; the green circle marks an example of a dominant T–G mode.

sections, $k_{H\perp}$ and $k_{T\perp}$ are now complex numbers, for which the imaginary part characterizes the radial damping of the two components, which can be very different. As a consequence, comparing the H and T–G amplitudes only reveals a potential dominance at $r = 0$, but it cannot be systematically extended to the whole r domain.

Comparing the k_z values at which the dominant H modes appear with the helicon approximation figure 11.4, we can see by eye a certain correlation for the lowest n modes, but the eigenvalues are nevertheless significantly shifted by the introduction of T–G waves. The correspondence is very similar when comparing the radial profiles of the magnetic field for these dominant H modes with the radial profiles of the helicon approximation case as shown in figure 11.11; the {H,T–G} solution clearly resembles the helicon approximation solution, with addition of a small amplitude T–G wave of much shorter wavelength. For all the other eigenvalues, the radial field profiles are completely dominated by the T–G wave, as shown in figure 11.12 for the green circled eigenvalue of figure 11.10.

All the eigenvalues appearing in figure 11.9 can propagate into the plasma column, and in this collisionless model there is no reason for one specific mode to dominate the others. In real experiments it is often claimed that the length of the source antenna determines the preferentially excited mode [15], which might be the case for radially bounded plasma, but not for experiments such as RAID (see chapter 12). As a matter of fact, the main wavelength measured in the RAID chamber has no relation to the antenna length, and is determined by the plasma density and the magnetic field strength [3]. Instead, the plasma density evolves continuously with the RF power level, and no 'spatial resonance' behaviour occurs when the wavelength measured downstream coincides with the antenna length (or multiples of it).

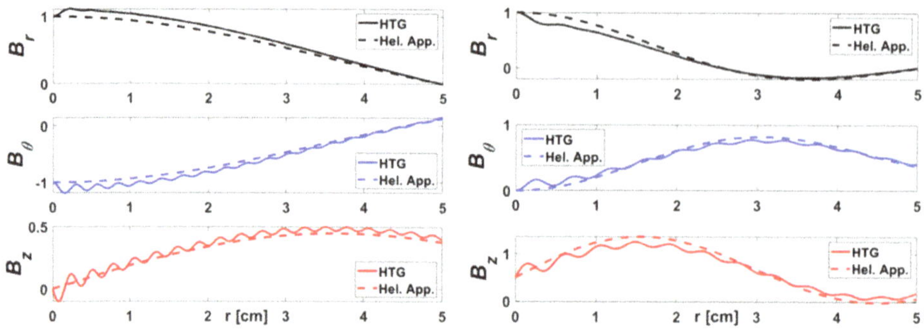

Figure 11.11. Radial profiles of the magnetic field components. (a) Modes $k_{z1,\,1}$ of the helicon approximation (dotted lines) compared to the dominant H mode of lowest n (highest k_z) in the {H,T–G} model (continuous lines). (b) Same as for (a), but for $k_{z1,\,2}$ of the helicon approximation, and corresponding dominant H mode of the {H,T–G} model.

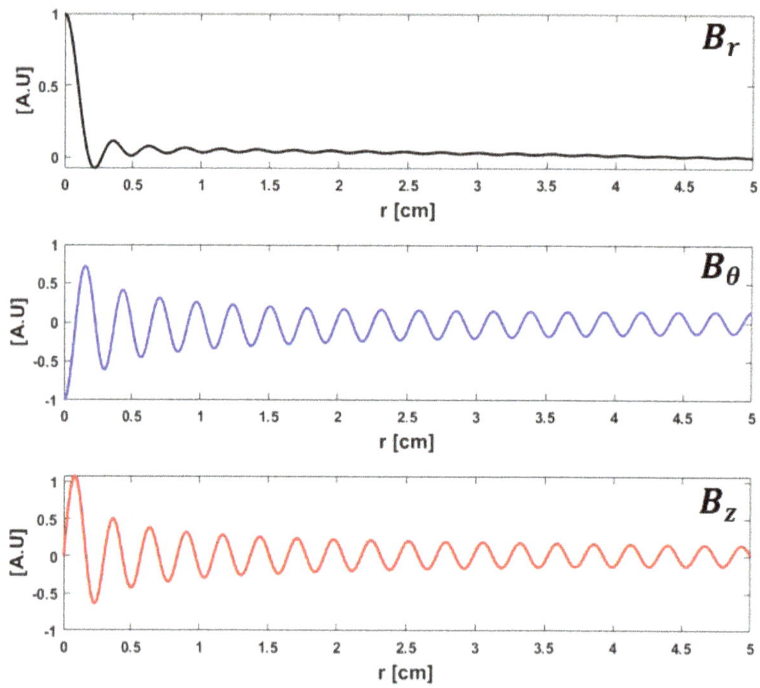

Figure 11.12. Radial profiles of the magnetic field components for a mode (green circle in figure 11.10) adjacent to a helicon-dominant mode (a blue circle in figure 11.10). The short-wavelength T–G waves are clearly dominant.

11.2.3 PEC at the plasma boundary, collisional {H,T–G} model

We now introduce some dissipation into the model by letting the electron–neutral effective collision frequency ν be non-zero. To fix the orders of magnitude, for a typical electron temperature of 3 eV, the collision frequency is approximately

10 MHz in argon and 23 MHz in hydrogen at 1 Pa pressure [16]. The operating pressure on RAID, chapter 12, is about 0.2–0.5 Pa, meaning $\nu_{Ar} \approx$ 2–3 MHz, and $\nu_{H_2} \approx$ 4–6 MHz. Therefore, roughly, a collision frequency $\leqslant$1MHz corresponds to a low pressure regime, 5 MHz would correspond to a typical pressure, and $\geqslant$10 MHz is a high pressure regime.

In the collisionless model, the wavenumbers were strictly real numbers. As collisions are introduced, $\hat{\omega} = \omega + j\nu$ is complex, and so complex wavenumbers arise, where the imaginary wavenumber component characterizes the wave damping in space. Therefore, the solutions of the equation Cond(k_z) = 0 are now to be found in the plane Re (k_z), Im (k_z), where the real and imaginary parts must be simultaneously zero, i.e. |Cond(k_z)| = 0. The expression (J.15) for Cond(k_z) in appendix J.2 is exactly the same as for the collisionless case, except that the values of $k_{\rm H}$ and $k_{\rm T}$, and therefore of k_z, are complex. Figure 11.13 is a contour plot for log|Cond(k_z)|, which is the complex plane equivalent of figures 11.3 and 11.9. The contour values are not labelled, but the pairs Re (k_z = 0) and Im (k_z = 0) which are solutions of Cond(k_z) = 0 (the eigenvalues) appear on the mapping as a concentration of lines around each singularity, as indicated by circles on the figure. Successive refinement of the mapping around the peaks confirms these solutions. The blue dashed lines are the same k_z(min), k_z(max) limits that appear in the framework of the collisionless model, figure 11.9, although these limits are only

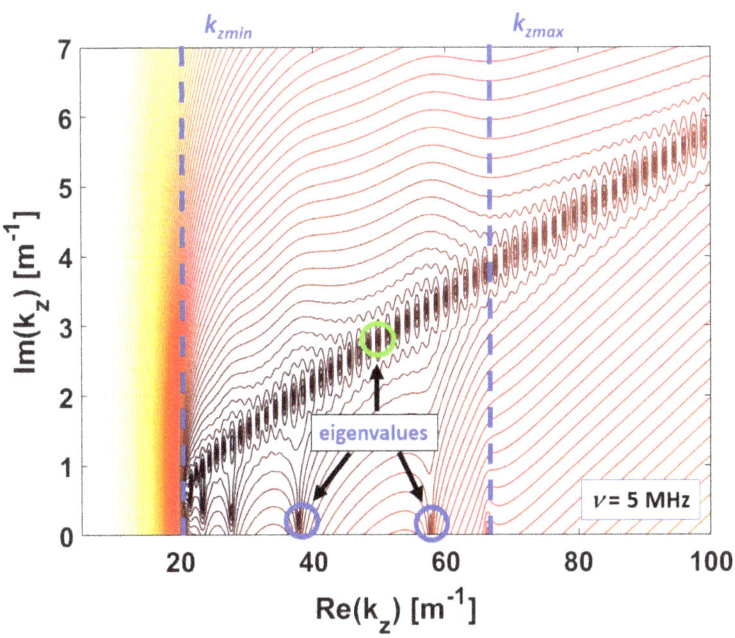

Figure 11.13. Collisional {H,T–G} case with PEC boundary: a contour map of log|Cond(k_z)| in the Re (k_z), Im (k_z) plane, for RAID standard conditions (11.44), m = 1, and assuming 5 MHz collision frequency. The function Cond(k_z) is given by (J.15) in appendix J.2. Blue circles mark examples of dominantly helicon modes; the green circle marks an example of a dominant T–G mode.

applicable when k_H and k_T are both real. It can be seen that the k_z(min) limit is apparently still valid because no eigenvalue is found for lower k values, but this is not the case for the upper limit, because the spectrum of possible modes is considerably extended to larger k (smaller wavelengths). Incidentally, this means that the numbering system of the eigenvalues starting from below the highest Re (k_z) value no longer holds in this dissipative approach. It will be seen below that the k_z(min) limit is likewise invalid in the case of non-uniform density profiles.

Considering figure 11.13, there are clearly two families of eigenvalues which can be distinguished according to their Im (k_z) values. The first one corresponds to waves characterized by long longitudinal damping lengths (small Im (k_z)), as for example the blue circled eigenvalues in figure 11.13. Note also that, within this family, the damping distance increases with Re (k_z). The major part of the eigenvalues belong to the second family, for example the green circle in figure 11.13, characterized by much shorter longitudinal damping lengths, and by the fact that these damping lengths are clearly decreasing with increasing Re (k_z). Considering the T–G to H amplitude ratio in this collisional case, it clearly appears that the first family of eigenvalues (blue), characterized by a low Im (k_z) value, corresponds to dominant H waves, while the second family (green) exhibits dominant T–G waves. This is illustrated in figure 11.14 which shows that the T–G/H amplitude ratio of the low Im (k_z) eigenvalue is very small.

The mapping shown in figure 11.15 is equivalent to the one of figure 11.13, but eliminating the regions for which either Im$(k_{\perp H})$ or Im$(k_{\perp T})$ are positive. The purpose is first to illustrate the fact that all the eigenvalues are retained in the remaining region for which both Im$(k_{\perp H})$ and Im$(k_{\perp T})$ are negative. The real parts of $k_{\perp H,T}$ are always positive (outward propagation). This means that both T–G and H components, when taken to propagate outward, have radially exploding amplitudes. But this is only due to the choice of sign, as fundamentally the transverse wavenumber is defined by the square relation $k_\perp^2 = k^2 - k_z^2$ and the positive root was chosen here, $k_\perp = +\sqrt{k^2 - k_z^2}$ in (11.20), thus defining outward propagating waves. Taking the negative root would correspond to inward propagating waves with damped

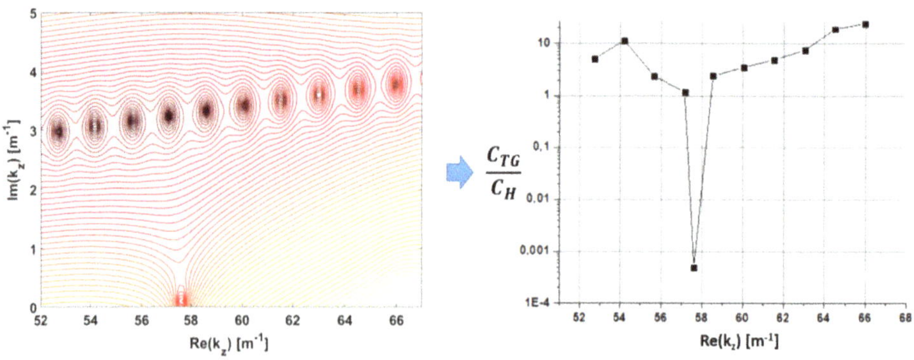

Figure 11.14. Left: Close-up of figure 11.13 on the lowest-n values (highest Re (k_z)). Right: T–G/H wave amplitude ratio for the eigenvalues shown on the left figure.

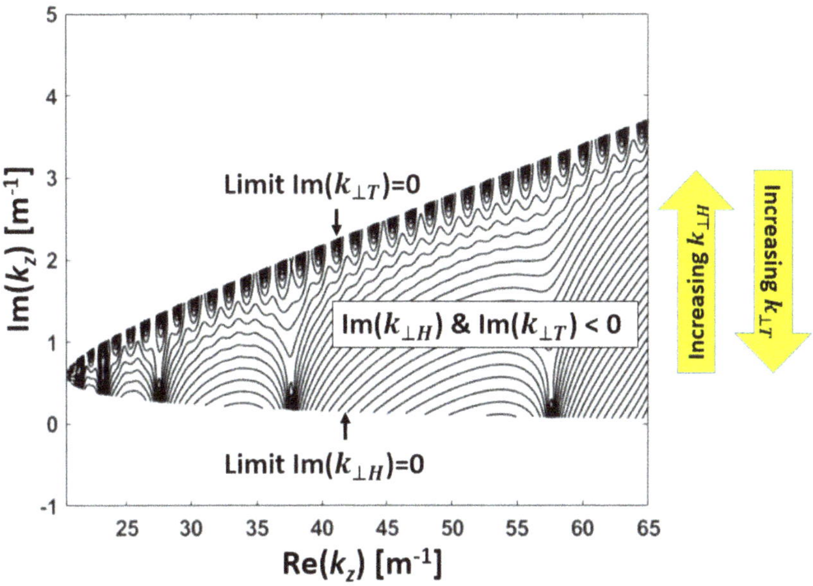

Figure 11.15. Mapping of Cond(k_z) in the (k_r, k_i) plane, as for figure 11.13, but with all the regions corresponding to either Im($k_{\perp H}$) > 0 or Im($k_{\perp T}$) > 0 removed.

amplitude, which corresponds typically to experimental conditions where an antenna surrounds the plasma column.

Considering the dominant H eigenvalues in figure 11.15, we can see that they are all close to the Im($k_{\perp H}$ = 0) limit, while the dominant T–G ones are close to the Im ($k_{\perp T}$ = 0) limit. Both Im($k_{\perp H}$) and Im($k_{\perp T}$) vary monotonically in the (k_r, k_i) plane, but with opposite trends as indicated in figure 11.15. Therefore, considering first the dominant H modes, these are characterized by small Im($k_{\perp H}$) values and high Im ($k_{\perp T}$), meaning that the H component of these modes is weakly damped radially, while the T–G component, on the contrary, is strongly damped. Taking the dominant T–G modes, we have the opposite situation of a radially strongly damped H component and a weakly damped T–G mode.

Figure 11.16 shows the evolution of the mapping of Cond(k_z) in the (k_r, k_i) plane with the value of the collision frequency. The main fact to be noted is that the longitudinal damping, Im(k_z), of dominant T–G modes is much more affected by a change in the collision frequency than for dominant H ones.

Finally, consider the radial profiles of dissipated power $\frac{1}{2}\mathrm{Re}\,[\mathbf{E}\cdot\mathbf{H}^*]$ (see appendix D.2.1) that can be expected for both dominant H and dominant T–G modes in RAID standard conditions (11.44) and m = 1. Figure 11.17 shows the dominant H field profiles and heating profile for a low collision frequency (1 MHz), while figure 11.18 corresponds to the dominant TG case. Considering figure 11.17, it appears that for a low collision frequency the T–G/H amplitude ratio is not that small, so that the mode is not strongly H dominant and the heating radial pattern shows a 'balanced' contribution of both H and TG components, i.e. a similar

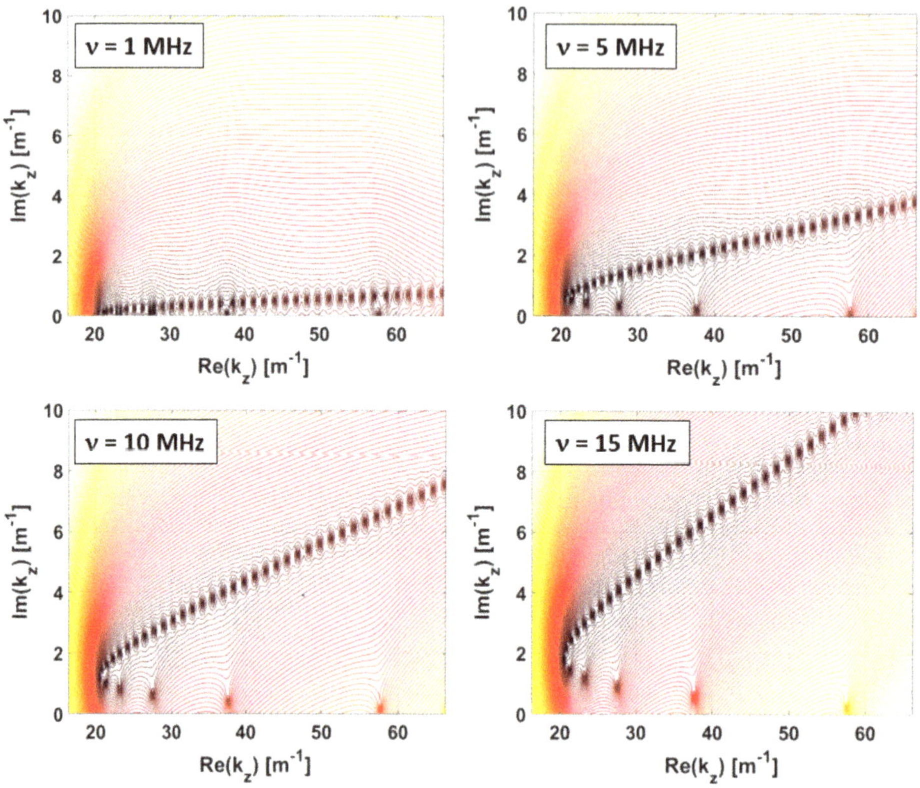

Figure 11.16. Collisional {H,T–G} case. Mapping of Cond(k_z) in the (k_r, k_i) plane, for RAID standard conditions (11.44), m = 1. Evolution with collision frequency ν.

heating contribution from long and short wavelengths, respectively. In fact, for such low collision frequencies, the dominant H eigenvalue is not sufficiently separated from the dominant TG group in the {Re(k_z),Im(k_z)} plane for a clear H dominance to occur, as can be seen in the figure 11.16 mapping. For higher collision frequencies, as shown in figure 11.19, the dominance of the H branch increases and the T–G dissipation appears as only a relatively small heating contribution at the edge of the plasma column. For dominant T–G modes, the heating radial profile is always strongly peaked on axis, and shows no major modification when varying the collision frequency from 1 MHz to 10 MHZ, as illustrated in figure 11.19.

11.2.4 Vacuum plasma boundary, collisionless {H,T–G} model

We now begin to consider the more interesting and relevant case of a vacuum/plasma boundary at $r = a$, and a PEC boundary at $r = b$, as sketched in figure 11.2. This situation naturally corresponds better to most experimental conditions. Without entering into detail here, we briefly summarize the collisionless situation. On the left part of figure 11.20 is shown the Cond(k_z) function for standard RAID conditions (11.44), with m = 1 and b = 20 cm. The expression for Cond(k_z) is

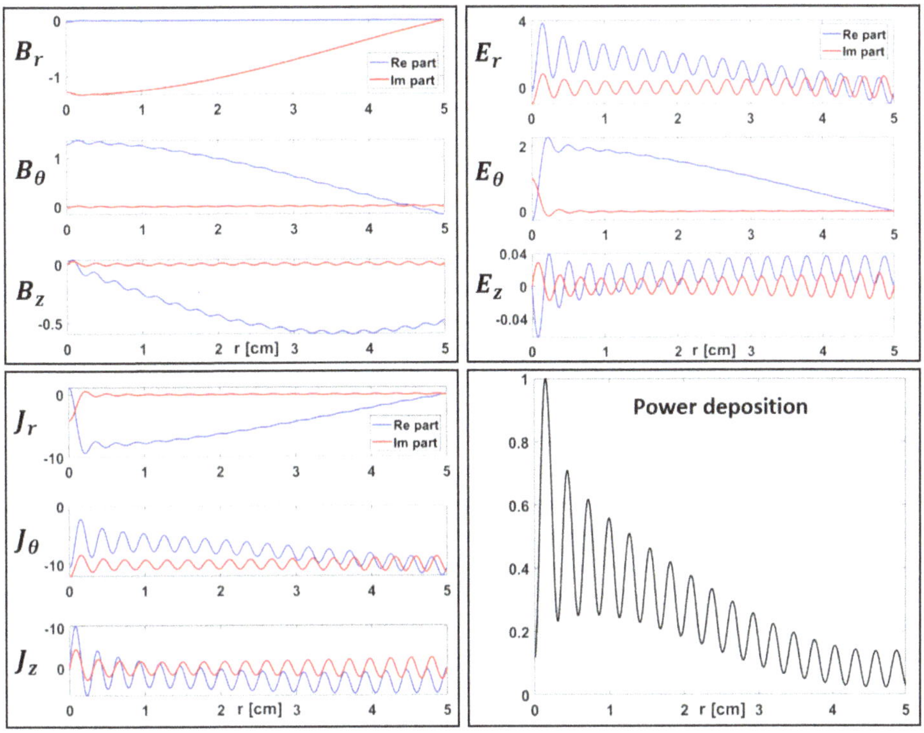

Figure 11.17. Collisional {H,T–G} model calculations of **B**, **E**, **J** radial profiles for the dominant H mode ($k_r \approx 57$ m^{-1} in figure 11.16) for a 1 MHz collision frequency.

obtained in a similar way to appendix J.2, although the algebra is considerably more cumbersome when six boundary conditions are to be satisfied. Similarly to the PEC case, there are many possible modes in the k_z(min) to k_z(max) range. The main difference with regard to the PEC boundary case concerns the T–G/H amplitude ratio for the different k eigenvalues. These are shown on the right part of figure 11.20, to be compared with the PEC result in the inset. It can be seen that there is now no dominant H mode. On the contrary, for all eigenvalues the T–G wave amplitude is dominant. It is then not possible to correlate this dielectric boundary case with the results from the helicon approximation, which posits helicon modes only—it would appear that the helicon approximation is valid only for the case of a PEC plasma boundary, which explains why there is no helicon approximation section for this case of vacuum boundary. If we were to stop at this stage, we would conclude the predominance of T–G modes over H ones, but we will see that things totally change with the introduction of both dissipation and edge density gradients.

11.2.5 Vacuum plasma boundary, collisional {H,T–G} model

Considering now the collisional situation, the vacuum boundary condition again leads to important differences with respect to the PEC case. Figure 11.21 shows a mapping of the log|Cond(k_z)| function in the {Re(k_z),Im(k_z)} plane, for RAID

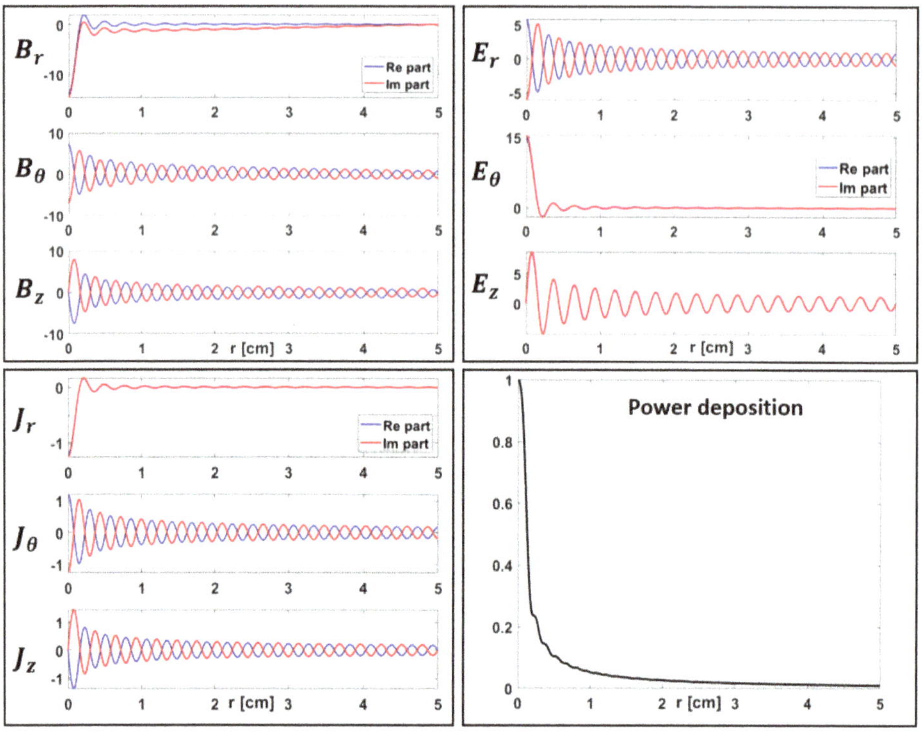

Figure 11.18. Collisional {H,T–G} model calculations of **B**, **E**, **J** radial profiles for the dominant T–G mode ($k_r \approx 57$ m^{-1} in figure 11.16) for a 1 MHz collision frequency.

standard conditions (11.44), with $m = 1$, assuming a 5 MHz collision frequency. This figure is to be compared with figure 11.13 which shows the PEC result for equivalent conditions. As for the PEC case, the collisionless k_z(min) limit is still valid but not so for k_z(max). The major difference in the present case concerns the separation of the low Im(k_z) family that was identified as dominant H modes in the PEC case. In figure 11.21 we now observe only one root, with comparatively strong axial damping, separated from the main T–G branch, whereas in the PEC case, figure 11.13, there were five clearly separated H roots with relatively little damping. The implication is that the boundary condition has a strong influence on the existence of the helicon-type modes.

Figure 11.22 shows the evolution of Cond(k_z) as a function of the collision frequency, which is to be compared with figure 11.16 for the PEC case. It can be seen that now there is no more rapid separation of low Im(k_z) and high Im(k_z) eigenvalues with the introduction of dissipation, as was the case for a PEC boundary. A first particular eigenvalue becomes well separated from the 'group', but for relatively high values of the collision frequency (see $\nu = 3$ MHz in figure 11.22), and at an Im(k_z) value (typically 2 m^{-1}) much higher than for the PEC boundary case (typically 0.1 m^{-1}). We can also note that, once separated, this particular eigenvalue is almost stationary while the collision frequency is further increased. Finally, as ν

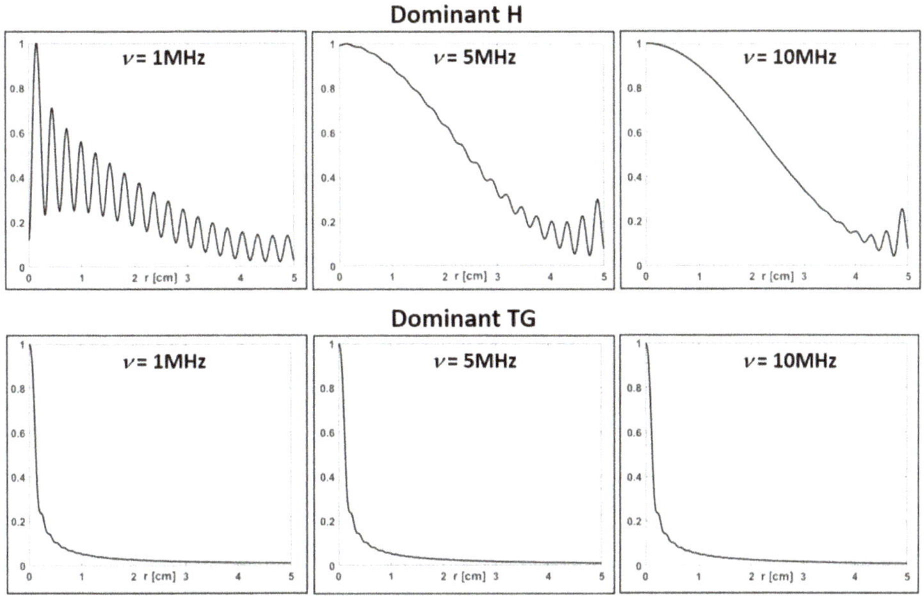

Figure 11.19. Collisional {H,T–G} model. Evolution of the radial energy dissipation profile as a function of the collision frequency, for dominant H modes (top), and dominant T–G modes (bottom).

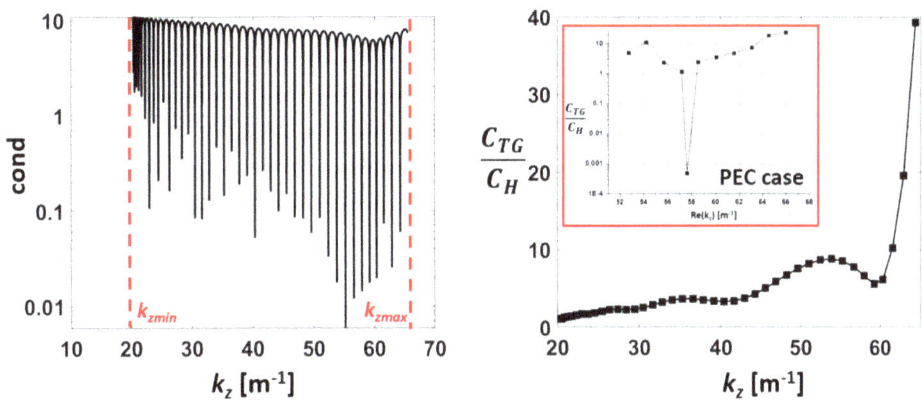

Figure 11.20. Collisionless {H,T–G} model with vacuum/plasma boundary. Left: function Cond(k_z) for standard RAID conditions (11.44) and $m = 1$, $a = 5$ cm, $b = 20$ cm. Right: T–G/H wave amplitude ratio for the k_z eigenvalues on the left figure, where the inset corresponds to the PEC boundary case.

continues to increase, some other eigenvalues become well separated from the others, as for example in the 15 MHz case of figure 11.22.

In the PEC situation, the rapidly separated eigenvalues of figure 11.16 all correspond to dominant H modes, while all the others are dominant T–G modes. In this dielectric (vacuum) boundary situation, it is less simple. For low collision frequencies, typically the 1–3 MHz cases of figure 11.22, all the eigenvalues correspond to dominant T–G waves. The roots that become separated from the

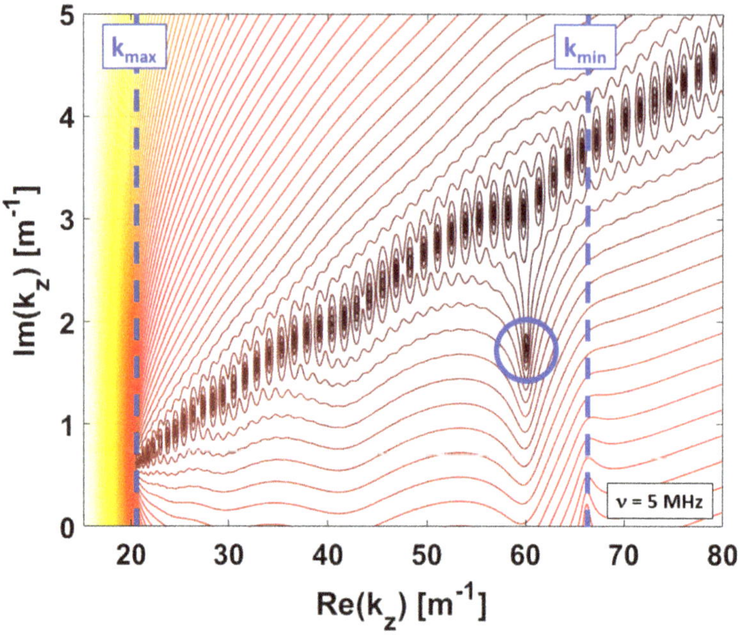

Figure 11.21. Collisional {H,T–G} case, vacuum boundary: a contour map of log|Cond(k_z)| in the {Re(k_z),Im(k_z)} plane, for RAID standard conditions (11.44), m = 1, a = 5 cm, b = 20 cm, and assuming a 5 MHz collision frequency.

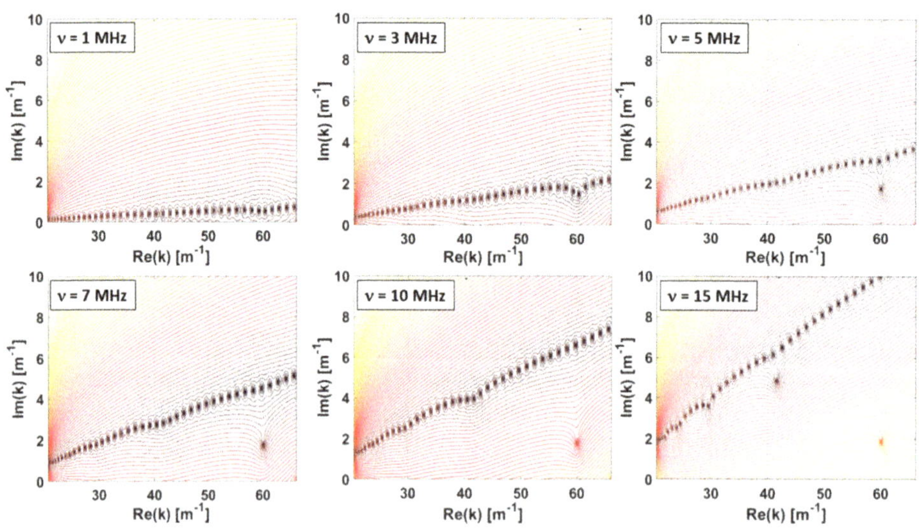

Figure 11.22. Collisional {H,T–G} case, vacuum boundary. Mapping of Cond(k_z) in the {Re(k_z),Im(k_z)} plane, for RAID standard conditions (11.44), m = 1, a = 5 cm, b = 20 cm: evolution with collision frequency.

group for higher collision frequencies can well be dominant H modes, but this dominance is only well established if the eigenvalue is largely isolated from the dominant T–G group, meaning for sufficiently high collision frequencies. For example, with $\nu = 3$ MHz, the separation only just begins (figure 11.22), and the considered eigenvalue still corresponds to a dominant T–G wave. This is illustrated in figure 11.23, where it can be seen that the T–G/H amplitude ratio for $\nu = 3$ MHz is about 3.5, and the dissipated power density profile is very peaked on the axis as it was also the case for the dominant T–G modes in the PEC boundary. When ν is increased to 10 MHz, the T–G/H amplitude ratio drops to $3 \cdot 10^{-4}$, so the mode is very dominantly H. But the dissipated power density profile is now very peaked at the edge of the plasma, where T–G modes are damped. This is very different from the PEC case for which the dissipated power density profile does not peak on the edge by T–G damping, but rather reveals a dominant H wave profile.

Up to now, the examples have only considered the mode $m = 1$, which is seen to be the most important for helicon sources, although it is not always easy to understand why. The point is that, at least for the RAID birdcage antenna (see chapter 12) which is built to dominantly excite the $m = 1$ modes, COMSOL simulations clearly indicate a very strong dominance of $m = 1$ modes, and the measured axial wavelengths are very consistent with a $m = 1$ spectrum, consistent

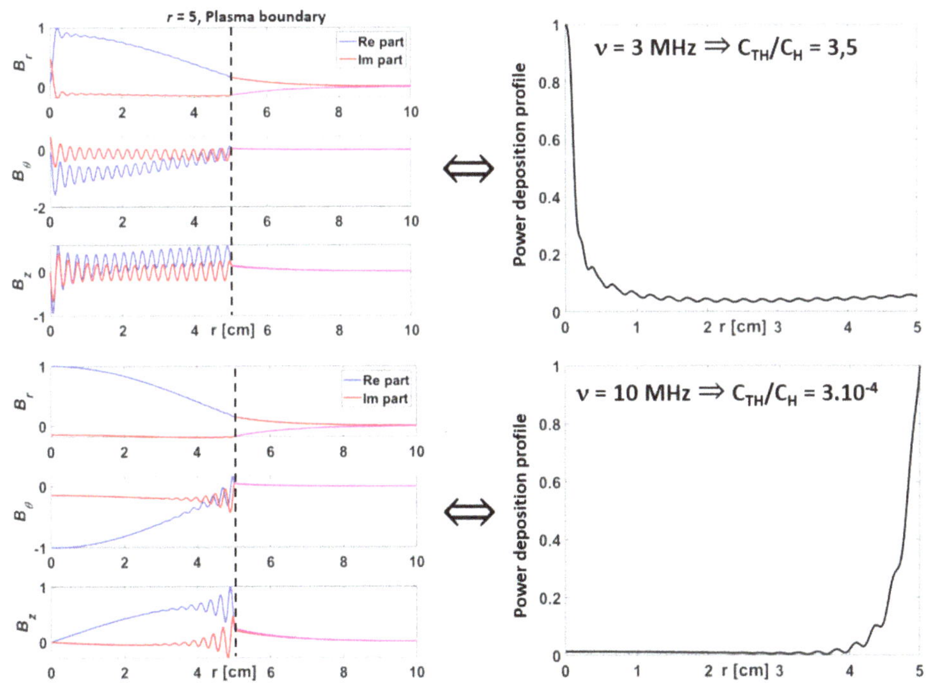

Figure 11.23. Collisional {H,T–G} case, vacuum boundary, RAID standard conditions (11.44), $m = 1$, $a = 5$ cm, $b = 10$ cm. Top left: radial profiles of the magnetic field components for the eigenvalue that becomes separated at $\nu = 3$ MHz. Top right: the corresponding power dissipation profile. Bottom figures, same as top but for $\nu = 10$ MHz.

with the birdcage current distribution. The fact is that the $m = 1$ mode shows some particularities, which we attempt to illustrate in the following. Figure 11.23 shows the mappings of the Cond(k_z) function in the {Re(k_z),Im(k_z)} plane, for several values of the collision frequency, and for RAID standard conditions (11.44), but now taking $m = 2$. This figure is to be compared with figure 11.22 which represents the $m = 1$ case. It can be seen that the behaviour of both mappings is similar. We also observe in this $m = 2$ case the separation of an eigenvalue as the collision frequency is increased. As for $m = 1$, once well separated, this eigenvalue corresponds to a dominant H mode, while all other solutions correspond to dominant T–G modes. However, it must be noticed that the clear separation of the first eigenvalue occurs for a collision frequency typically two times higher for the $m = 2$ case than for $m = 1$, and with an axial damping also typically two times higher. Further increasing m accentuates this effect (figure 11.24).

We now focus on this first dominant H eigenvalue that is separated, for different value of $m = \{-3, -2, -1, 0, 1, 2, 3\}$. In order to obtain a net eigenvalue separation for all m (dominant H condition), it is necessary to consider a quite high collision frequency of $\nu = 15$ MHz (still for RAID standard conditions (11.44)). Figure 11.25 shows the evolution of the longitudinal wavelength $\lambda_z = 2\pi/\text{Re}(k_z)$ and of the longitudinal damping distance $\delta_z = 2\pi/\text{Im}(k_z)$ of this dominant H root with the azimuthal mode number m. It can be seen that the $m = 1$ case corresponds to a minimal longitudinal wavelength and is the less axially damped mode. We further show the evolution of the H and T–G radial wavelengths $\lambda_{r\text{H,T}} = 2\pi/\text{Re}(k_{\perp\text{H,T}})$ and radial damping distances $\delta_{r\text{H,T}} = 2\pi/\text{Im}(k_{\perp\text{H,T}})$. Here again we see that the $m = 1$ case is particular because first, it maximizes the H transverse wavelength while minimizing the T–G transverse wavelength; and second, it minimizes the radial H damping while maximizes the T–G radial damping. Note that these observations are not

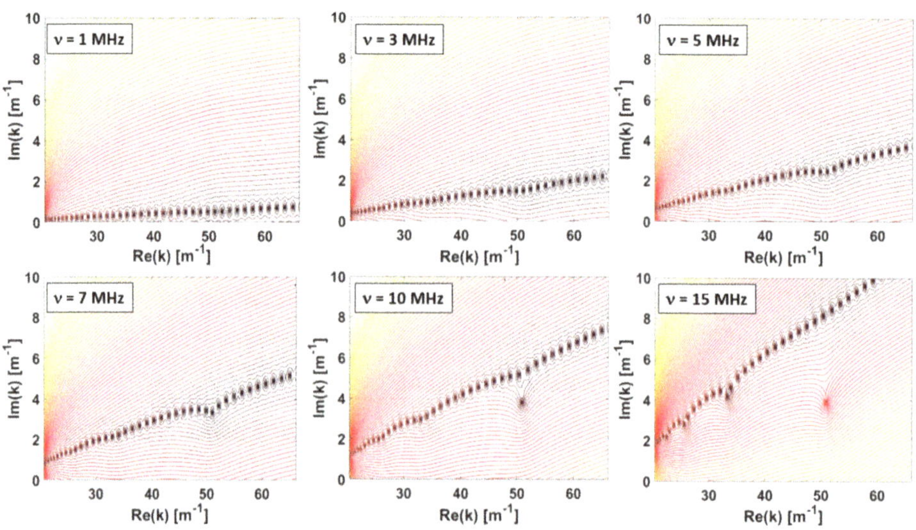

Figure 11.24. Collisional {H,T–G} case, vacuum boundary. Mapping of Cond(k_z) in the {Re(k_z),Im(k_z)} plane, for RAID standard conditions (11.44), $m = 2$, $a = 5$ cm, $b = 20$ cm: evolution with collision frequency.

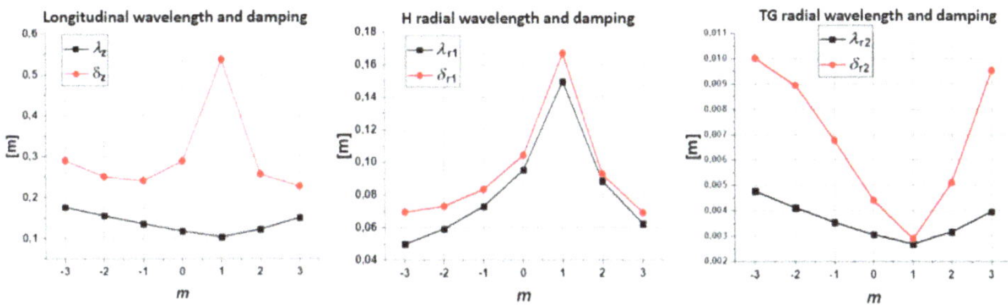

Figure 11.25. Wavelength and damping of the first separated root (dominant H wave at ν = 15 MHz) for different azimuthal mode number m.

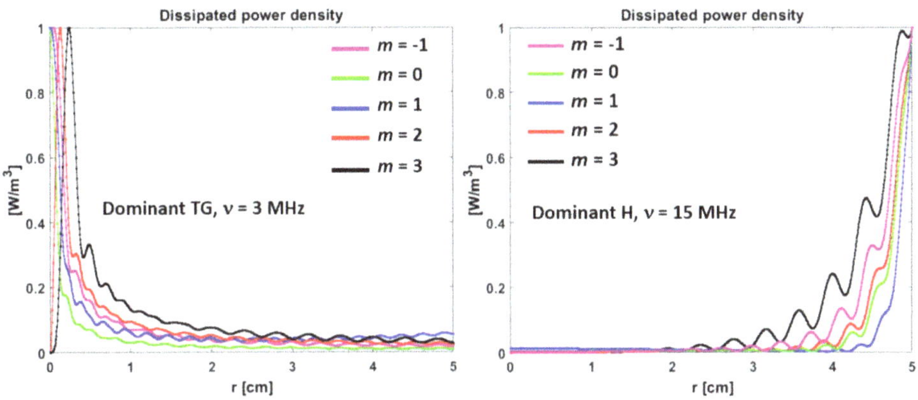

Figure 11.26. Power deposition profiles for different m.

specific to the dominant H mode, and are also retrieved at low collision frequency for dominant T–G modes.

With regard to the power deposition, we find the same overall behaviour regardless of the azimuthal mode number. This is illustrated in figure 11.26, where we can see that dominant H modes lead to a strong power deposition on the plasma edge, due to local damping of the T–G component, while dominant T–G modes lead to a power deposition peaked on axis. Going more into detail, only the $m = -1$, 0, 1 modes lead to profiles which are peaked exactly on axis. This is due to the fact that these modes are the only ones that exhibit even components of the fields (B_r for $m = \pm 1$, B_z for $m = 0$) that can be non-zero on the axis. For other m values, the field always vanishes at $r = 0$.

11.2.6 Intermediate conclusions from the uniform density models

The lesson to be drawn from this investigation into H and T–G modes on a uniform density plasma column is that their existence and damping characteristics depend sensitively on the model assumptions, and on the plasma collisionality and boundary conditions (PEC or vacuum boundary at the plasma edge). In the next section, the influence of the plasma density radial profile will also be estimated. All of these effects must be carefully evaluated before any meaningful comparison can be made between a model and magnetic field measurements in the RAID experiment of chapter 12.

11.3 Normal mode solutions for radially non-uniform plasma

The equations for a collisional plasma with radially non-uniform density, but no vacuum displacement current, were described in section 11.1.4. The resulting fourth order equation (11.32) can be solved numerically to study the influence of the shape of the density profile on the {H,T–G} mode properties. It was first verified that the fourth order numerical solution correctly reproduces the results for uniform density in section 11.2.

11.3.1 Collisional {H,T–G} model, non-uniform density

The main deficiency of the uniform density model of section 11.2 is that it imposes an abrupt drop of density at the plasma/vacuum boundary. In reality, the plasma density radial profile will vary from one experiment to another, but the density must always drop over a certain distance to zero at the edge, or, at least, to a sheath edge density. In experiments such as the RAID birdcage antenna [3], where the plasma column is not bounded by a close-fitting vacuum vessel, the measured plasma density typically exhibits a bell-shaped profile, well separated from the wall, as discussed in chapter 12. One of the aims of using the numerical approach is to understand the influence of the plasma density gradient on the helicon wave propagation and the dissipated power profile associated with the wave. Any number of mode studies can be calculated numerically, depending on how the non-uniformity is parameterized, and we do not intend to cover here a broad spectrum of cases. Nevertheless, we show below what we think to be the most notable effects of density gradients at the plasma edge.

In the continuity of the preceding chapters, we consider typical RAID-like conditions in the following example, but instead of a uniform density across the plasma column radius, we now define two regions, as illustrated in figure 11.27: a first region of uniform plasma density for radius $r \in [0, r_c]$, and a second region for $r \in [r_c, a]$, where the density drops to zero with a cosine shape. The choice of this function is motivated by the fact that it ensures vanishing density gradients at both $r = r_c$ and $r = a$, and smooth density transitions at these points.

The radius a of the plasma column is taken to be fixed, and we vary only the position of r_c. The sharpness of the density drop at the plasma edge can then be quantified by the ratio Δ/a.

Figure 11.28 shows a comparison between the eigenvalue spectrum (for $m = 1$ and $\nu = 5$ MHz) calculated for a uniform density, and for the profile of figure 11.27, which is parameterized by $\Delta/a = 0.4$. A first remark can be made concerning the dominant T–G modes, which appear to be not much affected, in terms of axial wavelength and damping, by the modification of the plasma radial profile. In contrast, the dominant H mode eigenvalues are strongly modified by the

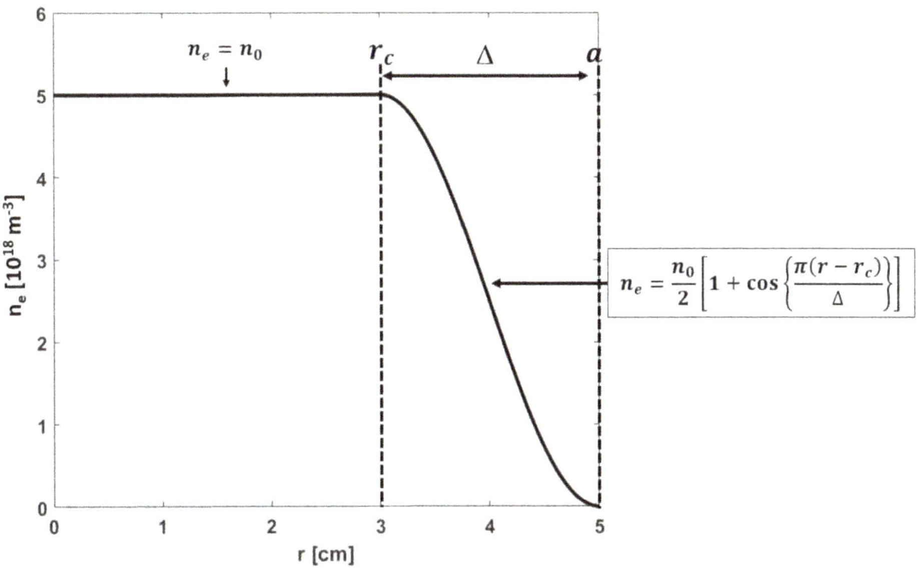

Figure 11.27. Definition of the ad hoc density profile used to investigate the effect of density profiles.

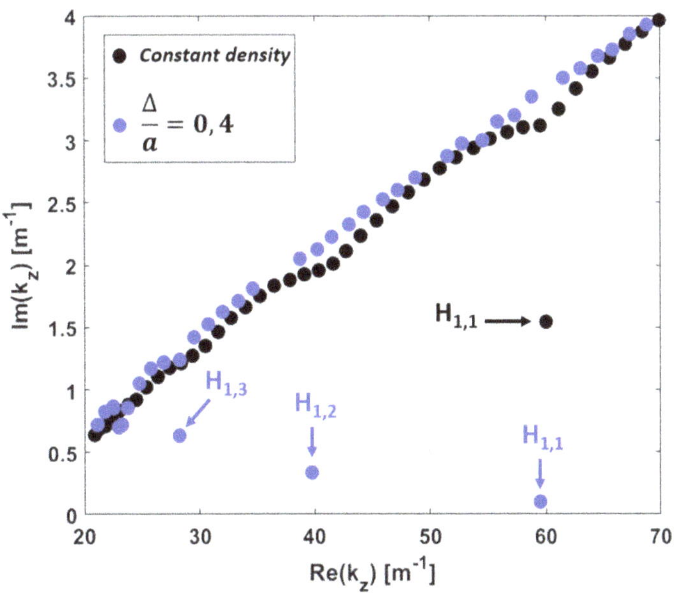

Figure 11.28. Comparison between the eigenvalues calculated for a uniform density, and for a varying profile characterized by $\Delta/a = 0.4$ ($\nu = 5$ MHz).

introduction of the edge gradient, especially in terms of damping. Typically, the axial damping distance for mode $H_{1,1}$ with a uniform plasma density is ten times smaller than for $\Delta/a = 0.4$. As a consequence of this damping reduction, three dominant H modes are now well separated from the dominant T–G family in the $\Delta/a = 0.4$ case, whereas only one H mode was clearly observable with a uniform density profile. We show in figure 11.29 the evolution of the real and imaginary parts of the $H_{1,1}$ axial wavenumber with the parameterization $\Delta/a = 0.4$.

It should be noticed that the modification of the eigenvalue spectrum shown in figure 11.28 is indeed an effect of the edge density gradient, and cannot be attributed to a variation of an effective radius of a uniform density column. As a matter of fact, a decrease in radius a in the framework of uniform densities has the opposite effect on the dominant H roots compared to that of the gradients, as illustrated in figure 11.30: there are no dominant H modes separated from the T–G family for

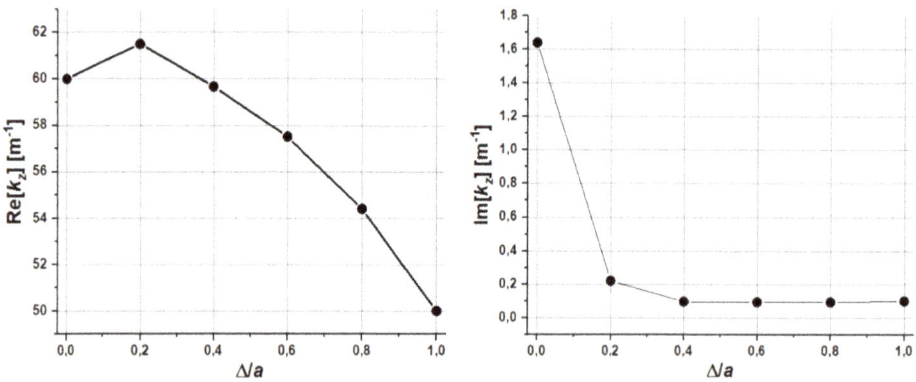

Figure 11.29. Evolution of the $H_{1,1}$ eigenvalue with Δ/a (ν = 5 MHz).

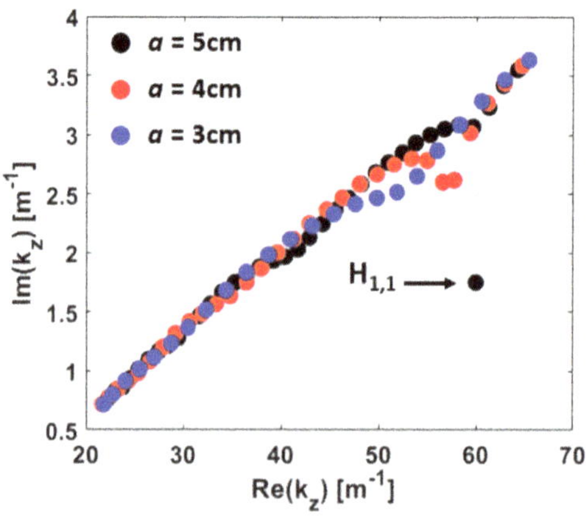

Figure 11.30. Evolution of the eigenvalues spectrum for a uniform density profile, for three values of the column radius a (ν = 5 MHz).

plasma radii smaller than $a = 5$ cm. For these smaller radii, well-separated dominant H modes would be recovered for higher collision frequency, see figure 11.22. Note also that the T–G family is once again only little affected by the variation of the plasma column width. This is a quite general observation, and it turns out that the T–G family range of wavelengths and damping seems to be essentially determined by the peak electron density and the value of the magnetic field.

Radial density gradients can have a very strong influence on the power deposition profile, at least for the dominant H modes. We show in figure 11.31 the normalized deposition profiles expected in four situations: a uniform density, a steep edge gradient ($\Delta/a = 0.1 \Rightarrow \Delta = 5$ mm), an intermediate gradient $\Delta/a = 0.4$, and a full-radius cosine variation ($\Delta/a = 1$). We recall that each eigenmode in cylindrical geometry can be seen as a superposition of pure H and pure T–G components. The $H_{1,k}$ roots have been designated as dominantly H eigenvalues, but it was shown that this dominance is really effective only when these roots are well separated from the dominant T–G family. For the uniform density case (or a steep gradient) in the present example, the $H_{1,1}$ separation is not very strong, and consequently, the mode presents a T–G mode contribution which is still comparable to the H mode (typically, 0.4 to 1). This appears clearly in the power deposition profile for these cases, which is concentrated on the edge of the plasma column where the T–G component is rapidly damped. Going towards more 'bell-shaped' profiles, we see that the power deposition profile progressively tends to become peaked on the axis, as the $H_{1,1}$ mode becomes more and more truly H dominant.

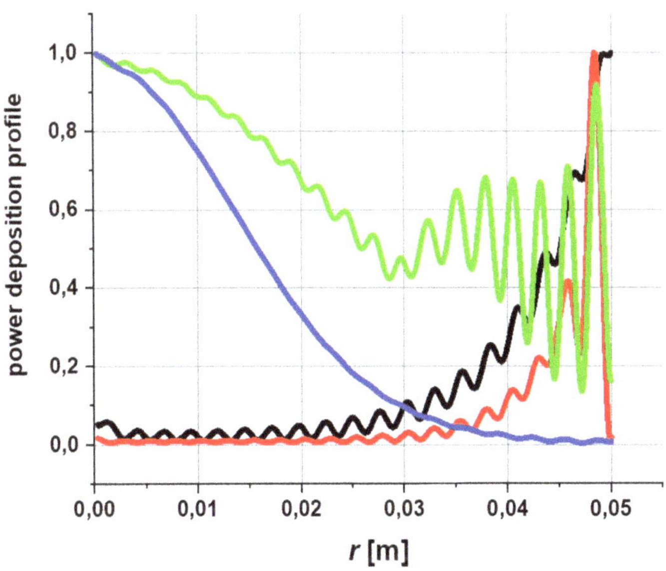

Figure 11.31. Calculated power deposition profiles for uniform density (black); $\Delta/a = 0.1$ (red); $\Delta/a = 0.4$ (green); and $\Delta/a = 1$ (blue).

There is finally one particular effect that appears only for 'gentle' radial variations of the plasma density (i.e. bell shapes), which concerns the prediction of a new family of eigenvalues. As a matter of fact, for a uniform density, or even for the $\Delta/a = 0.4$ case shown in figure 11.28, no roots are found below a Re (k_z) value of about 20 m^{-1}. Interestingly, this corresponds very well to the minimum Re (k_z) value defined by (11.47), although this equation was established in the framework of a collisionless approach. In these cases, we can identify only the dominant T–G and dominant H families. On the example of figure 11.32 we compare the eigenvalue spectra for a uniform density profile (RAID standard conditions (11.44)) and for a full cosine profile ($\Delta/a = 1$). It can be seen that some eigenvalues are now predicted to appear for Re (k_z) values below the 20 m^{-1} minimum corresponding to (11.47). We further designate this new family as the LTG mode group (standing for 'low T–G').

One notable particularity of these LTG modes concerns the associated power deposition profiles. Figure 11.33 shows the typical power radial profiles for the H, TG, and LTG families, for the example of figure 11.32. It can be seen that H and TG roots both lead to power deposition peaked on axis, while LTG modes are characterized by off-axis heating patterns.

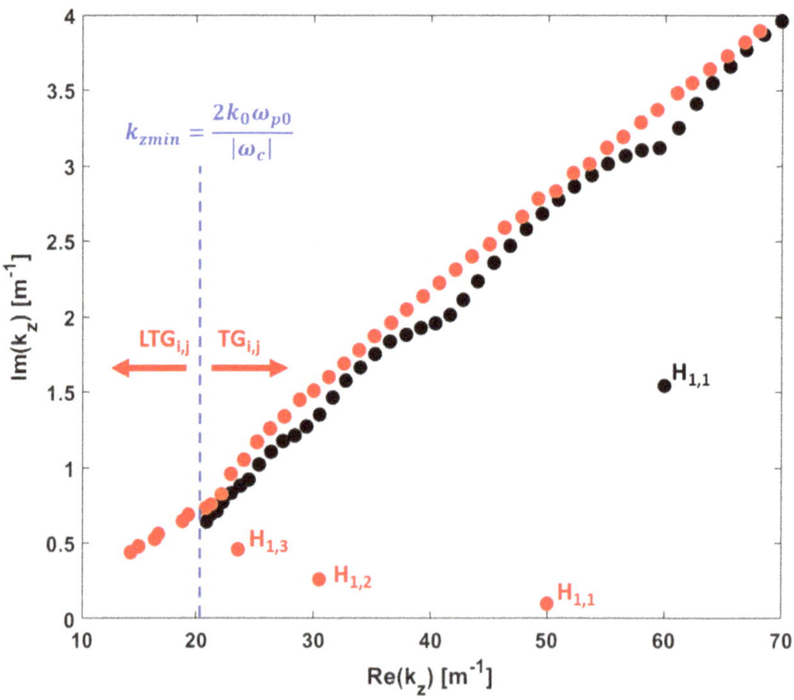

Figure 11.32. Comparison of the eigenvalue spectra for a uniform density (black dots) and a full cosine shape ($\Delta/a = 1$, red dots).

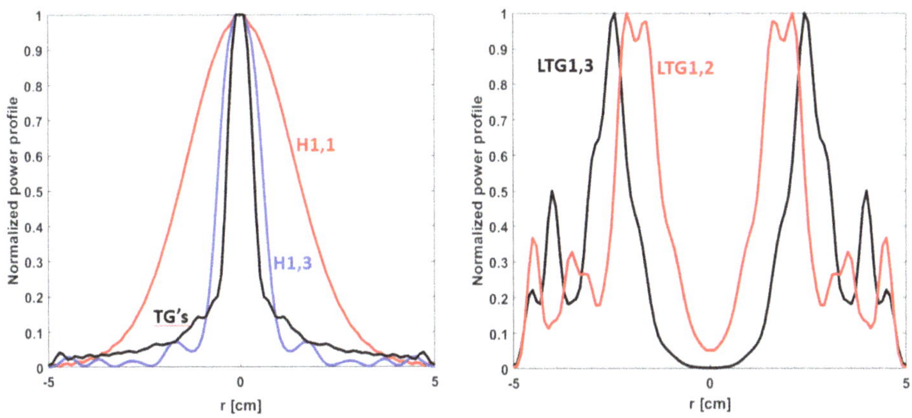

Figure 11.33. Typical radial power deposition profiles for different modes of figure 11.32. Left: peaked power profiles characteristic of H and TG families. Right: off-axis profiles characteristic of LTG modes.

11.4 Chapter summary for helicon modes in a magnetized plasma

This chapter considered the transformation of whistler plane waves into helicon and Trivelpiece–Gould hybrid modes in a bounded cylindrical plasma column with an axial magnetic field. Several different models were described, depending on the particular selection of approximations, and a mode study was performed for each class of equations. The model results, such as the power deposition profile, depend sensitively on the choice of plasma parameters, which include boundary conditions, collisionality, and the plasma density radial non-uniformity. In the latter case, a fourth order equation is appropriate for the experimental conditions of resonant birdcage antennas, and new modes of longer axial wavelength are predicted. Numerical simulation results are compared with measurements in the next chapter 12.

References

[1] Shinohara S 2022 *High-Density Helicon Plasma Science* (Singapore: Springer Nature) Springer Series in Plasma Science and Technology

[2] Trivelpiece A W and Gould R W 1959 Space charge waves in cylindrical plasma columns *J. Appl. Phys.* **30** 1784

[3] Guittienne Ph, Jacquier R, Pouradier Duteil B, Howling A A, Agnello R and Furno I 2021 Helicon wave plasma generated by a resonant birdcage antenna: magnetic field measurements and analysis in the RAID linear device *Plasma Sources Sci. Technol.* **30** 075023

[4] COMSOL Inc. https://www.comsol.com [accessed 30 January 2024]

[5] Lieberman M A and Lichtenberg A J 2005 *Principles of Plasma Discharges and Materials Processing* 2nd edn (Hoboken, NJ: Wiley)

[6] Klozenberg J P, McNamara B and Thonemann P C 1965 The dispersion and attenuation of helicon waves in a uniform cylindrical plasma *J. Fluid Mech.* **21** 545

[7] Chen F F and Arnush D 1997 Generalized theory of helicon waves. I. Normal modes *Phys. Plasmas* **4** 3411

[8] Abramowitz M and Stegun I A 1965 *Handbook of Mathematical Functions* (New York: Dover)

[9] Jin J-M 1998 *Electromagnetic Analysis and Design in Magnetic Resonance Imaging* (Boca Raton, FL: CRC Press)

[10] Arnush D and Chen F F 1998 Generalized theory of helicon waves. II. Excitation and absorption *Phys. Plasmas* **5** 1239

[11] Wolfram Research Inc 2022 *Mathematica, Version 13.1* (Champaign, IL: Wolfram Research Inc)

[12] Chen F F, Hsieh M J and Light M 1994 Helicon waves in a non-uniform plasma *Plasma Sources Sci. Technol.* **3** 49

[13] Sudit I D and Chen F F 1994 A non-singular helicon wave equation for a non-uniform plasma *Plasma Sources Sci. Technol.* **3** 602

[14] Light M, Sudit I D, Chen F F and Arnush D 1995 Axial propagation of helicon waves *Phys. Plasmas* **2** 4094

[15] Perry A J, Vender D and Boswell R W 1991 The application of the helicon source to plasma processing *J. Vac Sci. Technol.* B**9** *310*

[16] Hagelaar G J M and Pitchford L C 2005 Solving the Boltzmann equation to obtain electron transport coefficients and rate coefficients for fluid models *Plasma Sources Sci. Technol.* **14** 722

IOP Publishing

Resonant Network Antennas for Radio-Frequency Plasma Sources

Theory, technology and applications

Philippe Guittienne, Alan Howling and Ivo Furno

Chapter 12

Wave-sustained plasma

In the early 1970s, Boswell [1–3] showed that helicon waves can very efficiently generate high density plasma extending far away from the source along the magnetic field. Conventional helicon sources are based on specially shaped antennas, often partial-helix, made from interconnected metal straps. Preliminary evidence of helicon wave generation by an alternative antenna—a birdcage resonant network (figure 12.1)—was given in 2005 [4]. In fact, this was the very first use of an RF resonant network antenna as plasma source. In this experiment, the helicon regimes were confined in a long dielectric tube, leading to so-called 'bounded discharges'. In the first part of this chapter, section 12.1, we show a selection of results from this first campaign.

More recently, the Resonant Antenna Ion Device (RAID) experiment was developed [5–9] where a resonant birdcage antenna is used as a helicon plasma source [10, 11] in an initial step to investigate negative ion production for neutral beam heating of tokamaks. Apart from this aspect, the RAID experiment is well suited for fundamental studies of helicon wave propagation. A steady-state plasma column, typically 1.5 m long, is maintained in the main vacuum chamber where it can be conveniently measured, axially and radially, by various diagnostics. The notable difference with the 2005 experiment [4] is that the plasma column in RAID is unbounded (section 12.2), at least in the main chamber. We dedicate a large section of this chapter to results obtained on RAID.

Finally, in the last section 12.3 we present the very interesting case of 'helicon' discharges driven by planar networks. These are novel surface sources, which can drive a plasma column along the externally imposed magnetic field. They represent the 'magnetized' version of the planar inductively coupled antennas in chapters 6 and 7. These planar helicon sources can conveniently be mounted in air (outside the vacuum vessel), or inside the vacuum chamber (section 13.4).

doi:10.1088/978-0-7503-5296-3ch12

Figure 12.1. '#3: A birdcage containing a helicon and a confined whistler.' (Illustration by Alex Howlıng. Copyright 2023 Alex Howling.)

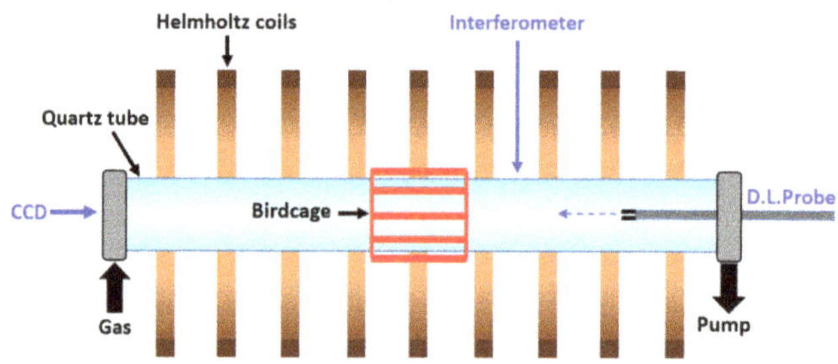

Figure 12.2. Schematic of a typical experimental set-up for a bounded helicon discharge experiment. The plasma, sustained by the birdcage antenna, is monitored by a CCD camera, and a double Langmuir probe (D L Probe), from opposite ends. (Adapted with permission from [4]. Copyright 2005 AIP Publishing.)

12.1 Bounded helicon discharge

In most experiments with helicon sources, plasmas are generated in long, thin dielectric tubes. The plasma column is closely radially bounded by walls, so the plasma density cannot gradually fall to zero as would be the case in an unbounded column, such as the RAID experiment that we present later in this chapter.

A typical experimental set-up for bounded helicon discharges is shown in figure 12.2. Here, a closed birdcage antenna (16 legs, 25 cm long) surrounds a 1.5 m long quartz tube, 10 cm diameter, which constitutes the discharge chamber. The plasma is magnetized by a set of 14 Helmholtz coils, generating a quite uniform field up to 1000 G all along the tube. In this experiment, the plasma density was measured by Langmuir probes and 33 GHz microwave interferometry. The plasma visible emission was photographed by a CCD camera along the discharge axis from one end of the quartz tube. As mentioned above, this experimental set-up was used already in 2005 for the first characterization of a birdcage antenna as helicon source [4]. It was shown that birdcage antennas are indeed very efficient for this kind of discharge, notably with an extended range of operating pressure, up to 25 Pa in these experiments.

Figure 12.3 shows an example of hysteresis behaviour that is typically observed in bounded discharges. The left figure shows a measurement of plasma density, 10 cm from the antenna, during an up and down RF power cycle. The plasma ignites (via the *E*-mode of chapter 5) at a quite low power of typically 80 W. Increasing the RF power from there, the plasma density remains relatively low, about 10^{17} m^{-3} in the inductively coupled *H*-mode, until a sudden and very strong jump occurs at 320 W RF power, to reach densities of typically $4 \cdot 10^{18}$ m^{-3}. In the literature [12–15], this is often called the transition from the *H*- to *W*-mode, where *W* signifies 'helicon wave-sustained mode'[1]. It can also be seen from the CCD camera pictures (figure 12.3 right), that this abrupt density jump is associated with a complete change of plasma shape: before the jump (image *a*) the plasma is located close to the current maxima of the birdcage (m = 1 azimuthal symmetry). The 'bean' shapes (m = 2 azimuthal periodicity) are typical of inductive coupling by a current with m = 1 azimuthal periodicity; see figure 8.23. In this regime, the plasma only diffuses from the antenna

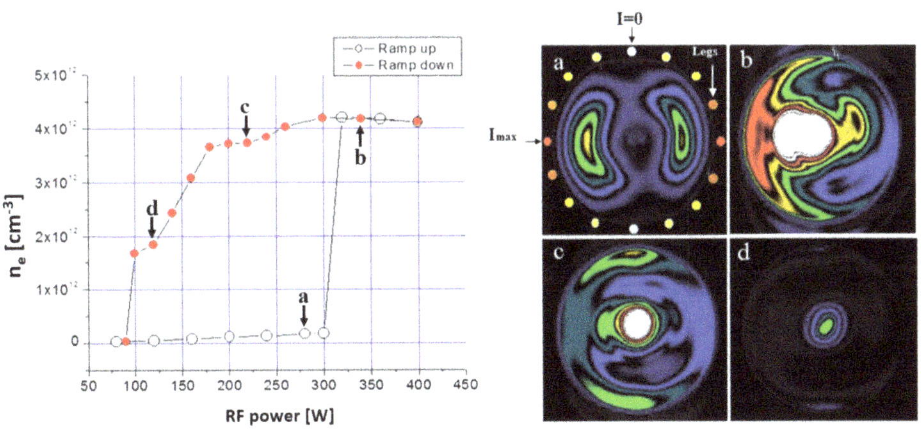

Figure 12.3. Left: measurement of peak density at 15 cm from the source in figure 12.2, during an up and down RF power cycle. Right: CCD pictures taken from one end of the discharge tube at different RF powers during the cycle, indicated by letters (*a*, *b*, *c*, *d*) on the left plot. (Adapted with permission from [4]. Copyright 2005 AIP Publishing.)

[1] According to Isayama *et al* [14]: 'There are believed to be three distinct modes of operation for a helicon plasma source: the capacitively coupled plasma (CCP or *E*-mode), the inductively coupled plasma (ICP or *H*-mode), and the helicon wave-sustained *W*-mode. The *E*–*H*–*W* mode transitions are observed with jumps in the plasma density.' Note that the ICP *H*-mode name is associated with the magnetic field strength *H*; this terminology is not to be confused with the Helicon H mode which is a component of the helicon wave-sustained mode, *W*.

and does not extend axially far away from it. After the density jump (image *b*), the plasma is now clearly centred on the tube axis. The transition is further associated with the creation of a plasma column, extending far from the source, up to the tube end for pressures of typically 1 Pa. A very interesting point concerns the observation of a hysteresis behaviour of the plasma density when ramping down the RF power, as can be seen in the left panel of figure 12.3. Centred and high density regimes are maintained down to 100 W, until an inductive regime is recovered. Successive abrupt mode transitions are also observed (images *c* and *d*), with corresponding plateaus in the density measurements.

Abrupt mode transitions have been observed in many other experiments, up to the point where they are sometimes considered as a characteristic of helicon discharges. Some authors also claim that these abrupt transitions are associated with certain ratios between the antenna length and the excited axial wavelength [16], but this remains uncertain. In experiments such as RAID [17] in section 12.2, helicon regimes occur while no sudden mode transition is observed. The main difference between the two systems is whether the plasma column is bounded or not, suggesting that mode transitions might well be characteristic of bounded helicon discharges.

The set-up of figure 12.2 was also used to check the ability of birdcages to excite a specific azimuthal wave mode *m*. To do this experiment, three different sets of capacitor assemblies were used; each set was designed to make the birdcage resonate at 13.56 MHz, but with different resonant mode numbers m = 1, 2, and 3. Figure 12.4 shows the typical CCD pictures of plasma discharges obtained for the three different mode excitations. Since the CCD camera images the plasma from one end of the quartz tube, there is naturally a certain viewing angle in these pictures.

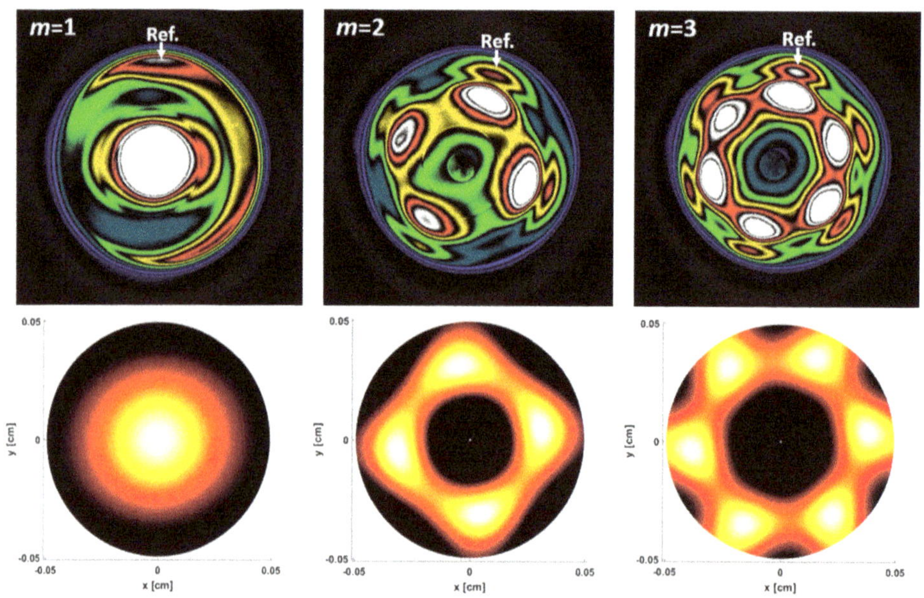

Figure 12.4. Top: CCD pictures of helicon discharges for m = 1, 2, 3 mode excitations of the birdcage antenna. Bottom: calculated mappings of power density deposition of pure $H_{m,1}$ theoretical modes.

Therefore, the outermost structures appearing on the images, indicated by white arrows labelled 'Ref.', are simply due to reflections of the plasma light on the internal wall of the quartz tube. Below the CCD pictures, the calculated power deposition mapping expected for each value of m is represented, according to a constant density model, but considering only the H component of the first dominant H mode ($H_{m,1}$), thereby excluding the T–G component. The excellent correspondence between the top and bottom representations clearly demonstrates a very effective selectivity of the azimuthal modes.

Interestingly, if we were to keep the T–G component for each $H_{m,1}$, the power deposition mapping would appear very different, totally peaked on the plasma edge, as explained in chapter 11 figure 11.26. But on the present CCD pictures there is no sign of such an edge localized enhanced power deposition, and on the contrary the plasma distribution really only reveals a H mode structure. It is then likely that the edge density gradients are sufficiently small (despite being a bounded plasma) in these experiments for T–G modes to play only a minor role in terms of power deposition.

12.2 Unbounded helicon discharges: the RAID experiment

The RAID experiment, shown schematically in figure 12.5, is a linear and axisymmetric plasma reactor which can sustain high power helicon discharges up to 10 kW [17]. In this set-up, the vacuum chamber is not a continuous dielectric tube as for the example of figure 12.2, but it is made up of two regions. The source region, on the extreme left of figure 12.5, is a double-walled, high-thermal-conductivity alumina tube with eight water-cooling channels. This cooled tube is 40 cm long, for an internal diameter of 9 cm and external diameter of 11 cm. A nine-leg open birdcage, figure 12.6, 13 cm diameter, 15 cm long, surrounds this dielectric discharge

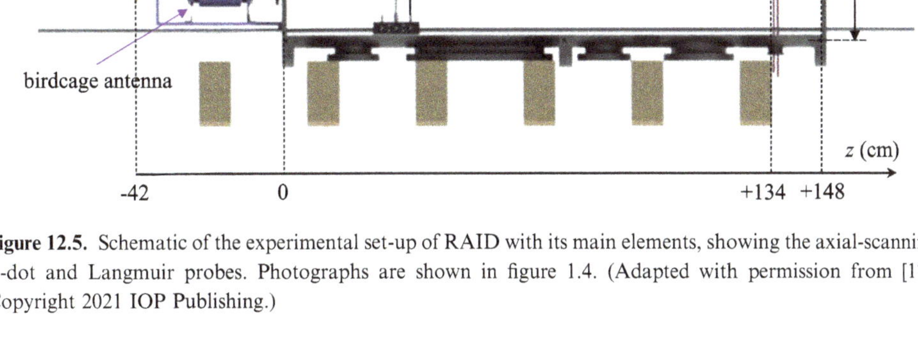

Figure 12.5. Schematic of the experimental set-up of RAID with its main elements, showing the axial-scanning B-dot and Langmuir probes. Photographs are shown in figure 1.4. (Adapted with permission from [17]. Copyright 2021 IOP Publishing.)

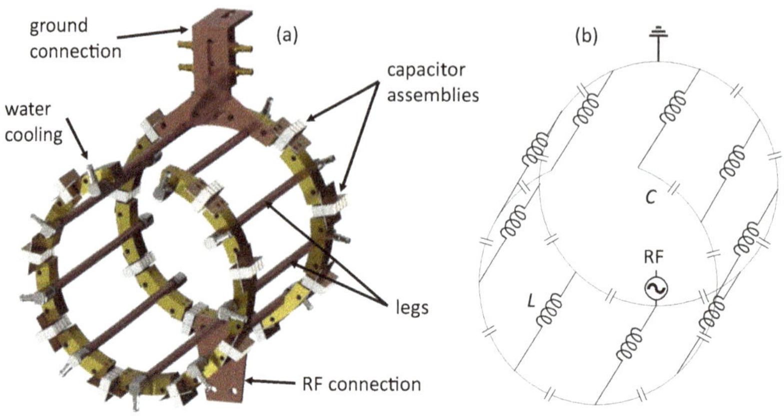

Figure 12.6. (a) Engineering drawing of the birdcage resonant antenna 13 cm diameter and 15 cm long, and (b) the equivalent electrical circuit. (Reproduced with permission from [17]. Copyright 2021 IOP Publishing.)

tube. The antenna is designed to present a m = 2 resonance at the 13.56 MHz operation frequency. We recall that in an open network, a m = 2 resonance corresponds to a $m = 1$ azimuthal variation of current amplitude into the antenna leg (see chapter 9); as such, the RAID antenna is dedicated to excite $m = 1$ helicon modes. Details about the construction and the RF properties of this source are given in chapter 13. Note that the dimensions of this birdcage are similar to partial-helix antennas used on many other helicon plasma sources [18–20]. A half-helical antenna was also tested on RAID, to compare with the birdcage design. It was found that the resonant network provides a wider range of operation parameters in terms of power (lower power modes) and higher pressure, with more stable discharges. In particular, helicon regimes can be difficult to ignite in hydrogen gas using conventional partial-helix antennas, and if so, the discharges are often very unstable [18, 21]. In contrast, the resonant antenna generates stable hydrogen discharges, which notably allowed the volumetric production of hydrogen negative ions to be studied intensively on RAID [8–11, 22].

The alumina discharge tube is mounted on the flange of a large stainless steel cylindrical chamber, 1.5 m long and 40 cm diameter, in which the helicon discharges, bright narrow columns of approximately 8 cm diameter (figure 1.4), are effectively unbounded. The great advantage of the RAID set-up is that it enables the plasma column properties in this main chamber to be thoroughly investigated by various types of diagnostics: axially and radially movable probes can be used to map either the plasma column density and electron temperature with double Langmuir probes, or the RF wave pattern by means of three-axis B-dot probes. Additionally, fixed diagnostics such as microwave interferometry and Thomson scattering are used for refined estimations of the plasma properties.

To end this general description, the plasma in RAID is magnetized using a set of five Helmholtz coils, generating a reasonably constant axial field in the main chamber region (figure 12.7), but not in the source region. The gradient of the magnetic field at the antenna location can be adjusted by means of the leftmost coil in figure 12.5 which is independently fed. Gradients of the magnetic field at the

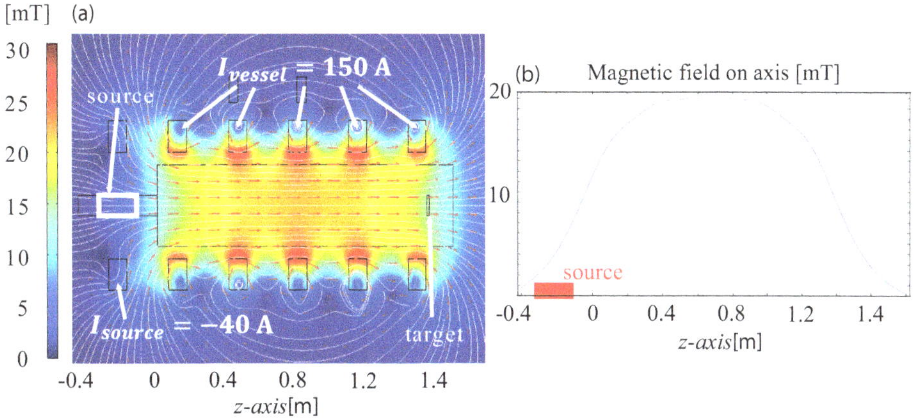

Figure 12.7. (a) Magnetic field mapping in standard conditions: −40 A in the source coil (counter field) and 150 A in the vessel coils. (b) Intensity of the magnetic field on axis for the same parameters and axial position of the antenna in red. At the maximum axial field strength near $z = 0.7$ m, the magnetic field is almost uniform over the vessel cross-section, falling by 1% at $r = 0.1$ m and by 10% at the wall, $r = 0.2$ m. Reproduced from [9].

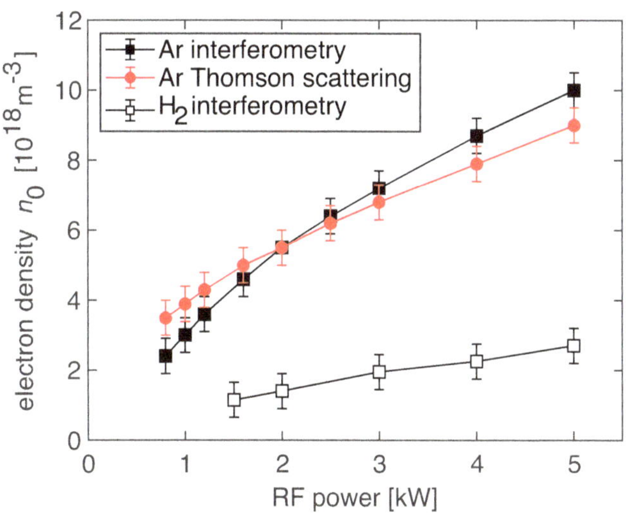

Figure 12.8. Peak electron density n_0 as a function of the input RF power (0.3 Pa, B_0 = 200 G) for argon and hydrogen discharges. (Reproduced with permission from [17]. Copyright 2021 IOP Publishing.)

antenna position were found to promote ignition and to stabilize the helicon regimes over a wider operational space, possibly thanks to a better confinement of slower electrons in the source region.

12.2.1 Plasma characterization

Figure 12.8 shows some measurements of the plasma column peak electron density performed by interferometry and by Thomson scattering on argon and hydrogen discharges, for a fixed magnetic field B_0 of 200 G, but varying the level of RF power.

The peak electron density in RAID discharges is generally in the 10^{18} – 10^{19} m^{-3} range, typical of helicon sources. Note that the density level for hydrogen plasma is significantly lower than for argon, because it is a molecular gas and a large part of the RF power is lost in dissociation and ro-vibrational excitation of molecules. The RAID experiment is not dedicated to low power operation, so it is difficult to properly characterize a possible inductive-to-helicon wave-sustained (H to W) regime transition. Nevertheless, in contrast to the bounded experiment of figure 12.2, no abrupt mode transition or hysteresis behaviour is observed in the unbounded plasma column of RAID.

In order to map the plasma properties, the axially/radially movable double Langmuir probes were used. Typical mappings for plasma density and electron temperature are shown in figure 12.9 for argon, 1500 W, 0.5 Pa, 520 G. The corresponding axial profiles are given in figure 12.10. It can be seen that the plasma density does not continuously decrease when moving away from the source, as would be the case for a purely diffusing plasma regime. On the contrary, the density first increases to reach a maximum approximately in the middle of the chamber, depending on conditions. This is a quite general characteristic of unbounded helicon discharges, which suggests that ionization indeed occurs within the plasma column, far from the antenna. On the other hand, the electron mean temperature continuously decreases with distance from the source.

The radial extent of the plasma column is quite dependent on the intensity of the DC magnetic field, as illustrated in figure 12.11 which shows the radial density profiles measured in argon discharges (600 W, 0.5 Pa) for a 65 G and a 520 G magnetic field strength. In contrast, the RF power level has only a small effect on the column width. Another interesting point is that the radial density profile is reasonably constant along the plasma column, although the peak value varies with z, typically by 20%–30%. As an example, we show in figure 12.12 the normalized radial density profiles taken between $z = 10$ cm and $z = 110$ cm (profiles measured every 10 cm) for the case of figure 12.9. The typical evolution of the plasma column full width half maximum (FWHM) with the magnetic field strength is shown in figure 12.13, for a fixed RF power. The error bars of figure 12.13 represent the maximum variation of the FWHM with z, and it can be seen that the radial density profile tends to become more constant along the column for high magnetic fields.

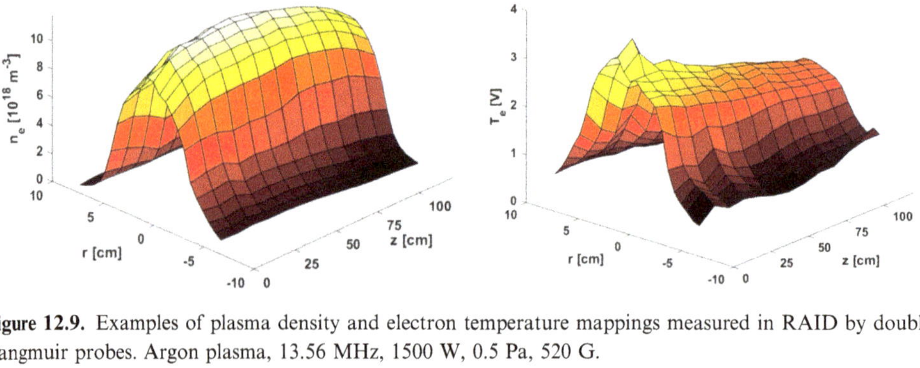

Figure 12.9. Examples of plasma density and electron temperature mappings measured in RAID by double Langmuir probes. Argon plasma, 13.56 MHz, 1500 W, 0.5 Pa, 520 G.

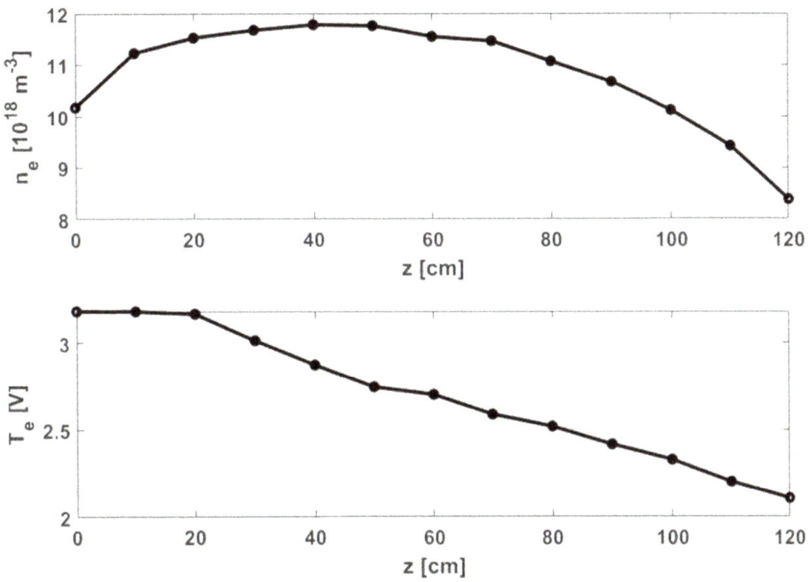

Figure 12.10. Axial plasma density and electron temperature profiles corresponding to the mappings of figure 12.9.

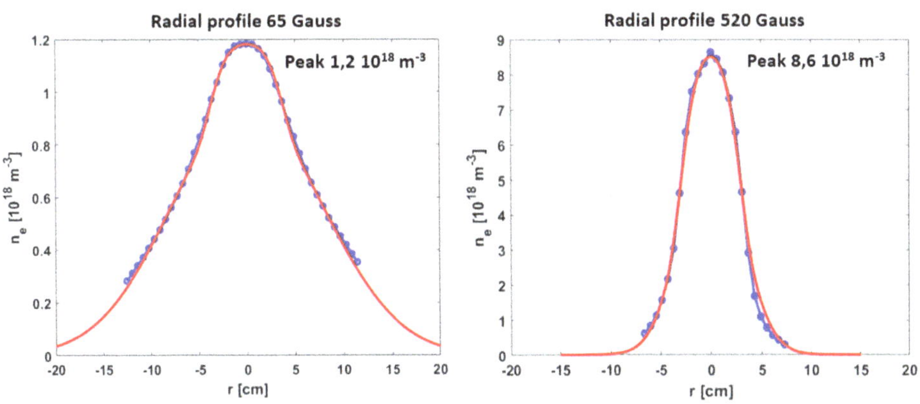

Figure 12.11. Examples of radial density profiles in RAID for a low and high magnetic field strength.

The magnetic field strength can also have a very strong influence on the mean electron temperature radial profile. To illustrate this point, figure 12.14 shows a comparison between the electron temperature profiles for 65 G and 520 G argon discharges, corresponding to the density profiles of figure 12.11. In these graphs, six radial temperature profiles are plotted for each case, corresponding to different z positions along the column. For the low magnetic field case, the temperature profile is very peaked at a radius of about 3 cm from the column axis. This hollow profile is particularly pronounced close to the antenna region, and tends to vanish after a few tens of centimetres.

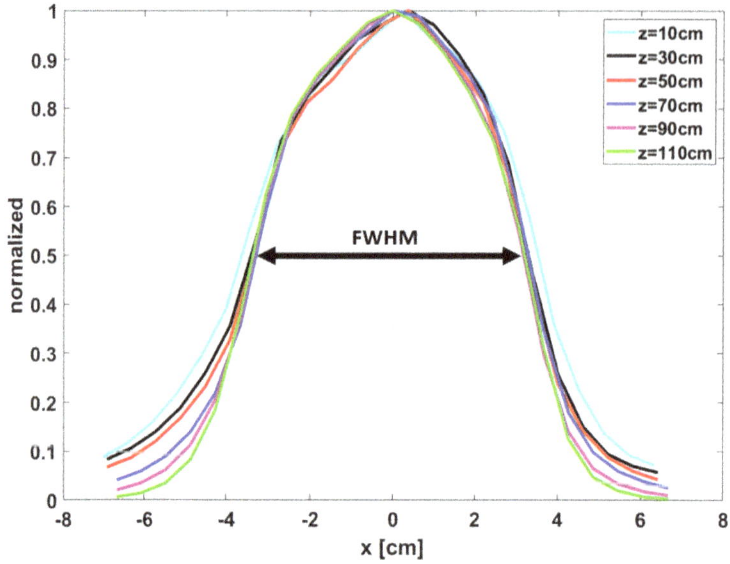

Figure 12.12. Normalized density profiles along the plasma column for the high field case of figure 12.11.

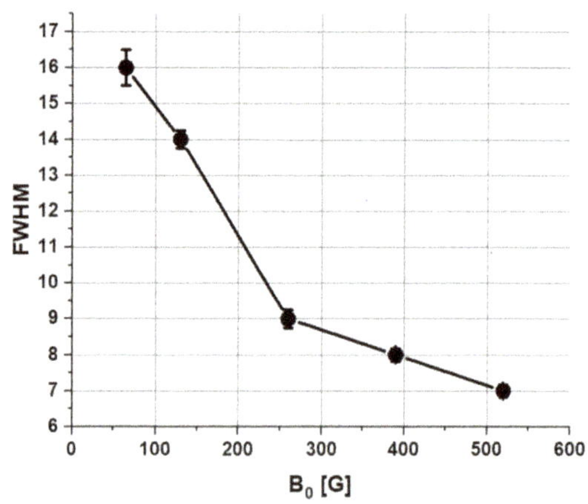

Figure 12.13. Evolution of the FHWM of plasma radial density profiles with plasma magnetization. The error bars represent the maximum variation of FHWM along the plasma column.

Considering the high field case, we observe an opposite situation where the electron temperature is now more peaked on the column axis. Furthermore, these observations appear to be indeed first related to the level of magnetic field, but not much to the RF power level (i.e. density), as illustrated in figure 12.15 which shows that the hollow profile associated with the low magnetic field in figure 12.14 is observable over a wide range of RF power.

The reasons for these observations are not fully clear at the present time, but it is probable that we see here a signature of phenomena occurring already in the source

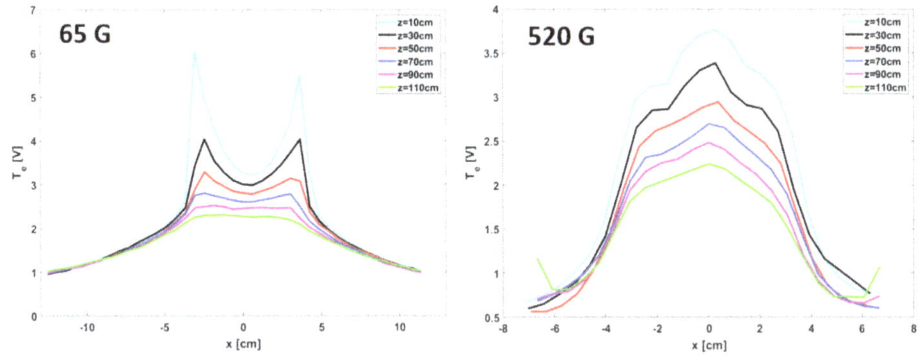

Figure 12.14. Radial electron temperature profiles for different z positions. Comparison between low and high magnetic field cases.

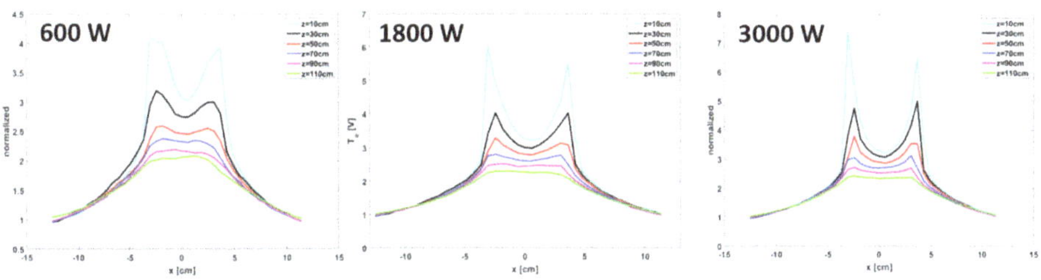

Figure 12.15. Radial electron temperature profiles for different z positions, for a low magnetic field case with different RF power levels.

region, which is unfortunately more difficult to characterize experimentally. Indeed, simulating the wave propagation for the two cases of figure 12.14, we clearly observe a net difference in terms of power dissipation distribution for the high and low magnetic field cases, as shown in figure 12.16. The high field case shows power deposition much more peaked on the column axis, which is consistent with experimental observation of blue cores with high magnetic fields. In the low field case, there is a dominant deposition of power off axis, especially in the source region, but remaining also in the main chamber over a few tens of centimeters. We show in figure 12.17 the radial power density profile for this low field case, taken at $z = 10$ cm from the main chamber flange, to be compared to the first T_e profile of figure 12.14.

12.2.2 Wave propagation

The B-dot probe in figure 12.18 measures the three Cartesian components of an oscillating B field using three orthogonal coils mounted on an alumina probe head. As for the Langmuir probe measurements shown in the preceding section, this B-dot probe was mounted on an axially/radially movable system for mapping the wave field across the plasma column. The interested reader will find more information about this diagnostic in [7].

The coils signals were triggered synchronously by a reference signal taken from the antenna input current. Figure 12.19 shows a typical measurement of the

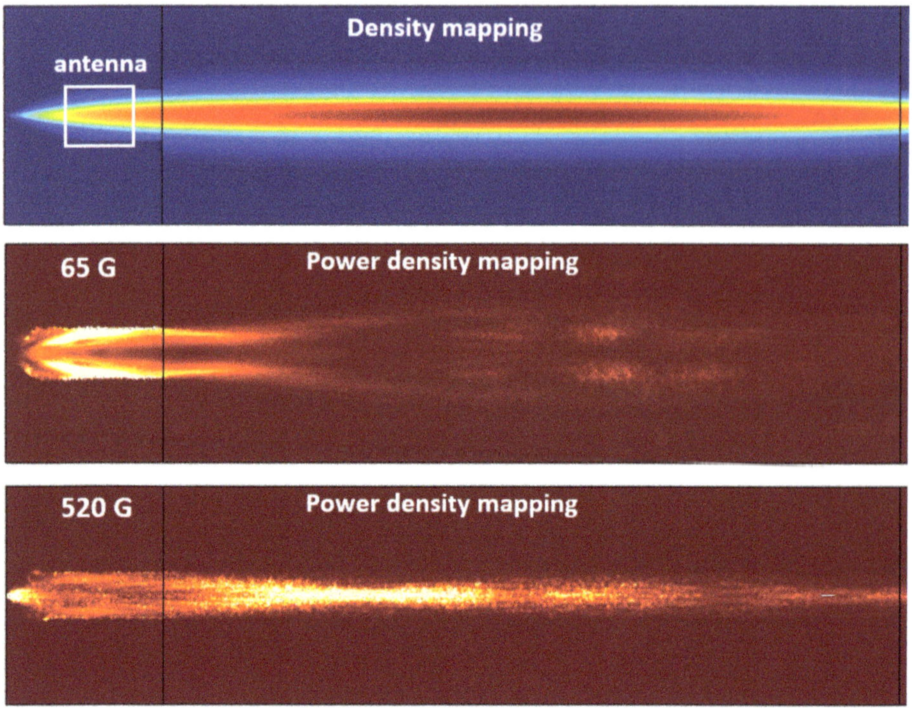

Figure 12.16. COMSOL [23] simulations of wave propagation, comparing the power deposition patterns for low and high magnetic field cases. The upper mapping shows the typical density distribution taken for the model, constructed in the main chamber region from Langmuir probe measurements, and extrapolated into the source region.

transverse magnetic wave components (B_x, B_y), for a given probe position, as a function of time. The black curve is a parametric plot of $B_x(t)$, $B_y(t)$ over hundreds of RF periods, and the red curve is the resulting mean over a single RF cycle. Note first that the field amplitude of ~1 G is typical of helicon waves and much smaller than the static field B_0, so wave equations can justifiably be linearized [2]. We further see that the wave is right and almost circularly polarized, as expected for R-waves (see chapter 10) when the axial wave vector k_z, and the DC magnetic field B_0, are in the same direction.

Figure 12.20 shows typical profiles of the transverse RF field measured along the plasma column axis z. The axial spatial structure of the transverse field is shown here at four equally spaced times in an RF half-period to show the rotation of the pattern. As for figure 12.19, the axial wave vector k_z and the DC magnetic field B_0 point in the same direction for this case. The wave has a left-handed helical pitch which rotates clockwise in time. By reversing the static field direction, the opposite orientations are observed. These characteristics are the same as for whistler R-waves in infinite uniform plasma in chapter 10, as shown schematically in figures 10.2, 10.5, 10.6, and 10.7. Comparison between the experimental measurements in figure 12.20 and the theoretical wave field structure in figure 12.21 shows

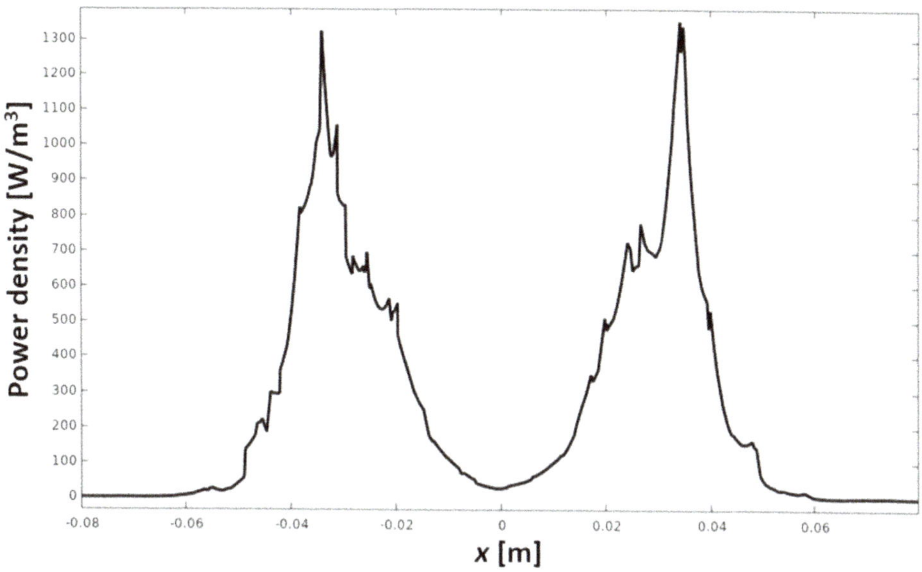

Figure 12.17. Radial profile of simulated power deposition for a low field case, taken from corresponding mapping of figure 12.16 at a distance of 10 cm from the chamber entrance.

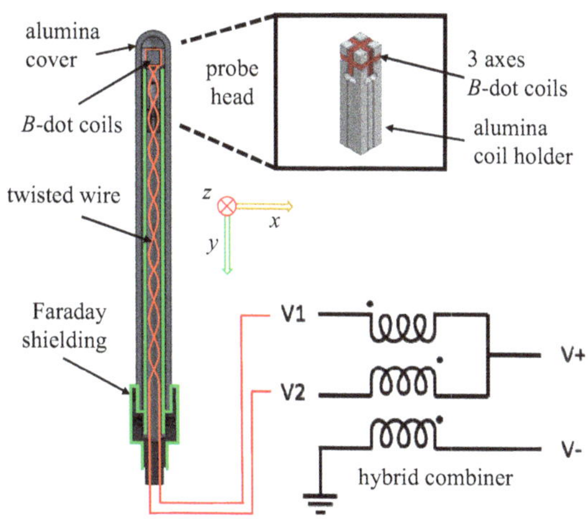

Figure 12.18. B-dot probes showing the probe head, cables, and hybrid combiner. The required magnetic induction signal is given by $V-$ (the capacitive pickup is given by $V+$). (Reproduced with permission from [17]. Copyright 2021 IOP Publishing.)

that the resonant birdcage antenna is generating a helicon wave-sustained plasma in RAID. This confirms, at least qualitatively, the characteristics expected for a helicon wave field generated by the resonant birdcage antenna. Quantitative agreement with the helicon dispersion relation is now investigated further.

Figure 12.22 shows measurements of the axial variation of one transverse component (B_y) of the magnetic field, at a given temporal phase, for different RF

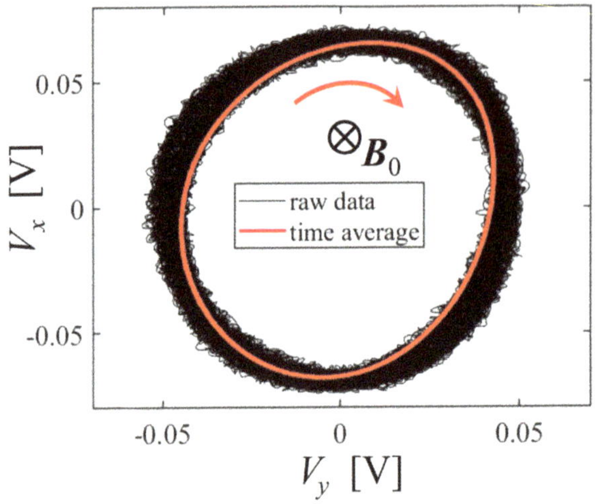

Figure 12.19. Lissajous figure showing the right-hand circular polarization of the measured transverse magnetic field with respect to the static magnetic field $\mathbf{B}_0$ along $+z$. The coil voltage signals are proportional to the oscillating magnetic field components B_x and B_y. The red line shows the time-averaged curve.

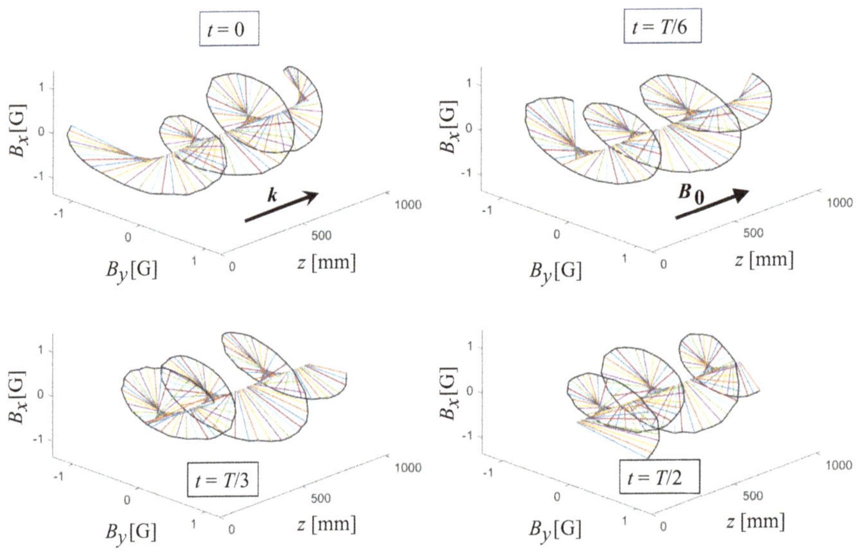

Figure 12.20. The transverse magnetic field raw data, measured on axis in Gauss, for three equal time intervals during an RF half-period (t = 0, T/6, T/3, T/2, where T is the RF period). The static magnetic field $\mathbf{B}_0$, and the wave vector **k** of propagation, are both oriented along the $+z$-axis as shown in the figure. Each chord is a perpendicular from the plasma column axis, at the z position where the measurement was made, to the [B_x, B_y] value measured at that position. The helix is the locus of these [B_x, B_y, z] points. The magnetic field spatial structure is seen to be a left-handed helix, rotating clockwise with respect to the direction of $\mathbf{B}_0$, in accordance with helicon theory (see chapters 10 and 11). Hydrogen at 0.3 Pa, B_0 = 200 G, P_{rf} = 1.5 kW. (Reproduced with permission from [17]. Copyright 2021 IOP Publishing.)

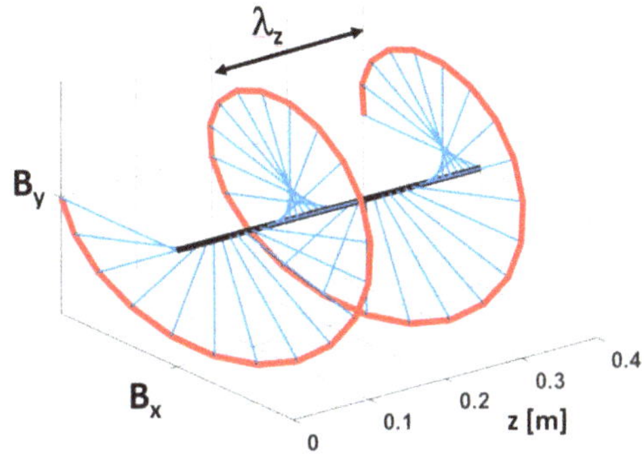

Figure 12.21. A schematic illustration of the helical structure of helicon waves which confirms the experimental measurements in figure 12.20. This figure represents the evolution of the transverse component of the wave field along the magnetic field axis z, for a fixed time, with an arbitrary value of λ_z. $\mathbf{B}_0$ and $\mathbf{k}$ are along the direction of the z-axis and the rotation is clockwise.

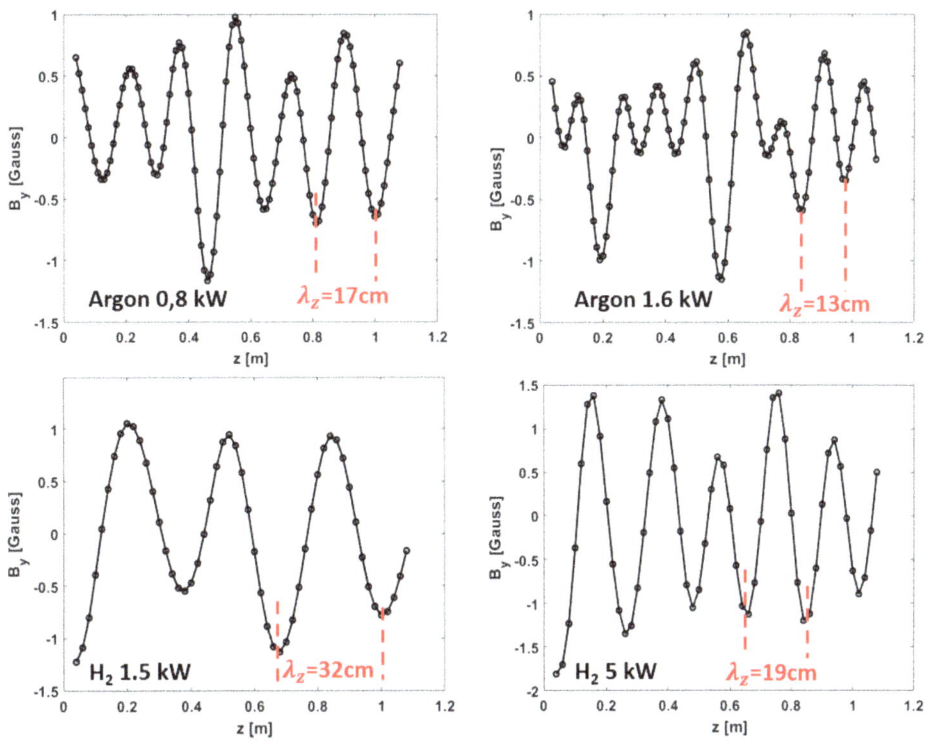

Figure 12.22. Examples of the B_y component of the measured field along the axis at a fixed temporal phase of an RF period, showing the measured wavelengths. The upper graphs correspond to argon plasmas at two different RF powers; the lower graphs are for hydrogen plasmas. Pressure = 0.3 Pa, B_0 = 200 G, for both gases.

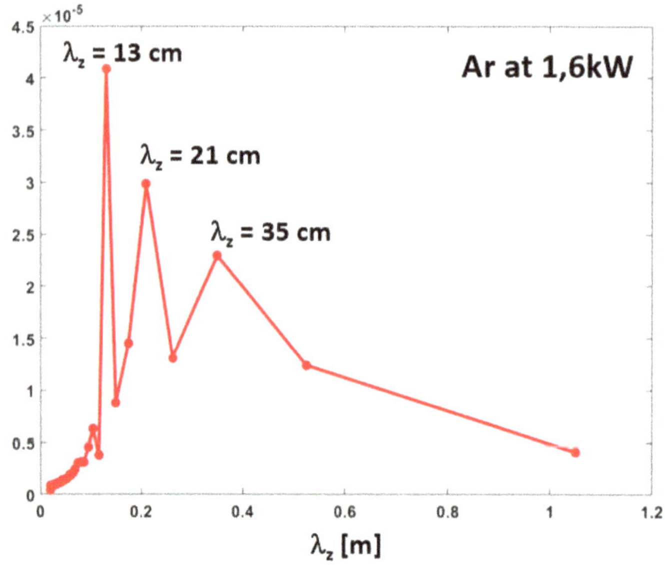

Figure 12.23. Spatial Fourier spectrum of the B_y axial variation shown in figure 12.22, for the argon 1.6 kW case.

power and gas conditions. At least one axial wavelength λ_z can clearly be extracted from these measurements by taking the distance between consecutive extrema. This axial wavelength can vary slightly (typically by ±1 cm) along the plasma column, presumably because of the axial change of plasma density shown in figure 12.10. These axial wavelength variations are nevertheless small enough so that a mean λ_z can be accurately estimated. More rigorously, axial wavelengths can clearly be identified in the spatial Fourier transforms of axial measurements, as shown for example in figure 12.23 which represents the spatial Fourier spectrum obtained for the 1.6 kW argon axial variation in figure 12.22. The main axial wavelength of 13 cm is clearly defined in this spectrum, but we also observe other higher-wavelength peaks, although not as well defined. The question arising is whether these secondary peaks are due to standing waves (reflections) or to a superposition of several modes propagating along the plasma column. Before considering this, we first focus in the following on the analysis of the shortest observable wavelength, and return later to this point.

12.2.3 Main axial wavelength analysis

In a first step, we therefore focus only on the predominant mode corresponding to the lowest wavelength extracted from axial measurements, which is the best defined mode in spatial Fourier spectra. We show in figure 12.24 the evolution of this wavelength, for a fixed static field of 200 G, as a function of the peak electron density. Note that this curve compiles the results obtained for both hydrogen and argon discharges, by using the relation $n_e(P_{\mathrm{RF}})$ measured for both gases (figure 12.8). The evolution of the wavelength is smooth and continuous over the whole density range, although the plasma composition changes from heavy ions to light ions,

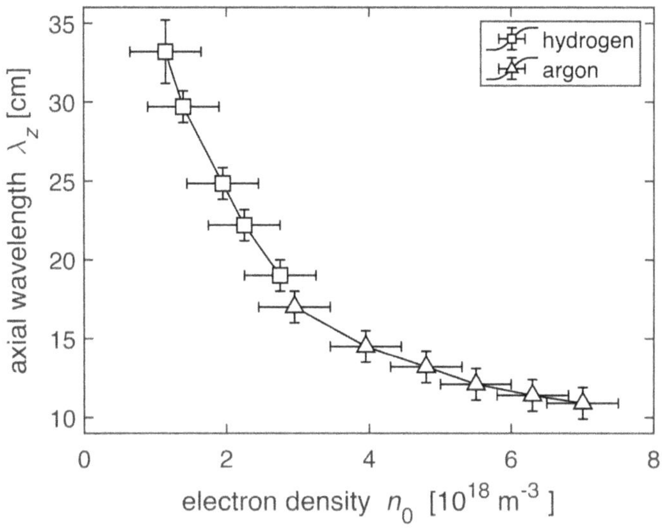

Figure 12.24. Measured axial wavelength λ_z as a function of plasma peak density n_0, at fixed magnetic field $B_0 = 200$ G. Squares correspond to hydrogen discharges, while triangles are for argon. These data are compared with theory in figure 12.25. (Reproduced with permission from [17]. Copyright 2021 IOP Publishing.)

which could involve different mechanisms of plasma heating. We shall conclude from this observation that the helicon wave axial wavelength is only, or at least dominantly, determined by the plasma density and the static magnetic field strength, regardless of the gas type, and consistent with helicon mode theory where ions are taken to negligibly contribute to the plasma dielectric tensor.

Notice that the measured axial wavelength in plasma, between 11 and 33 cm, is about two orders of magnitude smaller than the 13.56 MHz wavelength of 22 m in a vacuum. This is consistent with the 'super dielectric' nature of the whistler wave in magnetized plasma described in section 10.10 and figure 10.14, where the plasma effective relative permittivity for whistlers is $\varepsilon_p \sim 10^4$ for RAID standard conditions in table 10.2.

Another interesting fact concerns the non-observation in RAID of any particular phenomena associated with a match between axial wavelength and antenna length (or its multiples). In fact, the RAID antenna length is 15 cm, and axial wavelengths ranging between 11 cm and 33 cm are measured in figure 12.24, so that we would expect two possible 'resonances'. However, in RAID, the density evolution with power and magnetic field is always smooth, no abrupt mode transitions are observed which could be associated to a wavelength/antenna length match, and no hysteresis behaviour is seen. Once again, with regard to this aspect, RAID is very different from the bounded experiment of figure 12.2.

We now show in figure 12.25 the comparison between the experimental wavelengths of figure 12.24 and the predictions that can be obtained according to the different models introduced in chapters 10 and 11, as well as by COMSOL [23] numerical simulations of wave propagation. The models referred to in the legend of figure 12.25 are resumed in the following three sections.

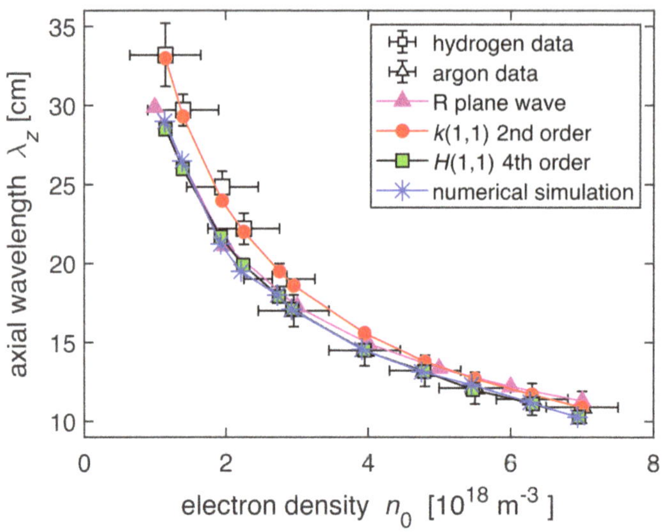

Figure 12.25. Variation of the axial wavelength λ_z as a function of the peak electron density n_0 for $B_0 = 200$ G, showing a comparison between experimental data, analytical, and numerical results. The open symbols (partly obscured) represent the measured axial wavelengths taken from the experimental data in figure 12.24. The magenta triangles indicate the plane wave dispersion relation from section 12.2.4 for an assumed average density $\bar{n}_e = 0.5n_0$. The full red circles correspond to the $k(1,1)$ eigenvalue arising from the integration of the second order equation (11.39) mentioned in section 12.2.5. The green squares correspond to the $H_{1,1}$ eigenvalue from integration of the fourth order equation also mentioned in section 12.2.5. The blue stars show the results of the COMSOL [23] numerical simulation of section 12.2.6. (Adapted with permission from [17]. Copyright 2021 IOP Publishing.)

12.2.4 The R plane wave

Assuming infinite, uniform plasma density, the plane wave whistler dispersion relation was determined analytically in chapter 10. According to (10.34), the principal electromagnetic electron R-wave propagating along B_0 [24–26] is given by

$$\frac{k^2}{k_0^2} = 1 - \frac{\omega_{\rm pe}^2}{\omega(\omega - \omega_{\rm ce})}. \tag{12.1}$$

In figure 12.25 a reasonable assumption of a radially averaged density $\bar{n}_e = 0.5n_0$ was necessary to obtain a good fit to the experimental data for RAID standard conditions in table 10.2.

12.2.5 Radially non-uniform density profile

Two models described in this book can predict axial wavelengths for the case of radially varying plasma density profiles. The first one, relying on the helicon approximation (11.39), implies the resolution of a second order differential equation, hence the label '2nd order' in figure 12.25. The most complete model, keeping dissipation and then T–G contributions, implies the integration of a fourth order equation (11.32). Naturally, in order to apply these models, it is first necessary to measure the radial variations of density, which was done here by using a Langmuir probe, typically as shown in figure 12.11. As explained above, for given conditions of RF power and static magnetic field, the radial density profiles in RAID are observed to be reasonably constant along the plasma column (figure 12.12), which is a necessary condition to apply these models.

As can be seen, the predicted wavelengths plotted in figure 12.25 for both models are all calculated for a $m = 1$ azimuthal mode, and correspond to the smallest axial wavelength of dominant H modes ($k_{1,1}$ in the helicon approximation denomination, and $H_{1,1}$ in the {H,T–G} collisional non-uniform denomination of chapter 11).

12.2.6 Numerical simulation

Finally, figure 12.25 also shows the wavelengths predicted by a 3D finite elements model (COMSOL [23] electromagnetic module) of wave propagation in the RAID experiment (figure 12.26). To match the model to the experimental conditions as closely as possible, the magnetic field of the external coils around the RAID chamber was computed [27]. For the wave calculations, the magnetized plasma was considered to be an axisymmetric cold-plasma-dielectric medium, of conductivity given by (10.14). The (r, z) mappings of density introduced into the numerical model

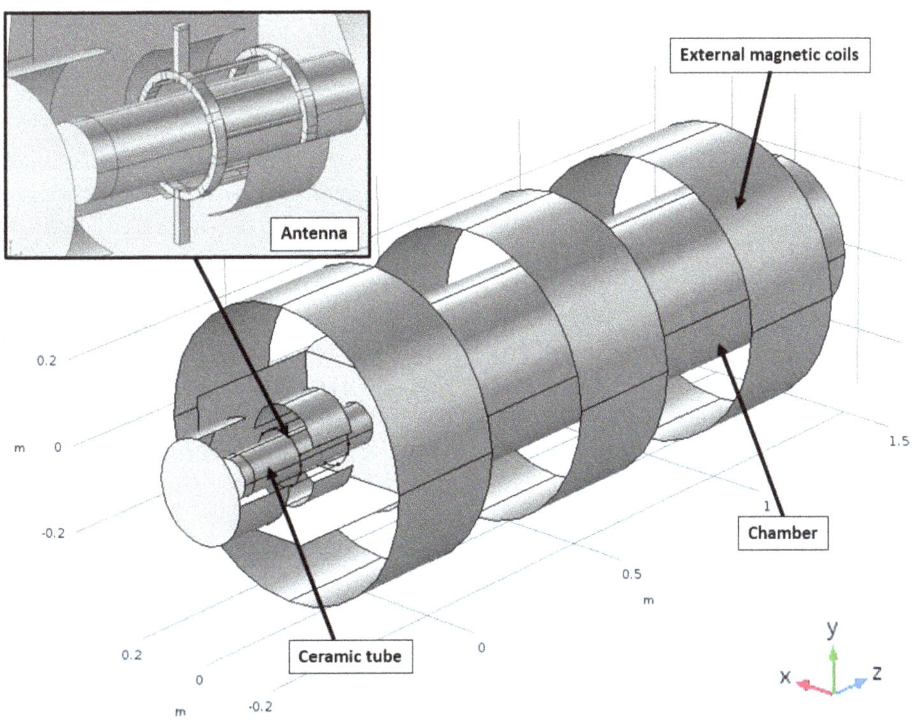

Figure 12.26. Numerical representation of the RAID experiment. The six coils in figure 12.5 coincide with the edges of the external magnetic coils in this figure. The inset shows the birdcage resonant antenna surrounded by two half-cylinder screens. (Reproduced with permission from [17]. Copyright 2021 IOP Publishing.)

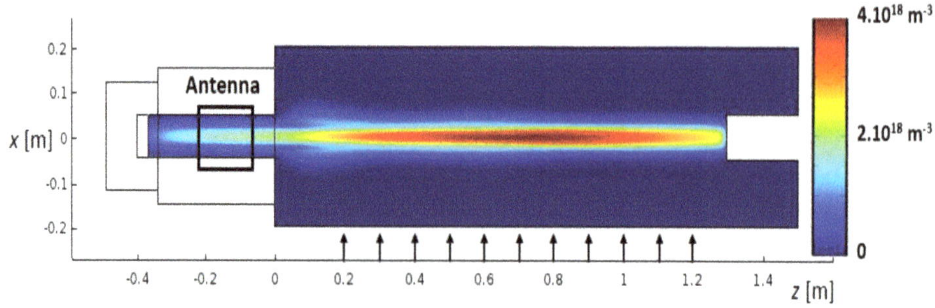

Figure 12.27. Mapping of electron density typically measured in RAID in a (x, z) longitudinal plane for hydrogen plasma. (Adapted with permission from [17]. Copyright 2021 IOP Publishing.)

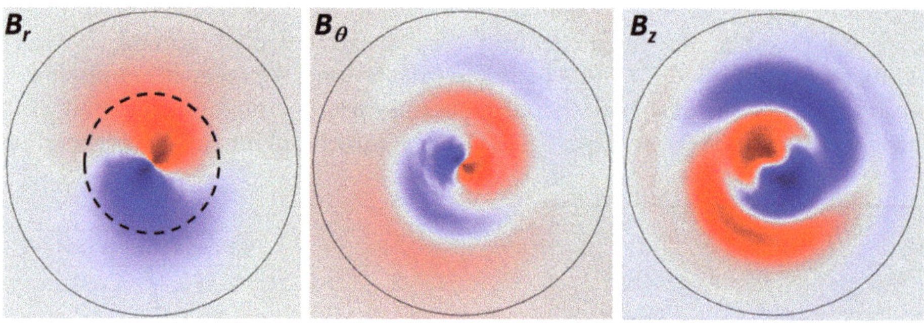

Figure 12.28. Azimuthal cross-section of the wave field components, calculated by the numerical simulation at $z = 0.5$ m in figure 12.27. The outside circles mark the approximate plasma radius of 5 cm; the black dashed circle at 1 cm radius is referred to in figure 12.29.

(example in figure 12.27) were taken from Langmuir probe measurements, with extrapolation for the source region. Therefore, only Maxwell's equations (D.3) and (D.4) are to be solved numerically. To estimate the local electron–neutral effective collision frequency ν, the Bolsig+ software [28] was used to obtain the T_e dependence of the collision rate, assuming Maxwellian electron energy distributions. Finally, the antenna structure (figure 12.6) is excited by the RF input voltage, and all the resultant antenna currents and wave fields are calculated self-consistently.

One particularly interesting result from numerical simulations is the demonstration that the birdcage antenna indeed excites very dominantly $m = 1$ waves into a plasma column. As an illustration, we show in figure 12.28 typical transverse (r, θ) mappings of the wave field components, in a middle position of the main RAID chamber ($z = 0.5$ m in figure 12.27). Following the azimuthal variation of these components at a given radius, indicated by the dashed circle in the B_r mapping of figure 12.28, one effectively sees that all the field components follow a quite pure $m = 1$ sinusoidal variation in figure 12.29.

To summarize, the shortest axial wavelength is apparently not determined by the antenna length, nor by standing waves in the cavity, but varies smoothly, depending

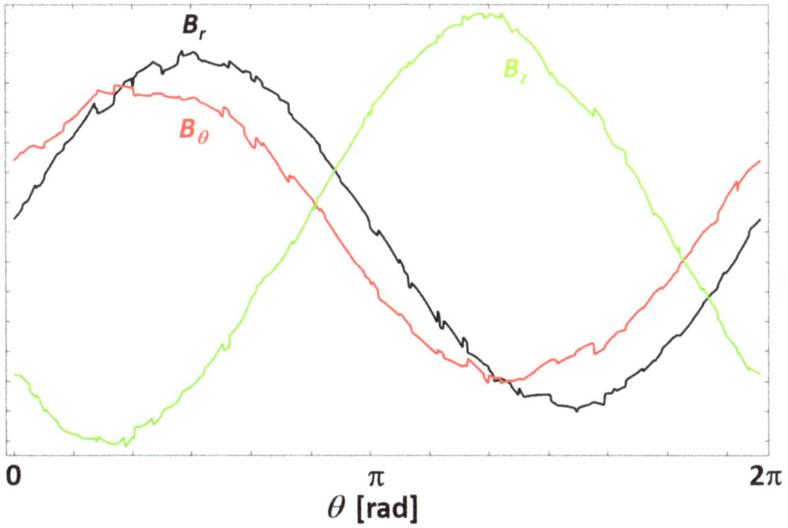

Figure 12.29. B_r, B_θ, and B_z components calculated around the circumference of the black dashed circle at 1 cm radius in figure 12.28, showing a dominant $m = 1$ symmetry. The magnetic fields are in arbitrary units.

on the plasma density and the magnetic field, satisfying the dispersion relation of the $m = 1$ helicon wave excited by the birdcage antenna. The fact that all four models for the axial wavelength present a good agreement with the experimental measurements in figure 12.25 shows that more detailed diagnostics of the wave field (such as fine-scale magnetic field measurements and multimode analysis) are necessary to determine which model gives the most accurate description of the wave physics in a magnetized plasma column.

12.2.7 Multimodal excitation

We now come back to the example of the spatial Fourier spectrum shown in figure 12.23. In the preceding section we showed that the shortest wavelength appearing in the spectra corresponds to a $m = 1$ mode. In the framework of the most sophisticated non-uniform model described at the end of chapter 11 (the {H,T–G} non-uniform density model in section 11.3), this shortest wavelength was clearly identified to correspond with the $H_{1,1}$ dominant helicon mode. We now return to the observation of the other modes in the measured wave field.

The Fourier spectrum of figure 12.23 is shown again in figure 12.30, along with the eigenvalues spectrum calculated from the {H,T–G} non-uniform density model. Clearly the second peak in the spatial Fourier transform, at 21 cm, would correspond very well to the $H_{1,2}$ root predicted by the model. This already suggests that we indeed observe a mode superposition in the plasma column. The third peak at 35 cm is very poorly defined, but it could well be associated with modes close to the branch separation point.

Figure 12.31 shows another example of comparison between the spatial Fourier spectrum obtained from the axial variations of the wave field components and

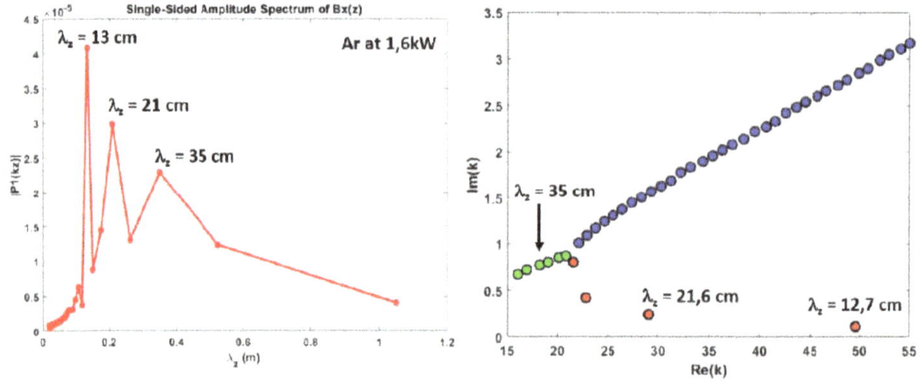

Figure 12.30. Comparison between an experimental spatial Fourier spectrum (argon 1.6 kW of figure 12.22) and theoretical predictions according to the {H,T–G} non-uniform density model with $m = 1$.

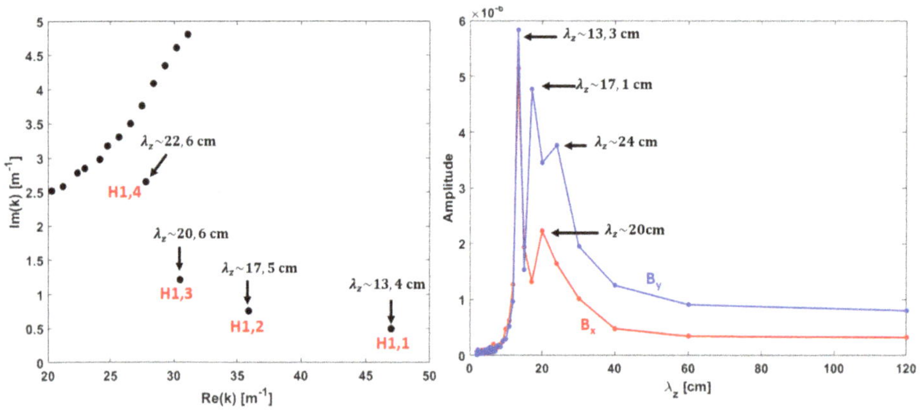

Figure 12.31. Comparison between an experimental spatial Fourier spectrum and theoretical predictions according to the {H,T–G} non-uniform density model with $m = 1$.

the predicted eigenvalues, for the corresponding plasma density radial profile (figure 12.11), according to the {H,T–G} non-uniform density model. This experiment was performed using argon, with a low magnetic field of 65 G and a low RF power of 600 W. The advantage of these soft conditions is that they allow a full (r, z) mapping of the wave field to be performed with the B-dot probe without overheating the probe head. Here again, the observed peaks in the Fourier transform correspond fairly well to the theoretical eigenvalue spectrum for the dominant H modes. Considering figure 12.31, the wave field appears to result from a superposition of at least the three first $H_{1,n}$ modes, with a moderate dominance of the $H_{1,1}$ mode in terms of amplitude.

Figure 12.32 shows a comparison between measured wave component mappings, in the (x, z) plane, and equivalent mappings predicted for the $H_{1,1}$ mode of figure 12.28. Clearly this single $H_{1,1}$ mode alone reproduces the experimental data

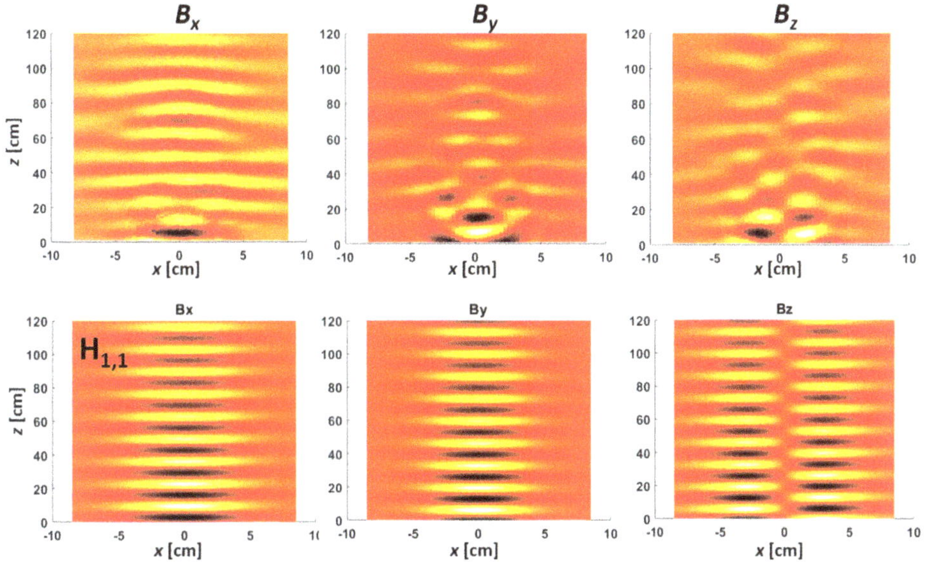

Figure 12.32. Top: mappings of the three components of the measured RF *B* field. Bottom: theoretical mappings of the $H_{1,1}$ mode, according to the {H,T–G} non-uniform model.

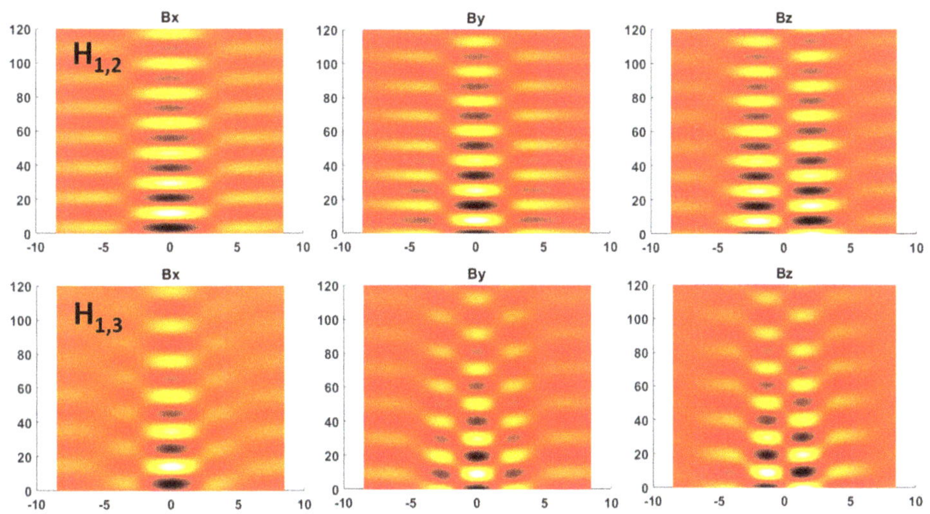

Figure 12.33. Theoretical mappings of modes $H_{1,2}$ and $H_{1,3}$, according to the {H,T–G} non-uniform model.

poorly. Additionally, we show in figure 12.33 the mappings for $H_{1,2}$ and $H_{1,3}$ individual modes, which appear to be even more different from the measurements.

Finally, figure 12.34 shows the comparison between the measured wave components mappings and a *superposition* of $H_{1,1}$, $H_{1,2}$, and $H_{1,3}$ modes, with relative amplitudes of 1, 0.75, and 0.75, respectively. These amplitudes are not finely adjusted by any fitting procedure, but are roughly representative of the relative amplitude of peaks in the spatial Fourier transform of figure 12.31. Clearly, by

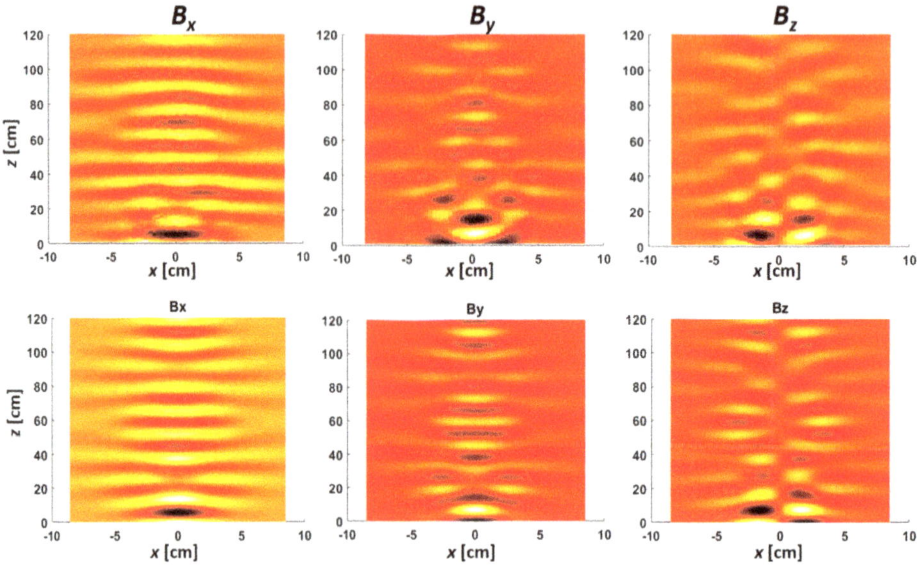

Figure 12.34. Top: mappings of the three components of the measured RF *B* field. Bottom: superposition of theoretical modes $H_{1,1}$ + 0.75 $H_{1,2}$ + 0.75 $H_{1,2}$.

summing these three modes, the resulting multimodal excitation provides a much better correspondence with the experiment.

12.2.8 Impedance measurements, antenna coupling

In this short section, we summarize the results of a set of experiments dedicated to investigate the coupling of the RAID birdcage, for different gases, and comparing inductive and helicon regimes. These experiments are based on impedance measurements of the source when coupled to plasma, which were performed by using a commercial impedance probe (Z-Scan™ [29] from Advanced Energy) for up to 5 kW, placed between the antenna and the matching network. The set-up measured the impedance over a frequency range of typically 1 MHz.

To begin with, figure 12.35 presents a comparison between the typical impedance peaks measured for inductive and helicon regimes in deuterium (2 kW, 1.5 Pa, 200 G). We observe essentially two important facts: inductive regimes lead to higher frequency shifts, but also to less reduction of the quality factor.

If we were to interpret these observations in a naive manner, we would first be tempted to attribute the higher frequency shift in the inductive regime to a higher density in the source region, and the stronger reduction of the quality factor in helicon regimes to a higher effective collision frequency.

In fact, this behaviour can be justified by the sole effect of plasma magnetization, for the same level of density or collision frequency. This was notably checked by COMSOL [23] simulations of RAID (figure 12.36). The 'inductive' and 'helicon' impedance curves of figure 12.36 were both obtained using the same density

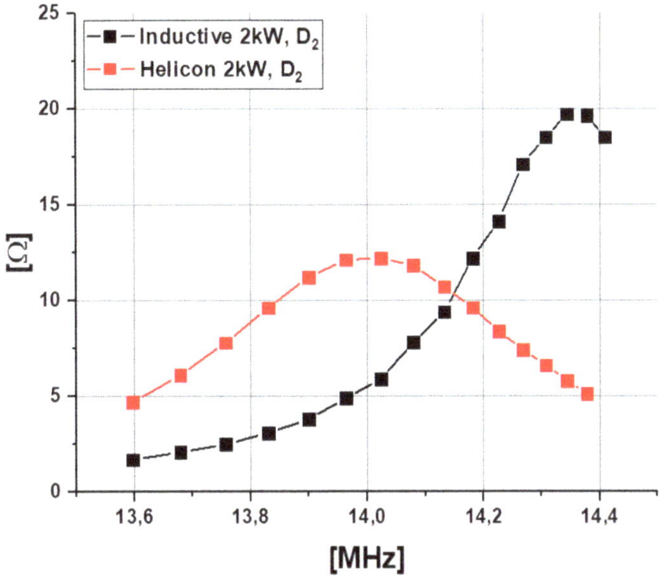

Figure 12.35. Impedance measurements of the loaded birdcage antenna, in inductive ($B_0 = 0$) and helicon regimes, over a frequency range of 1 MHz.

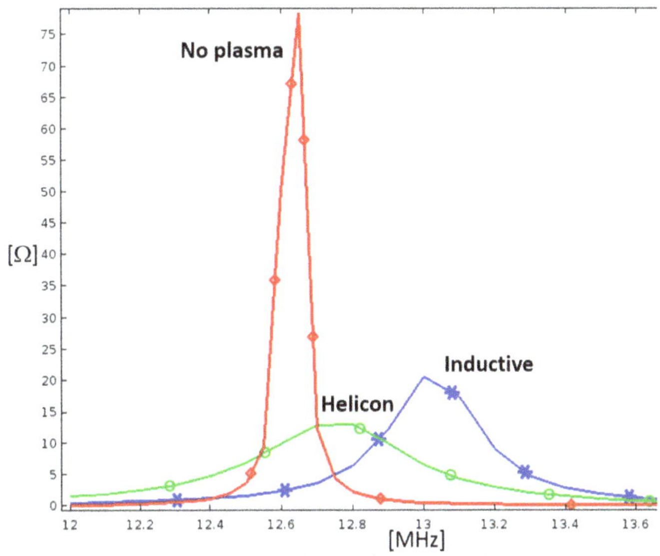

Figure 12.36. Simulated impedance of the RAID birdcage as a function of frequency, for a fixed mapping of plasma density (figure 12.27), but with (Helicon) and without (Inductive) plasma magnetization.

mapping of figure 12.27, and with the same collision frequency, but with or without the magnetic field. An interpretation for the stronger reduction of the quality factor in the helicon regime can be found in the fact that the antenna is effectively coupled to a larger plasma volume, thanks to wave propagation allowed by plasma

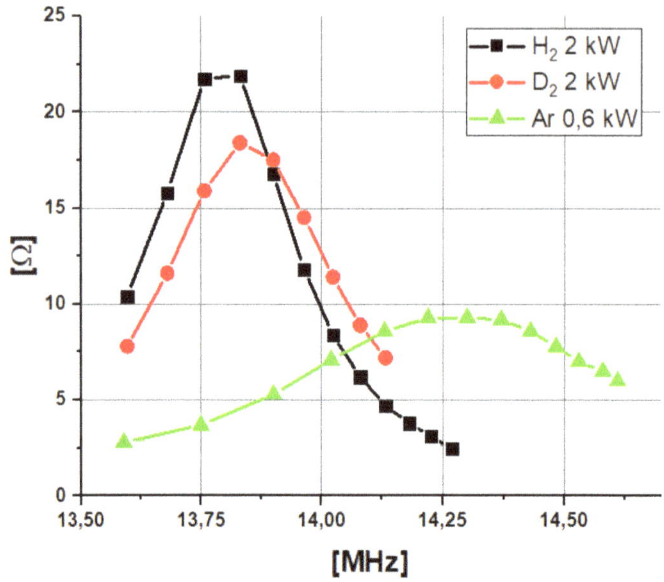

Figure 12.37. Measurements of the loaded birdcage impedance (real value) for different gases.

magnetization. As a matter of fact, without the external magnetic field, the fields are due only to inductive coupling by the antenna, and so are damped over a few skin depths, typically a few centimetres, and so the antenna does not interact with the rest of the column in this simulation. In terms of power transfer efficiency, the helicon regime shows an increase of about 5% compared to the inductively coupled regime, from typically 85% to 90%.

A second aspect concerns the influence of the gas type on the coupling. We show in figure 12.37 some impedance curves for H_2, D_2, and argon discharges, for the same magnetic field of 200 G. It can be seen that argon is indeed much more efficiently coupled to the antenna, showing lower quality factor and much higher frequency shifts, even for an RF power level typically three times lower than for H_2, D_2 discharges. Hence, when designing a resonant helicon source, the nature of the gas has to be taken into account, notably in order to anticipate the frequency shifts and to operate the coupled antenna as far as possible at its exact resonance. On RAID, this is the aim of the movable half screens (section 8.5.1 and 13.4.1) that surround the antenna, and enable its resonance frequency to be tuned over 1 MHz.

12.3 Planar helicon plasma source

Helicon sources are always associated with cylindrical geometry and, indeed, all the abundant literature on this topic describes results arising from the use of cylindrical sources, should they be ring type ($m = 0$), Nagoya, helical, double saddle, or even birdcage types. This leaves the pilgrim with the feeling that whistler wave heated discharges must intrinsically be of cylindrical symmetry.

This final section reports on the generation of whistler wave heated discharges using planar helicon sources (figure 12.38), a very different configuration from the helicon birdcage [30]. Planar resonant RF networks were previously described in part II, chapters 6 and 7, as inductively coupled sources of non-magnetized plasma. For the magnetized plasma in this part III, a static magnetic field is now applied perpendicular to the source plane, see figure 12.39. The RF source is placed in a large vacuum chamber, so the generated plasma is not bounded by a dielectric wall, similarly to the unbounded plasma column in RAID, section 12.2.

The plasma source in figure 12.39(a) is an $N = 11$ leg planar resonant network of size 20 cm by 20 cm [30]. For these experiments, the network capacitors were chosen to obtain a m = 2 resonance at 13.56 MHz. The antenna is placed in an open-top grounded metal box, and the empty spaces in the box below the antenna are filled with PTFE or silicone elastomer so that plasma is generated only in front of the antenna. A 1 mm thick glass plate was placed on the resonant network as a protective dielectric window between the plasma and the source.

This assembly is placed at one end of a large cylindrical vacuum chamber 1 m long, 70 cm in diameter, the plane of the antenna being perpendicular to the chamber axis in figure 12.39(b). A static magnetic field B_0 is applied using a pair of Helmholtz coils. Three types of diagnostics were installed to characterize the discharges and to evidence the excitation of electromagnetic waves. Ion saturation currents were measured on single tip Langmuir probes to obtain both axial and radial profiles of the plasma relative density. A 25 GHz interferometer was used to measure the absolute line-averaged electron density at a fixed distance of 25 cm from the source. Finally, a tri-axial B-dot probe was used to characterize the RF field along the chamber axis. All the results were obtained with a 30 sccm argon flow at an operating pressure of 1.1 Pa.

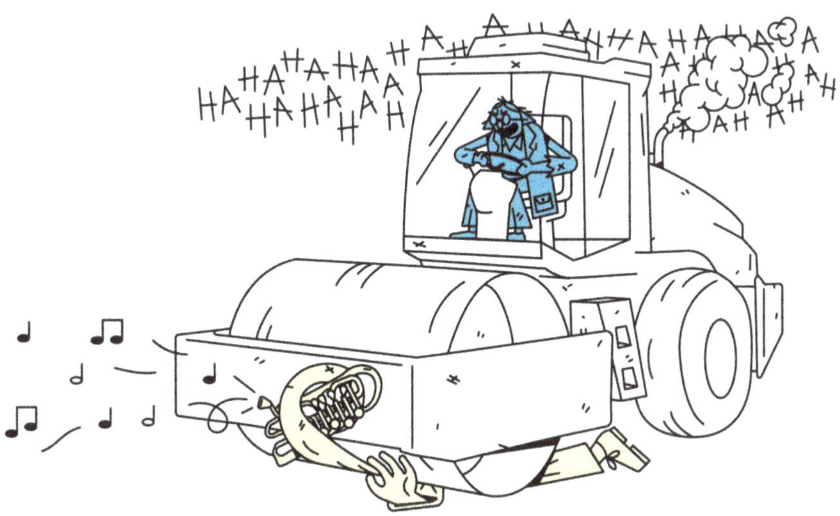

Figure 12.38. '#4: A planar helicon.' (Illustration by Alex Howling. Copyright 2023 Alex Howling.)

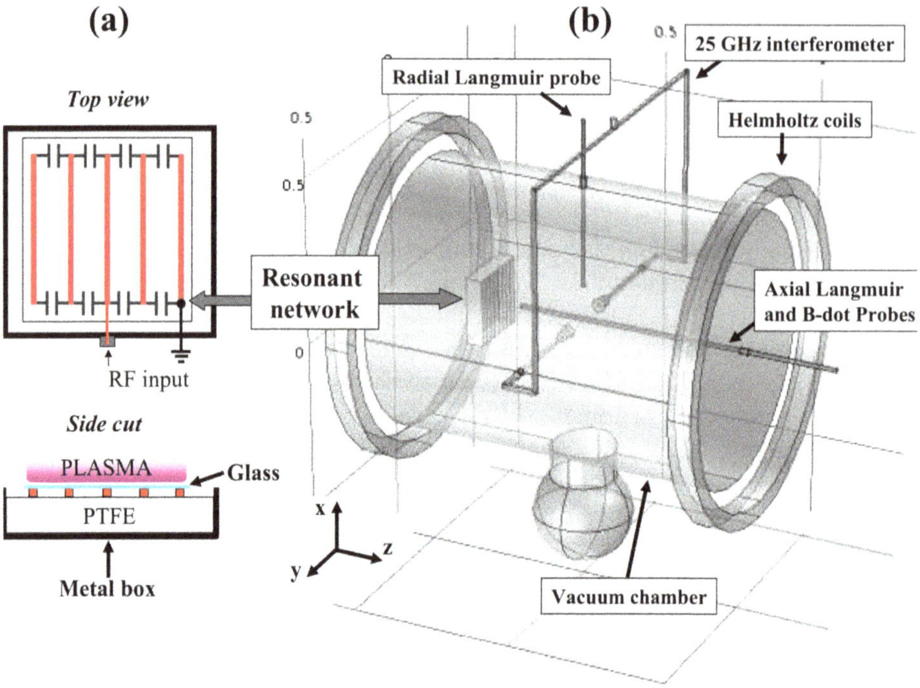

Figure 12.39. (a) Schema of the RF resonant network (only five legs represented). (b) Overall experimental set-up. A static magnetic field is applied perpendicular to the planar source by the Helmholtz coils. (Reproduced with permission from [30]. Copyright 2013 the American Physical Society.)

Figure 12.40 shows some typical ion saturation current profiles obtained with the axial Langmuir probe. Four of these profiles were measured at a fixed RF power (150 W) but for different static magnetic fields (B_0= 0, 13, 18, and 26 G), and the last profile is an example of a regime with higher RF power and field strength, 400 W and 34 G. At $B_0 = 0$ G, the plasma is non-magnetized and therefore generated by inductively coupling from the resonant network. Ionization mainly occurs in a narrow skin depth region a few centimeters from the source, and the plasma diffuses outwards into the chamber from there. When a small magnetic field is applied this diffusion profile spreads slightly along the z-axis because of electron confinement, up to a critical field value of about 15 G where a sudden regime transition is observed. This transition is characterized by an elongation of the axial density profile and is associated with the detection of RF magnetic fields far away from the source (see figures 12.41(c) and (d)). As shown in figure 12.40, with increasing RF power and magnetic field strength, flat axial density profiles extending typically 20 cm from the source up to the end of the vacuum chamber can be obtained. Diffusion from the skin depth region close to the antenna cannot account for such density profiles which, in fact, suggest that ionization persists quite far away from the RF planar network source, in a totally different way than it does for purely inductive ($B_0 = 0$ G) coupling.

A comparison of the transverse magnetic field patterns measured below ($B_0 = 0$ G) and above the regime transition ($B_0 = 18$ G) is shown in figure 12.41. The transition to

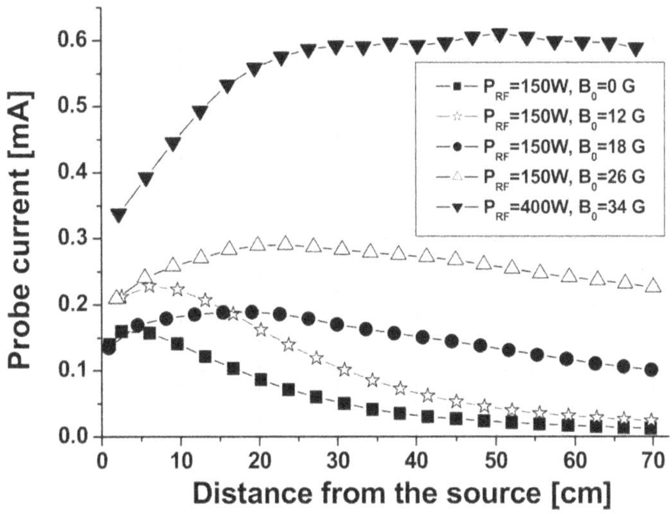

Figure 12.40. Axial ion saturation current profiles measured on a single tip (length 3 mm, diameter 0.6 mm) Langmuir probe biased at −30 V. (Reproduced with permission from [30]. Copyright 2013 the American Physical Society.)

a propagating wave regime is clear and the characteristics of these waves will be further described. In the inductive regime, figures 12.41(a) and (b), the field is linearly polarized along y and the spatial decay of B_y is well represented by an exponential, as expected in the framework of collisional damping. An estimation of the collision frequency for momentum transfer ν_m, using a Maxwellian electron energy distribution function with a 3 eV mean electron temperature in argon [28], leads typically to a 10 MHz collision frequency in argon at this pressure, which compared with $\omega = 85.2$ MHz indicates a moderately collisional plasma. The skin depth is measured here to be about 2 cm, which is 1.5 to 2 times shorter (for $\nu = 1$ MHz to 100 MHz) than expected according to a conventional 1D theoretical model of the field propagation into a collisional plasma [31], but it can be shown that effects due to the finite size of the plasma column can very well account for this reduced value.

The results of figures 12.40 and 12.41 were obtained for positive static field B_0, meaning oriented in the same direction as the wave vector k_z. In this case, we expect a right-handed polarization of the wave, which is effectively the case as shown in figures 12.42(a) and 12.43(a). Figure 12.42(a) shows the evolution of the measured polarization with the static field strength. For these measurements, the B-dot probe was placed 15 cm from the network, with a fixed RF power (300 W). Figure 12.43(a) shows the time evolution of the transverse field components (B_x, B_y), for the $B_0 = 60$ G case of 12.42. For $B_0 > 0$, the measured wave is reasonably circularly polarized, at least for the highest field values.

Interestingly, we observe a quite different behaviour for negative values of B_0, as illustrated by figures 12.42(b) and 12.43(b). First we note a left-handed polarization, as expected for k_z and B_0 in opposite directions. But furthermore, we see a very different evolution of the wave polarization with the intensity of the magnetic field in

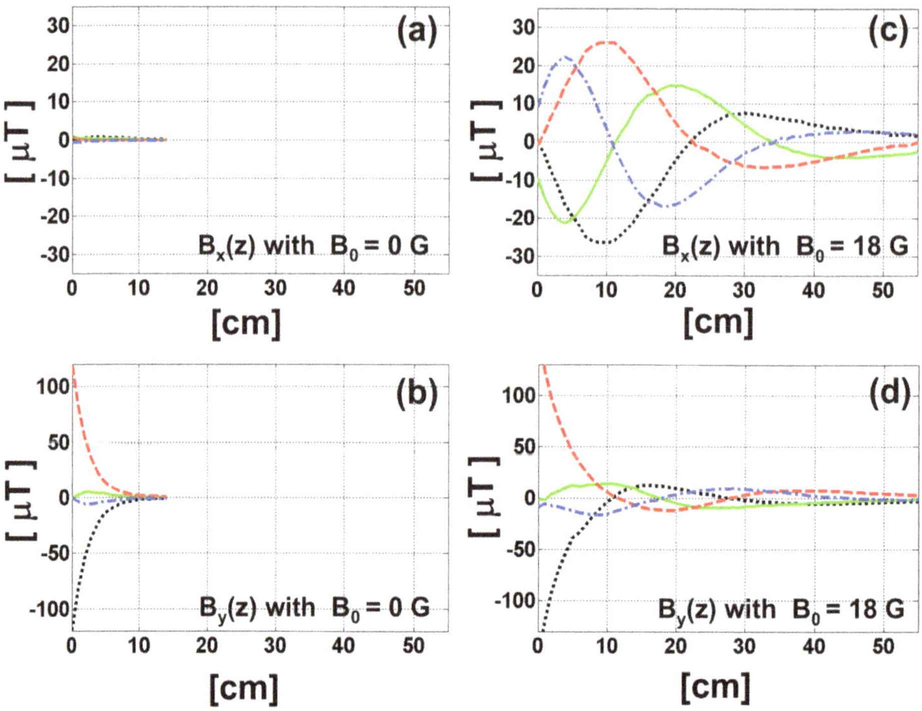

Figure 12.41. Measurements of the transverse field profile (components $B_x(z)$ and $B_y(z)$) along the chamber axis at four instants of an RF period τ ($t = 0$ (full line), $\tau/4$ (dotted line), $\tau/2$ (dash-dot line), $3\tau/4$ (dashed line)). The RF input power is 100 W. Graphs (a) and (b): $B_x(z)$, respectively, $B_y(z)$, without magnetic field (inductive coupling). Graphs (c) and (d) : $B_x(z)$, respectively, $B_y(z)$, with a 18 G static magnetic field. (Reproduced with permission from [30]. Copyright 2013 the American Physical Society.)

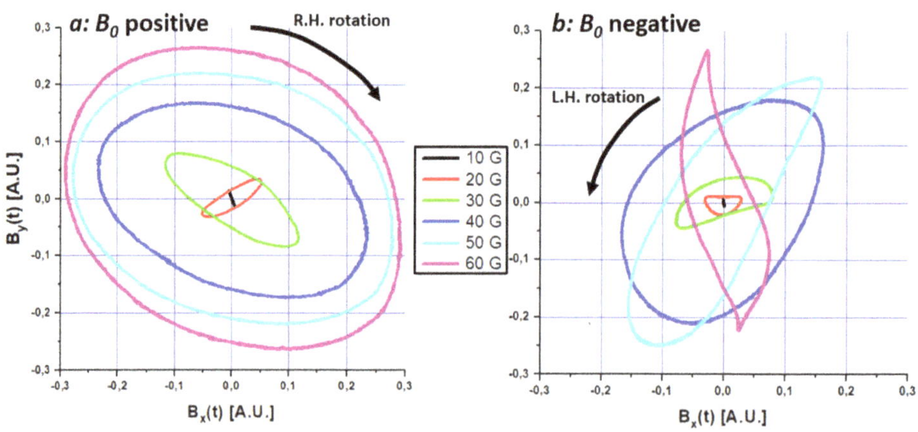

Figure 12.42. Comparison of the whistler wave polarization for (a) positive, and (b) negative static magnetic field B_0. The transverse field components are measured at a fixed distance from the source of 15 cm, for various values of B_0 (10–60 G).

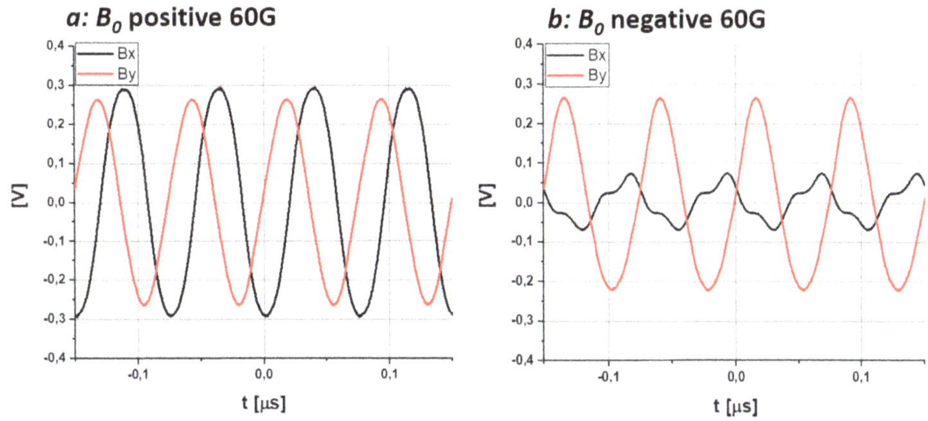

Figure 12.43. Temporal variations of the transverse field components for (a) positive, and (b) negative B_0, for the highest magnetic field value of figure 12.42 (60 G). The positive case shows a right-hand polarization (B_y leading B_x), while it is the opposite for $B_0 < 0$. The negative field case also exhibits a frequency doubling, which explains the very non-elliptic polarization shown in figure 12.42 observed in this case.

figure 12.42(b). Already for the small B_0 values, the wave appears barely elliptically polarized, with more bean-shaped Lissajou figures. For the highest value of 60 G, the polarization exhibits non sinusoidal variations, as can be seen in figure 12.43(b). A second harmonic frequency clearly appears in the measured signal, typically suggesting non-linear effects.

To end with the general properties of the measured waves, these clearly exhibit spatial helical structures, as expected for whistlers, with a left helical pitch for positive B_0, and right helical pitch for negative fields, as in the example of figure 12.44.

The radial density profiles, measured 25 cm away from the source, are peaked on the chamber axis, and can be approximated by a Gaussian with FWHM 25 to 30 cm (figure 12.45). These profile measurements combined with the line-averaged density measured with the 25 GHz interferometer at the same position are used to estimate the absolute density on the discharge axis. The peak plasma density as a function of the RF input power is shown in figure 12.46, for $B_0 = 16$ G and $B_0 = 32$ G. Densities up to $2 \cdot 10^{17}$ m^{-3} have been measured, and in the available power range the peak density varies approximately linearly with the input RF power level.

Finally, we have measured, as a function of the input RF power, the RF magnetic field wavelength λ_z along the z-axis, as well as its $1/e$ damping length δ_z (open symbols in figure 12.47). The linear fit n_{e0} (m^{-3}) $= 2.5 \cdot 10^{14} \times P_{\mathrm{RF}}(\mathrm{W})$ obtained for 16 G from figure 12.46 is used to set the horizontal scale in figure 12.47.

The full square symbols correspond to the expected values of λ_z and δ_z according to the simple relation of dispersion for planar waves (10.34). The only fitting parameter introduced here is the effective collision frequency ν, which negligibly affects λ_z but had to be set to typically 120 MHz in order to account for the measured damping constant δ_z, whereas for the 1.1 Pa operating pressure of this

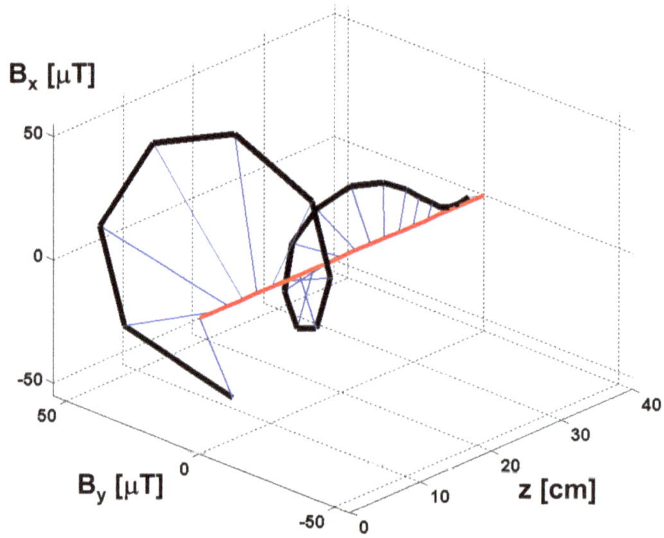

Figure 12.44. Transverse RF magnetic field (B_x, B_y) as a function of z for a given time. (Reproduced with permission from [30]. Copyright 2013 the American Physical Society.)

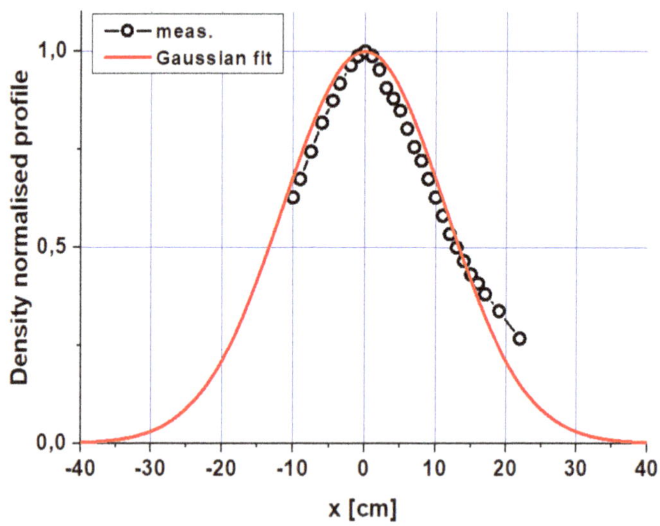

Figure 12.45. Radial density profile measured at 25 cm distance from the source plane by Langmuir probe (ion saturation current).

experiment, a 10–30 MHz electron–neutral collision frequency would instead be expected.

The star-shaped symbols correspond to wavelengths predicted by the {H,T–G} non-uniform model, taking the radial profile of figure 12.45, and assuming a cylindrical form for the plasma column. As also for the case of RAID, it is

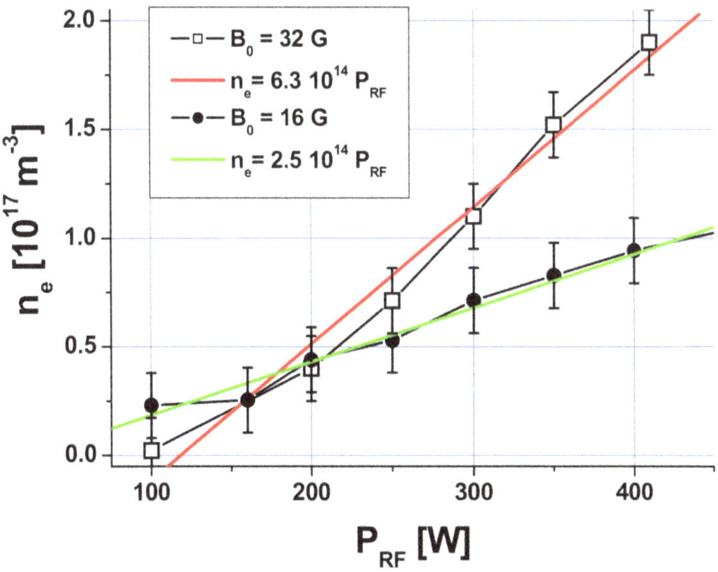

Figure 12.46. Electron peak density on the chamber axis at 25 cm from the RF network as a function of the RF input power (B_0 = 16 G and B_0 = 32 G). (Reproduced with permission from [30]. Copyright 2013 the American Physical Society.)

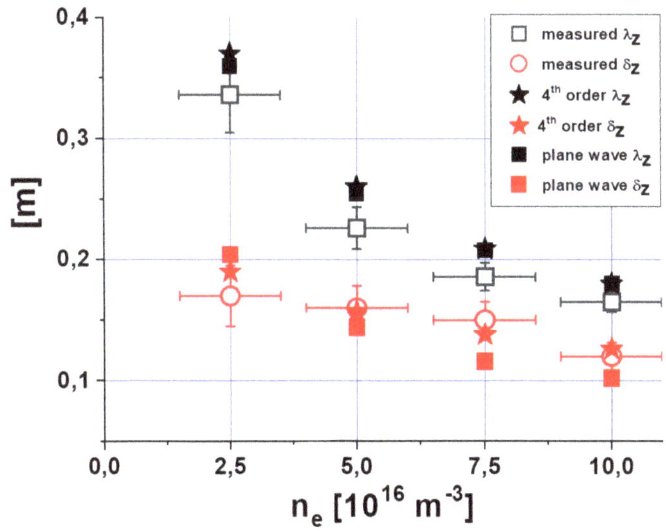

Figure 12.47. Open symbols: measurements of the wavelength $\lambda_z = 2\pi/\text{Re}(k)$ and damping constant $\delta_z = 1/\text{Im}(k)$ as a function of the peak electron density for B_0 = 16 G. Star symbols: calculations of λ_z and δ_z according to the {H,T–G} non-uniform model. Full squares: calculations of λ_z and δ_z according to the plane R-wave dispersion (10.34).

remarkable that the simplest and most sophisticated models both lead to almost the same results in terms of longitudinal wavelength predictions; see figure 12.25. The main difference concerns the damping prediction: the non-uniform model provides a noticeably better fit to experiment for the damping variation, but the most important fact is that the effective collision frequency has now to be set to only 50 MHz to justify for the level of damping. This is still a quite high value compared to the expected experimental value of 20–30 MHz, but more acceptable and potentially justifiable than 120 MHz. This remaining discrepancy could be explained either by invoking experimental uncertainties on the pressure measurement, for example, or due to a poor theoretical estimation of the effective electron collision frequency, for another example, by omitting electron–ion collisions.

12.4 Applications of birdcage helicon antennas

12.4.1 Application: helicon plasma thruster using a birdcage antenna

Based on a birdcage antenna, Romano *et al* [32] have designed an RF helicon plasma thruster for a space electric propulsion system. The electrodeless thruster shown in figure 12.48 includes a gas injector, a solenoid for the applied magnetic field up to 70 mT, and a discharge channel to accelerate the quasi-neutral RF plasma exhaust. The eight-leg birdcage operates at 40.68 MHz up to 1 kW. Advantages are that the thruster suffers no electrode erosion, and needs no on-board propellant nor plasma neutralizer. The scalable birdcage in figure 12.49 has a simple and efficient integrated matching circuit.

12.4.2 Application: negative ion sources using birdcage antennas

Beams of negative ions are of interest for a wide range of applications, with beam diameters from millimetres up to many centimeters. The smallest diagnostic beams are required in secondary ion mass spectroscopy (SIMS) for high resolution ion

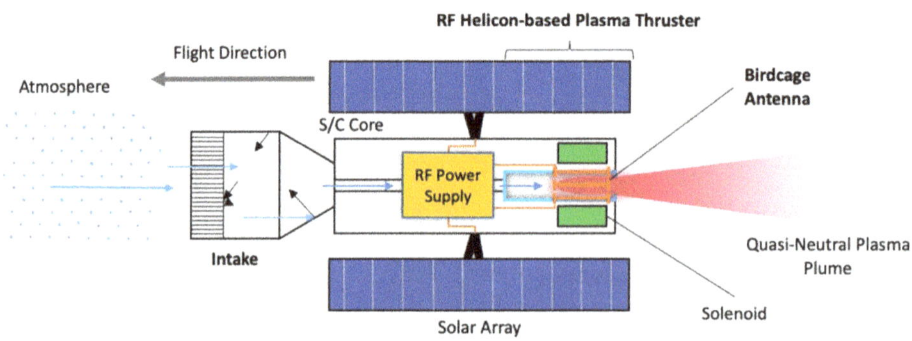

Figure 12.48. Schematic of an atmosphere-breathing electric propulsion system comprising a gas intake, a solar array for a solenoid and RF power supply, with a birdcage to generate a helicon plasma plume. (Reproduced with permission from [32]. Copyright 2020 Elsevier.)

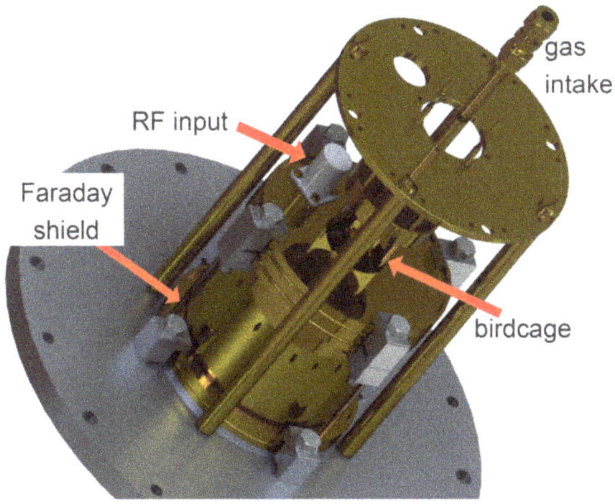

Figure 12.49. Engineering drawing of the helicon plasma thruster. The solenoid in [32] is removed to show the birdcage.

micro-probe isotope geochemistry. In this context, a negative atomic oxygen ion (O^-) source capable of long-term (days) steady-state operation is under development. The novel inductively coupled plasma source uses a birdcage antenna to produce positive or negative O^- beams. Changes in wall temperature and vacuum chamber dimensions influence the beam quality, indicating the critical role of surface reactions in the creation/destruction of negative oxygen ions [33].

RAID has also been extensively studied as a volume source for negative ions of atomic hydrogen and deuterium, H^- and D^-. Dedicated diagnostics have been applied, such as optical emission spectroscopy, cavity ring-down spectroscopy, and Langmuir probe-assisted laser photodetachment [9, 22]. H^- and D^- are principally concentrated in a halo around the plasma column with densities of $2 \cdot 10^{16}$ m^{-3} for only a few kW RF power in a Cs-free plasma. A hydrogen transport fluid code shows that RAID plasmas have a hot electron core favourable for ro-vibrational excitation and dissociation of H_2 molecules [34]. Indeed, dissociative attachment to ro-vibrationally excited H_2 molecules is the only significant source of H^- anywhere in the RAID volume.

Ultimately, the use of helicon sources in neutral beam injectors for fusion will crucially depend on the possibility of extracting negative ions from the plasma column. In the present configuration, radial extraction would entail a plasma grid in close proximity to the negative ion halo near 4 cm radius, with attendant issues of perturbation to the plasma column and modifications to the vacuum vessel. Furthermore, the plasma column would need to be uniform, and the ion extraction system would need pass between the magnetic field coils. Preliminary negative ion radial extraction experiments are presently underway on RAID.

12.5 Chapter summary for wave heated discharges

A resonant birdcage antenna was shown to be a suitable source for bounded and unbounded helicon plasmas, generating an intense, stable plasma column in argon or hydrogen up to 10 kW of continuous RF power. The magnetic wave field measured along the plasma axis exhibited the helical structure and polarization consistent with helicon waves. The dependence of axial wavelength on the plasma density and external DC magnetic field was smooth and continuous, showing no preferred dependence on the antenna length, nor the standing wave structure and no abrupt changes in the helicon mode structure or in the plasma density. Furthermore, heavy or light ions (Ar^+ or H_2^+) had no notable influence on the axial wavelength, which depended only on the electron density and the DC magnetic field as expected according to helicon theory.

The measured axial wavelengths were compared with a uniform plasma plane wave model, a non-uniform cylindrical second order 'helicon approximation' model, and a fourth order non-uniform helicon model including Trivelpiece–Gould modes—the {H,T–G} theory of chapter 11. All three analytical approaches compare well with a numerical simulation. The power density profile and magnetic field structure of the numerical simulation are reproduced by the {H,T–G} multimodal theory.

Planar resonant networks are introduced as a novel configuration for helicon wave-sustained plasma sources. The {H,T–G} theory is in fair agreement with the measured wavelength and $1/e$ damping length. Finally, two applications were briefly described, for plasma thrusters and negative ion sources.

In conclusion for part III, the birdcage and planar helicon sources present a new alternative to conventional partial-helix antennas, with advantages of wider parameter space and improved plasma stability, especially for light ion (hydrogen) plasmas.

References

[1] Boswell R W 1970 A study of waves in gaseous plasmas *PhD Thesis* School of Physical Sciences, Flinders University of South Australia

[2] Boswell R W 1970 Plasma production using a standing helicon wave *Phys. Lett.* A **33** 457

[3] Boswell R W 1984 Very efficient plasma generation by whistler waves near the lower hybrid frequency *Plasma Phys. Control. Fusion* **26** 1147

[4] Guittienne Ph, Chevalier E and Hollenstein Ch 2005 Towards an optimal antenna for helicon waves excitation *J. Appl. Phys.* **98** 083304

[5] Furno I, Agnello R, Fantz U, Howling A A, Jacquier R, Marini C, Plyushchev G, Guittienne Ph and Simonin A 2017 Helicon wave-generated plasmas for negative ion beams for fusion *EPJ Web Conf.* **157** 03014

[6] Marini C *et al* 2017 Spectroscopic characterization of H_2 and D_2 helicon plasmas generated by a resonant antenna for neutral beam applications in fusion *Nucl. Fusion* **57** 036024

[7] Jacquier R *et al* 2019 First B-dot measurements in the RAID device, an alternative negative ion source for DEMO neutral beams *Fusion Eng. Des.* **146** 1140

[8] Agnello R *et al* 2020 Negative ion characterization in a helicon plasma source for fusion neutral beams by cavity ring-down spectroscopy and Langmuir probe laser photodetachment *Nucl. Fusion* **60** 026007
[9] Agnello R 2020 Negative hydrogen ions in a helicon plasma source *PhD Thesis* no. 7817, Ecole Polytechnique Federale Lausanne (EPFL) Switzerland
[10] Agnello R *et al* 2018 Cavity ring-down spectroscopy to measure negative ion density in a helicon plasma source for fusion neutral beams *Rev. Sci. Instrum.* **89** 103504
[11] Agnello R, Andrebe Y, Arnichand H, Blanchard P, De Kerchove T, Furno I, Howling A A, Jacquier R and Sublet A 2020 Application of Thomson scattering to helicon plasma sources *J. Plasma Phys.* **86** 905860306
[12] Kinder R L, Ellingboe A R and Kushner M J 2003 H- to W-mode transitions and properties of a multimode helicon plasma reactor *Plasma Sources Sci. Technol.* **12** 561
[13] Sharma N, Chakraborty M, Neog N K and Bandyopadhyay M 2018 Development and characterization of a helicon plasma source *Rev. Sci. Instrum.* **89** 083508
[14] Isayama S, Shinohara S and Hada T 2018 Review of helicon high-density plasma: production mechanism and plasma/wave characteristics *Plasma Fusion Res.: Rev. Articles* **13** 1101014
[15] Shinohara S 2022 *High-Density Helicon Plasma Science* (Singapore: Springer Nature) (Springer Series in Plasma Science and Technology)
[16] Perry A J, Vender D and Boswell R W 1991 The application of the helicon source to plasma processing *J. Vac Sci. Technol.* B **9** 310
[17] Guittienne Ph, Jacquier R, Pouradier Duteil B, Howling A A, Agnello R and Furno I 2021 Helicon wave plasma generated by a resonant birdcage antenna: magnetic field measurements and analysis in the RAID linear device *Plasma Sources Sci. Technol.* **30** 075023
[18] Piotrowicz P A, Caneses J F, Green D L, Goulding R H, Lau C, Caughman J B O, Rapp J and Ruzic D N 2018 Helicon normal modes in Proto-MPEX *Plasma Sources Sci. Technol.* **27** 055016
[19] Shoji T, Sakawa Y, Nakazawa S, Kadota K and Sato T 1993 Plasma production by helicon waves *Plasma Sources Sci. Technol.* **2** 5
[20] Caneses J F and Blackwell B D 2016 Collisional damping of helicon waves in a high density hydrogen linear plasma device *Plasma Sources Sci. Technol.* **25** 055027
[21] Sakawa Y, Takino T and Shoji T 1999 Contribution of slow waves on production of high-density plasmas by m = 0 helicon waves *Phys. Plasmas* **6** 4759
[22] Furno I *et al* 2023 Helicon volume production of H^- and D^- using a resonant birdcage antenna on RAID *Physics and Applications of Hydrogen Negative Ion Sources* ed M Bacal (Cham: Springer) (Springer Series on Atomic, Optical, and Plasma Physics) vol 124 ch 9
[23] COMSOL Inc https://www.comsol.com [Accessed 31 January 2024]
[24] Chen F F 2016 *Introduction to Plasma Physics and Controlled Fusion* 3rd edn (Cham: Springer)
[25] Lieberman M A and Lichtenberg A J 2005 *Principles of Plasma Discharges and Materials Processing* 2nd edn (Hoboken, NJ: Wiley)
[26] Swanson D G 2003 *Plasma Waves* (Bristol: Institute of Physics Publishing) (Series in Plasma Physics) 2nd edn
[27] Thompson D, Agnello R, Furno I, Howling A A, Jacquier R, Plyushchev G and Scime E E 2017 Ion heating and flows in a high power helicon source *Phys. Plasmas* **24** 063517

[28] Hagelaar G J M and Pitchford L C 2005 Solving the Boltzmann equation to obtain electron transport coefficients and rate coefficients for fluid models *Plasma Sources Sci. Technol.* **14** 722

[29] Advanced Energy Industries Inc http://www.advanced-energy.com [accessed 31 January 2024]

[30] Guittienne Ph, Howling A A and Hollenstein Ch 2013 Generation of whistler-wave heated discharges with planar resonant RF networks *Phys. Rev. Lett.* **111** 125005

[31] Lieberman M A and Godyak V A 1998 From Fermi acceleration to collisionless discharge heating *IEEE Trans. Plasma Sci.* **26** 955

[32] Romano F *et al* 2020 RF helicon-based inductive plasma thruster (IPT) design for an atmosphere-breathing electric propulsion system (ABEP) *Acta Astronaut.* **176** 476

[33] Stoffels E, Stoffels W W, Kroutilina V M, Wagner H-E and Meichsner J 2001 Near-surface generation of negative ions in low-pressure discharges *J. Vac Sci. Tech.* A **19** 2109

[34] Agnello R, Fubiani G, Furno I, Guittienne P, Howling A, Jacquier R and Taccogna F 2022 A 1.5 D fluid–Monte Carlo model of a hydrogen helicon plasma *Plasma Phys. Control. Fusion* **64** 055012

Part IV

Technology, future developments, and appendices

IOP Publishing

Resonant Network Antennas for Radio-Frequency Plasma Sources

Theory, technology and applications

Philippe Guittienne, Alan Howling and Ivo Furno

Chapter 13

Technology of resonant network antennas

The purpose of this penultimate chapter is to enter into some practical considerations for the design and construction of a resonant plasma source. It is quite difficult to be completely general in this matter, because the design requirements depend on the final goal to be reached. For example, there is naturally a certain difference in terms of technical challenges between a source dedicated to operate at low pressure inside a vacuum chamber, as in section 13.4.2, or outside the vacuum chamber at atmospheric pressure (section 13.4.1). Second, the power level is a key parameter, notably with regard to heat management on the plasma side, but also for dimensioning the whole RF system elements, from the network's capacitor assemblies to the matching system. Finally, the network size is also critical for the design of the source, especially due to the level of RF potentials that can rise to high values for large area antennas.

This technology chapter is divided into section 13.1 on RF impedance matching, section 13.2 on RF capacitor assemblies, dimensioning of the RF elements in section 13.3, mechanical construction of antennas for high and low gas pressure in section 13.4, closing with high Q design in section 13.5.

13.1 Impedance matching of resonant network antennas

Almost all commercial RF power supplies are built to present a purely real, 50 Ω output impedance. Briefly, this means that such a generator delivers its RF power to a given load without reflection only if the load input impedance is also purely real at 50 Ω. Clearly, this is most certainly not the case for conventional circuits used for RF plasma sources—see section 5.5. For example, a three-turn ICP coil would typically exhibit an input impedance of $Z_{\text{in}} = (1 + j500)\ \Omega$, almost purely reactive, so that practically all of the RF power would be reflected back to the RF generator if it was directly connected to the coil. Therefore, for proper use of RF power supplies,

doi:10.1088/978-0-7503-5296-3ch13

the load input impedance must be adapted to bring it to a real, 50 Ω value. This 'impedance matching' is achieved by inserting an arrangement of purely reactive components (capacitors and inductors) between the generator and the load; it is commonly called the 'matching system' or 'matching box'.

13.1.1 Two element matching

Many commercial matching systems for plasma sources are based on a two-element arrangement, schematized in figure 13.1, in which Z_L is the load impedance and the reactive (imaginary) impedances jb and jc are the matching elements. However, this type of matching is generally not suitable for resonant network plasma sources, as we show below, because their impedance properties are very different from those of conventional ICP or CCP. Using basic relations for serial and parallel associations of impedances, the equivalent impedance Z_{in} for the arrangement sketched in figure 13.1 can be expressed as

$$Z_{\text{in}} = \left(\frac{1}{jb} + \frac{1}{jc + Z_L}\right)^{-1}. \tag{13.1}$$

Taking $Z_L = R + jX$, the real and imaginary parts of Z_{in} are

$$\begin{aligned} R_{\text{in}} &= \frac{b^2 R}{R^2 + (b + c + X)^2} \\ X_{\text{in}} &= \frac{b(R^2 + (c + X)(b + c + X))}{R^2 + (b + c + X)^2}. \end{aligned} \tag{13.2}$$

Defining the generator output impedance R_g to be real, the matching conditions $R_{\text{in}} = R_g$ and $X_{\text{in}} = 0$ lead, according to (13.2), to a pair of solutions for the two-element arrangement:

$$\begin{aligned} b_{1,2} &= \mp \frac{RR_g}{\sqrt{R(R_g - R)}}, \\ c_{1,2} &= \pm \sqrt{R(R_g - R)} - X. \end{aligned} \tag{13.3}$$

Negative values for b or c are obtained in practice by using capacitors, while positive values correspond to inductors. It can be seen from (13.3) that this matching solution is limited by the condition $R < R_g$, since b and c must be real. For conventional ICP or

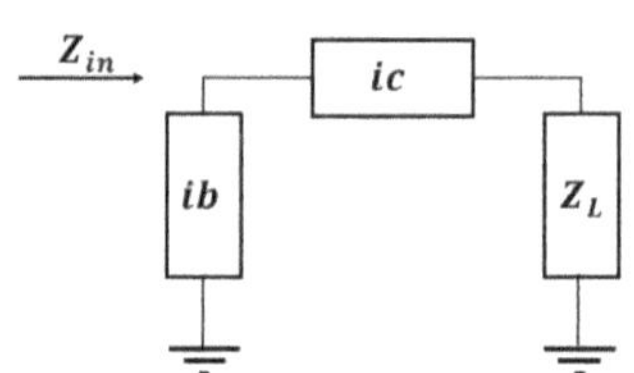

Figure 13.1. A two-element impedance matching circuit.

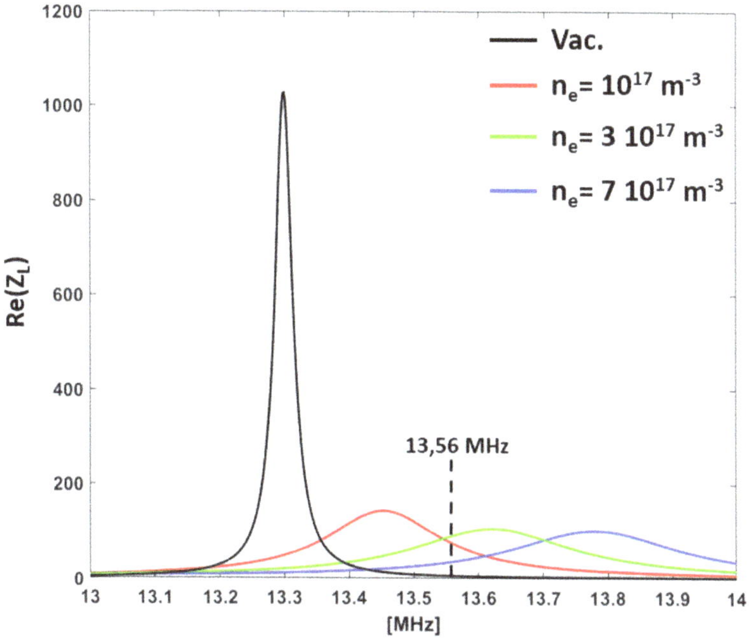

Figure 13.2. Real impedance model spectra for a nine-leg, 1 m long antenna, operated at 13.56 MHz in three different plasma densities, mode m = 5.

CCP, the real part of the reactor impedance, even when coupled to plasma, is always much smaller than $R_g = 50\ \Omega$, so this two-element system is well suited. This is generally not true for resonant networks which are characterized by a high and real input impedance in the vicinity of the resonance frequency. Furthermore the input impedance of a resonant plasma source varies quite strongly with the plasma coupling, first due to the lowering of the system Q factor, and second, due to the resonance frequency shift linked to plasma/antenna mutual inductance, as illustrated in figure 13.2.

Considering the example of figure 13.2, which is based on the measurements and models in figures 6.11, 7.13, and 7.14, the 'impedance path' to be matched during an RF power ramp at 13.56 MHz, starting with ignition at low plasma density and ending with a plasma density of $7 \cdot 10^{17}\ m^{-3}$, would typically be as shown in figure 13.3. X changes from negative to positive as the plasma load impedance curve traverses the resonance at 13.56 MHz, consistent with figure 2.4. It can be seen that both R and X vary strongly, and that a two-element matching system cannot be used because R_{in} crosses the 50 Ω limit.

13.1.2 T matching

To be able to cross the 50 Ω limit, three-element matching must be considered. Two configurations can be used, as illustrated in figure 13.4, so-called T and Π matching circuits. In the following we will focus only on the T solution, which is practically always applicable. Following the same procedure as for the two-element system, the T arrangement input impedance can be directly written from (13.2) as

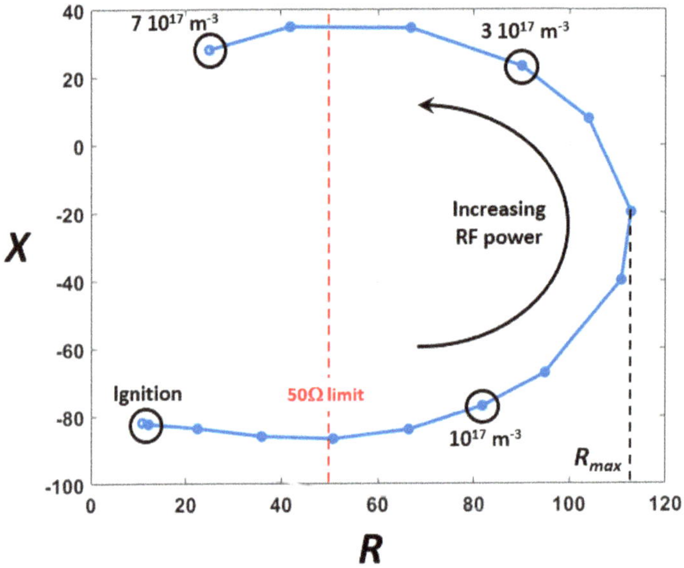

Figure 13.3. A representative antenna impedance path as the plasma power and plasma density increases, for a fixed frequency of 13.56 MHz.

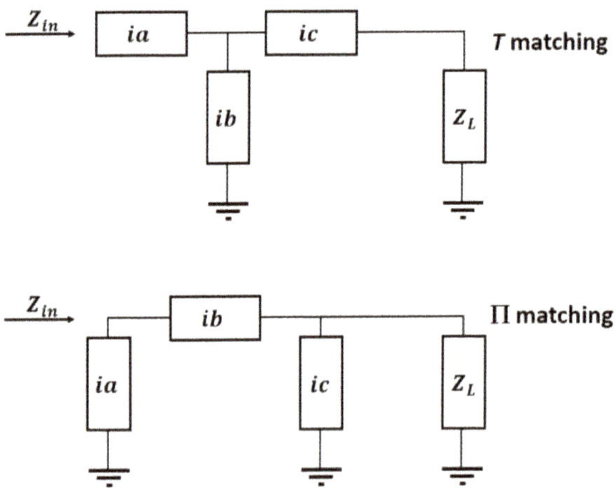

Figure 13.4. T and Π three-element matching circuits.

$$
\begin{aligned}
R_{\text{in}} &= \frac{b^2 R}{R^2 + (b + c + X)^2} \\
X_{\text{in}} &= \frac{b(R^2 + (c + X)(b + c + X))}{R^2 + (b + c + X)^2} + a,
\end{aligned}
\tag{13.4}
$$

where a is the third reactive element.

Two matching conditions are to be applied, $R_{\text{in}} = R_g$ and $X_{\text{in}} = 0$, while three matching elements are available. As a consequence, we can fix the impedance of one of these and use the two matching conditions to set the remaining ones. In practice, this means that one of the matching box elements can be either a fixed capacitor or a fixed inductor, while the other two elements are variable, to follow the load impedance variations.

From the expressions of R_{in} and X_{in} in (13.4), it can notably be seen that tuning the element a affects only X_{in}. Therefore, it is technically better to fix either b or c rather than a. In fact, if a is fixed, the tuning of b or c changes R_{in} and X_{in} simultaneously in (13.4), which significantly complicates the algorithms required for an automatic feedback control of the matching condition, as well as manual operation. If instead we choose to fix b, for example, then R_{in} is affected only by changes in the element c, which is then decoupled from the tuning of the imaginary part X_{in}, obtained by acting on element a.

For the following, we therefore set a fixed value for the element b. Solving for the same matching conditions $R_{\text{in}} = R_g$ and $X_{\text{in}} = 0$, with (13.4), we obtain another pair of solutions for the variable impedances a and c:

$$\begin{aligned} a_{1,2} &= \frac{-bR \pm \sqrt{RR_g(b^2 - RR_g)}}{R}, \\ c_{1,2} &= \frac{-R_g(b + X) \pm \sqrt{RR_g(b^2 - RR_g)}}{R_g}, \end{aligned} \tag{13.5}$$

which is self-consistent with the two-element solution (13.3) when a is set to zero. Because a and c must be real numbers, we first note from (13.5) that the value for b is constrained by the relation $b^2 - RR_g > 0$ which requires that $|b| > \sqrt{RR_g}$. This constraint only concerns the absolute value of the impedance b, so this element can be either a fixed capacitor or inductor. To set an appropriate value for b, one has to consider the maximum real impedance $R_{\max}$ to be matched. Considering the example of figure 13.3, the maximum value for R is $R_{\max} = 155\ \Omega$, so to satisfy the condition $|b| > \sqrt{RR_g}$ over the whole impedance path we must have $|b| > 76\ \Omega$, assuming that $R_g = 50\ \Omega$. If we choose the b element to be a fixed capacitor of capacitance C_b, its impedance $b = -1/(\omega C_b)$ should then be less than $-76\ \Omega$, which roughly gives $C_b < 150$ pF at 13.56 MHz. If the b element is chosen to be an inductor of inductance L_b, its impedance $b = \omega L_b$ should then be higher than 76 Ω, which gives $L_b > 900$ nH. The choice between an inductive or a capacitive element is mainly practical, depending on power, currents, and simply on the feasibility of the component itself. For example, if the necessary C_b capacitance were to fall in the 1–10 pF range, which is the order of magnitude of the stray capacitance in a metal box, it would be necessary to choose an inductor.

For the purposes of a quantitative analysis of currents and voltages, we now introduce an arbitrary, but realistic, relation between plasma density and RF power, as sketched in figure 13.5. We also set the fixed b element to be a 100 pF capacitor, easily allowing a matching solution over the full R range at 13.56 MHz. The element

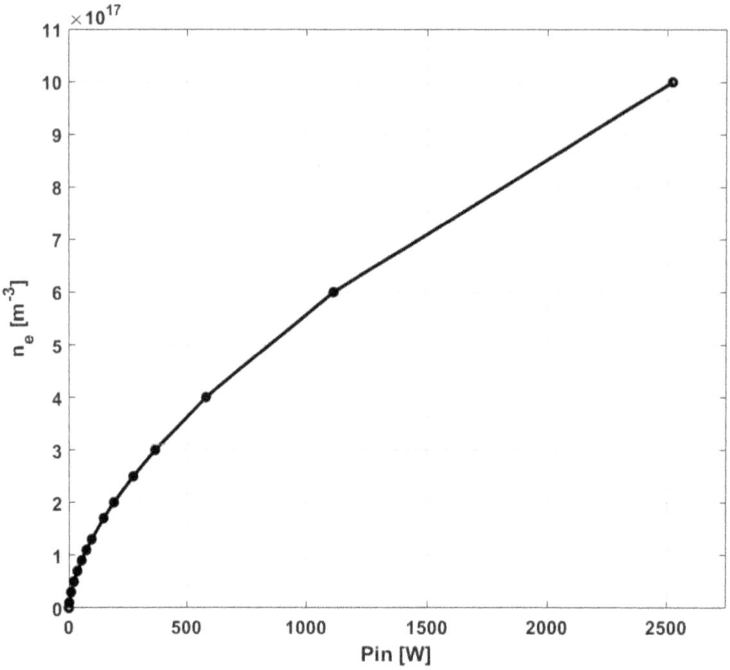

Figure 13.5. An arbitrary dependence of the plasma density on RF power.

impedances *a* and *c* required by (13.5) to match the whole impedance path of figure 13.3 are plotted in figure 13.6 as functions of the arbitrary RF input power. We now show how to choose between these two solutions, and how to design the physical hardware needed to provide these variable impedances.

In practice, a variable reactive impedance is almost always made by a series combination of a fixed inductor and a variable capacitor. The fixed inductor determines a positive offset value for the element total impedance, while the capacitor adds a variable negative contribution. Since the impedance of a capacitor is inversely proportional to its capacitance (figure 13.7), the range of impedance over which the variable element can be tuned depends essentially on the lower part of the capacitor range. Some variable capacitors can go down to a few picofarads (pF), but considering the unavoidable stray capacitance with the matching box housing, a 25 pF minimum value is more realistic. An upper capacitance value for variable capacitors is commonly a few hundred pF. Therefore, considering figure 13.7, we can roughly say that at 13.56 MHz, we can easily tune a matching element over typically 400 Ω. The required impedance variations for the elements *a* and *c* shown in figure 13.6 are slightly less than 200 Ω, which is well within the variable capacitor's range of impedance.

To define the value of the inductor that has to be connected in series with the variable capacitor, for a given matching element *a* or *c*, the simplest method is to consider first the minimum impedance to be reached for the given element. For example, taking the *a* element for the first matching solution in figure 13.6, the minimum necessary impedance

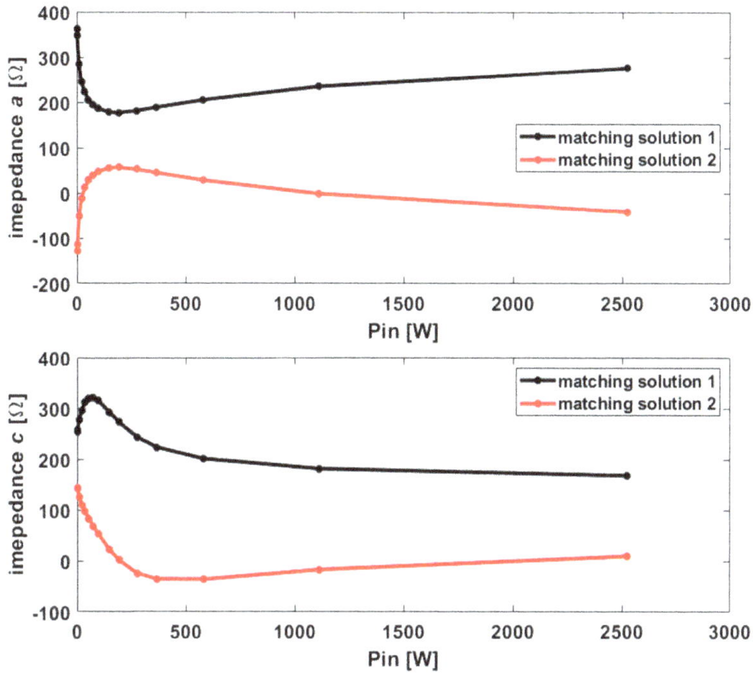

Figure 13.6. Variable impedances a and c for both matching solutions in (13.5), corresponding to the impedance path of figure 13.3.

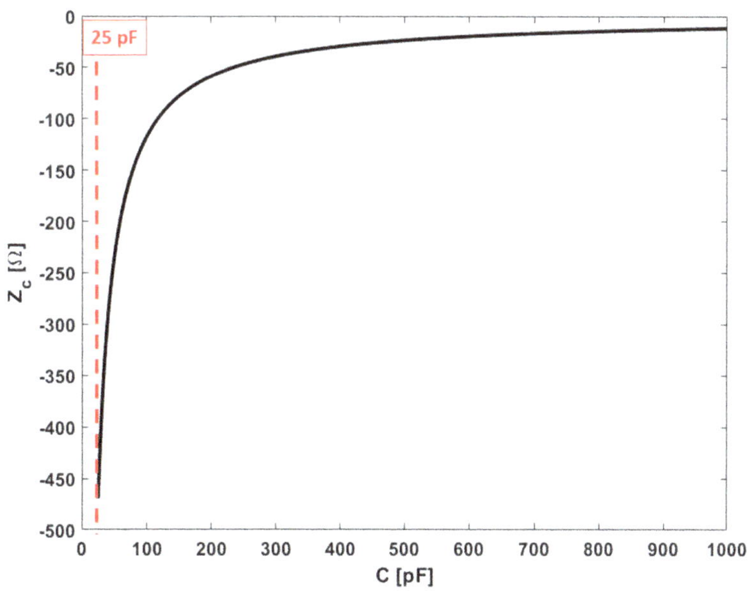

Figure 13.7. Impedance of a variable capacitor of capacitance C, at 13.56 MHz.

is about 170 Ω. This value should be obtained when the capacitor is in its lowest capacitance range, say, typically 30 pF, which roughly corresponds to an impedance of −390 Ω at 13.56 MHz. Therefore, the inductor impedance must be at *most* 170 + 390 = 560 Ω, which corresponds to an inductance of about 6.6 μH.

A minimum value for this inductor can now be determined by considering the maximum impedance to be reached by the *a* element, which is about 370 Ω (solution 1 for *a* in figure 13.6). This impedance should be obtained when the variable capacitor is in its high capacitance range. A typical value would be 250 pF, which corresponds roughly to a −50 Ω impedance at 13.56 MHz. Then, the inductor impedance must be at *minimum* of 370 + 50 = 420 Ω, which corresponds to an inductance of about 4.9 μH.

Finally, for the impedance path of figure 13.3, the matching solutions can be summarized as shown in figure 13.8, which presents the evolution of the $a_{1,2}$ and $c_{1,2}$ variable capacitances during the arbitrary power ramp, the value of the associated fixed inductor being given for each case.

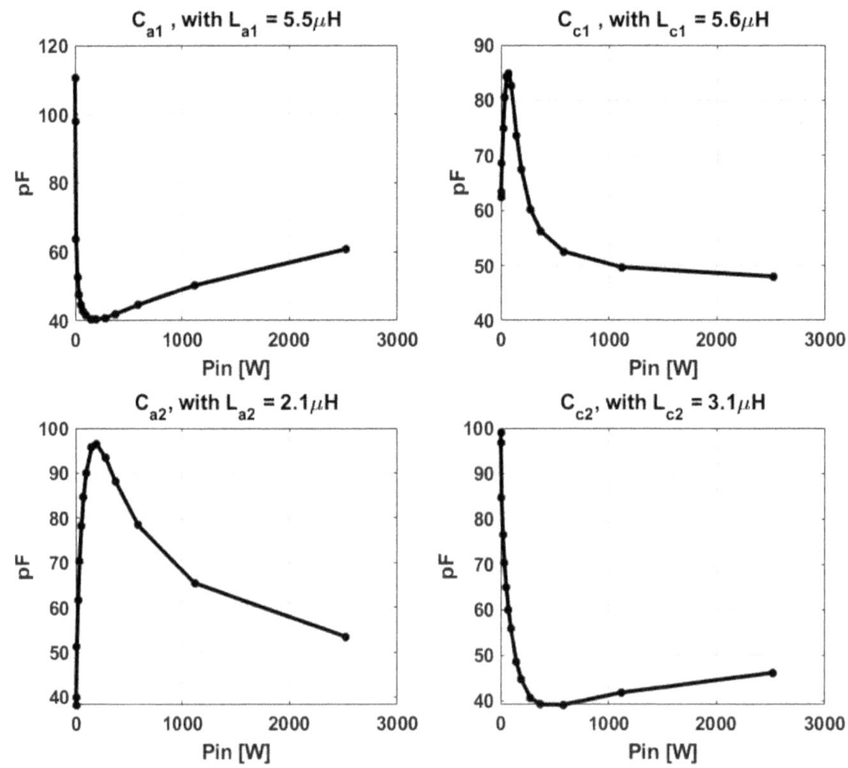

Figure 13.8. Example solutions for the variable matching capacitors C_a and C_c, combined with fixed inductors L_a and L_c, according to solutions 1 and 2 for *a* and *c* in (13.5), for the impedance path given as an example in figure 13.3.

Considering a given input power P_{in}, once the system is matched, simple and useful expressions for the voltages and currents (see figure 13.9) can be established. The matched system input rms voltage and current are

$$V_{50\Omega} = \sqrt{R_g P_{\text{in}}}, \quad I_{50\Omega} = \sqrt{P_{\text{in}}/R_g}, \tag{13.6}$$

where $R_g = 50\ \Omega$, and the load input rms voltage and current are

$$V_{\text{load}} = \sqrt{\frac{R^2 + X^2}{R} P_{\text{in}}}, \quad I_{\text{load}} = \sqrt{P_{\text{in}}/R}. \tag{13.7}$$

These results can be interpreted in terms of the high currents in CCP large area reactors, and the high voltages and currents in large area ICP systems, previously discussed in section 5.5.

The only difference between the two matching solutions is found in the potential and current expressions across the b element in figure 13.9. For solution 1 of (13.5):

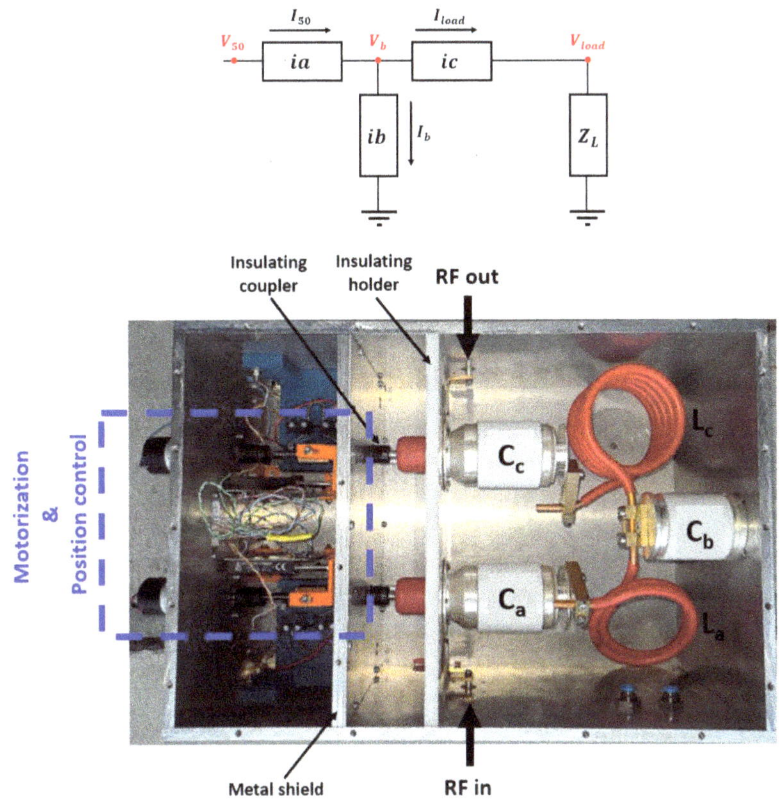

Figure 13.9. Equivalent circuit for estimating voltages and currents in matching conditions, and a practical realization of a corresponding matching box.

$$V_b = \sqrt{\frac{|b|(R_g + R) + 2\sqrt{R_g R(|b|^2 - R_g R)}}{R_g R}}\sqrt{|b|P_{\text{in}}},$$
$$I_b = \sqrt{\frac{|b|(R_g + R) + 2\sqrt{R_g R(|b|^2 - R_g R)}}{R_g R}}\sqrt{P_{\text{in}}/|b|}. \tag{13.8}$$

For solution 2 of (13.5):

$$V_b = \sqrt{\frac{|b|(R_g + R) - 2\sqrt{R_g R(|b|^2 - R_g R)}}{R_g R}}\sqrt{|b|P_{\text{in}}},$$
$$I_b = \sqrt{\frac{|b|(R_g + R) - 2\sqrt{R_g R(|b|^2 - R_g R)}}{R_g R}}\sqrt{P_{\text{in}}/|b|}. \tag{13.9}$$

It can be seen that the current I_b and potential V_b are expected to be lower for the second solution, due to the minus sign preceding the $2\sqrt{R_g R(|b|^2 - R_g R)}$ terms in the above expressions. Indeed, applied to our illustrative example, one finds $V_b = 1050$ V and $I_b = 9$ A for a 1 kW input power for solution 1, whereas the second solution leads to $V_b = 230$ V and $I_b = 2$ A. Then it is far preferable to opt for the second matching solution, and this is a quite general result. Another practical reason for why the second matching solution is preferable is to be found in the values for the fixed inductors indicated in figure 13.8. It can be seen that the first matching solution requires larger inductances, typically twice as big as for the second one. The point is that inductors are quite cumbersome parts to fit into matching boxes—see figure 13.9—and the reduced footprint of the second matching solution is highly preferable for practical reasons.

13.2 Capacitor assemblies for high RF power antennas

Since water-cooled copper tubes are essentially indestructible inductors, the onus is on the capacitors to withstand the high RF resonance currents circulating within an antenna network. The RF capacitors are the weak link in a resonant network antenna, limiting the maximum usable RF power before catastrophic overheating and explosive failure occurs. A good design of the capacitor assemblies is therefore one of the fundamental keys for achieving a reliable resonant plasma source. These assemblies must be realized with individual high Q, high power capacitors. Similarly to the quality factor for an inductor (2.9), the quality factor Q_C of a capacitor is defined as the ratio of its reactive impedance magnitude $Z_C = \frac{1}{\omega C}$ at the operating frequency, to its resistance [1]:

$$Q_C = 1/(\omega C R), \tag{13.10}$$

because the higher the value of Q_C, the better the approximation to an ideal lossless capacitor. R is the capacitor's ESR value which is the sum of its dielectric losses and resistive losses in the skin depth of its metal elements. This demands the highest quality porcelain ceramic dielectrics with low-loss tangent [1], and multi-layer soldered construction with non-magnetic, low resistivity metal leads [2].

At the time of writing, in 2023, ceramic and mica capacitors appear to be the best, and almost only, choices to build these assemblies. Typically, mica and ceramic capacitors with a maximum current at 13.56 MHz of about 15 A rms and a maximum voltage of 2 kV rms can easily be found on the market. Naturally, before designing the capacitor assembly and the RF system in general, it is necessary to estimate the expected maximum levels of current and voltage in the desired plasma source, using the successive models presented earlier in the book. If the capacitor assemblies are well dimensioned, they can operate for a long time (typically more than 1.5 year of continuous operation warranted by manufacturers) and will not heat up even in a low pressure plasma environment.

13.2.1 Series and parallel assemblies

An assembly is obtained by series and parallel associations of individual capacitors, as in figure 13.10. In this example, the assembly is made up of two parallel branches, each one incorporating three identical individual capacitors. All of the capacitors being equal, the assembly input current I is equally distributed between the two branches, so an individual capacitor has only to withstand one half of the total current. Additionally, in a given branch, the potential drop across an individual capacitor represents here only a third of the total assembly voltage ΔV. A practical example of a single stack assembly is shown in figure 13.11. Multiple capacitors in such assemblies can be mixed and matched to obtain high precision tolerance (1%) of the combined capacitance across all of the antenna stringers.

To generalize, we set out to design a source for which the maximum current expected in a capacitor assembly is $I_{\max}$, and the maximum voltage across the assembly is $\Delta V_{\max}$. We further consider this assembly to be made of identical individual capacitors that can withstand a maximum current $i_{\max}$ and a maximum voltage drop $\delta V_{\max}$. The minimum number of branches n for a safe assembly is then given by $n > I_{\max}/i_{\max}$, and the minimum number p of individual capacitors in each branch is given by $p > \Delta V_{\max}/\delta V_{\max}$. The assembly capacitance C_{ass} consists of np individual capacitances C_{ind} given by the simple relation $C_{\mathrm{ass}} = C_{\mathrm{ind}} n/p$.

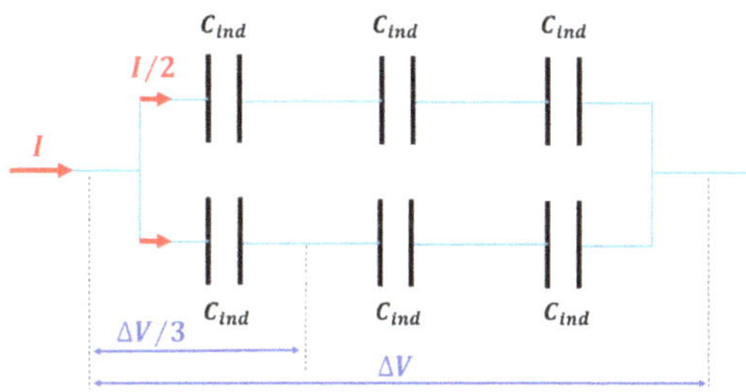

Figure 13.10. An example of current sharing and voltage division in a capacitor assembly.

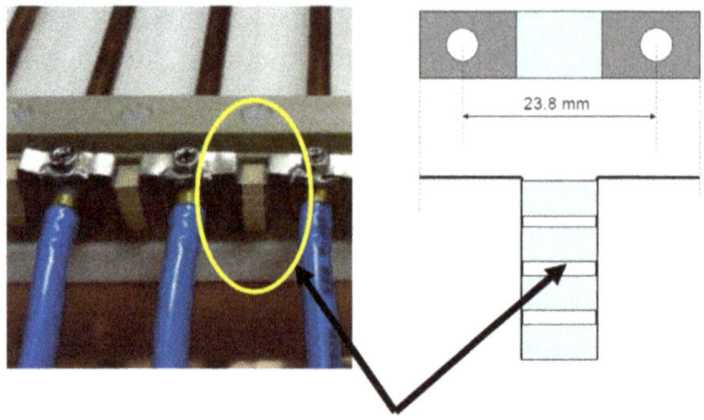

Figure 13.11. The photograph and sketch show a stringer capacitor assembly, $C = 2.6$ nF, consisting of four stacked, high Q RF capacitors, taken from point (d) in figure 3.1.

In practice, mica and ceramic capacitors have similar performance for capacitance above 400 pF or so. When the value of the individual capacitors is in this range, it is better to choose mica components for the assemblies, as these are cheaper and more rapidly available. But below 400 pF, the maximum current rating in mica capacitors falls rapidly to low values, typically down to around 1 A for a 100 pF capacitance. Ceramic capacitors do not suffer from this limitation, and so are preferable for individual capacitor values below about 400 pF. Naturally, the need for small individual capacitance values is primarily for high power sources, because high current assemblies require a large number n of branches, and secondarily to large networks for which the necessary assembly capacitance itself can be easily lower than 400 pF.

RF capacitors are the most fragile and expensive items in a resonant network antenna, since the rest of the antenna consists essentially of custom-built parts in copper and brass. Expensive capacitors with long delivery times are manageable for prototype antennas, but there is a natural motivation to manufacture all components in-house. In the case of the high-pressure antenna, section 13.5, the legs have much higher inductance because of the additional series inductance L_S, and therefore only need very small capacitor values of the order of 10^1 pF in order to resonate at, say, 13.56 or 40.68 MHz. For such small values it is conceivable to build capacitors as robust as the leg inductors by using parallel metal plates whose capacitance $C = \varepsilon_0\varepsilon_r A/d$ depends on the surface area A of a few cm^2 and dielectric thickness ~0.1–1 mm, for example. Preliminary tests nevertheless indicate that thermal expansion and self-resonance effects (due to inductance associated with the current flow over the capacitor plates) could be problematic, even for such a basic construction.

13.3 Dimensioning the RF system

One of the main parameters for designing a plasma source is the maximum level of RF power to be used. With regard to the RF system, this determines the choice and design of components for the matching circuit, the connecting cables, and the antenna itself. In this section, our aim is to give an overview of the analysis of a matched RF system to evaluate the level of currents and voltages in its different parts, and to dimension these for safe operation.

A typical RF system is sketched in figure 13.12. It can be seen that we allow for the possible use of a coaxial cable of length D for connecting the matching box to the antenna—this becomes necessary when the antenna is mounted inside a vacuum vessel, with the matching system outside. In the following, we consider the RF antenna only from the point of view of its input impedance, which is assumed to be known by measurement or well estimated from models.

13.3.1 The coaxial line

The coaxial transmission line in figures 13.12 and 13.13 consists of an inner conductor of radius r_{int} on the axis of an external conductor of radius r_{ext}. Both conductors are taken to have the same electrical conductivity σ_{m}. The space between them is filled with a dielectric material of relative permittivity ε_r and, to be general, electrical conductivity σ_d, although the latter can generally be neglected. The coaxial cable has a characteristic impedance Z_{c} which can be written as [1]

$$Z_{\text{c}} = \sqrt{\frac{\hat{R} + j\omega\hat{L}}{\hat{G} + j\omega\hat{C}}}, \tag{13.11}$$

where $\hat{R}$, $\hat{L}$, $\hat{G}$, and $\hat{C}$ are the per-unit-length resistance, inductance, conductance, and capacitance of the line. These can be expressed for a coaxial line as follows [1]:

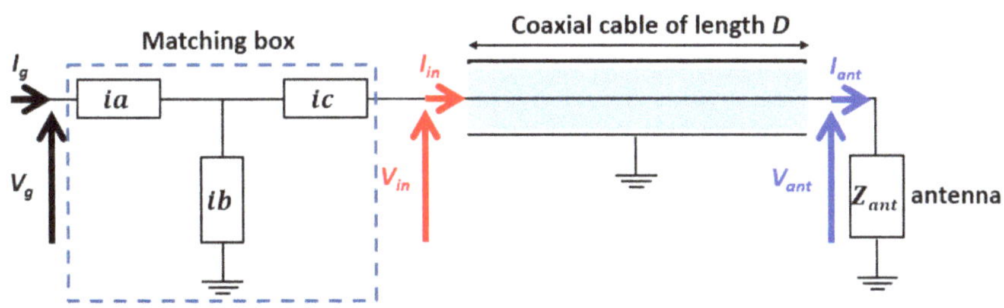

Figure 13.12. General schematic of an RF system for a plasma source of input impedance Z_{ant}.

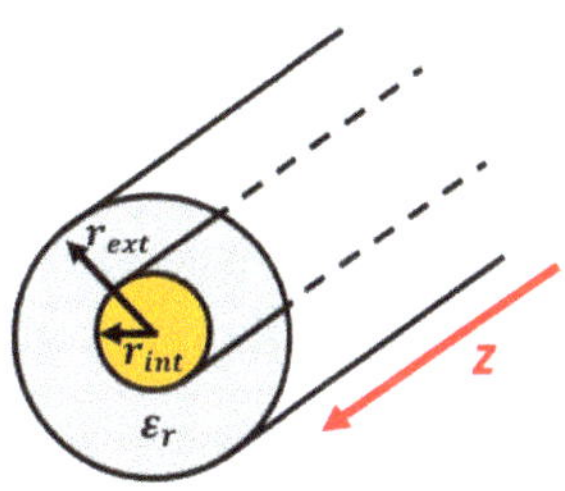

Figure 13.13. Schematic representation of a coaxial transmission line.

$$
\begin{aligned}
\hat{R} &= \frac{1}{2\pi\sigma_{\mathrm{m}}\delta}\left(\frac{1}{r_{\mathrm{int}}}+\frac{1}{r_{\mathrm{ext}}}\right), \\
\hat{L} &= \frac{\mu_0}{2\pi}\ln\left(\frac{r_{\mathrm{ext}}}{r_{\mathrm{int}}}\right), \\
\hat{G} &= \frac{2\pi\sigma_d}{\ln\left(\frac{r_{\mathrm{ext}}}{r_{\mathrm{int}}}\right)}, \\
\hat{C} &= \frac{2\pi\varepsilon_0\varepsilon_r}{\ln\left(\frac{r_{\mathrm{ext}}}{r_{\mathrm{int}}}\right)},
\end{aligned}
\tag{13.12}
$$

where $\delta = \sqrt{\frac{2}{\omega\mu_0\sigma_{\mathrm{m}}}}$ is the frequency-dependent skin depth for the line conductors at frequency $f = \frac{\omega}{2\pi}$. Note that a real characteristic impedance is possible only for a lossless cable ($\hat{R} = \hat{G} = 0$). Indeed, the characteristic impedance given for commercial cables, usually 50 Ω, assumes lossless transmission lines. By neglecting $\hat{R}$ and $\hat{G}$, we obtain

$$
Z_0 = \sqrt{\frac{\hat{L}}{\hat{C}}} = \frac{1}{2\pi}\sqrt{\frac{\mu_0}{\varepsilon_0\varepsilon_r}}\ln\left(\frac{r_{\mathrm{ext}}}{r_{\mathrm{int}}}\right), \tag{13.13}
$$

which relates the cable insulator's relative permittivity ε_r, and the inner and outer radii, to the real characteristic impedance Z_0.

Following the same analysis as in section 7.3 yields the voltage $V(z)$ and current $I(z)$ of a lossy transmission line along z:

$$
\begin{aligned}
V(z) &= e^{-\Gamma z}V^{+} + e^{\Gamma z}V^{-}, \\
I(z) &= \frac{1}{Z_{\mathrm{c}}}\left(e^{-\Gamma z}V^{+} - e^{\Gamma z}V^{-}\right),
\end{aligned}
\tag{13.14}
$$

where V^+ and V^- are the phasor amplitudes of the forward and backward waves, to be determined by boundary conditions at $z = 0$ (the line input) or at $z = D$ (the antenna input). The propagation factor is $\Gamma = \sqrt{(\hat{R} + j\omega\hat{L})(\hat{G} + j\omega\hat{C})}$.

13.3.2 Impedance matching at the cable input

We now consider the impedance matching of the antenna to the RF generator output impedance R_g, which is almost always 50 Ω. We set the antenna input impedance to be $Z_{\mathrm{ant}} = R_{\mathrm{ant}} + jX_{\mathrm{ant}}$. The transmission line transforms this impedance so that the input impedance Z_{in} at the transmission line input at $z = 0$ is [1]:

$$
Z_{\mathrm{in}} = Z_{\mathrm{c}}\frac{Z_{\mathrm{ant}} + Z_{\mathrm{c}}\tanh\Gamma D}{Z_{\mathrm{c}} + Z_{\mathrm{ant}}\tanh\Gamma D}, \tag{13.15}
$$

which is the impedance to be matched to the generator impedance $R_g = 50\ \Omega$.

A three-element T-matching circuit with three purely imaginary impedances ja, jb, jc (as sketched in figures 13.4, 13.9 and 13.12) is often used for resonant networks, and the matching condition is achieved as described in section 13.1, specifically for the preferred second solution in (13.5), namely

$$\begin{aligned} a &= \frac{-bR_{\text{in}} - \sqrt{R_{\text{in}}R_g(b^2 - R_{\text{in}}R_g)}}{R_{\text{in}}}, \\ c &= \frac{-R_g(b + X_{\text{in}}) - \sqrt{R_{\text{in}}R_g(b^2 - R_{\text{in}}R_g)}}{R_g}, \end{aligned} \tag{13.16}$$

for $b^2 > R_{\text{max}}R_g$, where R_{max} is the maximum value of R_{in} that can be matched.

We recall that, for conventional ICP antennas (typically solenoids), the real impedance R_{ant} is always very small ($<1\ \Omega$), allowing a two-element matching ($a = 0$), with preferred matching conditions given by the first solution of (13.3)

$$\begin{aligned} b &= -\frac{R_{\text{in}}R_g}{\sqrt{R_{\text{in}}(R_g - R_{\text{in}})}}, \\ c &= \sqrt{R_{\text{in}}(R_g - R_{\text{in}})} - X_{\text{in}}. \end{aligned} \tag{13.17}$$

13.3.3 Potentials and currents

For a three-element matching circuit, the potential V_{in} and current I_{in} at the matching circuit output (i.e. the transmission line input in figure 13.12) are given by

$$\begin{aligned} V_{\text{in}} &= \left[\left(1 + \frac{c}{b}\right)(R_g - a) - c\right]I_g, \\ I_{\text{in}} &= \left[1 - \frac{R_g - a}{b}\right]I_g, \end{aligned} \tag{13.18}$$

where I_g is the generator output rms current, which is straightforwardly given by $I_g = \sqrt{(P_g/R_g)}$ when matching conditions are achieved, for an RF power P_g delivered by the generator. For a two-element matching circuit, V_{in} and I_{in} are obtained from (13.18) by setting $a = 0$ in the expressions.

To be explicit, the matching system currents and voltages in figure 13.14 are given by

$$\begin{aligned} I_a &= I_g, \\ I_b &= \left(\frac{R_g - a}{b}\right)I_g, \\ I_c &= I_{\text{in}}, \end{aligned} \tag{13.19}$$

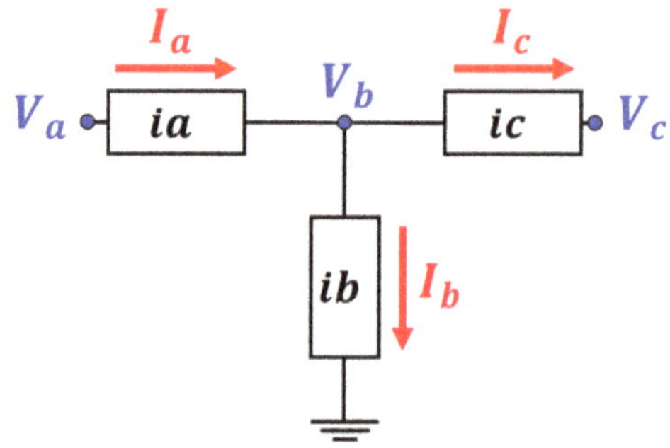

Figure 13.14. Definition of the variables in a three-element matching circuit used in (13.19) and (13.20).

and

$$\begin{aligned} V_a &= V_g, \\ V_b &= (R_g - a)I_g, \\ V_c &= V_{\text{in}}, \end{aligned} \tag{13.20}$$

remembering that $I_g = \sqrt{(P_g/R_g)}$ and $V_g = \sqrt{(R_g P_g)}$ when the system is matched.

The matching calculation relies on the assumption of purely reactive impedances (ja, jb, jc) for the matching elements. Of course, in real life, these always present a certain resistance $R_{(a,b,c)}$, but they are very small compared to the reactive impedances, so that the currents given by (13.19) are almost unaffected by introducing the element resistances into the calculation. The power $P_{(a,b,c)}$ dissipated in the matching elements can therefore be estimated correctly by the relation $P_{(a,b,c)} = R_{(a,b,c)} \cdot I^2_{(a,\, b,\, c)}$, where $I_{(a,b,c)}$ are given by (13.19). Naturally, the resistances $R_{(a,b,c)}$ have first to be well estimated, either from product datasheets, calculations, or directly by measurements, when possible, at 13.56 MHz.

V_{in} and I_{in}, (13.18), are the two boundary conditions needed to determine the V_+, V_- amplitudes of the RF wave along the coaxial line, which is done by setting $V(z = 0) = V_{\text{in}}$ and $I(z = 0) = I_{\text{in}}$ in (13.14). This leads to

$$\begin{aligned} V_+ &= \frac{1}{2}(V_{\text{in}} + Z_0 I_{\text{in}}), \\ V_- &= \frac{1}{2}(V_{\text{in}} - Z_0 I_{\text{in}}). \end{aligned} \tag{13.21}$$

The rms potential and current at the antenna input is finally determined from (13.14) with $z = D$. Once the potential and current are known, the time-averaged delivered power is expressed at any point by the relation

$$\bar{P} = \frac{1}{2}\left[VI^* + V^*I\right] = \text{Re}(VI^*). \tag{13.22}$$

We now undertake four example studies concerning the design of RF systems: (i) matching of resonant networks compared with classical solenoid ICPs in section 13.3.4; (ii) the choice of coaxial cable in section 13.3.5; (iii) the question of the number of antenna legs in section 13.3.6; and (iv) choosing the antenna mode number in section 13.3.7.

13.3.4 Study #1: resonant networks versus solenoid ICPs

One of the advantages of resonant networks lies in the fact that quite long transmission lines can often be used to connect them to the matching system. This is even possible in a low pressure environment, provided that the voltage levels in the system remain reasonably low. In the following, we apply the above formalism to justify this advantage which is entirely due to the impedance properties of resonant networks.

On the one hand, we consider the resonant source of figure 13.15, which is a cylindrical antenna, 15 cm long, producing plasma by inductive coupling in a 10 cm diameter tube. Note that in the framework of resonant networks, this can be considered as a 'small' antenna. On the other hand, we consider a solenoid antenna of four turns for comparison, extending over 10 cm, to be wrapped around the same discharge tube.

A typical measurement of the birdcage input impedance as a function of RF input power (0.5–5 kW) is shown below in figure 13.16. The variation of Z_{ant} with power is essentially due to the variation of the antenna resonance frequency with plasma density. It can be seen that, for this example, the birdcage resonance frequency coincides exactly with the 13.56 MHz of the RF generator for a power slightly below 1500 W. In the following we will consider a purely real input impedance for the birdcage case, of typically 20 Ω.

For the solenoid, the main contribution to R_{ant} comes from plasma coupling, leading to typical values of about 1 Ω. The inductance of the four-turn solenoid is about 3.5 μH, which, at 13.56 MHz, gives an imaginary part X_{ant} for this antenna of roughly 300 Ω, which is quite high compared to R_{ant}.

For the coaxial transmission line, we take copper conductors ($\sigma_{\mathrm{m}} = 5 \cdot 10^7$ S m^{-1}), with a PTFE dielectric ($\varepsilon_r = 2$, $\sigma_d \sim 10 - 20$ S m^{-1}), which is typical for low-loss cables. We set the inner conductor to be 2 mm diameter ($r_{\mathrm{int}} = 1$ mm), which gives a

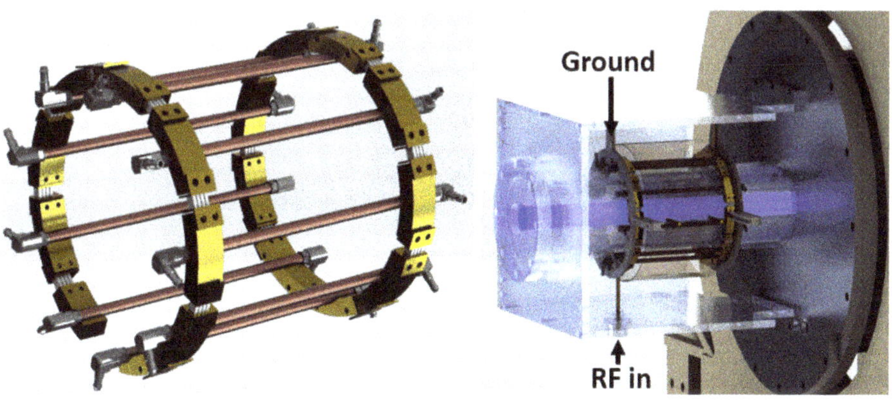

Figure 13.15. An open birdcage antenna (RAID, chapter 12) designed for high power operation (10 kW). The nine-leg antenna is 15 cm long, 13 cm in diameter, surrounding a water-cooled ceramic tube. This source is dedicated principally to helicon wave excitation, with mode m = 2.

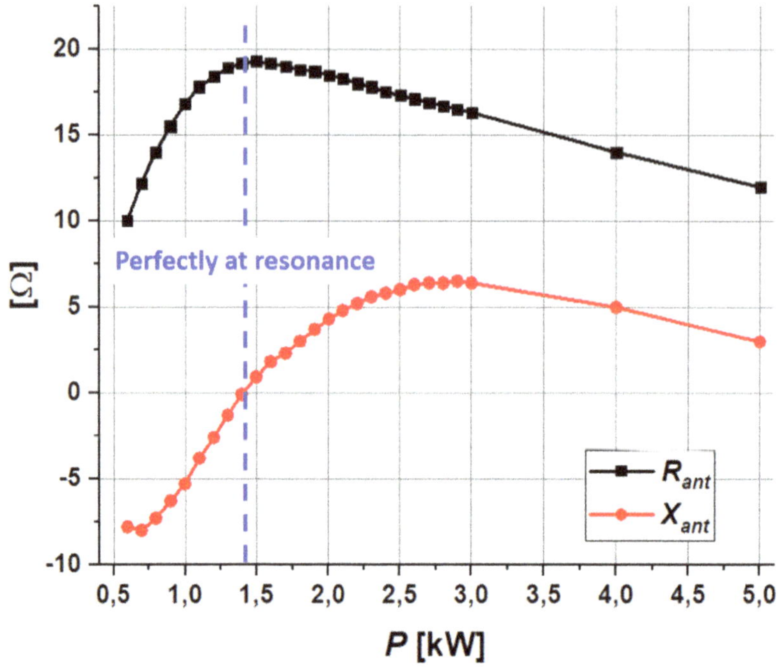

Figure 13.16. Dependence of the measured input impedance, Z_{ant}, on the RF input power of the birdcage antenna (figure 13.15), in inductive mode (no external magnetic field).

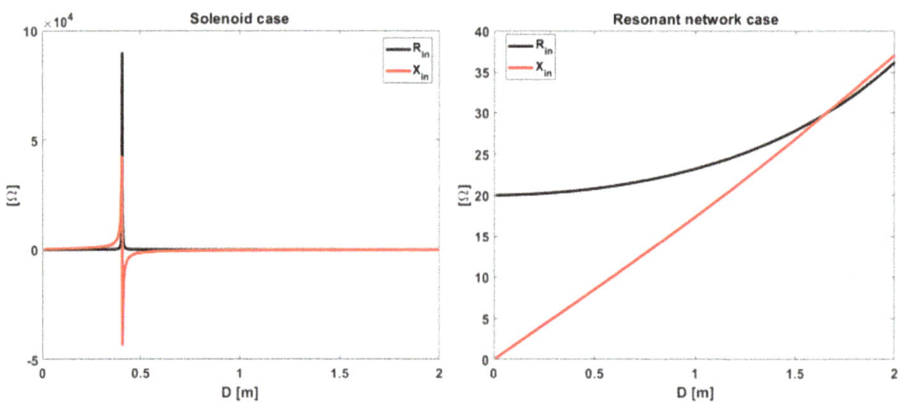

Figure 13.17. Dependence of the impedance to be matched (Z_{in}) on the coaxial cable length D, for the solenoid ICP case (left) and birdcage antenna (right).

diameter of about 7 mm for the outer conductor (r_{ext} = 3.5 mm) to obtain a 50 Ω characteristic impedance Z_0 according to (13.13) (this typically corresponds to coaxial cables of type RG 213).

To begin the comparison between the two systems, we show in figure 13.17 the evolution of the impedance Z_{in} (13.15) with the length D of the transmission line. It can be seen that a parallel resonance appears in the solenoid case for a cable length of about 0.4 m. Close to this resonance, it is technically impossible to match the system because the matching elements would be unachievable in practice. Then, with regard to matching, for the solenoid case one could imagine to use either a short coaxial cable, with a maximum length of typically 30 cm for the present example, or a sufficiently long one of at least 50 cm. For the case of the resonant network, the antenna impedance transformed by the transmission line varies much more gradually.

We continue our comparison by considering the voltages V_{in} and V_{ant} expected at the two ends of the transmission line, as shown in figure 13.18, for a moderate RF input power of 1 kW. We once again (see, for example, section 5.5) touch on probably the most important difference between the two types of plasma source—it can be seen that the cable voltage expected for the solenoid case is considerably higher than for the network. A 9.5 kV voltage, as obtained for small cable lengths, would require very special high voltage cables and connectors to prevent arcing. In contrast, the voltages for the birdcage network are typically 65 times lower, about 140 V only for short cable lengths. Things seem to improve for the solenoid case with long cables because the voltage levels fall significantly, but unfortunately this is not for good reasons, and the use of long transmission lines is hopeless for the solenoid ICP, as explained in the next paragraph.

Figure 13.19 shows the evolution of the power transfer ratio P_{ant}/P between the matching box and the antenna, as a function of the cable length. All the power which is not transmitted to the antenna is, in fact, dissipated by heating of the transmission line. With a resonant network, typically 1% of the power would be dissipated in a 1 m long cable, but ten times more with a solenoid antenna. This is already problematic and strongly limits the usable RF power level to a few hundred watts due to overheating of the cable. For a 2 m long cable, the situation becomes dramatic with more than 60% of the RF power lost into the cable for the solenoid, while it remains at only 2% with the RF birdcage network.

We see that, even for relatively small systems, using a transmission line to transfer the RF power from the matching box to a solenoid antenna is extremely

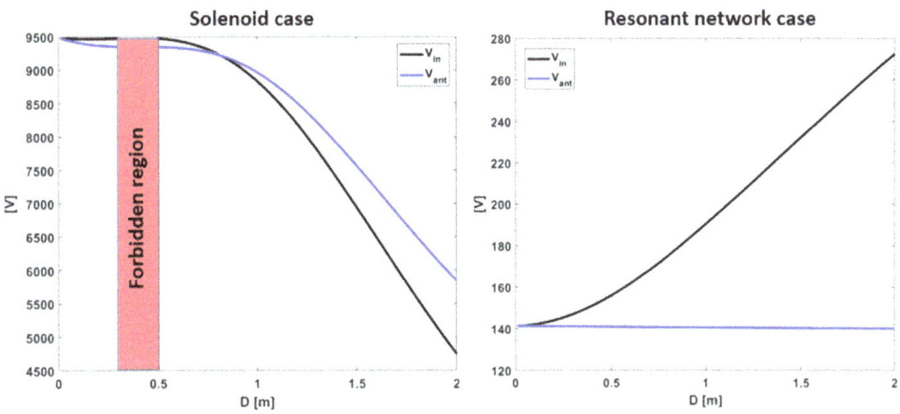

Figure 13.18. Voltages at the coaxial cable ends, V_{in} and V_{ant}, as a function of the cable length D, for 1 kW input RF power.

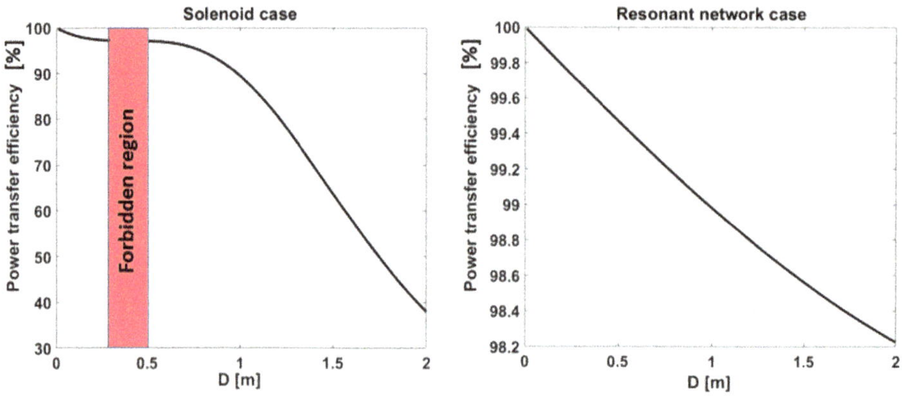

Figure 13.19. Dependence of the power transfer efficiency on the coaxial cable length *D*.

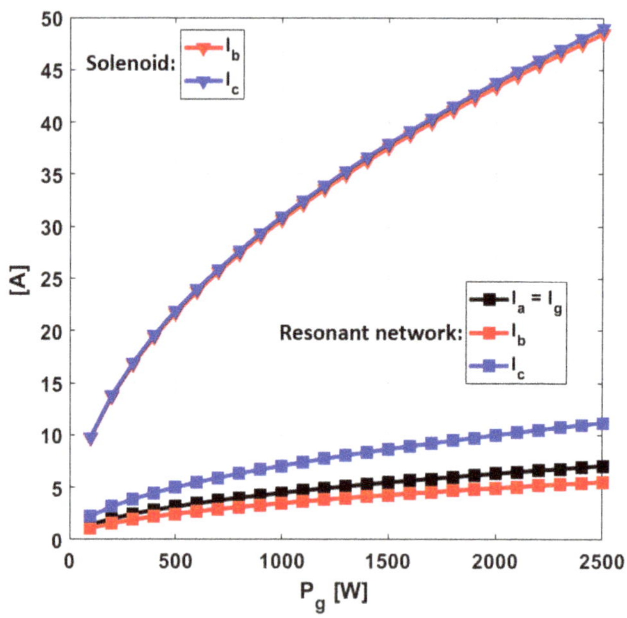

Figure 13.20. Currents expected in the matching circuit elements (*a*, *b*, *c*) as functions of the RF input power, for a solenoid antenna (triangles) and for a birdcage antenna (squares).

problematic, the main difficulty arising from the very high voltages causing arcs, notably in connectors, and from the heating of the transmission line itself. Indeed, for conventional ICPs, the matching box is always directly connected, in the shortest way, to the antenna. In contrast, the resonant network antenna, for this example and quite generally, can easily be operated with a quite long transmission line, permitting the whole RF power circuit to be well separated from the antenna itself.

To finish with this example, figure 13.20 compares the current expected in the matching elements for both sources, which is plotted here as a function of RF power.

For this calculation we have removed the transmission line from the system, to optimize the solenoid case in terms of power transmission. The current I_g is the same for both systems, as they are both matched to 50 Ω. We also recall that the solenoid antenna requires only two matching elements, *jb* and *jc*. It can be seen that the level of current in the matching elements is much lower for the resonant network than for the conventional solenoid ICP antenna. Hence, resonant networks are less demanding in terms of matching element maximum ratings, notably for the variable matching capacitors, and much less power is dissipated by these elements.

13.3.5 Study #2: choosing a coaxial cable

The aim of this small study is to emphasize the importance of the coaxial cable dimensions, in terms of internal and external radii, when used to connect the matching circuit to the plasma source. Once again, we apply the formalism to the example of the birdcage antenna of figure 13.15, which is assumed here to be connected to the matching system by a 0.5 m long coaxial cable. We already know the dependence of the source input impedance on the RF power (up to 5 kW) from the measurement shown in figure 13.16. From the previous study, section 13.3.4, we expect a good power transfer efficiency of the resonant network, but for a power of 5 kW, even a 1% loss (50 W) might not be insignificant, as we now show.

Consider the possible use of two typical types of commercial 50 Ω cables, as shown in figure 13.21. Cables of dimensions such as those on the right-hand photograph can be used up to at least 5 kW, according to datasheet specifications. But we have to be careful, as these specifications are only valid if the cable is used to feed a 50 Ω (matched) system. The commercial cable on the left is a high power cable, much bigger than the first one, and which allows a 22 kW power transmission, according to specifications. In the following, we assume both cables to be made of the same materials (copper and PTFE), and we concentrate on the influence of these different dimensions.

Figure 13.22 shows the expected dissipated power in both cable types for a 0.5–5000 W power ramp, taking into account the evolution of the antenna impedance with power shown in figure 13.16. It can be seen that about four times more power is lost in the small cable than in the big one.

Figure 13.21. Two examples showing the diameters of low-loss commercial coaxial lines. Left: A large cable; right: a smaller cable.

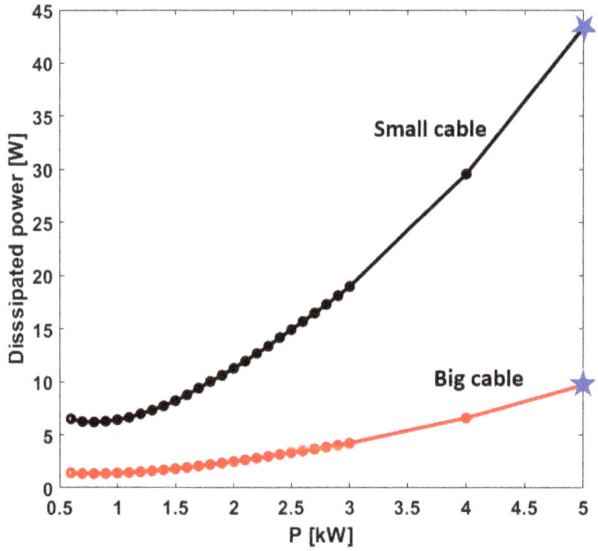

Figure 13.22. Power dissipated in the 0.5 m long cables as a function of RF input power for both cable types in figure 13.21. The two blue stars correspond to $P = 5$ kW transmitted power; see also figure 13.23.

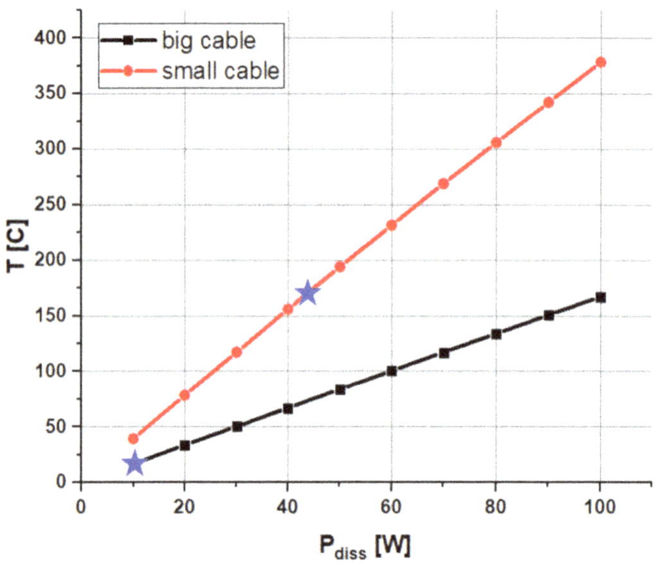

Figure 13.23. Evolution of the steady-state temperature of the inner conductor as a function of power dissipated into the large and small cables of figure 13.21. The two blue stars correspond to the 5 kW delivered power in figure 13.22.

To estimate the corresponding increase in temperature of the cables is not straightforward, but it can be quite easily carried out by means of numerical simulations with finite element solvers. Figure 13.23 shows the result of such a simulation for the present examples in air (the temperature rise is greater in a vacuum). The given temperature corresponds to the central conductor of the coaxial line, which is the hottest part. The stars in figure 13.23 correspond to the 5 kW condition of figure 13.22, so roughly 10 W dissipated in the big cable, and 44 W for the smaller one. We observe here a combination of favourable effects: using a bigger cable leads to a lower dissipated power for the same delivered RF power (figure 13.22), and, additionally, using a bigger cable leads to a lower temperature increase for the same dissipated power (figure 13.23).

Stating a maximum temperature of 50 °C of the inner conductor for safe operation, we see from figure 13.23 that we cannot allow more that 12 W of dissipated power in the smallest cable, which, according to figure 13.22, would correspond typically to a maximum delivered power of about 2 kW. The same 50 °C temperature increase is obtained for 30 W of dissipated power in the case of the bigger cable, which is out of the range of figure 13.22, but would typically be expected for about 10–15 kW of RF delivered power. The large cable type is therefore mandatory for high power antenna operation above 2 kW.

13.3.6 Study #3: choosing the number of antenna legs

One of the first questions that arises when designing a resonant network concerns the number of legs to be chosen. It is clear that a large number of legs makes networks more expensive, and globally more complicated with more connections, notably for water cooling, which entails a greater risk of leaks especially in a low pressure environment. On the other hand, a small number of legs often corresponds to a large spacing between them, which can degrade the resulting plasma uniformity (for a birdcage example, see figure 8.17) when the distance between two successive legs is larger than the typical plasma skin depth.

As a practical example for this discussion, we consider the construction of a 40 cm-by-40 cm planar source, to be operated at 13.56 MHz with a power of 1 kW, at the m = 6 mode resonance. Three designs are envisaged; for all of these, 40 cm long legs are used, but with different spacing between them of 1.25 cm, 2.5 cm, and 5 cm, respectively, giving 33, 17, and 9 leg networks. The m = 6 current distributions for these three networks are shown in figure 13.24, for the same input power of

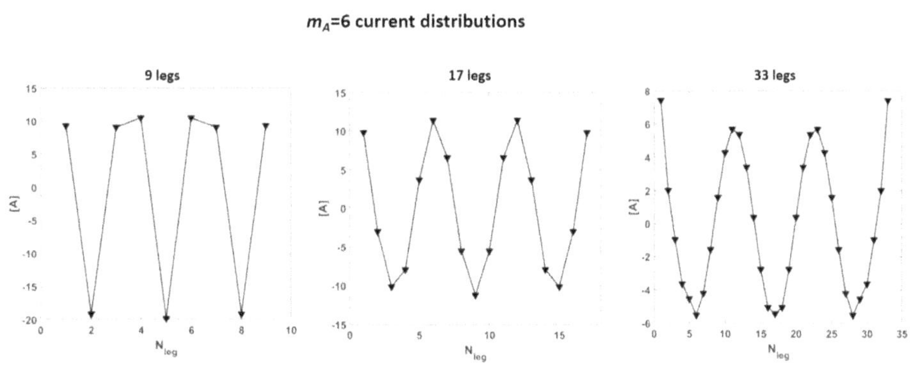

Figure 13.24. Current distributions for the three network designs for mode m = 6. Each symbol represents a leg position.

1 kW. From these distributions, we can calculate the mapping of power deposition in a plasma slab (5 cm thick, figure 13.25) for each network design. Figure 13.26 shows the profiles of power deposition (W m^{-3}) along a line 2 cm above the antenna plane, as indicated in figure 13.25, for the three different networks. It can be seen that the 17 and 33 leg networks provide similar power deposition profiles, but not the nine-leg one. In this nine-leg case, the separation between adjacent legs is too large. Therefore, for the present example, a 17-leg network would be a good compromise, minimizing the number of legs while keeping the expected heating pattern for a m = 6 current distribution.

Apart from the need of a certain 'leg spatial density', as shown above, the number of legs also influences the electrical properties of the networks. For our present example, table 13.1 resumes the impact of leg number on the most relevant parameters. First, a smaller number of legs implies a lower capacitor value, which can be seen as an issue to design the capacitor assembly for high powers. Interestingly, the level of maximum current into these capacitor assembles remains more or less independent of the number of legs. A smaller number of legs also always

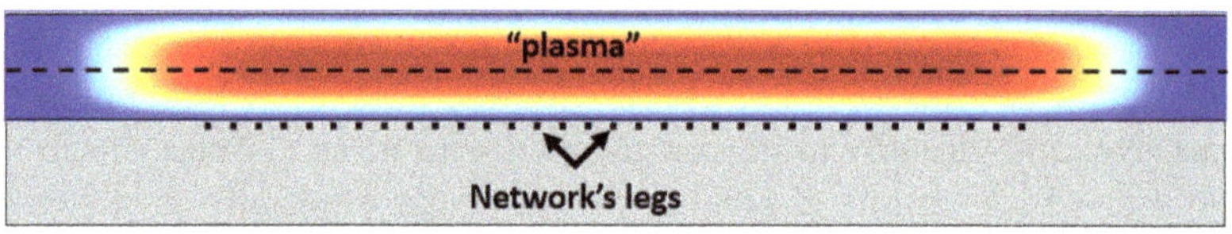

Figure 13.25. Plasma density distribution in slab geometry above the planar antenna.

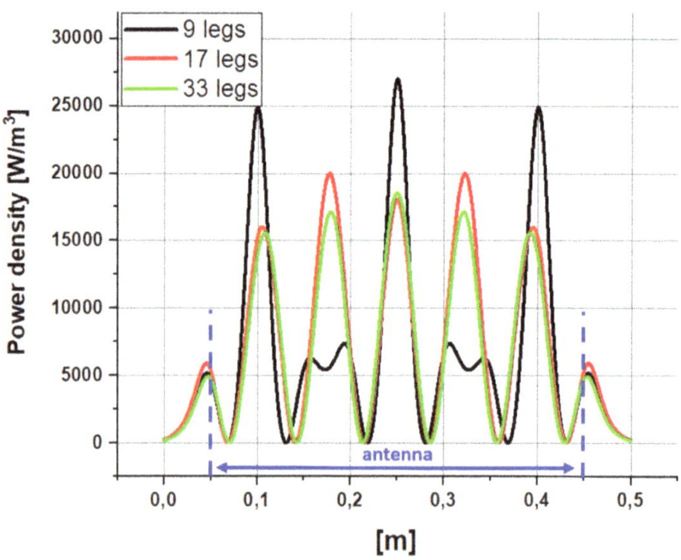

Figure 13.26. Power deposition profiles calculated along the dashed line represented in figure 13.25, 2 cm above the antenna plane.

Table 13.1. Summary of electrical parameters for 9, 17, and 33 legs in a planar antenna.

Number of legs	Assembly capacitance	Maximum capacitor assembly current	Maximum node voltage	Resonance frequency shift
9	470 pF	7 A	300 V	0.75 MHz
17	1100 pF	8 A	200 V	0.9 MHz
33	2450 pF	8 A	150 V	1 MHz

correspond to higher potentials at the antenna nodes, here typically a factor of two between the 33 and 9 leg examples. This can be an argument for choosing a larger number of legs, especially for sources in a low pressure environment. Finally, the shift of the antenna resonance frequency due to plasma coupling is greater for a higher number of legs, which can be an argument for minimizing the number of legs, especially for low mode numbers for which the frequency shifts are greater.

13.3.7 Study #4: choosing the antenna mode number

The choice of the resonant mode number for a given source is naturally dependent on the desired application and geometry. For example, it is clear that for a cylindrical birdcage antenna to be used as a helicon source, a m = 2 (for open birdcage) resonant mode is desired if one wants to dominantly excite the $m = 1$ helicon mode family (see section 12.1). On the other hand, considering a planar source for plasma processing, the most important point is often the plasma uniformity because it is directly connected to the process uniformity. The aim of this short section is to give some clues to guide the choice of the antenna mode number, especially for these types of applications.

Intuitively, it would seem logical that high mode numbers would be favourable for plasma uniformity, because the maxima of the sinusoidal current distributions—where we expect higher local density—are less spatially separated, and therefore more efficiently smoothed out by diffusion. This is correct, but only to a certain extent. To illustrate this point, we show as an example in figure 13.27 the results of 2D numerical models for a 17-leg planar antenna, where the separation between two successive legs is 2.5 cm, typical of a 40 cm-by-40 cm antenna. In these models the plasma density expected for any current distribution is computed, for argon at 1 Pa and with an RF power of about 500 W, and we show here the results for four increasing mode numbers, m = 2, 6, 10, 16. Additionally, we show in figure 13.28 the corresponding estimated density profiles at 2 cm above the antenna plane. It is clear that by increasing the mode number from 2 to 10, the density profiles become more uniform, but on going to higher m values, the plasma uniformity is degraded, as shown here for the m = 16 case which is the highest mode number for this 17-leg antenna. If we were to give a rule of thumb for the choice of the mode number with regard to plasma uniformity, we could recommend that medium values of m should be considered first.

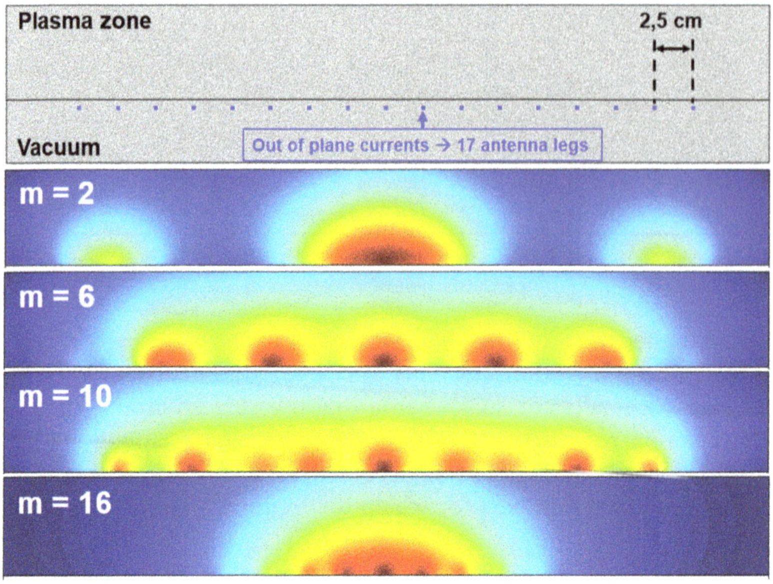

Figure 13.27. 2D mappings of density generated by a 17-leg planar antenna for different resonant mode numbers. 2D COMSOL [3] simulation for argon at 1 Pa, with 500 W RF power.

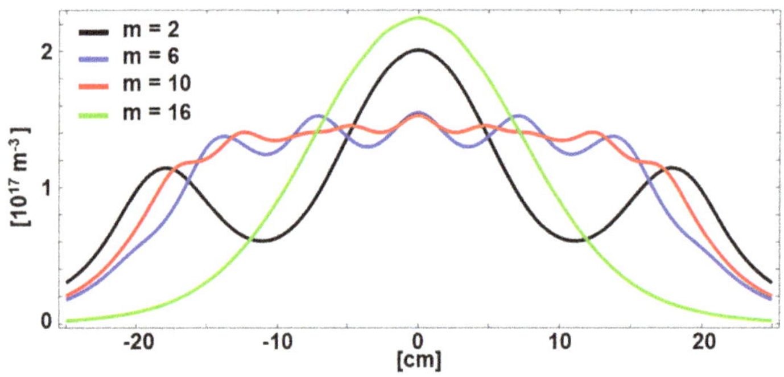

Figure 13.28. Density profiles extracted from mappings of figure 13.27, at 2 cm above the antenna plane.

Apart these arguments regarding the plasma density uniformity, RF considerations should also to be taken into account for the choice of the mode number:

Globally, the levels of current and voltage for a given RF input power tends to increase with m. For example, figure 13.29 shows a comparison of the node voltages and leg currents between the m = 6 and m = 10 cases of figure 13.28. The current level is typically two times higher for the m = 10 case than for m = 6. For high power applications, this will be an important criterion for the mode choice, because the capacitor assemblies will require two times more individual capacitors for m = 10 than for m = 6. Furthermore, the required capacitance for the assemblies to bring the desired resonance at the operation frequency, usually 13.56 MHz, is lower

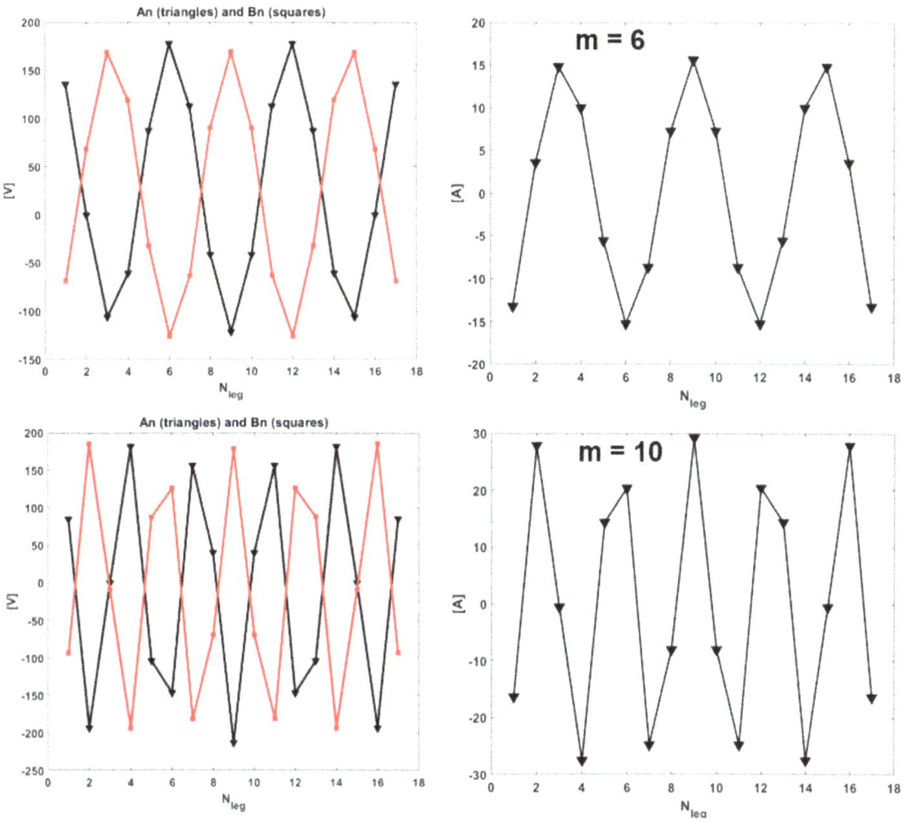

Figure 13.29. Comparison between the distributions of node voltages (left) and leg currents (right) on a 17-leg network, for modes m = 6 and m = 10 with identical RF power.

for higher modes (typically 700 pF for m = 10 in our example, 1200 pF for m = 6), which should also be considered for high power applications as it can affect the choice of the individual capacitor type.

It can also be seen in figure 13.29 that the maximum voltage level on the antenna nodes is also higher for the m = 10 case (400 Vpp) than for m = 6 (300 Vpp)[1]. For systems designed for operation in a low pressure environment, this will be an argument for choosing the lowest acceptable mode number, because high voltages are the main problem for these applications.

A final point concerns the shift in resonance frequency due to plasma loading, which is smaller for higher mode numbers. Taking the present example, for which a 10^{17} m^{-3} plasma density was assumed, the m = 6 mode leads to a 0.75 MHz frequency shift, compared to only 0.25 MHz for m = 10. For very high density applications, this might be an argument for choosing the highest possible mode number.

[1] In figure 13.29, the offset in the voltage distribution for m = 6 is because its grounded node at $n = 2$ is not a 'natural zero' of the sinusoidal distribution. For m = 10, the grounded node at $n = 3$ is closer to the 'natural zero' of the sinus, so the voltage is more symmetric about 0 V.

13.4 Antenna mechanical construction

The mechanical arrangement of an antenna's components depends on its physical environment: whether it is outside the vacuum vessel (i.e. in atmospheric air), or inside a vacuum vessel embedded in a dielectric, or exposed to low pressure gases inside the vacuum vessel. Several examples have already been mentioned throughout the book, and the detailed technology of atmospheric pressure and low pressure antenna sources are here explained in sections 13.4.1 and 13.4.2, respectively.

13.4.1 Antennas operating in an atmospheric pressure environment

Figure 13.15 shows a typical example of this kind of plasma source, taken from chapter 12. It is a birdcage antenna in air, enclosing a dielectric tube which acts as a vacuum vessel containing a low pressure plasma (note that the antenna operates in atmospheric air, but it is not a source of atmospheric pressure plasma!). This source type is notably used for helicon plasma generation, up to quite high levels of power of about 10 kW. It is a 9 leg antenna, 15 cm in length for a diameter of 13 cm, which is operated at 13.56 MHz with mode m = 2 (corresponding to a $m = 1$ wave excitation, see chapter 11).

In general, the RF network legs are made of copper tubes, which is convenient for fluid (usually water) cooling of the structure. The connectors for the capacitor assemblies at the extremities of the legs can be either brazed or tightly screwed. For RF compatibility, the source has to be enclosed inside a metal box (Faraday shielding). For the example of figure 13.15, we take advantage of this metal housing to mount the antenna around the discharge tube using a simple plastic holder. As a general rule, the dielectric properties of plastics used for parts in contact, or in the vicinity of, the antenna must be checked at the operating frequency. A 7/16 RF connector is mounted on the metal housing just in front of the antenna RF input node, which is connected to the central pin of this connector by a straight metal rod. The antenna grounded nodes are directly connected to the housing, also with thick metal rods. For high power operation, it is necessary to cool these rod connectors. Two half-cylindrical screens are placed around the antenna leg, whose distance can be adjusted to track the resonance frequency of the antenna, typically over a 1 MHz frequency range. This is especially useful for such a low mode system (m = 2), where the frequency shift due to plasma coupling is the more important.

It is generally preferable to connect the matching system (section 13.1) directly at the RF antenna input, to minimize power losses in connectors. For a conventional ICP, such as a solenoid antenna, we have seen that it is even mandatory to do so. But it is not necessarily the case for resonant networks, thanks to their impedance properties which permit the use of transmission lines, even quite long, between the matching box and the RF antenna (see section 13.3.4). It is often very convenient to have this possibility to move the matching circuit away from the source itself, and also especially for sources inside vacuum vessels which are described later in section 13.4.2.

As another example of an antenna operating in an atmospheric pressure environment, figure 13.30 shows a 20 cm-by-20 cm planar network, also used for helicon

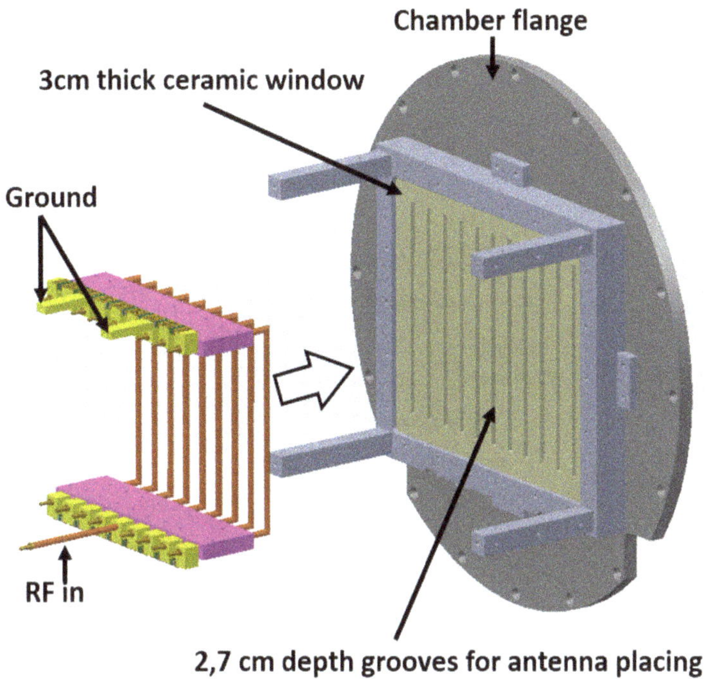

Figure 13.30. A planar antenna designed for 5 kW operation. The nine-leg antenna, area 20 cm-by-20 cm, is embedded within grooves machined into a thick dielectric window. This source is dedicated principally to helicon wave excitation, with mode m = 2.

plasma generation at high power in section 12.3. The antenna sustains the low pressure plasma via a dielectric (machinable ceramic) window, which therefore has to be strong enough to withstand atmospheric pressure. Considering the area of this window, a 3 cm thickness is necessary to ensure safe operation, which would bring the antenna quite far from plasma, leading to poor coupling and power transfer efficiency. Here, we take advantage of the network geometry to bring the legs to a distance of only 3 mm from the plasma, thanks to grooves in the machinable ceramic.

Similarly to the cylindrical source of figure 13.15, the planar network has to be enclosed by a metal housing. The network itself is mounted on the metal back plate of this housing (figure 13.31) using thick metal rods connecting the required grounded nodes, and a plastic holder at the RF feeding location (not shown in the figure). Once again, the resonance frequency of the network can be tuned by using a movable planar screen shown in the figure.

This solution of a grooved dielectric window is naturally only valid for reasonably small networks, for cost and even feasibility reasons. A solution for large sources is schematized in figure 13.32. Once again, we take advantage of the topology of planar systems. The principle is to have a 'grid' aperture in the vacuum chamber wall, instead of a large single one. A large window of quite small thickness (typically

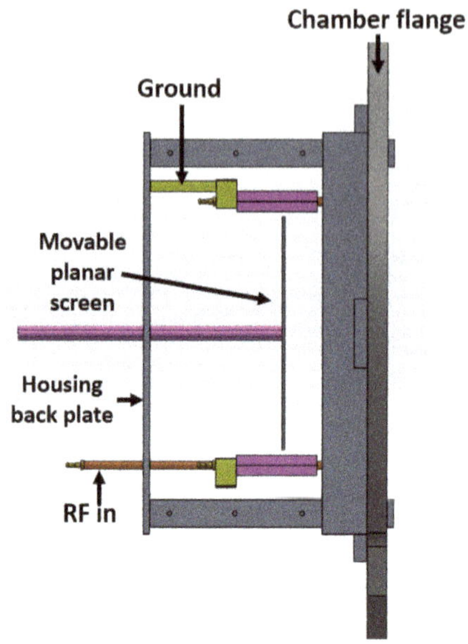

Figure 13.31. Side view of the planar source of figure 13.30 mounted on the chamber flange. The metal sidewalls of the housing are not shown.

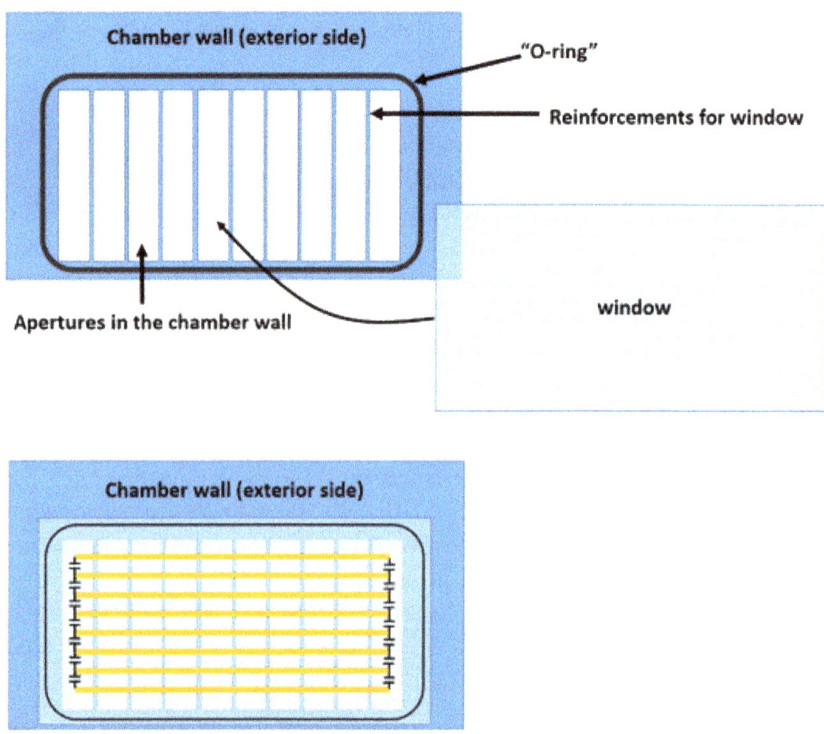

Figure 13.32. Mechanical arrangement for a large area planar network outside the vacuum vessel.

5 mm) could then be used owing to the mechanical reinforcement afforded by the regularly spaced grid bars. An RF network placed as represented in the bottom panel of figure 13.32 means that its legs are perpendicular to the reinforcement bars, and therefore totally uncoupled to the chamber grid because perpendicular conductors have no mutual inductance (see chapter 4).

13.4.2 Antennas operating in a low pressure environment

A resonant network antenna can be used in a low pressure environment much more easily than conventional ICPs, essentially because of the comparatively low voltage levels inherent to these kind of structures, as already mentioned. The main risk with high voltages in low pressures concerns, of course, the generation of arcs that can strongly damage the source itself, and also degrade the processes quality. In the following we show several examples of 'internal' planar sources that have been designed for use inside a vacuum vessel.

The principle for the construction of all these sources is always more or less the same, and is illustrated in figure 13.33. The RF network is placed in a grounded metal housing, and maintained by an insulating holder at a distance of typically a few centimeters from the housing's back plate. The RF connection is made through this back plate, directly opposite the designated antenna RF input node. The designated ground nodes of the network are connected directly to the housing back plate by straight rod connectors. For the design of figure 13.33, these ground connectors are not water cooled, but this may be necessary for high power operation. The legs of the network are made of copper tubes, with water pipe fittings brazed onto the ends. In the design of figure 13.33, the cooling water

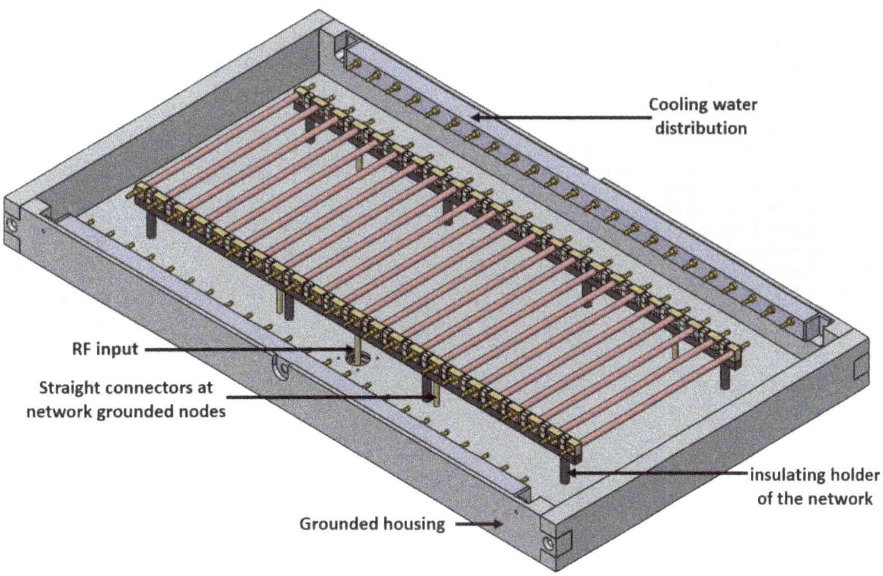

Figure 13.33. Schematic of the design for a planar source in a low pressure environment.

distribution is integrated into the grounded housing sidewalls, and the water flow is fed in to, and out of, each leg via plastic tubes—see the blue tubes in figures 3.1, 6.3, and 13.11. Note that the network water circulation very efficiently cools the legs and the stringer connectors for the capacitor assemblies, but not the assemblies themselves. These are only cooled by thermal conduction through their leads, which is generally not very efficient. However, provided that the current maximum rating for capacitors is conservatively respected, they will not heat up much, and even in a low pressure environment the heat removal by conduction through the capacitors leads is sufficient. The empty spaces within the source housing volume are filled with a dielectric material, as illustrated in figure 13.34, to preclude any ignition of parasitic plasma RF glow discharges in these zones—as a rule of thumb, any remaining interstices should be smaller than a few millimetres. If absolutely necessary, glass beads of sub-millimetre diameter can be used to fill any remaining interstices, or even to fill the whole volume; hollow spheres are a practical option for limiting the total weight. Lastly, the housing is closed (but not sealed, to allow pump down) by a protective dielectric window (figure 13.34), which prevents direct contact between the plasma and the dielectric embedment.

Although this design looks quite simple, there are some important points to watch out for. First, all the water connections must be totally vacuum compatible. Second, the properties of the dielectric embedding material must be carefully checked at the operating RF frequency. Dielectric foams are very interesting in this respect, as their

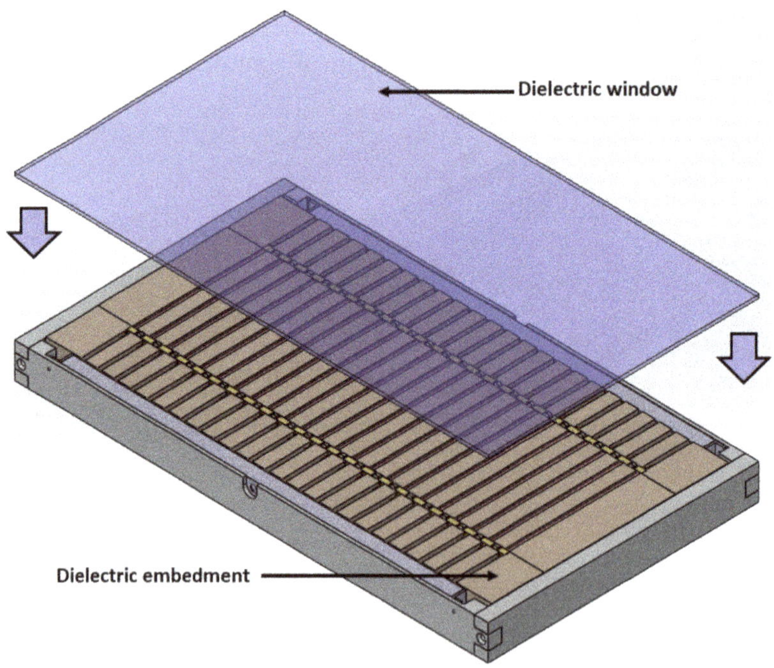

Figure 13.34. Schematic of a planar antenna plasma source in a low pressure environment, showing the network embedded in a dielectric medium, and the protective dielectric window.

effective dielectric constant is often very close to unity, with a very low-loss tangent. Additionally, the use of foams strongly minimizes the final weight of the source. Gas contamination and pump down time of the foam did not appear to be a problem in the roll-to-roll application of section 7.7.1.

Finally, a particularly important point concerns the management of the heat load on the dielectric window that arises from its contact with plasma. As a rule of thumb, in a planar geometry about half of the RF input power will be converted into heat load on the source dielectric window. This estimate is found to be more or less correct for atomic gases, such as argon, and constitutes an upper value for molecular gases for which a non-negligible part of the input power is used for dissociation and ro-vibrational excitation. The plasma heat load is naturally not uniformly distributed over the dielectric interface, but concentrated in the antenna region where the plasma is generated. Under these conditions, a single-piece window of regular glass rapidly bows to an alarming extent due to the higher thermal expansion of its central region, and can eventually break even for quite moderate RF power. A first solution to this issue consists in the use of tiles instead of a single-piece window. Another solution is to use a single-piece window made of ultra-low expansion glass, such as glass ceramics used for kitchen hobs (note that the dielectric properties at 13.56 MHz must always be checked when choosing a material to be in close vicinity to the RF antenna). In another direction, the heating of the dielectric window can be managed by a direct cooling of this part.

The choice of the dielectric embedment also depends on the strategy chosen for the plasma heat load management. As already mentioned, foams present some advantages concerning the dielectric properties and mass density, but they are also very good thermal insulators. As a consequence, the dielectric window heat losses are almost only radiative; as such, its temperature can rise to quite high values, depending on the power level. Foams are then to be considered for what could be called a 'hot system' strategy. Naturally, in this case the embedment foam must withstand the dielectric window temperature, which strongly limits the choice. SiOx/alumina foams are perfect with regard to this aspect, they are rigid and can be easily machined, although they are somewhat brittle. For moderate temperatures up to about 300 °C polyimide foams can also be used, with the advantage of being flexible. Figure 13.35 is a photograph of such a source with a polyimide foam embedment. The RF network covers a 60 cm-by-40 cm area, and operates at 13.56 MHz up to 1.5 kW. This source was used for deposition of micro-crystalline silicon thin films on glass substrates; see also section 6.5. For this process the substrate temperature is required to be at about 200 °C, which is entirely compatible with a 'hot system' design. Instead of a single piece for the dielectric interface, the antenna is isolated from the plasma by alumina tiles (figure 13.36) with overlapping, stepped edges. Alumina was preferable in this case notably with regard to process contamination. The source placed inside the process vacuum chamber is shown in figure 13.37. It can be seen that the RF power is transmitted from the chamber wall to the RF network simply by a robust commercial coaxial cable, comparable to the big coaxial cable of figure 13.21, preferentially with high voltage connectors.

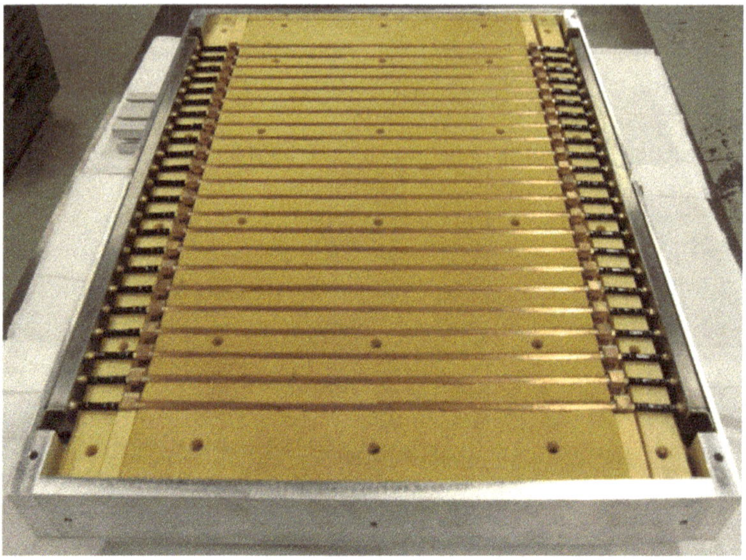

Figure 13.35. Planar antenna embedded in polyimide dielectric foam.

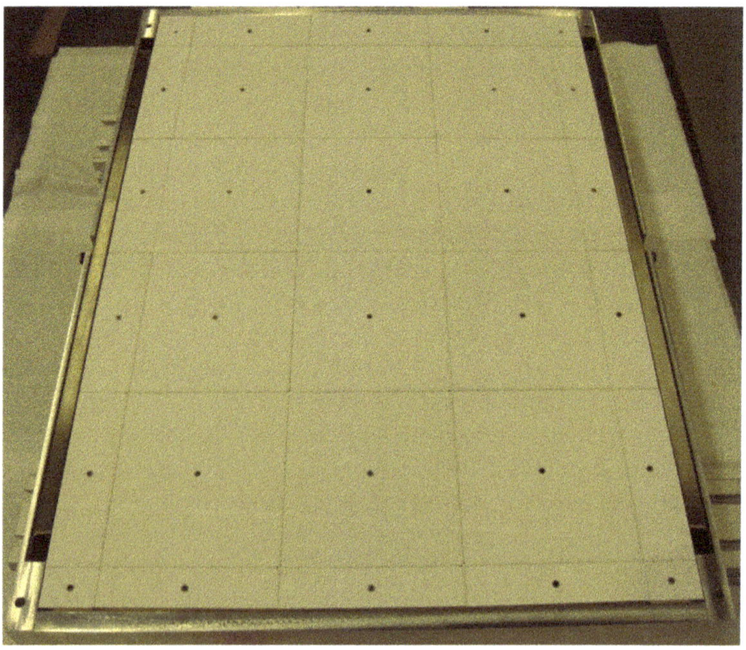

Figure 13.36. Planar antenna protected by a dielectric window of alumina tiles.

Another *in situ* photograph of a 'hot system', but this time with a SiOx/alumina foam embedment, is shown in figure 13.38. Here the antenna size is 25 cm-by-55 cm, for a 13.56 MHz frequency up to 3 kW. This type of source was installed in a roll-to-roll vacuum chamber for thin film deposition on plastic webs, 50 cm wide, as

Figure 13.37. The complete planar antenna assembly mounted inside the process vacuum chamber.

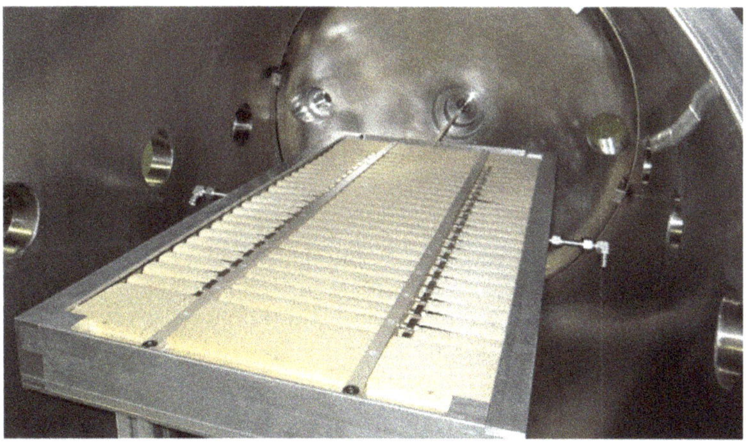

Figure 13.38. Photograph of a planar antenna *in situ* with SiOx/alumina foam embedment.

sketched in figure 7.17(b). For this kind of roll-to-roll application described in section 7.7.1, the antenna has a quite large aspect ratio (long and thin), and the question arises of the choice between two possible network configurations, which we denote here to be the 'short leg' and 'long leg' configurations, represented in figure 13.39. Both designs have been tested and give the same overall performance in terms of average plasma density and process rates and film quality. The short leg design is a lower impedance/lower voltage system, which is always an advantage at

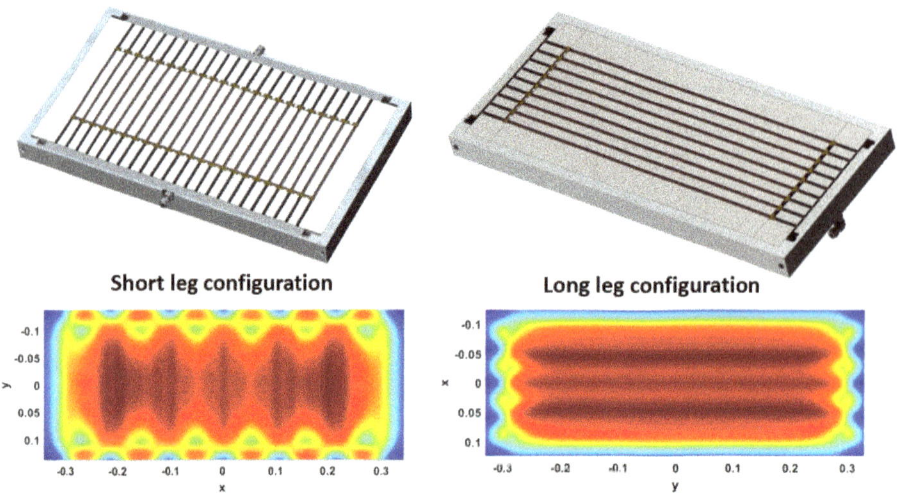

Figure 13.39. Top: 'short leg' and 'long leg' designs, for the same network area. Bottom: Corresponding mappings of induced magnetic field amplitude in a plane 2 cm above the network surface.

low pressures, and allows higher power operation. On the other hand, the short leg configuration requires 23 legs and 44 capacitors assemblies, whereas the long leg version has only 9 legs and 18 assemblies, making the latter antenna cheaper and of lower risk with regard to water leaks or capacitor defects, for example. To finish with this question, we also show in figure 13.39 some mappings of the RF induced magnetic field amplitude for both configurations, in a plane 2 cm above the antenna surface. The sinusoidal current distribution of the networks (m = 6 for short legs, m = 4 for long legs) clearly shows through in these mappings, and, as a consequence, the plasma generated by these antennas will present a certain non-uniformity following these patterns. The non-uniformity associated with the current distribution strongly depends on the process pressure and on the distance to the antenna, but to give an order of magnitude, it is typically about 5%–10% for a 1–10 Pa pressure range. It can be seen from the long leg mapping that the plasma uniformity along the legs can be expected to be fairly uniform. In the specific case of a roll-to-roll machine, the web scrolls parallel to the legs direction for the short leg configuration, while it is perpendicular for the long leg one. Therefore, the plasma non-uniformity due to the sinusoidal current distribution in the long leg configuration will not affect the process uniformity, which makes this design preferable for this kind of application.

Finally, the now-familiar design previously presented in figures 3.1 and 6.3 shows another design tested for this roll-to-roll application. It is equivalent to the design of figure 13.38, but instead of a foam, a thermally conductive silicone elastomer was used to embed the antenna (not shown in figures 3.1 and 6.3 because it is opaque). The silicone completely fills the housing volume, and the dielectric window is directly placed onto its surface, which has to be sufficiently flat to ensure a good window/silicone thermal contact. The back plate of the housing is made of copper, for high thermal conductivity, and a copper pipe circuit is soldered on it for cooling water circulation, clearly visible in figure 6.3. Finally, the silicone block is then cooled by this rear plate, as well as by the antenna legs. This solution was found to be very efficient to maintain the whole source at a low (room) temperature. However, one disadvantage is the large weight of the silicone, which is much greater than with foams. For very large networks this can be a problem, and embedment solutions combining foam and thermally conductive silicone will be considered, the latter embedding essentially only the complicated shape of the network itself. Moreover, thermally conductive silicones have to be cured typically at 80 °C to 100 °C for polymerization in an oven. For medium to large sources, it can be a real problem just to find these. Finally, the pouring of the silicone must be carried out very carefully to avoid trapped bubbles in this very viscous bulk material, before curing. The best solution consists in placing the whole system into a vacuum chamber, and to very gently pump down to a few mbar. Bubbles will be efficiently removed by this process, but it is quite delicate, and we strongly recommend to keep an eye on the bubbling silicone in the chamber as the pressure falls. In fact, the bubbling intensity can increase dangerously, especially at the beginning of the procedure, and the pressure ramp-down must be halted immediately. To do so, briefly opening a manual valve is very efficient, and has the advantage that the small pressure shock will help to explode bubbles.

To finish with 'internal' sources, we consider now the case of long-dimension devices, as shown for example in figure 13.40. This is a long-leg system, 180 cm-by-40 cm, with an SiOx/alumina foam embedment, dedicated to operation in a low pressure environment. In fact, this source resembles the long leg source of figure 13.39, for a power of typically 3 kW, but with an area approximately 5 times larger. As a rule of thumb, the plasma density for a planar source is given by the injected RF power per unit area of network (W m^{-2}). Then, to have the same plasma density with a 180 cm-by-40 cm antenna (Figure 13.40) as for the 25 cm-by-55 cm antenna (figure 6.3), the RF power for the former must be about five times higher than for the latter, i.e. up to 15 kW.

We suppose for now that the source is to be operated at 13.56 MHz, the same as for the small source of figure 6.3. A first general point that distinguishes small and large antennas concerns the choice of the grounded nodes. For a small system, this choice mainly sets the input impedance of the antenna (see chapter 3), and we have shown that it is often advantageous to maximize it by choosing the appropriate ground connection, for a given antenna mode. For a big antenna, the ground node positions not only define the antenna input impedance, but also has a clear influence on the resonance frequency shift induced by plasma coupling. We illustrate this effect in figure 13.41, which represents the expected real impedance peaks, for the $m = 4$ resonance of the network of figure 13.40, with three different grounding configurations: A3–A7, B1–B9, and B4–B6 (for a typical plasma density of $2 \cdot 10^{17}$ m^{-3}). For all of them, the RF feeding is set on the central node A5. It can be seen that the frequency shift varies from 0.2 MHz for the B1–B9 configuration up to 0.8 MHz for the B4–B6 one. In terms of impedance, the A3–A7 grounding solution

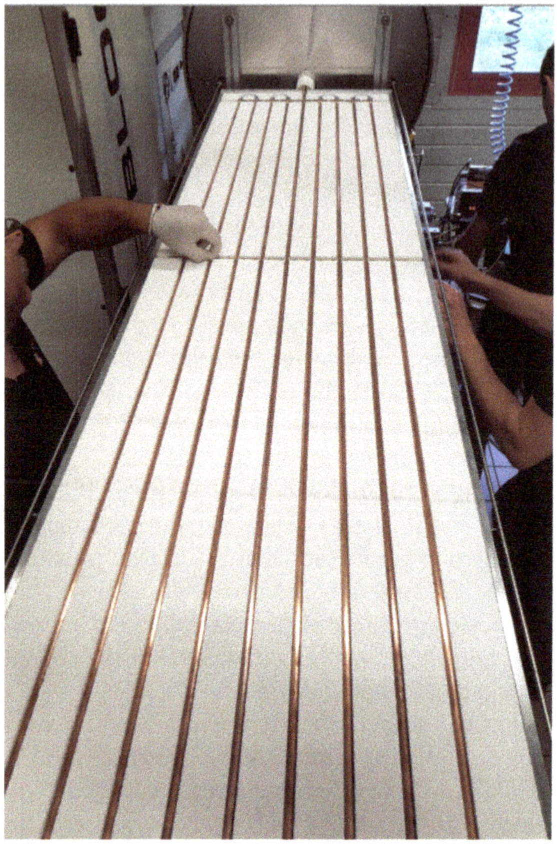

Figure 13.40. Example of a long-dimension source: 180 cm-by-40 cm long leg network

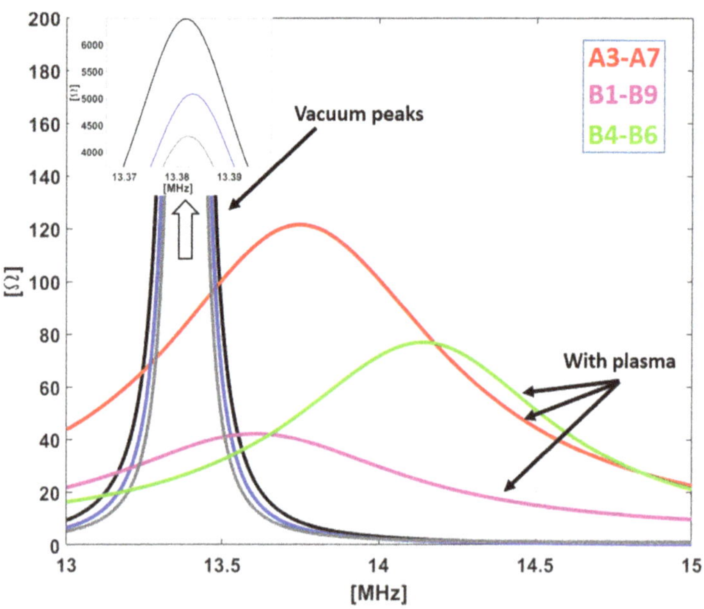

Figure 13.41. Effects of the grounding configuration on the network resonance properties.

gives the highest value, and the frequency shift is only 0.35 MHz, hence this configuration could be considered as a good choice for this example.

Note that, to obtain a m = 4 resonance with this network, a typical 165 pF value for the capacitor assemblies is necessary. The current into the capacitors at 15 kW, with a good plasma coupling, is not expected to exceed 20 A to 25 A, but, as explained in section 13.2, for a small capacitance value such as 165 pF, it will be generally necessary to choose ceramic capacitors, more expensive and less rapidly available than mica capacitors.

In reality, the main problem for this source in a low pressure environment arises from the high voltage on the network. At 15 kW, one can expect a 1500–2500 V voltage on the antenna nodes, depending on the plasma coupling. These values are unfortunately too high for an 'internal' source, and destructive arcs will probably occur.

A very simple way to sidestep this limitation is to abandon the totemic 13.56 MHz, and go for lower frequencies. Taking a 3.39 MHz (i.e. 13.56/4 MHz) excitation frequency for example, the voltage at 15 kW is now expected to be about 750 V–1200 V, so typically two times lower than for 13.56 MHz. Additionally, the required capacitor assembly for a resonance at 3.39 MHz is about 2800 pF, which easily allows mica capacitors to be used. Finally, with this lower frequency, we come back to the situation for which the grounding configuration mainly defines the antenna input impedance, and has only a small influence on the plasma induced frequency shift. On the other hand, the skin depth for RF field into plasma is inversely proportional to the square-root of the operating frequency. For good plasma coupling, the plasma slab thickness should be at least two times this skin depth. Then, if a typical 5 cm plasma gap is appropriate at 13.56 MHz, it should be about 10 cm when reducing the frequency to 3.39 MHz. Finally, a small drawback connected to the reduction of the excitation frequency can be found in the design of the matching circuit inductors. In fact, the required inductance for these parts varies inversely proportional to the frequency. Then, physically, the matching inductors for 3.39 MHz operation will be four times bigger than at 13.5 6 MHz, which leads to a more cumbersome RF system.

Another way to limit the voltage level on the network is to increase the number of legs. As shown in section 13.3.6, taking a 17-leg network instead of a 9-leg one, so with a 2.5 cm inter-leg distance instead of 5 cm, would reduce voltages by one third.

A final remark concerns the heat management with large area 'internal' sources. For the above-mentioned 'hot' designs, there is no cooling of the structure apart from the antenna itself. It was found empirically that, for the same level of RF power per unit area, the overall temperature of a large system stabilizes at a significantly higher value (say, 500 °C) compared to a small antenna (about 250 °C), which can be explained by the higher surface-to-volume ratio of big sources. For example, the 25 cm-by-55 cm antenna dielectric window typically heats up to 200 °C–250 °C, whereas the 180 cm-by-40 cm antenna source can rise to more than 500 °C. A temperature of 500 °C is too hot for a robust, industrial system, and supplementary cooling solutions must be implemented, for example, a thermally conducting dielectric embedment, or a cooled dielectric window, etc.

13.5 High Q design

The most straightforward design that comes to mind for resonant antennas consists of a direct connection, as short as possible, between the capacitors and the ends of the legs, as shown in figures 3.1, 3.2, 6.3, and explicitly in figure 13.42. In the following, we will refer to this configuration as the *direct design*. But there are two situations for which an alternative configuration, a so-called *high Q design*, must be considered.

The first case arises essentially when 'high-pressure' discharges are to be driven by the resonant network—because the plasma resistivity is proportional to the electron–neutral collision frequency, increasing the gas pressure implies that the resistive load on the coupled antenna will also increase (although not linearly), up to the point where the Q factor of a direct design network falls to such low values that the source can no longer be said to resonate (typically $Q < \frac{1}{2}$, see section 2.2.2). The range of pressure for which a direct configuration can be used depends on the network topology, dimensions, and number of legs, but to give an order of magnitude we could say that this design is well adapted for pressures up to few tens of Pa.

The second case for which a high Q design might be required is when the shift in resonance frequency of the antenna, induced by plasma coupling, is considered to be too large. This shift is notably higher for low mode numbers, and for small antennas for which the plasma/antenna capacitive coupling (negative shift) is not sufficient to counterbalance the inductive coupling (positive shift)—see section 7.6.3, for example. The resonance frequency of a network can be tuned over a certain range of frequencies (typically 1 MHz) by acting on the distance between metallic screens and the antenna elements, but the frequency shift due to plasma can easily exceed this tuning range and it can sometimes be useful to be able to minimize it.

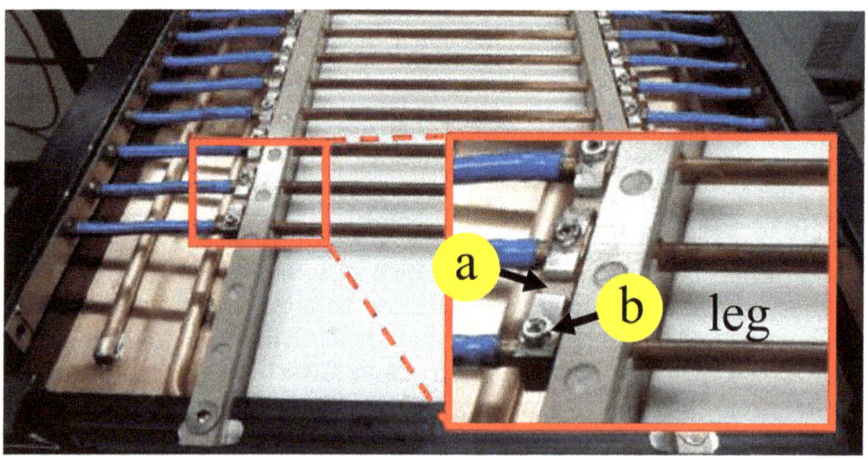

Figure 13.42. An example of the typical direct design, taken from figure 6.3. The zoom rectangle shows (a) a ceramic capacitor assembly, and (b) a screw connection directly on to the end of a copper leg. The transverse grey strip is the PVC mechanical support for the legs held by nylon screws, and the blue plastic tubes carry the cooling water from the sidewall distributor channel into each leg tube.

To briefly explain the principle of a high Q configuration, let us first consider a N-leg direct design network, as schematized in figure 13.43, coupled to a plasma slab. For sake of simplicity, we consider here a simplified system for which the stringers' inductance L_{leg} and resistance r are neglected. We further apply a simple dissipative model (section 3.5.6) and consider the network equivalent parallel resonant R_{eq}, L_{eq}, C_{eq} circuit for a given mode number m. We thereby implicitly ignore mutual inductance for this discussion. In the circuit sketched in figure 13.43, R and L are to be taken to be the individual leg resistance and inductance when coupled to plasma. For this Ddirect arrangement, the leg effective resistance R is almost solely due to plasma coupling.

With these simplifications, the quality factor Q_{Direct} of the equivalent parallel R_{eq}, L_{eq}, C_{eq} circuit for mode m is simply expressed as

$$Q_{\text{Direct}} = \frac{\omega_{\text{m}} L}{R'} = \frac{1}{R}\sqrt{\frac{L}{C\left(1 - \cos\left[\frac{\text{m}\pi}{N}\right]\right)}}, \tag{13.23}$$

using (3.54), (3.47), and (3.50) with negligible stringer inductance L_{leg} and resistance r.

We now consider the design represented in figure 13.44. It can be seen that the network legs are now made up of two parts: a first part consisting of a straight

Figure 13.43. Plasma coupling to the whole length of each leg in the direct design of a resonant ladder antenna.

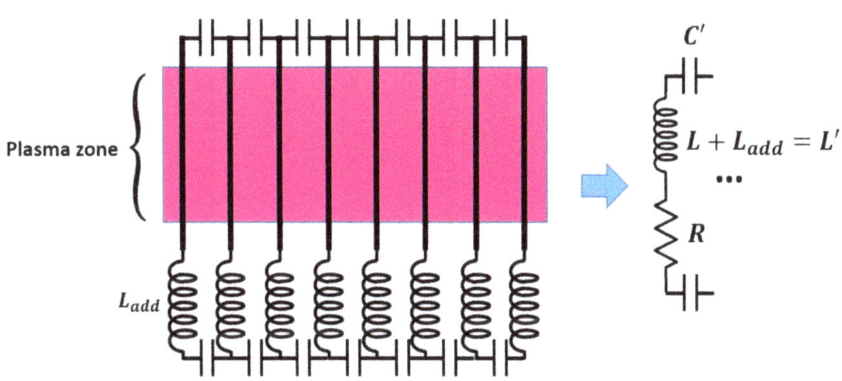

Figure 13.44. Plasma coupling to a partial length of each leg, in series with an additional inductor L_{add}, in the high Q design of a resonant ladder antenna.

conductor as for the direct design which is coupled to plasma, while the second part is simply an additional inductor of inductance L_{add} which is not coupled to plasma.

Let us state that the total leg inductance $L' = L + L_{\text{add}}$ in figure 13.44 is n times higher than L. In order to have the same resonance frequency for both networks in figures 13.43 and 13.44, as a rule of thumb we must have $L'C' = LC$, so C' should be typically n times smaller than C. On the other hand, the value of the effective leg resistance remains approximately the same as R, because this is related to the plasma-coupled part of the leg which is unchanged with regard to figure 13.43. Now expressing the quality factor Q_{HQ} for the high Q design of figure 13.44, we find

$$Q_{\text{HQ}} = \frac{1}{R}\sqrt{\frac{L'}{C'\left(1 - \cos\left[\frac{\text{m}\pi}{N}\right]\right)}} = \frac{1}{R}\sqrt{\frac{n^2 L}{C\left(1 - \cos\left[\frac{\text{m}\pi}{N}\right]\right)}} = nQ_{\text{Direct}}. \quad (13.24)$$

To illustrate the advantages (and disadvantages) of the high Q configuration, we consider below the example of a 23-leg planar network as shown in the photographs of figures 3.1 and 6.3 (20 cm legs). In this direct design with 1.4 nF capacitors[2], the m = 6 resonant mode is excited at 13.56 MHz. Figure 13.45 shows the network impedance spectrum typically expected for a discharge pressure in the pascal range. It can be seen that the antenna resonates with a quite high quality factor, and uniformly distributed currents flow across the whole antenna. Figure 13.46 is analogous to figure 13.45, but for 50 times higher pressure. It is clear that the resonance peaks are not even observable, and that the current distribution in the antenna is strongly damped close to the RF central input.

We next consider a high Q configuration for this 23-leg network, where we have added one inductor per leg, uncoupled from the plasma, of four times the leg

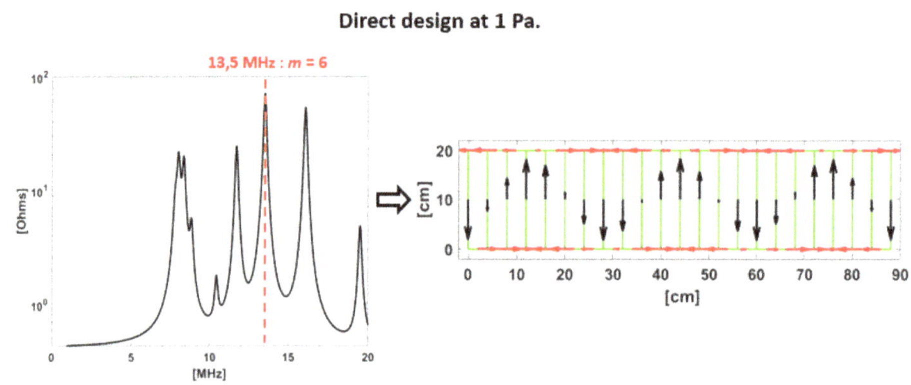

Figure 13.45. Direct design ladder resonant antenna operated at 1 Pa pressure; the calculated loading curve impedance spectrum, and the corresponding currents circulating uniformly in the antenna. Mode m = 6 at 13.56 MHz.

[2] This is different from the 2.6 nF capacitors in section 3.1 because the stringers' inductance and resistance are neglected, and all mutual inductances are ignored, for simplicity.

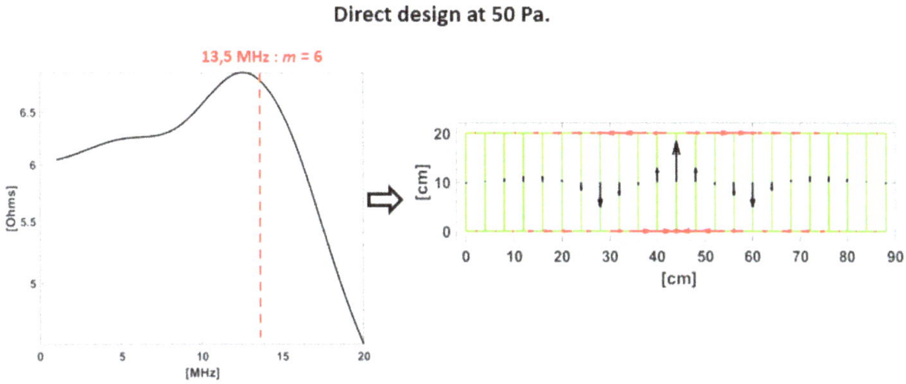

Figure 13.46. Direct design ladder resonant antenna operated at 50 Pa pressure; the calculated loading curve impedance spectrum, and the corresponding currents circulating in the antenna. Mode m = 6 at 13.56 MHz. Compared to figure 13.45, the resonant modes cannot be distinguished, and the current is strongly damped near to the RF central input.

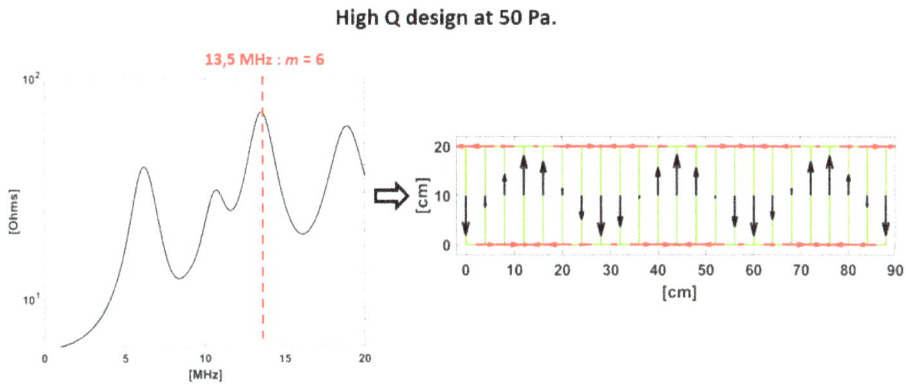

Figure 13.47. High Q design ladder resonant antenna operated at 50 Pa pressure; the calculated loading curve impedance spectrum, and the corresponding currents circulating in the antenna. Mode m = 6 at 13.56 MHz. Compared to figures 13.45 and 13.46, the resonant modes are partially recovered, and the current is again more uniform across the antenna length.

inductance, i.e. $L_{\text{add}} = 4L$, so that the total inductance is $5L$. The m = 6 resonance frequency is kept at 13.56 MHz by reducing the antenna capacitors' value to 0.31 nF, a factor 4.5 smaller than the 1.4 nF of the direct design. This is not exactly a factor of 5, as suggested above by rule of thumb, because mutual inductance between the coupled legs is taken into account here. The network impedance spectrum, calculated for the 50 Pa range in this high Q configuration, is shown in figure 13.47. It can be seen that the resonance properties are partly retrieved, with a quality factor maintained around a value of about $Q = 10$. The main penalty to pay when going from a direct design to a high Q design is a system with globally higher impedance and hence higher voltages, for a given antenna current. This is naturally due to the higher impedance of the network's individual components, namely, lower capacitance $C' < C$, and higher leg inductance $L' > L$. As a matter of fact, for a given input power, the current amplitudes are almost identical for both direct and high Q configurations, because the higher impedances have higher potential drops across them. This is illustrated in figure 13.48, which shows the current and voltage distributions expected for both configurations, for a 100 W discharge in the 1 Pa pressure range.

Note that the increased level of potentials of high Q configurations does not necessarily imply a higher antenna-to-plasma capacitive coupling, because this depends on the physical layout of the network. For example, consider the photograph in figure 13.49, showing a high Q birdcage source designed to operate in the mbar (10^2 Pa) pressure range. It can be seen that two additional inductors per leg are inserted, one at both ends of each leg. These are short coils oriented perpendicularly to the discharge tube, so that the high potential antenna nodes are several centimeters away from the plasma, resulting in a very low capacitive coupling with these parts. In fact, in this configuration the plasma experiences mainly the potentials at the extremities of the straight part of the leg, which are identical for both direct and high Q designs.

In practice, the main limitation of the high Q configuration lies in the fact that the network capacitors' value rapidly becomes much smaller than for the direct design. This can lead to difficulties because high power, low capacitance capacitor assemblies are difficult to manufacture. In the current state-of-the-art, only ceramic-based technologies can provide high current, low capacitance capacitors. This is, in fact, the same problem as for large antennas, where reducing the frequency below the traditional 13.56 MHz to a few MHz allows for most of the issues to be solved.

We now illustrate the second situation for which a high Q design can be preferred to a direct design in order to minimize the resonance frequency shift due to plasma coupling. To do so, we consider the example of a nine-leg birdcage antenna to be excited at mode m = 1, as often employed for a birdcage helicon source (chapter 12). The high Q configuration for this example is typically as shown in the photograph of figure 13.50. It can be seen that it is comparable in principle to the high Q design of

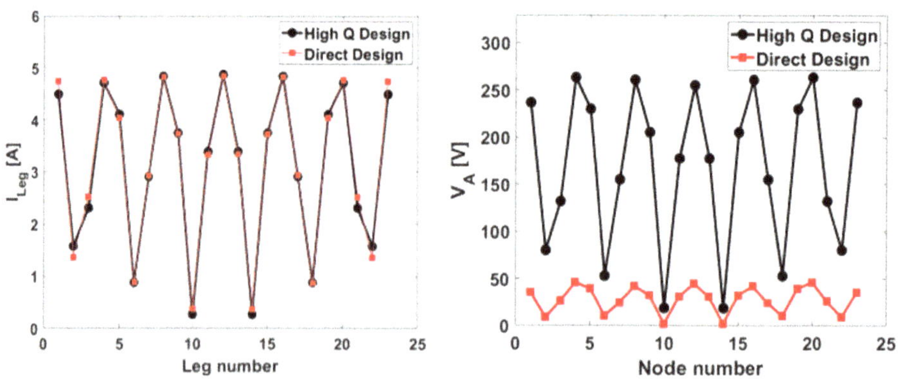

Figure 13.48. Comparison of the leg currents and node voltages for the direct and high Q designs, for the conditions of figure 13.45.

Figure 13.49. A practical example of a high Q design for a birdcage antenna, incorporating short solenoid 'hair-curler' inductors at the end of each leg.

Figure 13.50. Another example of a high Q design for a birdcage antenna, but incorporating short, straight inductors at the end of each leg.

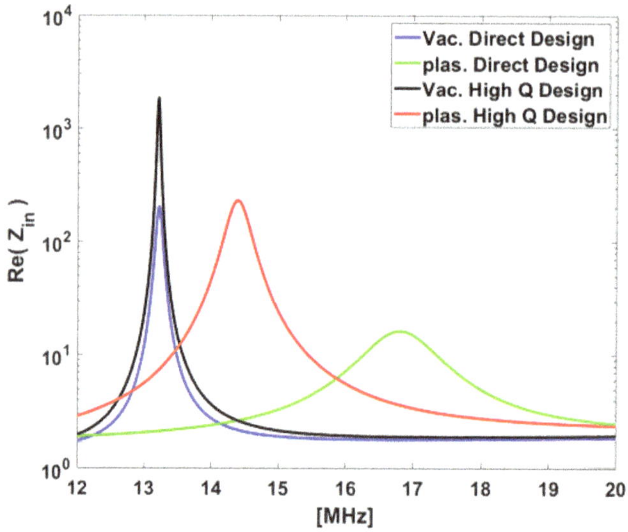

Figure 13.51. The frequency shift from vacuum loading to plasma coupling for direct and high Q designs. The frequency shift is almost three times smaller for the high Q design.

figure 13.49, but instead of short solenoids, the additional inductors of figure 13.50 are just straight portions perpendicular to the coupled legs. Stating the length of these added straight portions to be typically half of the coupled lengths, the total leg inductance would be typically twice the value of a direct design configuration. The Q factor variation between the two configurations is only about a factor of 2, but here the point is to minimize the frequency shift, not to facilitate high resistive coupling.

Finally, figure 13.51 shows the expected shift of the resonance frequency with plasma loading for the two designs. For the direct network, the frequency shift is about 3.5 MHz for a typical plasma density of $5 \cdot 10^{17}$ m^{-3}, while it is reduced to about 1.2 MHz for the high Q design. To summarize, the L_{add} inductors in the high Q design mean that the plasma interacts with only part of the antenna legs, so the antenna's normal mode vacuum properties (high Q and uniform current, in chapter 3) are not strongly perturbed by the plasma, which causes only a small frequency shift. The plasma is exposed to similar currents and voltages as the direct connection design, while being separated from the higher voltage nodes of the high Q design.

13.6 Chapter summary for the technology of resonant network antennas

Several technical issues are explained to be of practical help to the reader (figure 13.52). These concern RF impedance matching and RF capacitor assemblies, as well as coaxial cable types, the number of antenna legs, and the choice of mode number. The mechanical construction of antennas is outlined for operation at high and low gas pressures, and for antennas inside or outside the vacuum chamber,

Figure 13.52. 'Tech info.' (Illustration by Alex Howling. Copyright 2023 Alex Howling.)

paying attention to the limits of current and voltage. A novel 'high *Q*' design is introduced to maintain the network resonance properties and to limit the frequency shift in the case of strong plasma loading. These design considerations enable the advantages listed in section 1.4 to be realized in practice.

References

[1] Bleaney B I and Bleaney B 1976 *Electricity and Magnetism* (London: Oxford University Press) 3rd edn p 1
[2] ATC American Technical Ceramics https://rfs.kyocera-avx.com/ Accessed 31 January 2024
[3] COMSOL Inc https://www.comsol.com Accessed 31 January 2024

Chapter 14

Future developments and applications

In this final chapter, we briefly list several new and potential applications of resonant network antennas, whether in non-magnetized plasma (complementary to part II) or in magnetized plasma (complementary to part III), to exploit the advantages of resonant network antennas listed in section 1.4 to the full.

14.1 Hybrid design

Electromagnetic solutions in chapter 7 show that standing waves along the legs begin to impair the uniformity in large area (~1.4 m^2) reactors, especially in the presence of plasma (section 7.6.5). Probably the most efficient way to reduce the standing wave effect is to lower the RF frequency, whenever possible. Here we consider an alternative strategy employing hybrid networks. This design was already mentioned in section 7.7.2, in a completely different context of RF networks as plasma diagnostics.

As an example, we consider here a nine-leg antenna with 1.8 m-long legs, as on the photograph of figure 13.40, using the 'direct' (straight leg) high-pass design of previous chapters in this book. The wavelength at 13.56 MHz in vacuum would be about 22 m, from which we could expect a 3% variation of current amplitude between the centre and the end of a leg, which can be considered as small. However, as explained in chapter 7, the proximity of plasma significantly reduces the propagating wavelength, as illustrated in figure 14.1, which shows a calculation of the current along a leg expected for our example antenna. Here, a quite high plasma density (10^{18} m^{-3}) was used to enhance the standing wave amplitude (typically 20% non-uniformity).

The principle of a hybrid antenna entails placing capacitors in the middle of each leg, as illustrated in figure 14.2. If the correct value is chosen for these capacitors (unfortunately only empirically determined from models), the leg current consists of

doi:10.1088/978-0-7503-5296-3ch14

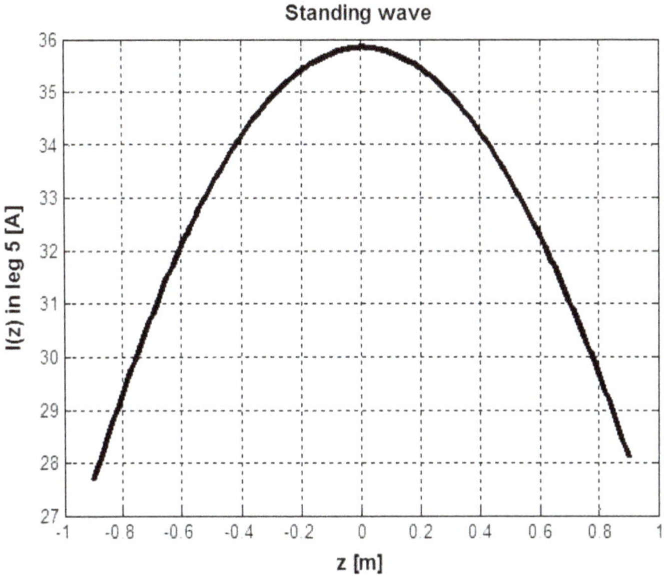

Figure 14.1. Typical standing wave expected in a 180 cm direct high-pass antenna configuration in the presence of a high plasma density of 10^{18} m^{-3}.

Figure 14.2. Schematic representation of a hybrid configuration.

two consecutive standing waves (figure 14.3), thus reducing the current non-uniformity along the leg. In the example given here the non-uniformity is reduced from 22% down to typically 4.5% by inserting 300 pF capacitors in the middle of each leg.

Naturally, the drawback is a discontinuity in terms of voltage in the middle of the leg, as shown in figure 14.4. This will lead to a local capacitive coupling with plasma (capacitive coupling is driven by voltage gradients), and then probably to a local non-uniformity.

The hybrid design is relevant only if large-amplitude standing waves are expected. The standing waves are expected to be more pronounced for very long systems, but also for high frequencies and high densities. Additionally, the standing wave

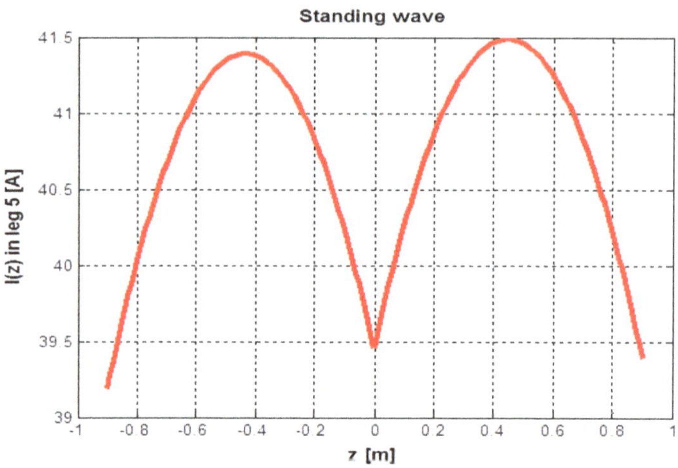

Figure 14.3. Effect of the inserted capacitor on the standing wave in the hybrid configuration.

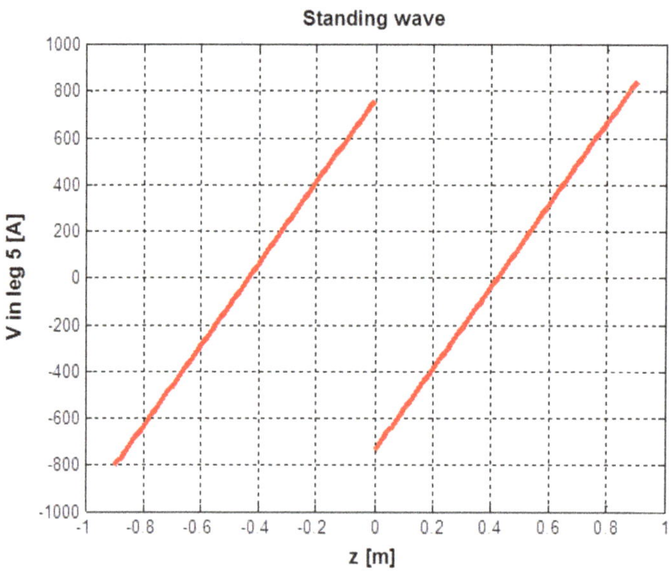

Figure 14.4. Potential along a leg in the hybrid configuration. The capacitor is located at $z = 0$.

amplitude is quite dependent on the possibility for a capacitive ground return current through the plasma to occur, which depends on the particular application. Figure 14.5 shows a schematic of the distributed capacitance on an antenna leg in a typical experimental environment. We suppose here that a substrate is to be processed by plasma. Behind the substrate, a ground plane is schematized, representing a vacuum chamber wall, a substrate holder, or a drum for roll-to-roll applications (section 7.7.1), for example. The capacitors C_s represent the distributed capacitance due to plasma sheaths, which typically corresponds to a 1 mm vacuum

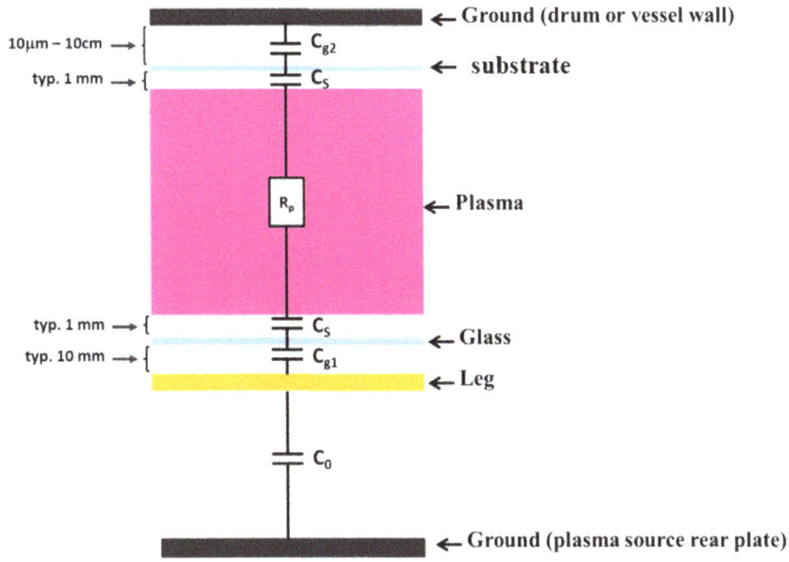

Figure 14.5. Schematic representation of distributed capacitive coupling on a leg.

Figure 14.6. Schematic representation of a hybrid-high Q network.

gap. The values for C_{g1} and C_{g2} depend on the gap distances between plasma and leg, for the former, and between plasma and the grounded top plate for the latter. The ground return current through the plasma is therefore determined by the series combination of these four capacitors. It is clear that for a large substrate/ground gap, of a few centimeters, the capacitance C_{g2} becomes very small, so that almost no capacitive current returns to ground through the plasma gap. The opposite situation would typically occur in a roll-to-roll application, with a grounded drum, for which the substrate/ground gap would virtually vanish. Typically, one could expect the current non-uniformity along an antenna leg to be doubled between the cases of a large, and a vanishing, substrate/ground gap.

The hybrid concept might be even more interesting when combined with the high Q design (section 13.5), as illustrated in figure 14.6, where additional inductors reduce the frequency shift. On the other hand, additional inductors lead to higher voltages on the antenna, because the total inductance of the legs is higher than for a

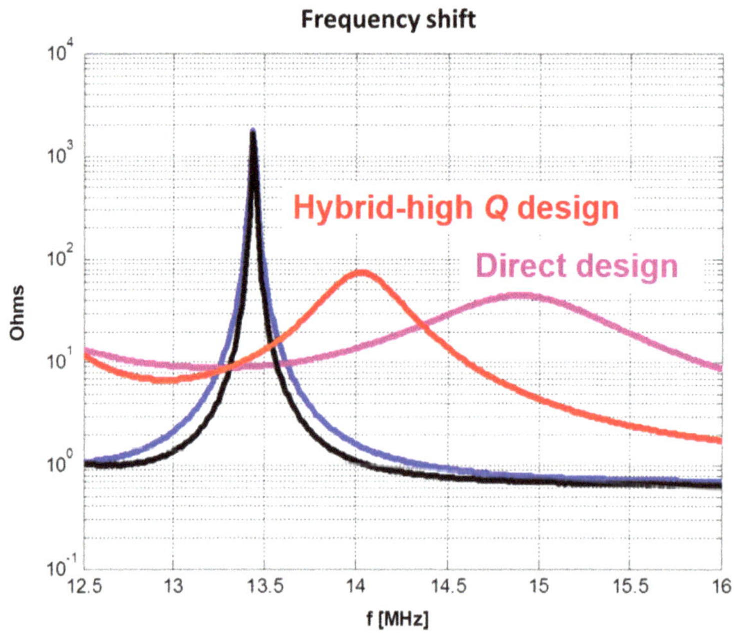

Figure 14.7. Frequency shifts for direct high-pass (magenta) and hybrid-high Q configurations (red), with respect to to the vacuum impedance curves (blue and black).

straight antenna. By placing a capacitor in the middle of the leg, its effective impedance can be brought back to a lower value. It may seem rather ridiculous to do so, because we seem to only compensate the effect of the additional inductor with this capacitor. This would be true with short legs for which capacitive effects can be neglected, but not when standing waves are considered. As a matter of fact, the potential distribution along the leg is very different in the hybrid design (figure 14.4) with regard to other configurations. The capacitive RF current to ground is higher in the hybrid configuration, which is equivalent to a negative contribution in terms of frequency shift.

As an example, the following figures show the main results obtained with a hybrid-high Q design for our example antenna. A $5 \cdot 10^{17}$ m^{-3} plasma density was considered for this calculation, and the ground return current through the plasma was maximized (vanishing substrate/ground gap in figure 14.5). The value for the capacitors in the middle of the legs was chosen to compensate the standing wave (300 pF). In order to compare the results with those of a direct, straight leg, high-pass antenna design, the added inductance value was chosen so that the required stringer capacitor remains the same (166 pF) in both configurations. Hence the added inductances are only about 0.165 times the straight leg value. We first show in figure 14.7 the difference in terms of frequency shift between the two configurations. It can be seen that the hybrid-high Q design strongly diminishes the impact of plasma coupling. The cutting of the standing wave is shown in figure 14.8. Finally, figure 14.9 shows the comparison between the node voltages expected for both

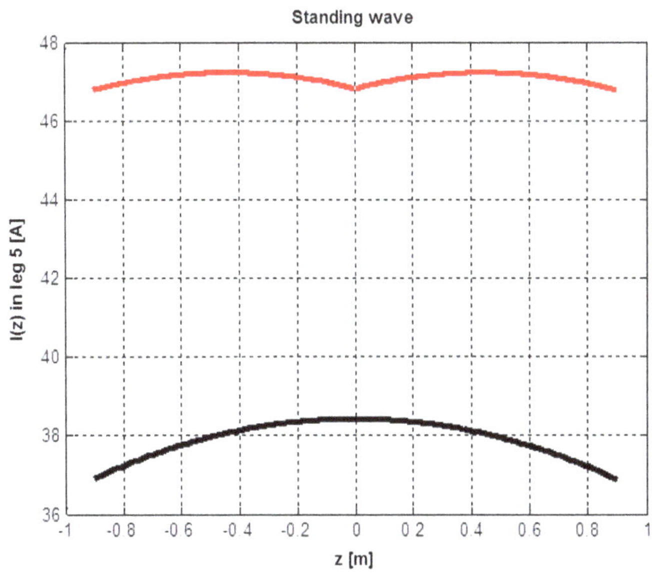

Figure 14.8. Standing waves of current in leg 5 for direct (black) and hybrid-high Q (red) configurations.

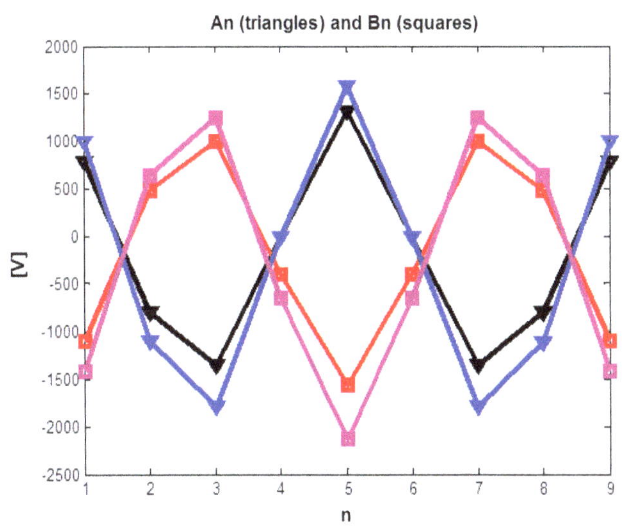

Figure 14.9. Node voltages of the nine legs for direct (black and red) and hybrid/high Q (blue and magenta) configurations.

configurations. As explained, the stringer capacitors having the same value for both configurations, the voltage levels are very close in both cases.

14.2 Two-dimensional resonant network antennas

Two-dimensional resonant antennas bring the book full circle back to the first figure, figure 1.1. Indeed, since that figure, the great majority of the book has been based on

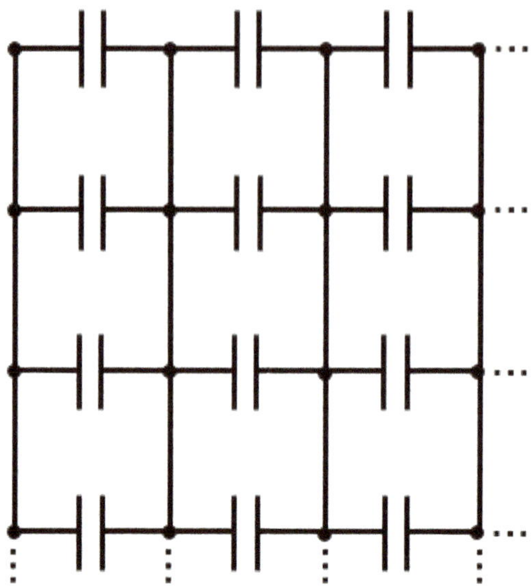

Figure 14.10. Building a 2D network. The straight-wire sections represent inductances.

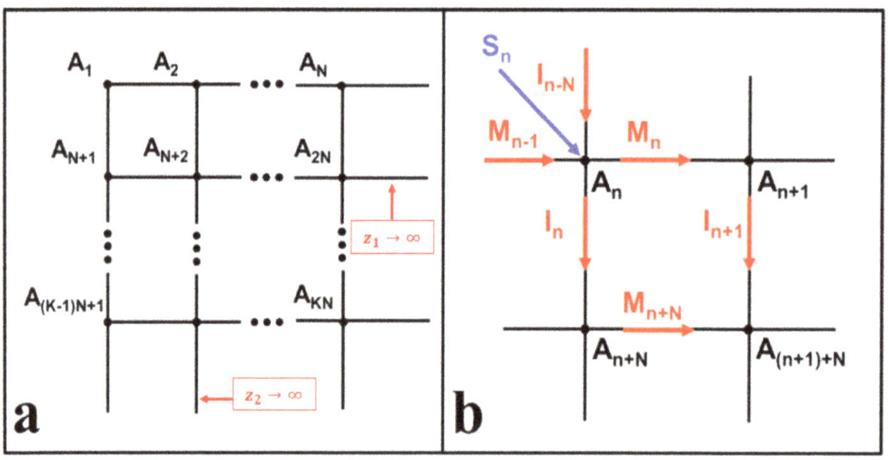

Figure 14.11. (a) Convention for numbering the nodes, indicating the hypothetical additional branches of infinite impedance allowing a simple matrix formalism to be used. (b) Representation of one elementary mesh with definition of the currents.

the narrow sub-category of one-dimensional (1D), high-pass antennas. Clearly, there remains a rich potential for novel configurations by considering two-dimensional networks, as typically sketched in figure 14.10.

Figure 14.11 gives the basis of a 2D matrix formulation for network analysis. An N-by-K network is shown in figure 14.11(a), with a line-by-line numbering of the antenna nodes. Figure 14.11(b) details one elementary mesh. The voltage at the nth node is A_n, the nth horizontal current is denoted by M_n, while the nth vertical current is I_n. S_n is an externally driven source current imposed at the nth node, or a return current (with negative sign) for a grounded node. The impedance of each branch can be inductive, capacitive, or any combination, including resistance, depending on whether the impedances are lossless, dissipative, or have mutual partial inductance. Figure 14.11 is the 2D manifestation of figures 3.5, 3.11, or 4.13. Instinctively, the 2D antenna has been drawn as a rectangular mesh in figure 14.11(a), where the perpendicular elements have zero mutual partial inductance, although any arbitrary orientation could be imagined, exploiting mutual inductive coupling between all elements. Variable component values or flexible mesh orientations of 2D (or possibly 3D) resonant networks could perhaps be conceived as feedback sensors, or even as adaptive artificial neural networks.

14.2.1 Matrix formalism

The matrix formulation for treating the 2D network problem is, in fact, an extension of what has already been presented for 1D networks in appendix B, but it is necessary to pay attention in particular to the boundary conditions. We recall here the fundamental steps.

Considering the elementary mesh drawn in figure 14.11(b) we have

$$M_n = M_{n-1} - I_n + I_{n-N} + S_n. \tag{14.1}$$

As a practical example we define here the Ohm's laws for both horizontal and vertical branches in the framework of rectangular elementary meshes: only parallel segments are mutually coupled. We therefore write

$$\begin{aligned} A_n - A_{n+1} &= \sum_{p=1}^{K\cdot N} z_1(n, p) M_p \\ A_n - A_{n+N} &= \sum_{p=1}^{K\cdot N} z_2(n, p) I_p. \end{aligned} \tag{14.2}$$

The impedances $z_{1,\,2}(n, p)$ characterize the coupling between the nth and the pth segments. To switch to a matrix formulation we define the column vectors $\mathbf{A} = (A_1, \ldots, A_{N\cdot K})^T$, $\mathbf{M} = (M_1, \ldots, M_{N\cdot K})^T$, $\mathbf{I} = (I_1, \ldots, I_{N\cdot K})^T$, and $\mathbf{S} = (S_1, \ldots, S_{N\cdot K})^T$. We further define the $(K \cdot N)$-by-$(K \cdot N)$ matrix shift operators $\bar{S}_1$ and $\bar{S}_N$ such that

$$\begin{aligned} [\bar{S}_1\mathbf{X}]_n &= [\mathbf{X}]_{n-1}, \\ [\bar{S}_N\mathbf{X}]_n &= [\mathbf{X}]_{n-N}. \end{aligned} \tag{14.3}$$

It can be verified that we then also have

$$\begin{aligned} [\bar{S}_1^T\mathbf{X}]_n &= [\mathbf{X}]_{n+1} \\ [\bar{S}_N^T\mathbf{X}]_n &= [\mathbf{X}]_{n+N}. \end{aligned} \tag{14.4}$$

According to these definitions, (14.1) can be generalized as

$$(\bar{1} - \bar{S}_1)\mathbf{M} + (\bar{1} - \bar{S}_N)\mathbf{I} = \bar{U}_1\mathbf{M} + \bar{U}_N\mathbf{I} = \mathbf{S}, \tag{14.5}$$

$\bar{1}$ being the identity matrix, $\bar{U}_1 = (\bar{1} - \bar{S}_1)$, and $\bar{U}_N = (\bar{1} - \bar{S}_N)$. Equations (14.2) can now be summarized as

$$\begin{aligned} \bar{U}_1^T \mathbf{A} &= \bar{Z}_1 \mathbf{M}, \\ \bar{U}_N^T \mathbf{A} &= \bar{Z}_2 \mathbf{I}, \end{aligned} \tag{14.6}$$

where we have introduced the $(K \cdot N)$-by-$(K \cdot N)$ impedance matrices $\bar{Z}_1$ and $\bar{Z}_2$ constituted by the $z_1(n, p)$ and $z_2(n, p)$ elements. Note that, to ensure a consistent numbering (**M**, **I**, and **A** must have the same size), we have added some hypothetical branches of 'infinite' impedance at the right and lower boundaries of the network of figure 14.11(a), leading to vanishing current, and therefore respecting the boundary conditions.

Provided that $\bar{Z}_1$ and $\bar{Z}_2$ can be inverted, we obtain from (14.6)

$$\begin{aligned} \mathbf{M} &= \bar{Z}_1^{-1} \bar{U}_1^T \mathbf{A}, \\ \mathbf{I} &= \bar{Z}_2^{-1} \bar{U}_N^T \mathbf{A}, \end{aligned} \tag{14.7}$$

which, introduced into (14.5), finally leads to the solution for the node voltages **A**:

$$\mathbf{A} = [\bar{U}_1 \bar{Z}_1^{-1} \bar{U}_1^T + \bar{U}_N \bar{Z}_2^{-1} \bar{U}_N^T]^{-1} \mathbf{S}. \tag{14.8}$$

14.2.2 Verifying the boundaries

Current conservation (14.1) and Ohm's law (14.1) are relations between adjacent currents or node voltages. But with our line-by-line numbering convention, starting from the left, the nth node voltage/current is not systematically adjacent to the $(n + 1)$ th nor the $(n - 1)$th one, as for example nodes A_N and A_{N+1} in figure 14.11(a). Therefore, in order to express our shift operators $\bar{S}_1$ and $\bar{S}_N$, we have to pay attention to the finite size of the network. To explicitly illustrate the form of the operators, we take as an example a three-by-three network in figure 14.12. In the above formulation the shift operators $\bar{S}_1$ and $\bar{S}_N$ are introduced

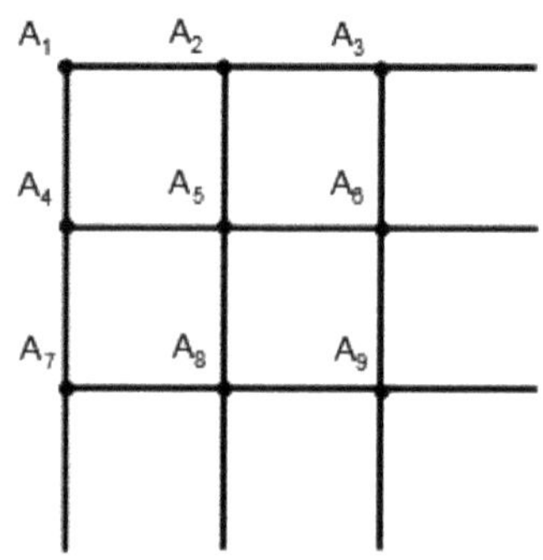

Figure 14.12. A three-by-three network.

to generalize the current conservation expression for all the nodes. We recall that these operators are defined from (14.3) as

$$\begin{aligned}[\bar{S}_1\mathbf{X}]_n &= [\mathbf{X}]_{n-1}, \\ [\bar{S}_N\mathbf{X}]_n &= [\mathbf{X}]_{n-N}.\end{aligned} \tag{14.9}$$

If we take p to be an integer (zero included), we must clearly have

$$[\bar{S}_1\mathbf{X}]_{pN+1} = 0, \tag{14.10}$$

which expresses the finite size of the network on its left boundary. For a three-by-three network this gives

$$\bar{S}_1 = \begin{pmatrix} 0&0&0&0&0&0&0&0&0 \\ 1&0&0&0&0&0&0&0&0 \\ 0&1&0&0&0&0&0&0&0 \\ 0&0&0&0&0&0&0&0&0 \\ 0&0&0&1&0&0&0&0&0 \\ 0&0&0&0&1&0&0&0&0 \\ 0&0&0&0&0&0&0&0&0 \\ 0&0&0&0&0&0&1&0&0 \\ 0&0&0&0&0&0&0&1&0 \end{pmatrix} \Rightarrow \bar{S}_1\mathbf{A} = \bar{S}_1\begin{pmatrix} A_1 \\ A_2 \\ A_3 \\ A_4 \\ A_5 \\ A_6 \\ A_7 \\ A_8 \\ A_9 \end{pmatrix} = \begin{pmatrix} 0 \\ A_1 \\ A_2 \\ 0 \\ A_4 \\ A_5 \\ 0 \\ A_7 \\ A_8 \end{pmatrix}. \tag{14.11}$$

In the same manner we must have $[\bar{S}_N\mathbf{X}]_{n\leqslant N} = 0$, which expresses the finite size of the network on its top boundary. For a three-by-three network this gives:

$$\bar{S}_N = \begin{pmatrix} 0&0&0&0&0&0&0&0&0 \\ 0&0&0&0&0&0&0&0&0 \\ 0&0&0&0&0&0&0&0&0 \\ 1&0&0&0&0&0&0&0&0 \\ 0&1&0&0&0&0&0&0&0 \\ 0&0&1&0&0&0&0&0&0 \\ 0&0&0&1&0&0&0&0&0 \\ 0&0&0&0&1&0&0&0&0 \\ 0&0&0&0&0&1&0&0&0 \end{pmatrix} \Rightarrow \bar{S}_N\mathbf{A} = \begin{pmatrix} 0 \\ 0 \\ 0 \\ A_1 \\ A_2 \\ A_3 \\ A_4 \\ A_5 \\ A_6 \end{pmatrix}. \tag{14.12}$$

If we now take the transpose of these operators, we have first for $\bar{S}_1$:

$$\bar{S}_1^T = \begin{pmatrix} 0&1&0&0&0&0&0&0&0 \\ 0&0&1&0&0&0&0&0&0 \\ 0&0&0&0&0&0&0&0&0 \\ 0&0&0&0&1&0&0&0&0 \\ 0&0&0&0&0&1&0&0&0 \\ 0&0&0&0&0&0&0&0&0 \\ 0&0&0&0&0&0&0&1&0 \\ 0&0&0&0&0&0&0&0&1 \\ 0&0&0&0&0&0&0&0&0 \end{pmatrix} \Rightarrow \bar{S}_1^T\mathbf{A} = \begin{pmatrix} A_2 \\ A_3 \\ 0 \\ A_5 \\ A_6 \\ 0 \\ A_8 \\ A_9 \\ 0 \end{pmatrix}. \tag{14.13}$$

satisfying $[\bar{S}_1^T \mathbf{X}]_n = [\mathbf{X}]_{n+1}$, from (14.4), respecting the condition $[\bar{S}_1^T \mathbf{X}]_{pN} = 0$ (p being an integer, 0 excluded), which expresses the finite size of the network on its right boundary.

Second, we have for $\bar{S}_N$

$$\bar{S}_N^T = \begin{pmatrix} 0 & 0 & 0 & 1 & 0 & 0 & 0 & 0 & 0 \\ 0 & 0 & 0 & 0 & 1 & 0 & 0 & 0 & 0 \\ 0 & 0 & 0 & 0 & 0 & 1 & 0 & 0 & 0 \\ 0 & 0 & 0 & 0 & 0 & 0 & 1 & 0 & 0 \\ 0 & 0 & 0 & 0 & 0 & 0 & 0 & 1 & 0 \\ 0 & 0 & 0 & 0 & 0 & 0 & 0 & 0 & 1 \\ 0 & 0 & 0 & 0 & 0 & 0 & 0 & 0 & 0 \\ 0 & 0 & 0 & 0 & 0 & 0 & 0 & 0 & 0 \\ 0 & 0 & 0 & 0 & 0 & 0 & 0 & 0 & 0 \end{pmatrix} \Rightarrow \bar{S}_N^T \mathbf{A} = \begin{pmatrix} A_4 \\ A_5 \\ A_6 \\ A_7 \\ A_8 \\ A_9 \\ 0 \\ 0 \\ 0 \end{pmatrix}, \tag{14.14}$$

which correctly satisfies the condition $[\bar{S}_N^T \mathbf{X}]_n = [\mathbf{X}]_{n+N}$, again from (14.4), respecting the condition $[\bar{S}_N^T \mathbf{X}]_{n>(K-1)N} = 0$, which expresses the finite size of the network on its bottom boundary.

We now focus on the impedance matrices $\bar{Z}_1$ and $\bar{Z}_2$. As mentioned above we have added some hypothetical branches at the right and bottom boundaries of the network, which ensure a consistent numbering of the system. A necessary boundary condition is that no currents flow into these hypothetical conductors, which can be expressed as

$$\begin{aligned} M_{pN} &= 0, \quad p = [1, 2, \ldots, K], \\ I_{n>(K-1)N} &= 0. \end{aligned} \tag{14.15}$$

To satisfy these conditions we can set the impedance of these branches to tend toward infinity, which is a good representation of the reality. In the impedance matrix $\bar{Z}_1$ we set $z_1(pN, pN) \to \infty$. In the impedance matrix $\bar{Z}_2$ we set $\forall n > (K-1)N$, $z_2(n, n) \to \infty$.

14.2.3 Modes in 2D networks

In 1D networks, the normal modes correspond to sinusoidal distributions of current and potentials within the structure. We recall that an N-leg open network presents $N - 1$ possible modes. For a N-by-K 2D network, one can intuitively anticipate that all combinations of sinusoidal distributions in both dimensions will be obtained.

For the 1D networks in this book, we have essentially considered the high-pass configuration, where the capacitors are inserted in the stringers of the ladder structure. The low-pass configuration corresponds to the situation for which the

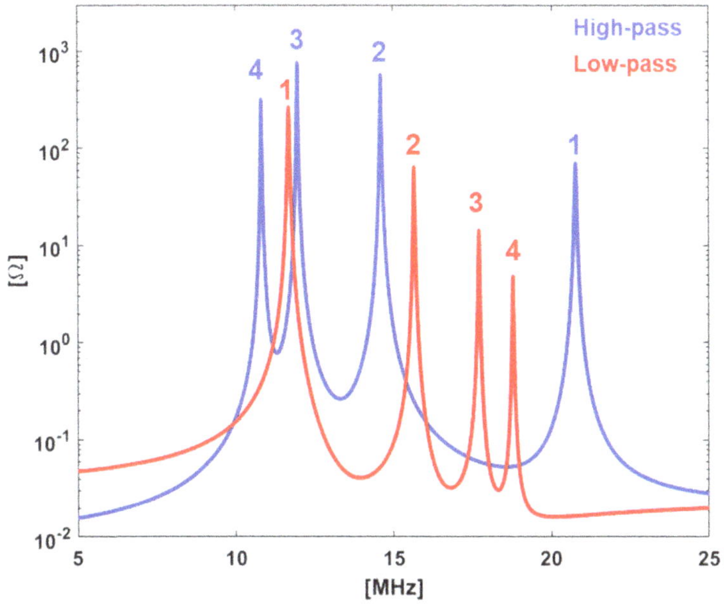

Figure 14.13. Comparison between the resonance spectra (real impedance) of high-pass and low-pass configuration. Example given for a five-leg network.

capacitors are inserted into the antenna legs. Both types lead to the same number of resonant modes, with the same sinusoidal distributions of current, but an important point is that the ordering of these modes in terms of frequency is inverted. For a high-pass network, the resonance frequencies decrease with the mode number m, while it is the opposite for a low-pass configuration. As an illustration, we show in figure 14.13 some typical real impedance spectra expected for a low-pass and a high-pass five-leg network. In addition to the fact that the mode ordering is inverted, we also note that the 'frequency separation' between the different modes is also different.

Considering now the 2D network general construction in figure 14.10, we see that it has a high-pass structure for modes developing horizontally, but a low-pass structure for modes developing vertically. We can then expect a different behaviour for vertical and horizontal modes, notably in terms of resonance frequency ordering.

As an illustrative example, we consider below a five-by-five 2D network, each elementary mesh being square with a 5 cm side. A 1 nF value for capacitors was used, bringing the resonance spectrum, shown in figure 14.14, into the 10 MHz range. We can observe 15 modes in this spectrum, combinations of the four vertical (low-pass) and the four horizontal (high pass) sinusoidal distributions. We denote

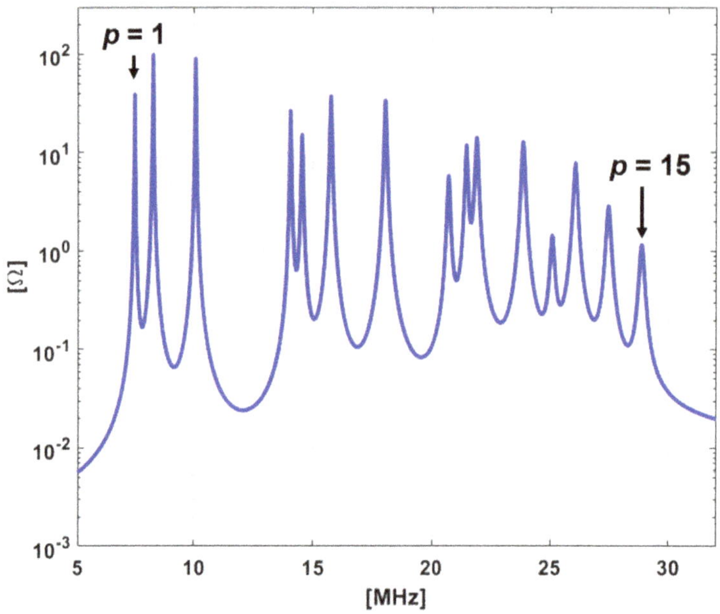

Figure 14.14. Resonance spectra (real impedance) of a five-by-five 2D network.

here these different modes by the index p, starting from the lowest resonance frequency of the spectrum.

The left couple of plots in figure 14.15 represent the current distributions corresponding to the two extreme modes, $p = 1$ and $p = 15$. The green lines in these plots show the network itself, and the blue and black arrows represent the current flowing in each branch. The dotted lines are guides for the eye representing the global sinusoidal variation of the current distribution in both dimensions. Clearly these two modes result in an equivalent combination of m = 1 and m = 4 distributions (low-pass m = 1 and high-pass m = 4 for $p = 1$, low-pass m = 4 and high-pass m = 1 for $p = 15$). The mappings on the right of figure 14.15 represent the calculated amplitude of the RF magnetic induction field generated by these current distributions, in a plane 2.5 cm above the network, which are taken to be associated with plasma generation. It is very clear that both modes are equivalent, one being just a 90° rotated version of the other, but with very different resonance frequencies (7 MHz and 29 MHz).

Just by considering the above examples of modes $p = 1$ and $p = 15$, one could hope that a logical occurrence of high-pass and low-pass sinusoidal distributions would follow, but it is not the case. We show in figure 14.16 the induction field mappings for all 15 modes of the five-by-five network. Equivalent high-pass/low-pass combinations can be very well identified (pairs 2 and 13, or 3 and 9, for example), but it can be seen that the order in which they appear does not follow any

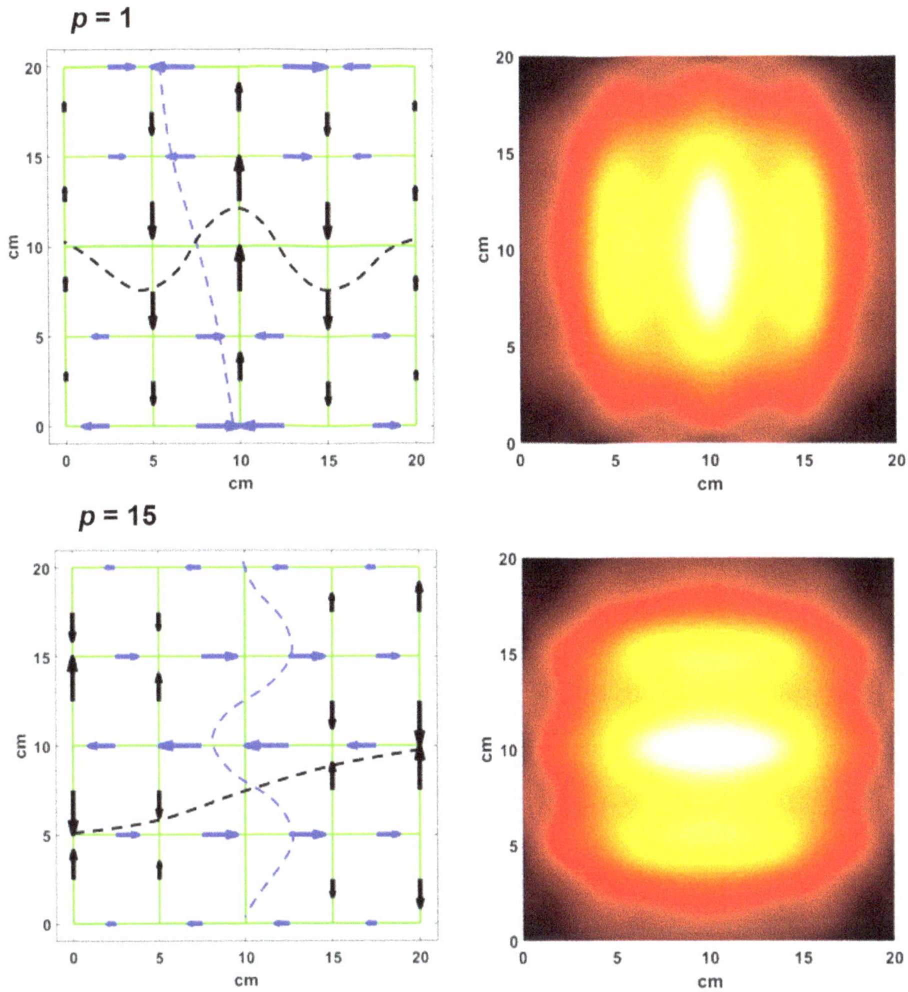

Figure 14.15. Left: representations of the five-by-five network (green lines) with current distributions (arrows), for modes $p = 1$ and $p = 15$ of figure 14.14. Right: mappings of the magnetic induction field magnitude 2.5 cm above the network plane.

obvious logic. The more interesting modes for plasma sources, at least in this square configuration, would probably be the ones combining high-pass and low-pass distributions of identical m, for instance modes $p = 4, 7, 10, 12$ in the present example.

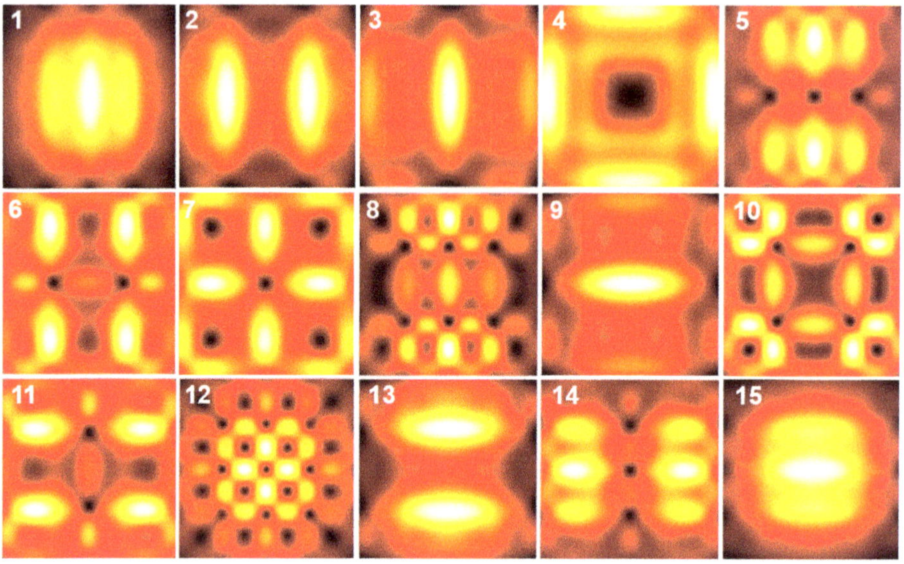

Figure 14.16. Mappings of the RF induction field magnitude 2.5 cm above the antenna plane, for all 15 modes of figure 14.14.

14.3 Phased antennas

We have seen in the preceding chapters that helicon waves are characterized by circular, or at least elliptical polarizations. These can be right or left handed, depending on the sign of the RF field azimuthal wave number m. Experiments, such as RAID in chapter 12, show that positive m modes are very dominantly excited and are responsible for ionization far from the source, sustaining a plasma column or beam. The positive mode numbers are always associated with a right handed polarization, while negative modes are left handed.

The birdcage antenna was shown to produce an RF field which is spatially favourable for helicon excitation, but which nevertheless remains temporally linearly polarized. We have already mentioned before that a linear polarization can be seen as a superposition of right- and left-handed circular polarizations of equal amplitude in figures 10.15 and 9.5. It can therefore be intuited that half of the antenna power will be used in the excitation of right-handed components (positive m), which is desirable, and the other half in left-handed ones, which is seen to be less useful for plasma generation—see also the end of section 10.10. There would then be an interest in building sources that produce the proper right-handed circular polarization in order to put all the antenna power into positive m excitation.

A circular polarization can be constructed by summing two spatially perpendicular linear polarizations, which must furthermore be temporally 90° out of phase (phase quadrature). To be clear, in the conventional ($\hat{\mathbf{x}}$, $\hat{\mathbf{y}}$, $\hat{\mathbf{z}}$) right-handed orthogonal coordinate triad, a general unitary circular polarization vector $\mathbf{C}$ can be generally expressed as:

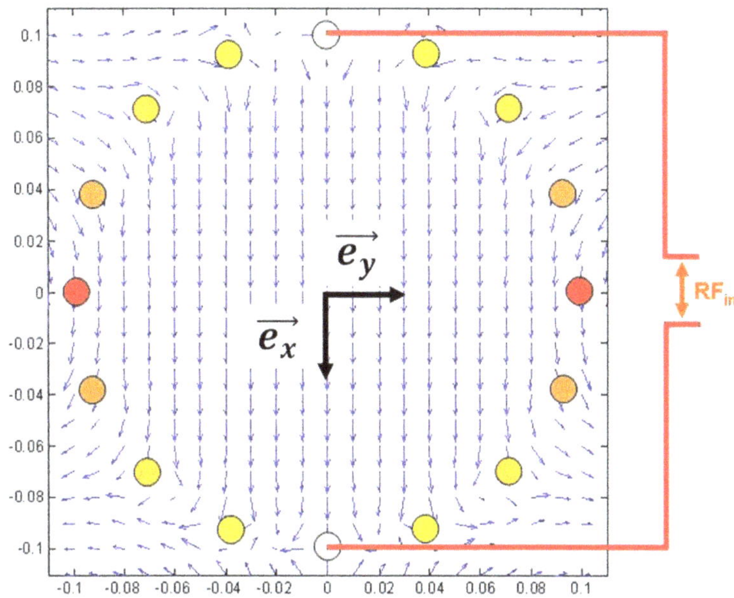

Figure 14.17. A closed, 16-leg birdcage generating a uniform RF linearly polarized field along the x-axis.

$$\mathbf{C} = \hat{\mathbf{x}}\cos(\omega t) \pm \hat{\mathbf{y}}\sin(\omega t). \tag{14.16}$$

The right-handed polarization is obtained using the plus sign in (14.16), as in (10.3); the negative sign in (14.16) corresponds to the left-handed polarization, as in (10.5). See also figures 10.3 and 9.5.

We can imagine two configurations allowing such a polarization to be obtained from cylindrical networks. The first configuration is a simple closed birdcage antenna. Closed birdcage normal modes are degenerate in sine and cosine current distributions as discussed in section 9.2.1. The degeneracy is removed once connections have been imposed at given antenna nodes, thus determining a spatial phase reference, as shown for example in figure 14.17 for a 16-leg closed birdcage, with mode m = 1. In this configuration, the m = 1 current distribution of the birdcage generates a quite uniform transverse field inside the coil, linearly polarized, and aligned with the x-axis of figure 14.17. Then, with this basic configuration we generate the $\hat{\mathbf{x}}\cos(\omega t)$ component of a circular polarization as expressed in 14.16 (with arbitrary temporal phase reference), but, as on the left-hand side of figure 10.15, this field is not rotating. Now consider the set-up sketched in figure 14.18. Here the same 16-leg closed birdcage is fed by two different RF inputs, spatially placed at 90° one from the other. This does not affect the resonance properties of the antenna. At the same resonance frequency, each source generates its own m = 1 sinusoidal distribution of currents in the legs. In the situation represented in figure 14.18, 'RF input 1' generates a transverse linearly polarized RF field oriented along the x-axis, as in figure 14.17, while 'RF input 2' generates a

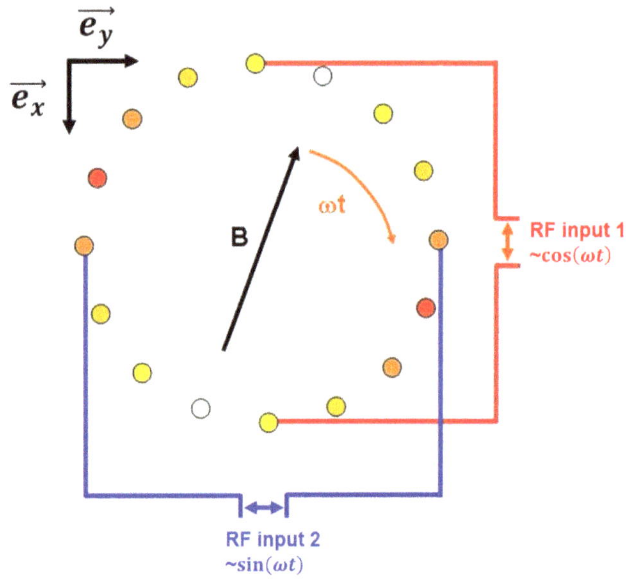

Figure 14.18. The same 16-leg closed birdcage as in figure 14.17, but driven by two separate RF power supplies in spatial and temporal phase quadrature.

transverse linearly polarized RF field oriented along the y-axis. This configuration corresponds to a spatial phase quadrature feeding of the antenna, because we superpose two sinusoidal distributions of currents azimuthally shifted by a quarter of a wavelength ($m = 1$). As long as the two injected signals are temporally in phase, the total RF field $\mathbf{B}_{tot} \propto (\hat{\mathbf{x}} + \hat{\mathbf{y}})\cos(\omega t)$ remains linearly polarized. But, by further imposing a $\pm 90°$ temporal phase shift between RF1 and RF2 (temporal phase quadrature), the total RF field now has the desired circularly polarized structure $\mathbf{B}_{tot} \propto \hat{\mathbf{x}}\cos(\omega t) \pm \hat{\mathbf{y}}\sin(\omega t)$, right- or left-handed depending on whether a $-90°$ or a $+90°$ phase shift is imposed. The problem for realizing this source design in practice resides precisely in the generation of this quadratic excitation. Using two different RF generators on the same antenna is not straightforward, in terms of generator protection as well as in terms of impedance matching.

A possible way to solve this problem could be to consider the second configuration, relying on the same general principle, but where, instead of taking a single closed antenna, two separate open antennas are used, interleaved as schematized in figure 14.19. Here, each antenna has six legs, which is the only configuration allowing a regular spacing between adjacent legs of the two antennas. Each antenna is excited by its own RF generator, with RF feeding spatially at 90° from each other (spatial phase quadrature), and with a 90° temporal phase shift. Note that this configuration does not completely solve the problem of decoupling the two RF sources RF1 and RF2, because it is clear that the two antennas are still coupled by their mutual inductance.

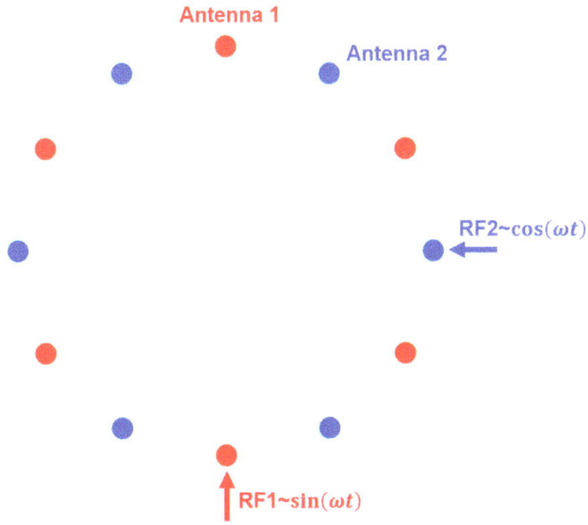

Figure 14.19. Two interleaved, six-leg, open birdcage antennas, driven by two separate RF power supplies in spatial and temporal phase quadrature.

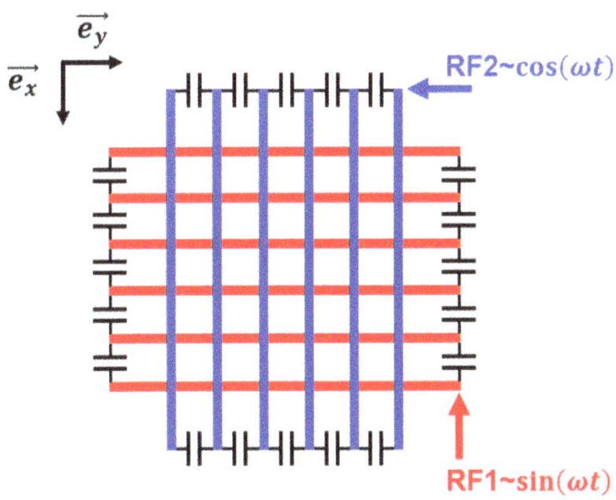

Figure 14.20. Two superposed, planar resonant network antennas, driven by two separate RF power supplies in temporal phase quadrature.

To end with this section dedicated to phased antennas, we now consider the case of planar networks for wave-sustained sources in section 12.3. The goal is the same as for the cylindrical (birdcage) geometry, meaning that we aim to generate a right-handed rotating field with the antenna, for a better coupling to positive *m* helicon modes. We recall that a planar network generates an RF magnetic induction field which is perpendicular to its legs, and linearly polarized. Consider the configuration sketched in figure 14.20, which shows two superposed planar networks, placed perpendicular to each the other. Here again, each antenna is excited by its own generator, and with a 90° temporal phase shift. The total field generated by this structure will therefore be circularly polarized. Naturally, because one network has to be physically behind the other, it will be less strongly coupled to the plasma, but which could be compensated by adjusting the RF power level. The great advantage of this set-up is that the coupling between the two antennas will be quite small, because perpendicular conductors are not inductively coupled, making the quadratic excitation by two different RF generators almost straightforward.

14.4 Toroidal plasma generated by a birdcage antenna

TORPEX is a toroidal plasma device used for fundamental physics experiments in simple magnetized toroidal configurations [1], a paradigm for the tokamak scrape-off layer, as well as in more complex magnetic geometries, such as closed field lines and X-point configurations, of direct relevance for fusion. Plasmas of different gases are created by using a magnetron that injects microwaves at 2.45 GHz in the electron cyclotron frequency range. In hydrogen plasmas, electron temperatures $T_e \sim 5 - 15\text{eV}$ and densities $n_{e0} \sim 1 - 5 \cdot 10^{16}\ \text{m}^{-3}$ are observed. Experimental techniques on RAID, chapter 12, are being expanded to TORPEX to investigate helicon waves in toroidal geometry.

A resonant birdcage antenna is under development, figure 14.21(a), comprising 15 copper legs connected via 1820 pF capacitors. The birdcage is 25 cm long and 32.5 cm in diameter, mounted around a 30 cm diameter glass tube that is inserted into the 40 cm diameter toroidal main vessel via dedicated flanges shown in figures 14.21(a) and (b). Powered at 13.56 MHz, the antenna will excite helicon waves either in a background magnetron plasma or by directly generating the

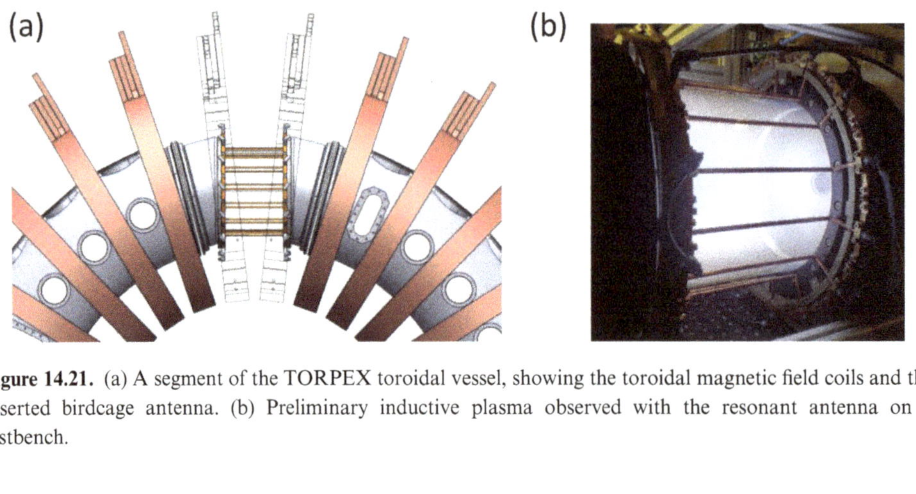

Figure 14.21. (a) A segment of the TORPEX toroidal vessel, showing the toroidal magnetic field coils and the inserted birdcage antenna. (b) Preliminary inductive plasma observed with the resonant antenna on a testbench.

plasma itself, to reach higher electron densities in hydrogen or argon. With this upgrade, TORPEX will investigate toroidal helicon waves by characterizing the wave field over the full plasma volume, thanks to the extensive set of diagnostics available on TORPEX, and by comparing it with full wave numerical simulations of the helicon modes. A better understanding of helicon physics in toroidal geometry will be the starting point to investigate their interaction with fast ions in TORPEX. These fundamental studies will offer valuable input for future experimental and numerical work on helicon current drive in tokamaks [2].

14.5 Multiple birdcage antennas along a plasma column

Multiple antennas strung out along a single plasma column could be used to improve the plasma axial uniformity. To take one example, the next generation particle accelerator, AWAKE, requires a plasma at least 10 m long of axial density $\sim 10^{20}$ m^{-3} with 0.1% uniformity [3–5]. These stringent specifications could possibly be met by multiple aligned birdcage antennas. The interplay between helicon waves generated by adjacent sources in a multiple alignment is being studied on RAID using double-ended helicon sources to symmetrize the plasma boundary conditions [6]. Figure 14.22 shows how the electron temperature profile is more symmetric when both antennas are powered, even though some spatial modulation remains on the temperature and the plasma density profiles. Depending on the relative phase between the phase-locked birdcage antennas, the plasma emission is seen to beat at low frequency (a few Hz), probably due to interference of the two antenna waves. Magnetic field diagnostics combined with laser induced fluorescence (LIF), Thomson scattering, and microwave interferometry will give a better understanding of helicon wave propagation and power dissipation involved in helicon plasma generation.

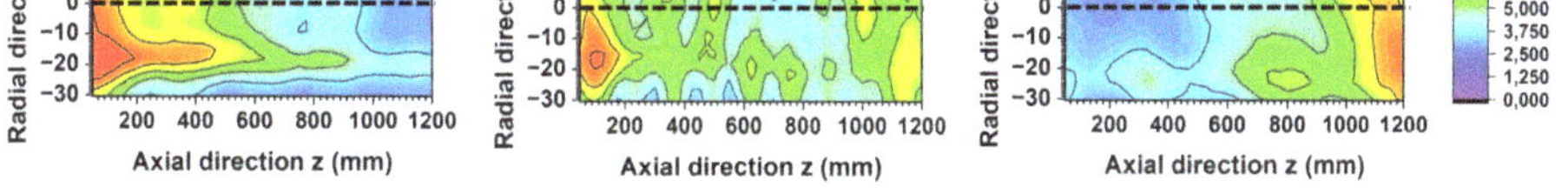

Figure 14.22. Electron temperature profiles measured along the axis (top graphs) and 2D mappings across the RAID diameter (bottom graphs). Left: Left-hand birdcage alone at 3 kW. Middle: Both birdcages at 1.5 kW each. Right: Right-hand birdcage alone at 3 kW. H_2 plasma at 0.3 Pa, 200 G on axis, 13.56 MHz. (Adapted with permission from [6]. Copyright 2023 Elsevier.)

14.6 Matchless antennas

The convention for maximum power transfer from a source of output impedance Z_{out} to a device of input impedance Z_{in} is that the source and load impedance should be complex conjugates of each other, i.e. $Z_{\text{out}} = Z_{\text{in}}^*$; this is called conjugate matching. For reflectionless matching, the impedances should be equal. Conjugate and reflectionless impedance matching are identical when both impedances are purely resistive, usually set equal to 50 Ω which is the output impedance of conventional RF generators. The source and load can then be connected via a coaxial cable of 50 Ω characteristic impedance to avoid reflections in the cable.

As discussed in section 5.5, the impedance of conventional ICP and CCP reactors is dominantly imaginary, hence a matching box is inserted at the input terminals to transform the load input impedance to 50 Ω at the expense of large currents and/or voltages circulating within the matching box. In contrast, resonant network antennas can be designed for dominantly real input impedance; a matching box described in section 13.1 can be used in a similar way for final adjustment to 50 Ω.

A completely different scenario, a so-called 'matchless antenna', proposes to do away with the RF power output impedance circuitry, and hence also with the matching box and 50 Ω cable, and instead, connect the RF output directly to the antenna irrespective of any consideration of 50 Ω in the power transfer. A reactive circuit can thereby directly drive a plasma exhibiting a dynamic impedance, without using a dynamic matching network [7]. This technique is being evaluated for driving resonant antennas.

14.7 Conclusions

A great diversity of resonant network antenna configurations and plasma applications remains to be explored. The first three parts of this book provide the basic principles of antenna normal modes, their inductive coupling to non-magnetized plasma, and their excitation of helicon modes in magnetized plasma. These principles can be combined and re-combined in many imaginative ways to discover new applications; some examples are given in this final chapter.

The following appendixes set out various mathematical steps to accompany the derivations in the main text. A link is provided in appendix K to many of the programs used to generate the figures shown throughout the book, so that the reader can adapt the parameters to their own particular resonant network antennas (figure 14.23).

Figure 14.23. 'A link is given in the last appendix K to a library of programs used to generate many of the figures in this book. We hope that this may be helpful for industrial development of resonant network antennas for RF plasma sources.' (Illustration by Alex Howling. Copyright 2023 Alex Howling.)

References

[1] Fasoli A, Furno I and Ricci P 2019 The role of basic plasmas studies in the quest for fusion power *Nat. Phys.* **15** 872

[2] Van Compernolle B *et al* 2021 The high-power helicon program at DIII-D: gearing up for first experiments *Nucl. Fusion* **61** 116034

[3] Agnello R, Andrebe Y, Arnichand H, Blanchard P, De Kerchove T, Furno I, Howling A A, Jacquier R and Sublet A 2020 Application of Thomson scattering to helicon plasma sources *J. Plasma Phys.* **86** 905860306

[4] Gschwendtner E *et al* 2019 Proton-driven plasma wakefield acceleration in AWAKE *Phil. Trans. R. Soc.* A **377** 20180418

[5] Buttenschön B, Fahrenkamp N and Grulke O 2018 A high power, high density helicon discharge for the plasma wakefield accelerator experiment AWAKE *Plasma Phys. Control. Fusion* **60** 075005

[6] Jacquier R *et al* 2023 A double-ended helicon source to symmetrize RAID plasma *Fusion Eng. Des.* **192** 113614

[7] Pribyl P 2006 Plasma production device and method and rf driver circuit with adjustable duty cycle *US Patent* US7,100,532 B2

IOP Publishing

Resonant Network Antennas for Radio-Frequency Plasma Sources

Theory, technology and applications

Philippe Guittienne, Alan Howling and Ivo Furno

Appendix A

Expansions near resonance for the dissipative antenna

A.1 Modified characteristic equation

Concerning the dissipative antenna of chapter 3, section 3.5, the modified characteristic equation (3.46) for a dissipative antenna is now

$$\begin{aligned}\cosh\gamma &= \left(1+\frac{Z_{\mathrm{str}}}{Z_{\mathrm{leg}}}\right),\\ &= 1+\frac{r+\dfrac{1}{j\omega C}+j\omega L_{\mathrm{str}}}{R+j\omega L_{\mathrm{leg}}},\\ &= 1+\frac{\left(r+\dfrac{1}{j\omega C}+j\omega L_{\mathrm{str}}\right)(R-j\omega L_{\mathrm{leg}})}{R^2+\omega^2 L_{\mathrm{leg}}^2}.\end{aligned} \tag{A.1}$$

Neglecting the square of small resistance in the denominator, $R^2 \ll \omega^2 L_{\mathrm{leg}}^2$, we have

$$\begin{aligned}\cosh\gamma &\simeq 1+\frac{1}{\omega^2 L_{\mathrm{leg}}^2}\left[rR+\omega^2 L_{\mathrm{leg}}L_{\mathrm{str}}-\frac{L_{\mathrm{leg}}}{C}+j\left(\omega L_{\mathrm{str}}R-\omega L_{\mathrm{leg}}r-\frac{R}{\omega C}\right)\right],\\ &= \left[1+\frac{rR}{\omega^2 L_{\mathrm{leg}}^2}+\frac{L_{\mathrm{str}}}{L_{\mathrm{leg}}}-\frac{1}{\omega^2 L_{\mathrm{leg}}C}\right]-j\left[\frac{R'}{\omega^3 L_{\mathrm{leg}}^2 C}\right],\\ R' &= R\left[1+\omega^2 L_{\mathrm{str}}C\left(\frac{rL_{\mathrm{leg}}}{RL_{\mathrm{str}}}-1\right)\right].\end{aligned} \tag{A.2}$$

Our arbitrary arrangement of R' is a combined effective resistance of R and r in the network.

From section 3.3, $\cosh\gamma = \cosh(\alpha + j\beta)$, where α is the attenuation constant per section, and β is the phase change per section (α and β are both real). Using trigonometrical expansions and assuming weak attenuation per section,

$$\begin{aligned}\cosh(\alpha + j\beta) &= \cosh(\alpha)\cosh(j\beta) + \sinh(\alpha)\sinh(j\beta),\\ &= \cosh(\alpha)\cos(\beta) + j\,\sinh(\alpha)\sin(\beta)\\ &\simeq \cos\beta + j\alpha\sin\beta, \quad (\alpha \ll 1),\end{aligned} \tag{A.3}$$

assuming $\alpha \ll 1$ so that $\cosh\alpha \approx 1$ (for the parameters in figure 3.12(a), $\alpha = 0.0015$ and $\cosh\alpha = 1$ to six decimal places).

Equating real and imaginary parts, we obtain

$$\begin{aligned}\cos\beta &\simeq \left[1 + \frac{rR}{\omega^2 L_{\text{leg}}^2} + \frac{L_{\text{str}}}{L_{\text{leg}}} - \frac{1}{\omega^2 L_{\text{leg}} C}\right],\\ &\simeq \left[1 + \frac{L_{\text{str}}}{L_{\text{leg}}} - \frac{1}{\omega^2 L_{\text{leg}} C}\right],\end{aligned} \tag{A.4}$$

$$\alpha\sin\beta \simeq -\frac{R'}{\omega^3 L_{\text{leg}}^2 C}, \tag{A.5}$$

where another product of small resistances, $rR \ll \omega^2 L_{\text{leg}}^2$, has again been neglected. As in section 3.4.1, using the boundary condition $\sin(\beta N) = \sin(\text{m}\pi) = 0$, (A.4) is the same as for the normal mode frequencies. The neglected resistive terms R^2 and rR make only a very small difference to the driven resonance frequencies compared to the normal mode frequencies. This is reminiscent of the similar very small effect of resistance in the parallel resonant circuit, section 2.2.2.

A.2 Input impedance near resonance

We begin from the exact voltage difference (3.5.4), namely

$$V_{\text{rf}} = \left[Z_{\text{str}}(N_f - N_g)I_{\text{rf}} - Z_{\text{leg}}(I_{N_f} - I_{N_g})\right]/2, \tag{A.6}$$

and the exact expression for input impedance (3.44),

$$\begin{aligned} Z_{\text{in}} &= \frac{Z_{\text{str}}}{2}(N_f - N_g) + Z_{\text{eq}},\\ \text{with}\quad Z_{\text{eq}} &= Z_{\text{leg}} F_{(\gamma,N)} G_{(\gamma,N_g,N_f,N)},\end{aligned} \tag{A.7}$$

where Z_{eq} is the input impedance with the parallel chain of stringer impedances, Z_{str}, removed. By comparison with the current in the limit of weak dissipation, (3.41), we can see that $G_{(\gamma,N_g,N_f,N)} \simeq (-1)^{\text{m}} D^2$. Substituting also for Z_{leg} and $F_{(\gamma,N)}$, gives

$$Z_{eq} = j\omega L_{leg} \cdot \frac{\tanh(\gamma/2)}{2\sinh(\gamma N)} \cdot (-1)^m D^2. \tag{A.8}$$

First we re-write the numerator of $F_{(\gamma,N)}$, neglecting α

$$\tanh\frac{\gamma}{2} \simeq \tanh\frac{j\beta}{2} = j\tan\frac{\beta}{2} = \frac{j(1-\cos\beta)}{\sin\beta}. \tag{A.9}$$

Normal modes excited in a lossless antenna present a discrete frequency spectrum whereas driven currents in a dissipative antenna exhibit a continuous spectrum, with peaks of real impedance at the normal mode resonance frequencies to first order in α, as in (A.4). For a dissipative antenna, therefore, the mode frequency has continuous values and we consider a small variation in angular frequency $d\omega$ around the normal mode frequency ω_m, so that $\omega = \omega_m + d\omega$. This corresponds to a mode variation dm about mode number m (where $d\text{m} \ll 1$). We can therefore obtain an expression for $\sin\beta$ by differentiating (A.4) with respect to m:

$$-\sin(\beta) \cdot \frac{d\beta}{d\text{m}} = \frac{2}{\omega^3 LC} \cdot \frac{d\omega}{d\text{m}}. \tag{A.10}$$

Furthermore, the boundary condition $(\beta N) = (\text{m}\pi)$ can also be differentiated to give $\frac{d\beta}{d\text{m}} = \frac{\pi}{N}$. Substituting, we have

$$\sin\beta = -\frac{2N}{\pi\omega^3 LC} \cdot \frac{d\omega}{d\text{m}}. \tag{A.11}$$

Hence the numerator (A.9) can be expressed in terms of circuit components using (A.4) and (A.11).

Now we turn to the denominator of $F_{(\gamma,N)}$ and consider the trigonometric expansion of $\sinh(\gamma N)$:

$$\begin{aligned}\sinh(\gamma N) &= \sinh(\alpha N + j\beta N)\\ &= \sinh(\alpha N)\cos(\beta N) + j\cosh(\alpha N)\sin(\beta N).\end{aligned} \tag{A.12}$$

The boundary conditions are such that $\cos(\beta N) = \cos(\text{m}\pi) = (-1)^\text{m}$, and in the neighbourhood of a normal mode, $\sin(\beta N) = \sin((\text{m} + d\text{m})\pi) = \sin(\text{m}\pi)\cos(\pi d\text{m}) + \cos(\text{m}\pi)\sin(\pi d\text{m}) \simeq (-1)^\text{m}\pi d\text{m}$. Substituting in (A.12):

$$\sinh(\gamma N) \simeq (\alpha N + j\pi d\text{m})(-1)^\text{m}. \tag{A.13}$$

Finally, we can replace all the factors in (A.8) using (A.9), (A.13), (A.4), (A.5), (A.11), as follows:

$$
\begin{aligned}
Z_{\text{eq}} &= j\omega L_{\text{leg}} \cdot \frac{\tanh(\gamma/2)}{2\sinh(\gamma N)} \cdot (-1)^{\text{m}} D^2. \\
&= j\omega L_{\text{leg}} \cdot \frac{j(1-\cos\beta)}{\sin\beta} \cdot \frac{(-1)^{\text{m}} D^2}{2(\alpha N + j\pi d\text{m})(-1)^{\text{m}}} \\
&= -\,\omega L_{\text{leg}} \cdot (1-\cos\beta) \cdot \frac{D^2}{2} \cdot \frac{1}{N\alpha\sin\beta + j\pi\sin(\beta) d\text{m}} \\
&= (1-\omega^2 MC) \cdot \frac{D^2}{2N} \cdot \frac{1}{\frac{R'}{\omega^2 L_{\text{leg}}^2}\left[1 + \frac{2j\omega L_{\text{leg}}}{R'}\frac{d\omega}{\omega}\right]} \\
&= \frac{\text{fac}}{\frac{R'}{\omega^2 L_{\text{leg}}^2}\left[1 + \frac{2j\omega L_{\text{leg}}}{R'}\frac{d\omega}{\omega}\right]}; \quad \text{fac} = (1-\omega^2 MC) \cdot \frac{D^2}{2N}, \\
&= \frac{\text{fac}}{\frac{R'}{\omega^2 L_{\text{leg}}^2}\left[1 + 2jQ\frac{d\omega}{\omega}\right]}; \quad Q = \frac{\omega L_{\text{leg}}}{R'}, \\
Z_{\text{eq}} &= \frac{1}{Y_{\text{eq}}}; \quad Y_{\text{eq}} = \frac{1}{\text{fac}} \cdot \frac{R'}{\omega^2 L_{\text{leg}}^2}\left[1 + 2jQ\frac{d\omega}{\omega_{\text{m}}}\right],
\end{aligned}
\tag{A.14}
$$

where $Q = \omega_{\text{m}} L_{\text{leg}}/R'$ is the quality factor of the antenna at the resonance frequency of the mth mode, and Y_{eq} is the admittance of a parallel resonant $RL\|C$ circuit which is recognized from the introduction to parallel resonant circuits, equation (2.18) in section 2.2. Finally, the complete expression for the network input impedance is

$$
Z_{\text{in}}^{\text{eq}} \simeq \frac{Z_{\text{str}}}{2}(N_f - N_g) + \frac{1}{Y_{\text{eq}}}, \tag{A.15}
$$

as explained and discussed in section 3.5.6.

Appendix B

Impedance matrix calculations

With reference to figure 4.13 in chapter 4, the antenna node voltages, leg currents, stringer currents, and source currents are, respectively, written as column vectors:

$$\mathbf{A} = \begin{pmatrix} A_1 \\ A_2 \\ \vdots \\ A_N \end{pmatrix}, \quad \mathbf{I} = \begin{pmatrix} I_1 \\ I_2 \\ \vdots \\ I_N \end{pmatrix}, \quad \mathbf{J} = \begin{pmatrix} J_1 \\ J_2 \\ \vdots \\ J_N \end{pmatrix}, \quad \mathbf{S}^J = \begin{pmatrix} S_1^J \\ S_2^J \\ \vdots \\ S_N^J \end{pmatrix}, \tag{B.1}$$

and similarly for node voltages $\mathbf{B}$, stringer currents $\mathbf{K}$, and source currents $\mathbf{S}^K$. For consistent numbering, the stringer currents $\mathbf{J}$ and $\mathbf{K}$ are extended to N values by setting very large values for the impedance of the Nth stringers, which effectively sets $J_N = 0$ and $K_N = 0$ as required by the open network boundary conditions.

Current conservation at the network nodes in figure 4.13 is

$$J_n = J_{n-1} + I_n + S_n^J, \tag{B.2}$$

$$K_n = K_{n-1} - I_n + S_n^K, \tag{B.3}$$

where the source currents are included. Introducing the lower shift matrix $\bar{U}_L$, which has ones on the subdiagonal and zeroes elsewhere, and writing $\bar{U} = \bar{1} - \bar{U}_L$, where $\bar{1}$ is the identity matrix, current conservation expressed in matrix form becomes

$$\bar{U}\mathbf{J} = \mathbf{I} + \mathbf{S}^J, \tag{B.4}$$

$$\bar{U}\mathbf{K} = -\mathbf{I} + \mathbf{S}^K. \tag{B.5}$$

Ohm's law for the legs in matrix form, using (4.32) and (4.33) is

$$\mathbf{B} - \mathbf{A} = \bar{Z}_{\text{leg}}\mathbf{I}, \tag{B.6}$$

doi:10.1088/978-0-7503-5296-3ch16

where $\bar{Z}_{\text{leg}} = R\bar{1} + j\omega\bar{M}_{\text{leg/leg}}$, using $\bar{M}_{\text{leg/leg}}$ for the combined mutual partial inductance matrix for the legs, accounting for image currents in the screen (see figure 4.17) and plasma (see figure 6.14), so that

$$\bar{M}_{\text{leg/leg}} = \begin{pmatrix} M_{11} & M_{12} & \cdots & M_{1N} \\ M_{21} & M_{22} & \cdots & M_{2N} \\ \vdots & \vdots & \ddots & \vdots \\ M_{N1} & M_{N2} & \cdots & M_{NN} \end{pmatrix}, \tag{B.7}$$

where $M_{nq} = M_{qn} = M_{nq}^{\text{leg/leg}} - M_{nq}^{\text{leg/screen}} - M_{nq}^{\text{leg/plasma}}$, for $n, q = 1, \ldots, N$. The terms $M_{nn}^{\text{leg/leg}} = L_{\text{leg}}$ represent the self partial inductance of the legs.

Ohm's law for the stringers is more complicated than for the legs because of the contribution from the in-line stringers as well as from the opposing stringers, as mentioned in section 4.5.2. It can be written as

$$A_n - A_{n+1} = \left(r - \frac{j}{\omega C}\right) J_n + j\omega \sum_{q=1}^{N} M_{nq}^{\text{line}} J_q + j\omega \sum_{q=1}^{N} M_{nq}^{\text{opp}} K_q, \tag{B.8}$$

$$B_n - B_{n+1} = \left(r - \frac{j}{\omega C}\right) K_n + j\omega \sum_{q=1}^{N} M_{nq}^{\text{line}} K_q + j\omega \sum_{q=1}^{N} M_{nq}^{\text{opp}} J_q, \tag{B.9}$$

where $\bar{M}_{\text{line}}$ is the combined mutual partial inductance matrix of the elements M_{nq}^{line} for the in-line stringers, subtracting the mutual partial inductances for image currents in the screen and plasma, and similarly for $\bar{M}_{\text{opp}}$ regarding the elements M_{nq}^{opp} for the opposing stringers. Recognizing that the upper shift matrix is the transposition of the lower shift matrix, Ohm's law in matrix form for the stringers is

$$\bar{U}^T \mathbf{A} = \left[\left(r - \frac{j}{\omega C}\right) + j\omega \bar{M}_{\text{line}}\right] \mathbf{J} + j\omega \bar{M}_{\text{opp}} \mathbf{K}, \tag{B.10}$$

$$\bar{U}^T \mathbf{B} = \left[\left(r - \frac{j}{\omega C}\right) + j\omega \bar{M}_{\text{line}}\right] \mathbf{K} + j\omega \bar{M}_{\text{opp}} \mathbf{J}, \tag{B.11}$$

whose difference gives

$$\bar{U}^T(\mathbf{B} - \mathbf{A}) = \bar{Z}_{\text{str}}(\mathbf{K} - \mathbf{J}), \tag{B.12}$$

where we define a stringer impedance matrix as $\bar{Z}_{\text{str}} = \left(r - \frac{j}{\omega C}\right)\bar{1} + j\omega\left(\bar{M}_{\text{line}} - \bar{M}_{\text{opp}}\right)$.

We now proceed to solve for the leg current vector $\mathbf{I}$, first by eliminating the node voltage difference using (B.6) and (B.12) to give

$$\bar{U}^T \bar{Z}_{\text{leg}} \mathbf{I} = \bar{Z}_{\text{str}}(\mathbf{K} - \mathbf{J}), \tag{B.13}$$

and then using (B.5) minus (B.4) to eliminate the stringer current difference in (B.13), to finally obtain

$$\mathbf{I} = \left(\bar{U}\ \bar{Z}_{\text{str}}^{-1}\ \bar{U}^{T}\ \bar{Z}_{\text{leg}} + 2\bar{1}\right)^{-1}(\mathbf{S}^{K} - \mathbf{S}^{J}). \tag{B.14}$$

This defines the leg currents in terms of the network impedances and the source currents, which are known for symmetrical connections, or for a single ground connection.

The node voltages are found by adding the two pairs (B.4), (B.5), and (B.10), (B.11), and using (B.6) to give

$$\mathbf{A} = \frac{1}{2}\left[(\bar{U}^{T})^{-1}Z_{\text{str}}\ \bar{U}^{-1}(\mathbf{S}^{K} + \mathbf{S}^{J}) + Z_{\text{leg}}\mathbf{I}\right], \tag{B.15}$$

$$\mathbf{B} = \frac{1}{2}\left[(\bar{U}^{T})^{-1}\ \bar{Z}_{\text{str}}\ \bar{U}^{-1}(\mathbf{S}^{K} + \mathbf{S}^{J}) - \bar{Z}_{\text{leg}}\mathbf{I}\right]. \tag{B.16}$$

The matrices $\bar{U}$, $\bar{Z}_{\text{leg}}$, $\bar{Z}_{\text{str}}$, and their combinations, arc always invertible and give physically reasonable values for the antenna currents and voltages. Finally, the antenna input impedance is given by the input voltage divided by the input current, i.e.

$$Z_{\text{in}} = (A_{\text{input}} - A_{\text{ground}})\big/S_{\text{input}}^{J}, \tag{B.17}$$

which can be compared with measurements. See appendix K for a link to an example.

IOP Publishing

Resonant Network Antennas for Radio-Frequency Plasma Sources

Theory, technology and applications

Philippe Guittienne, Alan Howling and Ivo Furno

Appendix C

Electron–molecule energy transfer fraction

Chapter 5, specifically section 5.1.1, makes reference to electron collisions in relation to plasma chemistry. Some basic notions [1] are described here.

C.1 Elastic collisions

For simplicity, we consider a head-on elastic collision in figure C.1; this gives the maximum energy transfer fraction. Momentum is always conserved, and because the collision is elastic, the kinetic energy is also conserved:

$$\begin{cases} \text{Conservation of momentum} & mV = mv_1 + Mv_2 \\ \text{Conservation of kinetic energy} & \dfrac{1}{2}mV^2 = \dfrac{1}{2}mv_1^2 + \dfrac{1}{2}Mv_2^2. \end{cases} \tag{C.1}$$

The first equation gives $v_1 = (mV - Mv_2)/m$, which, when substituted into the second equation, yields

$$v_2 = \frac{2mV}{M+m}. \tag{C.2}$$

The maximum fraction of energy transferred to M is then

$$\delta_{\max} = \frac{\text{Energy transferred to M}}{\text{Initial energy of m}} = \frac{\frac{1}{2}Mv_2^2}{\frac{1}{2}mV^2} = \frac{4mM}{(M+m)^2}. \tag{C.3}$$

We calculate two cases:

- For an electron striking an atom ($m \ll M$) head-on:

$$\delta_{\max} \simeq \frac{4m}{M} << 1 \qquad (\sim 1/18360 \text{ for argon}), \tag{C.4}$$

doi:10.1088/978-0-7503-5296-3ch17

Figure C.1. Schematic of a head-on collision between masses m and M. Initially, M is stationary and m impacts at velocity V.

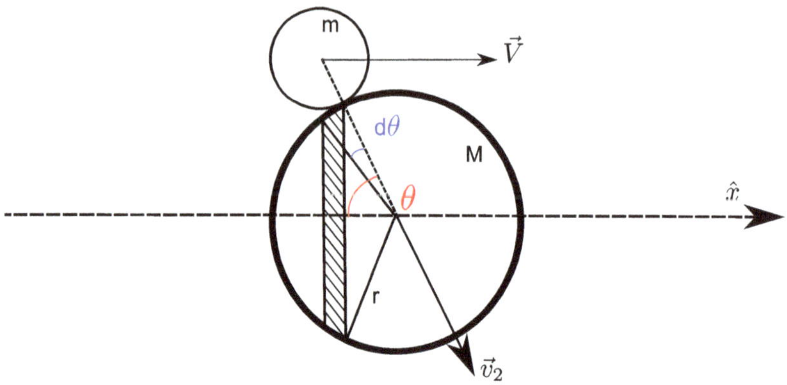

Figure C.2. Sketch of the collision in figure C.1, but for the impact at a general angle θ with respect to the centre of M.

which implies that electrons are inefficient in heating the gas. The electron ricochets off the massive molecule which hardly moves. On averaging over the impact parameter for these hard sphere collisions, the average electron forward momentum is zero after the elastic collision in the limit $m \ll M$.

- For an ion striking an atom head-on ($m = M$ in (C.3)):

$$\delta_{\max} = 1, \tag{C.5}$$

so ions and atoms thermalize efficiently ($T_i \approx T_{\text{gas}}$). For the gas in thermal equilibrium with the walls, we have $T_i \approx T_{\text{gas}} \approx T_{\text{wall}}$.

Elastic collisions at an arbitrary impact angle

We can also consider a general impact angle θ as shown in figure C.2. To compute the energy transfer fraction corresponding to the angle θ, we can proceed in the same way as point (a), keeping in mind that now in the equation of the momentum conservation along the line of centres, i.e. along the direction of v_2, we have the projection of the initial velocity of the incoming particle, $V \cos\theta$. It then follows:

$$\delta(\theta) = \frac{4mM}{(m+M)^2}\cos^2\theta \tag{C.6}$$

To compute the average of this quantity, we should first calculate the probability $p(\theta)$ of a collision with a certain angle θ (fraction of collisions corresponding to an angle θ). This can be obtained as

$$p(\theta) = \frac{dA}{A} = \frac{2\pi r \sin\theta\cos\theta\, d\theta}{\pi r^2} = 2\sin\theta\cos\theta\, d\theta, \tag{C.7}$$

where $dA = 2\pi r \sin\theta \cos\theta\, d\theta$ is the impact area between the angle θ and $\theta + d\theta$, while $A = \pi r^2$ is the total available impact area. Then the average energy transfer fraction is

$$\begin{aligned}\bar{\delta} &= \int_0^{\pi/2} \delta(\theta)p(\theta) = \frac{4mM}{(m+M)^2}\int_0^{\pi/2} 2\sin\theta\cos^3\theta\, d\theta \\ &= \delta_{\max} 2\left(-\frac{\cos^4\theta}{4}\right)\Bigg|_0^{\pi/2} = \frac{\delta_{\max}}{2},\end{aligned} \tag{C.8}$$

which is just half of the maximum in the head-on collision in (C.4). We see that for an electron striking an argon atom ($m \ll M$), the average energy transfer fraction is

$$\bar{\delta} \simeq \frac{2m}{M} = \frac{1}{9180} \ll 1. \tag{C.9}$$

Elastic collisions with an infinite mass target

The average momentum in the original direction for the situation of a target particle of infinite mass can be obtained in a similar way, first calculating the projected momentum along the impact direction for a general angle θ, and then averaging. In figure C.3, the projected momentum along the impact direction ($\hat{x}$) is

$$Q_x(\theta) = -mV\cos(2\theta). \tag{C.10}$$

The average is then

$$\begin{aligned}\bar{Q} &= \int_0^{\pi/2} Q_x(\theta)p(\theta) = -mV\int_0^{\pi/2} \cos(2\theta)\sin(2\theta)\, d\theta \\ &= -mV\left[\frac{\sin^2(2\theta)}{4}\right]\Bigg|_0^{\pi/2} = 0.\end{aligned} \tag{C.11}$$

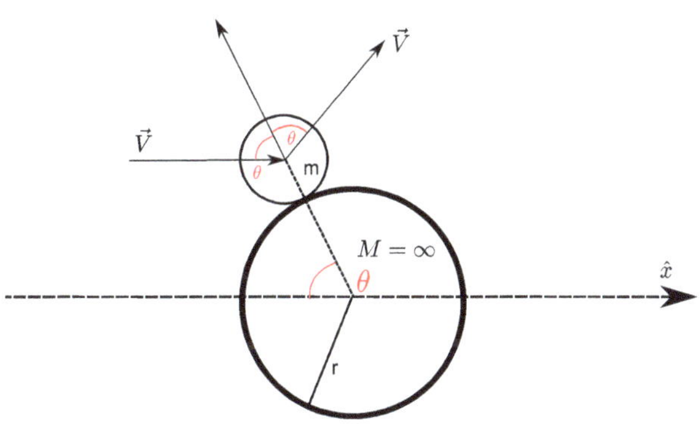

Figure C.3. Sketch for an elastic collision of m with an infinite mass target particle.

This means that on average, all the forward momentum is lost in a collision with an infinitely heavy scattering centre.

C.2 Inelastic collisions

We again consider a head-on collision, to maximize the possible energy transfer, by an electron, mass m in an inelastic collision with M. In the previous example, the target molecule M was taken to be a billiard ball, but now we consider that the molecule has some internal structure that can be modified (excited) by gaining some energy ΔU. The schematic is the same as in figure C.1, except that now the kinetic energy is not conserved because of the simultaneous transfer of ΔU in the collision. The conservation equations are now

$$\begin{cases} \text{Conservation of momentum} & mV = mv_1 + Mv_2 \\ \text{Conservation of total energy} & \frac{1}{2}mV^2 = \frac{1}{2}mv_1^2 + \left(\frac{1}{2}Mv_2^2 + \Delta U\right) \end{cases}. \tag{C.12}$$

There are two equations but three unknowns. To obtain a third condition, we first eliminate v_1, as before, to obtain

$$0 = -2mMVv_2 + M^2v_2^2 + mMv_2^2 + 2\,m\Delta U, \tag{C.13}$$

and then we choose to maximize ΔU with respect to v_2 using $\partial \Delta U/\partial v_2 = 0$. This gives

$$v_2 = \frac{mV}{m+M} \quad \Rightarrow \quad \Delta U_{\max} = \frac{1}{2}mV^2\left(\frac{M}{M+m}\right). \tag{C.14}$$

The maximum energy transfer fraction is therefore

$$\delta_{\max} = \frac{\Delta U_{\max}}{\frac{1}{2}mV^2} = \frac{M}{m+M} \approx 1, \tag{C.15}$$

which shows that an electron can transfer almost all of its kinetic energy in an inelastic collision with an atom or molecule having a suitable internal excited state of comparable energy. The simultaneous kinetic energy transfer fraction is

$$\delta_{\max} = \frac{\frac{1}{2}Mv_2^2}{\frac{1}{2}mV^2} = \frac{Mm}{(m+M)^2} \approx \frac{m}{M} \ll 1. \tag{C.16}$$

This means that high energy electrons can strongly modify the molecular internal energy, such as its chemical bonds, without heating the molecule. This is a key interest of plasma chemistry, as outlined in section 5.1.1.

Reference

[1] Raizer Y P 1991 *Gas Discharge Physics* (Berlin: Springer)

IOP Publishing

Resonant Network Antennas for Radio-Frequency Plasma Sources

Theory, technology and applications

Philippe Guittienne, Alan Howling and Ivo Furno

Appendix D

Maxwell's equations, plasma permittivity, and skin depth

This appendix provides some background to the introductory remarks on inductively coupled plasma (ICP) in chapter 5. We begin by revising Maxwell's equations and their quasistatic approximations which correspond to the *E*-mode (CCP reactors) and the *H*-mode (ICP reactors). Since part II is concerned with non-magnetized plasma, we then consider electron motion in an RF electric field in presence of electron–neutral collisions, which leads to the plasma permittivity. The Helmholtz wave equation then shows how a wave's amplitude is damped when propagating through the plasma dielectric, over a $1/e$ distance called the skin depth. This is the basis of inductive coupling in plasma.

D.1 Maxwell's equations

For free charges in vacuum, Maxwell's equations can be written as follows [1]:

$$\varepsilon_0 \nabla \cdot \mathbf{E} = \rho, \tag{D.1}$$

$$\nabla \cdot \mathbf{B} = 0, \tag{D.2}$$

$$\nabla \times \mathbf{E} = -\frac{\partial \mathbf{B}}{\partial t}, \tag{D.3}$$

$$\nabla \times \mathbf{H} = \mathbf{J} + \varepsilon_0 \frac{\partial \mathbf{E}}{\partial t}, \tag{D.4}$$

where ε_0 is the permittivity of vacuum, ρ is the volume density of free charge, and $\mathbf{J}$ is the conduction current density of free charge flowing across unit cross-sectional area. In absence of any ferromagnetic materials in this work, $\mathbf{B} = \mu_0 \mathbf{H}$, where μ_0 is the vacuum permeability. Self-consistent solutions give rise to electromagnetic waves

doi:10.1088/978-0-7503-5296-3ch18

(section D.4) and electromagnetic *EM*-mode solutions (section 5.2.3 and chapter 7) for the currents, charges, and the **E** and **B** fields in all plasma reactors, for any dimensions.

D.1.1 Two quasistatic approximations

These Maxwell equations feature two time derivatives, one for the magnetic flux density **B** in (D.3), and one for the electric field intensity **E** in (D.4). Quasistatic approximations of these equations are easier to solve and the solution accuracy is acceptable when frequencies are 'low', and reactor dimensions are 'small'; in practice, this means 'when an electromagnetic wave can cross the reactor before anything much changes in the system'. More quantitatively, this is when $L/c \ll 1/f_{\rm rf}$, i.e. quasistatic conditions hold when $L \ll \lambda_{\rm rf}$, where $\lambda_{\rm rf}$ is the RF wavelength in the system. This is therefore the same as the definition for lumped element circuits in section 2.3.

It is important to note, however, that a 'quasistatic approximation' does not automatically mean that *both* of the time derivatives are simultaneously zero [2]. Two different situations arise, depending on which time derivative is neglected, as described in the following sections.

The electro-quasistatic approximation

Also called the quasi-electrostatic approximation—applies when the magnetic induction term is neglected in (D.3), namely, $\nabla \times \mathbf{E} \approx 0$ (so electromagnetic wave solutions are excluded), and the time-varying magnetic field is effectively zero, but the vacuum displacement current $\varepsilon_0 \frac{\partial \mathbf{E}}{\partial t}$ is retained in (D.4). Electric fields and potentials dominate in the static limit, and it is natural to denote this the electro-quasistatic limit even though the excitation electric potentials are (slowly) time-varying [2]. Clearly, this electro-quasistatic approximation is associated with the *E*-mode of the CCP reactor in section 5.2.1. The ratio of the electric error field to the quasistatic electric field is of the order $(L/\lambda_{\rm rf})^2 \ll 1$ [2]. The electro-quasistatic fields are resumed as follows [2]:

$$\varepsilon_0 \nabla \cdot \mathbf{E} = \rho, \tag{D.5}$$

$$\nabla \cdot \mathbf{B} = 0, \tag{D.6}$$

$$\nabla \times \mathbf{E} \approx 0, \tag{D.7}$$

$$\nabla \cdot \mathbf{J} + \frac{\partial \rho}{\partial t} = 0. \tag{D.8}$$

The fourth equation (D.8), representing current continuity, is obtained from the divergence of $\nabla \times \mathbf{H} = \mathbf{J} + \varepsilon_0 \frac{\partial \mathbf{E}}{\partial t}$. The magnetic field is thus effectively eliminated from the equation set, but it can be recovered, if required, using this latter equation.

The magneto-quasistatic approximation

Also called the quasi-magnetostatic approximation—applies when the vacuum displacement current is neglected in (D.4), namely, $\nabla \times \mathbf{H} \approx \mathbf{J}$ (so electromagnetic wave solutions are excluded), but the magnetic induction term $-\partial\mathbf{B}/\partial t$ is retained in (D.3), so the induced electric field is, in fact, not zero. Magnetic fields and currents dominate in the static limit, and it is natural to denote this the magneto-quasistatic limit even though the magnetic fields and currents are (slowly) time-varying [2]. Clearly, this magneto-quasistatic approximation is associated with the H-mode of the ICP reactor in section 5.2.2. The ratio of the magnetic error field to the quasistatic magnetic field is of the order $(L/\lambda_{\rm rf})^2$, the same negligible order of magnitude as in the electro-quasistatic approximation [2]. The magneto-quasistatic fields are resumed as follows [2]:

$$\varepsilon_0\nabla\cdot\mathbf{E}=\rho, \tag{D.9}$$

$$\nabla\cdot\mathbf{B}=0, \tag{D.10}$$

$$\nabla\times\mathbf{E}=-\frac{\partial\mathbf{B}}{\partial t}, \tag{D.11}$$

$$\nabla\times\mathbf{H}\approx\mathbf{J}. \tag{D.12}$$

Note that the induced electric field is retained, representing Faraday's law in (D.11).

D.2 Electron fluid conductivity in RF electric fields

The linearized, force balance equation (momentum conservation) for an electron fluid in a cold, uniform, unmagnetized plasma, neglecting pressure, is [3]

$$m_e n_{e0}\frac{d\mathbf{u}}{dt}=-qn_{e0}\mathbf{E}-m_e\nu n_{e0}\mathbf{u}, \tag{D.13}$$

where q is the magnitude of the electron charge, ν the electron–neutral collision frequency, $\mathbf{u}$ the electron fluid velocity, m_e the electron mass, and n_{e0} the time-constant electron number density. This is essentially the Drude model for electron collisions in a gas in presence of an oscillating electric field. The collisional force term in (D.13) applies in the sense that, on average, a collision reduces the average electron forward momentum to zero, as noted at the end of appendix C.1. The force on the left-hand side is the mass density times the acceleration, representing Newton's second law. The acceleration is the convective derivative of the fluid velocity $\mathbf{u}$; it comprises two terms: $\frac{d\mathbf{u}}{dt}=\frac{\partial\mathbf{u}}{\partial t}+(\mathbf{u}\cdot\nabla)\mathbf{u}$, where $\frac{\partial\mathbf{u}}{\partial t}$ is the acceleration due to an explicitly time-varying $\mathbf{u}$, and $(\mathbf{u}\cdot\nabla)\mathbf{u}$ is a so-called inertial term due to a spatial variation in $\mathbf{u}$. The inertial term is non-linear in $\mathbf{u}$, and in the absence of significant plasma flow in the RF plasma sources of this book, it can be neglected compared to first order terms [3], so that $\frac{d\mathbf{u}}{dt}\approx\frac{\partial\mathbf{u}}{\partial t}$ to all intents and purposes. Finally, the force equation (D.13) is the linearized first velocity moment of Boltzmann's equation where the pressure gradient has been neglected (cold plasma theory [4]), and products of first order terms are small enough to be ignored compared with zero order and first order terms.

Putting $\mathbf{E} = \mathrm{Re}(\mathbf{E}_0 e^{-j\omega t})$, as throughout this book according to section 2.1.1, then $\frac{d}{dt} = -j\omega$ for all the first order phasor flow variables such as $\mathbf{u}$ and the electron current $\mathbf{J}_e$, and all of the electromagnetic field phasor variables such as $\mathbf{E}$, $\mathbf{H}$, and $\mathbf{B}$. The electron density n_{e0} is the background, time-constant plasma density term (zeroth order in frequency ω) which contributes no temporal variation to first order in the RF frequency ω, because any product with an electron density perturbation is of second order at least. Hence (D.13) can be written as

$$\mathbf{u} = -\frac{q}{m_e}\frac{1}{(\nu - j\omega)}\mathbf{E}, \tag{D.14}$$

and the electron current is

$$\mathbf{J}_e = -n_{e0}q\mathbf{u} = \frac{n_{e0}q^2}{m_e}\frac{1}{(\nu - j\omega)}\mathbf{E}. \tag{D.15}$$

Hence the plasma conductivity $\sigma_p = \mathbf{J}_e/\mathbf{E}$ is generally a complex value because of electron inertia:

$$\sigma_p = \frac{n_{e0}q^2}{m_e\nu}\left[\frac{1}{1 - j\omega/\nu}\right] = \sigma_{\mathrm{dc}}\left[\frac{1}{1 - j\omega/\nu}\right]; \quad \sigma_{\mathrm{dc}} = \frac{n_{e0}q^2}{m_e\nu}, \tag{D.16}$$

where σ_{dc} is the real DC plasma conductivity in a static electric field. The complex plasma conductivity σ_p can also be re-written using the electron plasma frequency squared, $\omega_{\mathrm{pe}}^2 = \frac{n_{e0}q^2}{\varepsilon_0 m_e}$, to give

$$\sigma_p = \frac{\varepsilon_0\omega_{\mathrm{pe}}^2}{\nu - j\omega}. \tag{D.17}$$

Equally, the plasma resistivity $\rho_p = \mathbf{E}/\mathbf{J}_e = 1/\sigma_p$ is also complex:

$$\rho_p = \frac{m_e\nu}{n_{e0}q^2}(1 - j\omega/\nu) = \rho_{\mathrm{dc}}(1 - j\omega/\nu); \quad \rho_{\mathrm{dc}} = \frac{m_e\nu}{n_{e0}q^2}, \tag{D.18}$$

where ρ_{dc} is the real DC plasma resistivity in a static electric field.

D.2.1 Ohmic power dissipation in an unmagnetized RF plasma

The ohmic power dissipated per unit volume in the plasma as a function of time, following section 2.1.1, is $P(t) = \mathrm{Re}(\mathbf{E}(t))\mathrm{Re}(\mathbf{J}(t))$. An expression for the time-averaged electrical dissipated power density $\bar{P}$ can be developed as follows [5]:

$$\begin{aligned} P(t) &= \mathrm{Re}\,[\mathbf{E}_0 e^{-j\omega t}] \cdot \mathrm{Re}\,[\mathbf{J}_0 e^{-j\omega t}] \\ &= \frac{1}{2}[\mathbf{E}_0 e^{-j\omega t} + \mathbf{E}_0^* e^{j\omega t}] \cdot \frac{1}{2}[\mathbf{J}_0 e^{-j\omega t} + \mathbf{J}_0^* e^{j\omega t}], \end{aligned} \tag{D.19}$$

using the complex number general result $\mathrm{Re}[z] = \frac{1}{2}(z + z^*)$. In taking the time average, the oscillatory factors $e^{\pm 2j\omega t}$ average out, leaving

$$\bar{P} = \frac{1}{4}[\mathbf{E}_0 \cdot \mathbf{J}_0^* + \mathbf{E}_0^* \cdot \mathbf{J}_0] = \frac{1}{2}\,\mathrm{Re}\,[\mathbf{E}_0 \cdot \mathbf{J}_0^*], \tag{D.20}$$

noting that $\mathbf{E}_0^* \cdot \mathbf{J}_0 = (\mathbf{E}_0 \cdot \mathbf{J}_0^*)^*$, and using the same complex number result in reverse. The time-averaged ohmic power density, as a function of the electric field amplitude E_0, is therefore

$$\begin{aligned}\bar{P} &= \frac{1}{2}\,\mathrm{Re}(\mathbf{E}_0 \cdot \mathbf{J}_0^*)\\ &= \frac{1}{2}\,\mathrm{Re}\left(\mathbf{E}_0 \cdot \left[\frac{n_{e0}q^2}{m_e}\right]\frac{1}{(\nu + j\omega)}\mathbf{E}_0^*\right)\\ &= \left[\frac{n_{e0}q^2}{2m_e}\right]\left(\frac{\nu}{\nu^2 + \omega^2}\right)E_0^2\\ &= \frac{1}{2}\sigma_{\mathrm{dc}}\left(\frac{\nu^2}{\nu^2 + \omega^2}\right)E_0^2.\end{aligned} \tag{D.21}$$

Ohmic power dissipation depends on ν as stated in section 5.1.1: continuous power transfer occurs due to the phase-randomizing collisions relative to the electric field oscillation. Electrons oscillating in collisionless conditions, $\nu = 0$, do not gain power continuously. We can find the collision frequency for maximum power density at a given electric field amplitude E_0 and frequency ω, by setting $\frac{\partial \bar{P}}{\partial \nu} = 0$, which gives $\nu = \omega$. Hence, the maximum ohmic power transfer by electron elastic collisions is

$$\bar{P}_{\max} = \left[\frac{n_{e0}q^2E_0^2}{4\nu m_e}\right] = \frac{1}{4}\sigma_{\mathrm{dc}}E_0^2. \tag{D.22}$$

To give an example relevant to section 5.1, for an RF discharge at 13.56 MHz in argon, ν is equal to ω at 3.33 Pa gas pressure [3] for maximum ohmic power dissipation on the basis of this simple volume model, which does not account for sheaths or other phenomena such as stochastic heating.

The time-averaged power density can also be calculated as a function of the current density amplitude, J_0, and the plasma resistivity:

$$\begin{aligned}\bar{P} &= \frac{1}{2}\,\mathrm{Re}(\mathbf{E}_0 \cdot \mathbf{J}_0^*)\\ &= \frac{1}{2}\,\mathrm{Re}\left(\rho_p \mathbf{J}_0 \cdot \mathbf{J}_0^*\right)\\ &= \frac{1}{2}\,\mathrm{Re}\left(\rho_p J_0^2\right)\\ &= \frac{1}{2}\rho_{\mathrm{dc}}J_0^2,\end{aligned} \tag{D.23}$$

obtained directly from (D.18); this ohmic power expression[1] is apparently simpler than (D.21) because the complex denominator in (D.15) and (D.16) appears in the numerator of the resistivity (D.18). The equality of the power expressions (D.21) and (D.23) is recognized by arranging the magnitude of the initial equation (D.15) as follows:

$$
\begin{aligned}
J_0^2 &= \left[\frac{n_{e0}q^2}{m_e}\right]^2 \frac{1}{\nu^2+\omega^2} E_0^2, \\
J_0^2 &= \left[\frac{n_{e0}q^2}{m_e\nu}\right]^2 \frac{\nu^2}{\nu^2+\omega^2} E_0^2, \\
\left[\frac{m_e\nu}{n_{e0}q^2}\right] J_0^2 &= \left[\frac{n_{e0}q^2}{m_e\nu}\right] \frac{\nu^2}{\nu^2+\omega^2} E_0^2, \\
\text{i. e.}\quad \frac{1}{2}\rho_{\mathrm{dc}} J_0^2 &= \frac{1}{2}\sigma_{\mathrm{dc}}\left(\frac{\nu^2}{\nu^2+\omega^2}\right) E_0^2.
\end{aligned}
\tag{D.24}
$$

D.3 Unmagnetized plasma permittivity

The electrical response of a plasma can be modelled as an electric conduction current of free charges moving in a low pressure gas with an overall electrical conductivity σ_p, or as an effective dielectric medium with complex relative permittivity ε_p. These two descriptions are equivalent as shown using the following Maxwell equation

$$
\begin{aligned}
\nabla \times \mathbf{H} &= \mathbf{J} + \varepsilon_0 \frac{\partial \mathbf{E}}{\partial t} = \sigma_p \mathbf{E} - j\omega\varepsilon_0 \mathbf{E} \\
&= (\sigma_p - j\omega\varepsilon_0)\mathbf{E} \quad \text{(for charges moving in vacuum)}
\end{aligned}
\tag{D.25}
$$

$$
\begin{aligned}
&= -j\omega\varepsilon_0\left(\frac{j\sigma_p}{\omega\varepsilon_0} + 1\right)\mathbf{E} \\
&= -j\omega\varepsilon_0\varepsilon_p\mathbf{E}, \quad \text{(for effective dielectric medium)}
\end{aligned}
\tag{D.26}
$$

$$
\text{where}\quad \varepsilon_p = \left(\frac{j\sigma_p}{\omega\varepsilon_0} + 1\right),
$$

remembering that σ_p itself is generally complex because of electron inertia in section D.2. By substitution for σ_p, (D.17), in (D.26), the standard expression for plasma relative permittivity ε_p is [3, 6]

$$
\varepsilon_p = 1 - \frac{\omega_{\mathrm{pe}}^2}{\omega(\omega + j\nu)},
\tag{D.27}
$$

where $\omega_{\mathrm{pe}}^2 = \frac{n_{e0}q^2}{\varepsilon_0 m_e}$ is the square of the electron plasma frequency first considered in section 5.1.2.

[1] This same expression is also obtained for RF magnetized plasma in (10.72).

D.4 The wave equation in unmagnetized plasma

The electric and magnetic fields incident on a uniform, non-magnetized plasma are assumed to vary as

$$\mathbf{E}, \mathbf{H} \propto e^{j(\mathbf{k}\cdot\mathbf{r}-\omega t)}, \tag{D.28}$$

which are plane waves because $\mathbf{k} \cdot \mathbf{r} =$ constant (at a given time), which is the equation of a plane perpendicular to the wave vector $\mathbf{k}$. The propagation of waves in a plasma can be treated from the point of view of the plasma as an effective dielectric of relative permittivity ε_p (D.26). Maxwell's equations in a uniform, non-ferromagnetic ($\mu_r = 1$) dielectric can be written generally, and for the specific case of the plane harmonic wave, as

$$\nabla \cdot \mathbf{E} = 0; \qquad j\mathbf{k} \cdot \mathbf{E} = 0, \tag{D.29}$$

$$\nabla \cdot \mathbf{B} = 0; \qquad j\mathbf{k} \cdot \mathbf{B} = 0, \tag{D.30}$$

$$\nabla \times \mathbf{E} = -\frac{\partial \mathbf{B}}{\partial t}; \quad j\mathbf{k} \times \mathbf{E} = j\omega\mathbf{B}, \tag{D.31}$$

$$\nabla \times \mathbf{H} = \varepsilon_0\varepsilon_p\frac{\partial \mathbf{E}}{\partial t}; \quad j\mathbf{k} \times \mathbf{H} = -j\omega\varepsilon_0\varepsilon_p\mathbf{E}. \tag{D.32}$$

In this description of plasma as an effective dielectric, all of the current is carried by the total effective displacement current, $-j\omega\varepsilon_0\varepsilon_p\mathbf{E}$. The first two equations show that $\mathbf{B} = \mu_0\mathbf{H}$ and $\mathbf{E}$ are both perpendicular to the direction of propagation $\mathbf{k}$, i.e. the plane wave fields are transverse. Furthermore, the third and fourth equations each show that $\mathbf{E}$ and $\mathbf{B}$ are perpendicular to each other, as well as to the direction of propagation. Hence, in an unmagnetized plasma, an electromagnetic electron plane wave is simply a light wave (propagating along $\mathbf{E} \times \mathbf{B}$) modified by the plasma relative permittivity. The wave has an oscillating magnetic field component $\mathbf{B}$, but there is no static magnetic field in an unmagnetized plasma by definition. Combining the last two Maxwell equations by taking the curl of the third and substituting the fourth, gives the Helmholtz wave equation for unmagnetized plasma:

$$\begin{aligned}\nabla \times (\nabla \times \mathbf{E}) = \nabla \times \left(-\frac{\partial \mathbf{B}}{\partial t}\right) &= -\mu_0\frac{\partial}{\partial t}(\nabla \times \mathbf{H}) = -\mu_0\varepsilon_0\varepsilon_p\frac{\partial^2 \mathbf{E}}{\partial t^2},\\ \nabla^2\mathbf{E} = \mu_0\varepsilon_0\varepsilon_p\frac{\partial^2 \mathbf{E}}{\partial^2 t},&\end{aligned} \tag{D.33}$$

and the same for $\mathbf{H}$, using the vector identity $\nabla \times (\nabla \times \mathbf{a}) \equiv \nabla(\nabla \cdot \mathbf{a}) - \nabla^2\mathbf{a}$. For general geometry, but the specific case of a single angular frequency ω, the Helmholtz equation becomes

$$\nabla^2\mathbf{E} = -\omega^2\mu_0\varepsilon_0\varepsilon_p\mathbf{E} = -\frac{\omega^2}{c^2}\varepsilon_p\mathbf{E} = -k_0^2\varepsilon_p\mathbf{E} = -k_d^2\mathbf{E}, \tag{D.34}$$

where k_d is the wavenumber in the effective dielectric. The plane wave formulation for (D.33) is

$$j\mathbf{k} \times (j\mathbf{k} \times \mathbf{E}) = j\mathbf{k} \times (j\omega\mu_0\mathbf{H}) = -j\omega(j\omega\mu_0\varepsilon_0\varepsilon_p\mathbf{E}),$$
$$k^2\mathbf{E} = \frac{\omega^2}{c^2}\varepsilon_p\mathbf{E} = k_0^2\varepsilon_p\mathbf{E}, \tag{D.35}$$

using $\mathbf{k} \times \mathbf{k} \times \mathbf{E} \equiv (\mathbf{k} \cdot \mathbf{E})\mathbf{k} - k^2\mathbf{E} = -k^2\mathbf{E}$ here, and $k_0 = \frac{\omega}{c}$ is the wavenumber of light in vacuum. A non-zero electric field exists only if

$$k = k_0\sqrt{\varepsilon_p}, \tag{D.36}$$

which is the dispersion relation for transverse electromagnetic electron waves.

D.5 Plasma skin depths for inductive plasmas

D.5.1 General complex skin depth

Taking a plane wave to propagate along z, then $\mathbf{k} = k\hat{\mathbf{z}}$, and $\mathbf{k} \cdot \mathbf{r} = kz$, so the fields propagate as $\mathbf{E}, \mathbf{H} \propto e^{j(kz-\omega t)}$. Inspecting the unmagnetized plasma dispersion relation (D.36) and the plasma relative permittivity (D.27) which is complex, we anticipate that the wavenumber k will have real and imaginary parts. We choose to write these as

$$k = K + \frac{j}{\delta}, \tag{D.37}$$

where $K = \mathrm{Re}(k)$ and $(1/\delta) = \mathrm{Im}(k)$ are purely real, because the wave is then seen to propagate as

$$H = H_0e^{j(kz-\omega t)} = H_0[e^{-z/\delta}]e^{j(Kz-\omega t)}, \tag{D.38}$$

where δ, the real skin depth, is the $1/e$ spatial decay constant as the wave amplitude is damped, or attenuated, during propagation with real wavenumber K into the plasma.

Alternatively, we could equally well choose to introduce a complex skin depth p, defined as follows for plane waves [7]:

$$H = H_0[e^{-z/p}]e^{-j\omega t}, \tag{D.39}$$

where, by equating the factors for $(-z)$ between (D.38) and (D.39), we have

$$\frac{1}{p} = \frac{1}{\delta} - jK, \tag{D.40}$$

so that $\mathrm{Re}(1/p) = (1/\delta)$ and $\mathrm{Im}(1/p) = -K$. Comparison of (D.37) with (D.40) gives the simple relation between the complex wavenumber and the complex skin depth:

$$k = \frac{j}{p}, \quad \text{or} \quad p = \frac{j}{k}. \tag{D.41}$$

To avoid disappointment, note that the real part of the complex skin depth is *not* equal to the skin depth! Instead, from (D.40), $p = \frac{1}{1/\delta - jK} = \frac{1/\delta + jK}{1/\delta^2 + K^2}$ so that

$\mathrm{Re}(p) = \frac{1/\delta}{1/\delta^2 + K^2} = \frac{\delta}{1 + K^2\delta^2} \neq \delta$. The real parts do equate for their reciprocals: $\mathrm{Re}(1/p) = (1/\delta)$, as stated for (D.40). The usefulness of the complex skin depth p will become apparent in appendix E and section 6.4.

D.5.2 Collisional skin depth

For collisional RF plasma, $\omega \ll \nu$, ω_{pe} and the plasma conductivity σ_p in (D.16) approaches the DC value σ_{dc} because collisions dominate the force balance in (D.13). Hence the plasma permittivity in (D.26) is (the subscript c denotes collisional)

$$
\begin{aligned}
&\varepsilon_p \approx \frac{j\sigma_{\mathrm{dc}}}{\omega\varepsilon_0},\\
&k_c = k_0\sqrt{\varepsilon_p} \approx k_0\sqrt{\frac{j\sigma_{\mathrm{dc}}}{\omega\varepsilon_0}} = \sqrt{j\omega\mu_0\sigma_{\mathrm{dc}}} = (1+j)\sqrt{\frac{\omega\mu_0\sigma_{\mathrm{dc}}}{2}},\\
&\delta_c \approx \sqrt{\frac{2}{\omega\mu_0\sigma_{\mathrm{dc}}}}; \quad K_c \approx \sqrt{\frac{\omega\mu_0\sigma_{\mathrm{dc}}}{2}},\\
&k_c = (1+j)\frac{1}{\delta_c};\\
&p_c = \frac{j}{k_c} = (1+j)\frac{\delta_c}{2},
\end{aligned}
\tag{D.42}
$$

remembering that $\sqrt{j} = (e^{j\pi/2})^{1/2} = e^{j\pi/4} = \cos(\pi/4) + j\sin(\pi/4) = (1+j)/\sqrt{2}$, where p_c is the complex skin depth in the collisional limit, $\nu \gg \omega$. In this case, the collisional skin depth δ_c is the same as the familiar direct current (DC) resistive skin depth in any conventional resistive conductor [1], because the frequent electron–neutral collisions dominate any significant influence of electron inertia in (D.13), so the plasma conductivity (D.16) and (D.17) is essentially real, $\sigma_p = \sigma_{\mathrm{dc}}$. The incident wave energy is dissipated principally by heating the electrons via electron–neutral collisions, until the wave energy is entirely absorbed. The wave propagates with its wavenumber K_c equal to the inverse skin depth; its amplitude decreases by a factor $1/e$ for each collisional skin depth δ_c travelled with wavelength $2\pi\delta_c$. In this collisional case, $\mathrm{Re}(p_c) = \delta_c/2$.

D.5.3 Collisionless skin depth

For collisionless RF plasma, $\nu \ll \omega \ll \omega_{\mathrm{pe}}$ according to section 5.1.2, so the plasma relative permittivity (D.27) is negative (the subscript nc denotes non-collisional, or collisionless),

$$
\begin{aligned}
&\varepsilon_p \approx -\frac{\omega_{\mathrm{pe}}^2}{\omega^2},\\
&k_{nc} = k_0\sqrt{\varepsilon_p} \approx jk_0\frac{\omega_{\mathrm{pe}}}{\omega} = j\frac{\omega_{\mathrm{pe}}}{c},\\
&\delta_{nc} \approx \frac{c}{\omega_{\mathrm{pe}}}; \quad K_{nc} \approx 0.
\end{aligned}
\tag{D.43}
$$

The wave is reflected for frequencies below ω_{pe}, physically because electrons can respond fast enough to cancel the wave's electric field, and their motion is not interrupted by collisions. The vibrating electrons collectively act as dipoles which coherently re-emit the incident radiation; this is similar to the familiar case of light reflection from the surface of metals by the conduction electrons [8]. Since it cannot propagate through the plasma, the wave energy is reflected, not dissipated as in the case of collisional skin depth because there are no collisions. The collisionless skin depth δ_{nc} equals the distance that light travels during an electron plasma oscillation period, i.e. the time it takes for the electrons to respond and reflect the wavefield. There is no propagation ($K_{nc} = 0$) and no power dissipation because $\nu = 0$; only evanescent penetration and total reflection. Clearly, collisionless conditions, by this definition, are not suited for sustaining an inductive plasma. Anomalous skin depth is not considered here [3].

D.5.4 Propagation, or evanescence and reflection, in collisionless plasma

The propagation or evanescence of a wave into an unmagnetized plasma can perhaps be understood from a physical point of view by comparing the conduction and displacement currents associated with the incident wave. Continuing with collisionless plasma as the simplest case, and treating the plasma as charges moving in vacuum, the electron conduction current density in the wave electric field **E**, from (D.15), is

$$\mathbf{J}_e = -n_{e0}q\mathbf{u} = +j\left(\frac{n_{e0}q^2}{m_e\omega}\right)\mathbf{E}. \tag{D.44}$$

The vacuum displacement current density of the wave is

$$\mathbf{J}_d = \varepsilon_0\frac{\partial \mathbf{E}}{\partial t} = -j(\omega\varepsilon_0)\mathbf{E}. \tag{D.45}$$

Both currents are in quadrature ($j = e^{j\pi/2}$) to the electric field, but note that the conduction current $\mathbf{J}_e$, caused by the wave electric field, is opposite to (in anti-phase with) the vacuum displacement current $\mathbf{J}_d$, so the sign of the resultant current will depend on which is dominant[2]. Their algebraic sum is

$$\begin{aligned}\mathbf{J}_d + \mathbf{J}_e = \mathbf{J}_d\left(1 + \frac{\mathbf{J}_e}{\mathbf{J}_d}\right) &= \mathbf{J}_d\left(1 - \frac{|\mathbf{J}_e|}{|\mathbf{J}_d|}\right)\\ &= -j\omega\varepsilon_0\left(1 - \frac{n_{e0}q^2}{\varepsilon_0 m_e\omega^2}\right)\mathbf{E}\\ &= -j\omega\varepsilon_0\left(1 - \frac{\omega_{\text{pe}}^2}{\omega^2}\right)\mathbf{E},\end{aligned} \tag{D.46}$$

[2] This is independent of the charge sign; it is not because the electrons have negative charge that the conduction current opposes the vacuum displacement current, as can be seen by re-calculating the force balance equation (D.13) for positive charges. The ion plasma frequency is much lower because of the heavy ion mass, so the conduction current is, in any case, dominated by the electrons.

which is nothing other than the effective total dielectric displacement current $(-j\omega\varepsilon_0\varepsilon_p)\mathbf{E}$ from (D.26), where the collisionless plasma relative permittivity is $\varepsilon_p = \left(1 - \frac{\omega_{\mathrm{pe}}^2}{\omega^2}\right)$. By comparing equations (D.44)–(D.46), two cases emerge:

- When the conduction current dominates ($|\mathbf{J}_d| < |\mathbf{J}_e|$, so ε_p is negative), the wavefield vacuum displacement current loses out to the opposing conduction current. The plasma has the characteristics of a metal (even if the effective conductivity is small), where vibrating conduction electron dipoles collectively cancel the wave electric field and the wave is reflected [5, 8]. Wave penetration is purely evanescent. This corresponds to $\omega < \omega_{\mathrm{pe}}$, which is the usual case for RF excitation of low temperature collisionless, unmagnetized plasma.
- When the vacuum displacement current dominates ($|\mathbf{J}_d| > |\mathbf{J}_e|$, so ε_p is positive), the wavefield vacuum displacement current wins out over the opposing conduction current (even though the effective relative permittivity is less than one, but still greater than zero), the plasma behaves as a dielectric, and the wave continues to propagate through the plasma with a modified wavenumber $k = k_0\sqrt{\varepsilon_p}$, which is real. This corresponds to $\omega > \omega_{\mathrm{pe}}$, appropriate for the transmission of microwave radiation. The cut-off occurs when $\omega = \omega_{\mathrm{pe}}$, as resumed in figure D.1.

This approach may seem trivial, but it will come in useful for interpreting R- and L-wave propagation or evanescence in magnetized plasma in section 10.10.

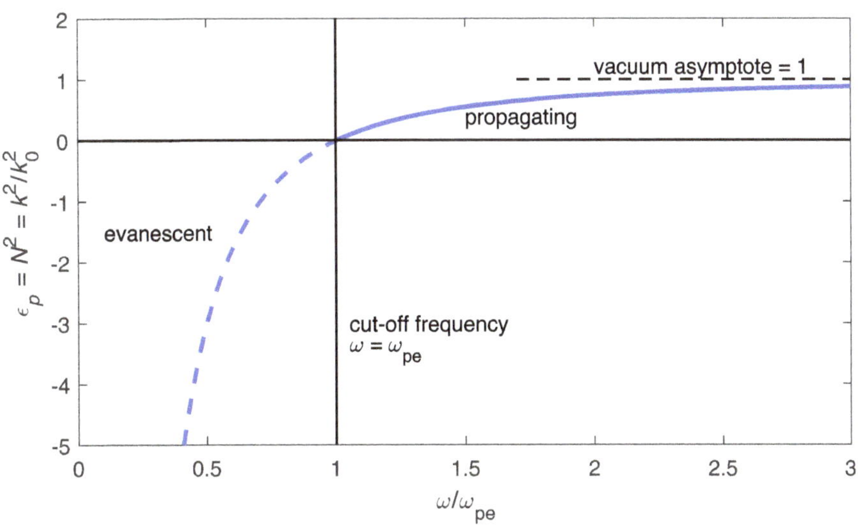

Figure D.1. The plasma relative permittivity ε_p in a collisionless unmagnetized plasma, as a function of the wavefield frequency ω normalized to the electron plasma frequency ω_{pe}. For $\varepsilon_p < 0$, the wave is reflected by the dominant conduction current and penetration is purely evanescent. Above the cut-off frequency ω_{pe}, where $\varepsilon_p > 0$, the wave propagates as a dominant displacement current. In the limit of high frequency, $\omega \gg \omega_{\mathrm{pe}}$, the electron inertia prevents them from following the wave electric field, and the wave propagates as in a vacuum, $\varepsilon_p \to 1$.

References

[1] Bleaney B I and Bleaney B 1976 *Electricity and Magnetism* 3rd edn (London: Oxford University Press)

[2] Haus H A and Melcher J R 2022 Electromagnetic fields and energy *Massachusetts Institute of Technology: MIT OpenCourseWare* http://ocw.mit.edu (Accessed: 22 April 2022) Creative Commons BY-NC-SA

[3] Lieberman M A and Lichtenberg A J 2005 *Principles of Plasma Discharges and Materials Processing* 2nd edn (Hoboken, NJ: Wiley)

[4] Chen F F 2016 *Introduction to Plasma Physics and Controlled Fusion* 3rd edn (Cham: Springer)

[5] Jackson J D 1999 *Classical Electrodynamics* 3rd edn (New York: Wiley)

[6] Chabert P and Braithwaite N 2011 *Physics of Radio-Frequency Plasmas* (Cambridge: Cambridge University Press)

[7] Howling A A, Guittienne Ph, Jacquier R and Furno I 2015 Complex image method for RF antenna–plasma inductive coupling calculation in planar geometry. Part I: basic concepts *Plasma Sources Sci. Technol.* **24** 065014

[8] Weisskopf V F 1968 How light interacts with matter *Sci. Am.* **219** 60

IOP Publishing

Resonant Network Antennas for Radio-Frequency Plasma Sources

Theory, technology and applications

Philippe Guittienne, Alan Howling and Ivo Furno

Appendix E

Theory of the complex image method

The approximations behind the complex image method used in chapter 6, in particular section 6.4, are derived in this appendix. For the case of an antenna current next to a uniform plane plasma, we calculate the magnetic field of the induced plasma current to see if it can be approximated by an antenna current image.

The complex image method has previously been successfully applied to the fields of telecommunications and power transmission [1–6], geophysics [7–9], and the microelectronics industry [10–12] to obtain solutions for the skin depth effect on inductive coupling in various resistive media. A few papers combine the complex image method with the partial inductance approach; some examples are to be found in the field of microelectronics [10–12]. However, to our knowledge, the complex image method combined with mutual partial inductance calculations in plasma processing (as introduced all-too-briefly in [13]) has not been applied before.

E.1 Diffusion equation for the magnetic vector potential

To start with, the theory is only concerned with magnetic fields, currents, and mutual partial inductance between conductors separated by distances much smaller than the RF wavelength—there is no regard for the conductors' capacitances, potentials and electric vacuum displacement currents. The first approximation to be made is therefore the magneto-quasistatic approximation (appendix D.1.1), whence

$$\nabla \times \mathbf{B} \approx \mu_0 \mathbf{J}, \tag{E.1}$$

$$\nabla \times \mathbf{E} = -\frac{\partial \mathbf{B}}{\partial t}. \tag{E.2}$$

We will require the self-consistent combination of the magnetic field generated by source currents, and also by the currents driven by the induced electric field in resistive media. This will be obtained from the solution of these simultaneous equations. One approach is to express both **E** and **B** in terms of a single parameter, the magnetic vector potential **A**, previously discussed in section 4.3.4. Substituting $\mathbf{B} = \nabla \times \mathbf{A}$ in (E.1) and (E.2), respectively, gives

$$\nabla \times (\nabla \times \mathbf{A}) = -\nabla^2 \mathbf{A} = \mu_0 \mathbf{J}, \tag{E.3}$$

$$\nabla \times \mathbf{E} = -\frac{\partial}{\partial t}(\nabla \times \mathbf{A}) = -\nabla \times \left(\frac{\partial \mathbf{A}}{\partial t}\right), \tag{E.4}$$

using the vector identity $\nabla \times (\nabla \times \mathbf{a}) \equiv \nabla(\nabla \cdot \mathbf{a}) - \nabla^2 \mathbf{a}$, and with $\nabla \cdot \mathbf{A} = 0$ for the Coulomb gauge [14] in (E.3). The solution of (E.4) is found by grouping terms and then integrating as follows [15]:

$$\begin{aligned} \nabla \times \left(\mathbf{E} + \frac{\partial \mathbf{A}}{\partial t}\right) &= 0, \\ \left(\mathbf{E} + \frac{\partial \mathbf{A}}{\partial t}\right) &= -\nabla V, \end{aligned} \tag{E.5}$$

where $-\nabla V$ is a constant of integration because $\nabla \times (\nabla V)$ is identically zero. As stated in the first paragraph, potentials and electric vacuum displacement currents are neglected in the magneto-quasistatic approximation, hence the ∇V term in (E.5) will be neglected. The starting equations (E.1) and (E.2) can therefore be re-written as

$$\nabla^2 \mathbf{A} = -\mu_0 \mathbf{J}, \tag{E.6}$$

$$\mathbf{E} = -\frac{\partial \mathbf{A}}{\partial t}. \tag{E.7}$$

The same equation (4.8) is solved for a current filament in section 4.3.4, but here we are concerned with spatially distributed current density. Substituting Ohm's law for a resistive medium, $\mathbf{J} = \sigma \mathbf{E}$, we obtain the required self-consistent equation

$$\nabla^2 \mathbf{A} = \mu_0 \sigma \frac{\partial \mathbf{A}}{\partial t}, \tag{E.8}$$

which is a diffusion equation for the magnetic vector potential in a resistive medium, subject to a time-varying magnetic field. This is the equation to be solved for the complex image method.

E.2 Line current above a semi-infinite plasma

Figure E.1 depicts a geometry relevant to a straight antenna wire above the plane surface of a semi-infinite plasma. The current flows along the filament and returns via the plasma. The single-harmonic current is directed along the z direction and, therefore, from section 4.3.4, the only component of **A** is A_z. Using (E.8) and (2.1), the governing equations for the two regions are

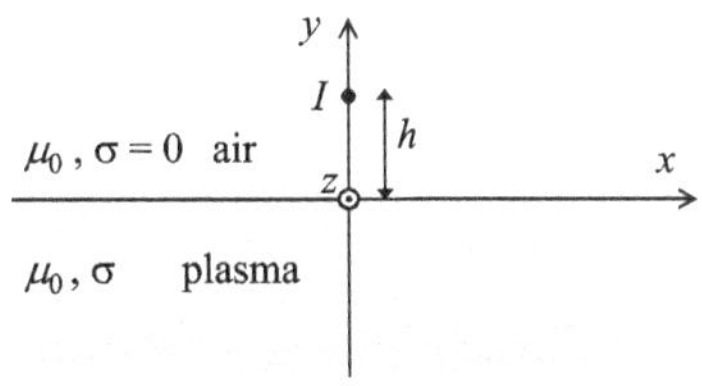

Figure E.1. AC line current source I parallel to the z-axis (out of the page), height h above and parallel to the surface of a semi-infinite, uniform plasma of electrical conductivity σ.

$$\nabla^2 A_z = 0 \qquad \text{in air,} \quad y > 0, \tag{E.9}$$

$$\nabla^2 A_z = -j\omega\mu_0\sigma\, A_z = -k_c^2 A_z \qquad \text{in plasma,} \quad y < 0, \tag{E.10}$$

where $k_c^2 = j\omega\mu_0\sigma$ is the square of the complex wavenumber, from (D.42) in the collisional plasma of appendix D.5.2; we see that diffusion of the magnetic vector potential is linked to the plasma collisional skin depth δ_c. The first equation is Laplace's equation because the electrical conductivity in air is $\sigma = 0$. The line current source I in air will be considered later, along with the boundary conditions, to find the complete solution. For now, we go on to solve these two equations, following [10, 16], using the separation of variables by setting $A_z = X(x)Y(y)$. We note that [16] gives a blow by blow account of the solution for the eddy currents, from which the complex image method is deduced by [10]. For a first reading, the reader may find this approach easier to understand than much earlier very thorough, but very mathematical, theories such as [2, 17].

Separation of variables for air
From (E.9),

$$\begin{aligned}\nabla^2 A_z = \frac{\partial^2 A_z}{\partial x^2} + \frac{\partial^2 A_z}{\partial y^2} = Y\frac{d^2X}{dx^2} + X\frac{d^2Y}{\partial y^2} = 0,\\ \frac{1}{X}\frac{d^2X}{dx^2} + \frac{1}{Y}\frac{d^2Y}{dy^2} = 0,\end{aligned} \tag{E.11}$$

$$\text{hence} \quad \frac{1}{X}\frac{d^2X}{dx^2} = -k^2; \qquad \frac{1}{Y}\frac{d^2Y}{dy^2} = k^2, \tag{E.12}$$

where k is a real separation constant (the wavenumber in air) because the two latter equations are independent of each other. The sign of $-k^2$ was chosen with regard to the symmetry of figure E.1 in order for $X(x) \propto \cos(kx)$ to be an even function. Solving (E.12) also for $Y(y)$ in terms of exponentials in y gives

$$A_z(k) = X(x)Y(y) = (Q_k e^{ky} + R_k e^{-ky})\cos(kx), \tag{E.13}$$

where Q_k and R_k are constants to be determined for each k. Because of a lack of boundary conditions along the infinite x direction, k can be any positive real wavenumber, so we must now integrate (instead of summing) over all possible $k = 0 \to \infty$ to obtain the general solution for A_z in air:

$$A_z(x, y) = \int_0^\infty (Q_k e^{ky} + R_k e^{-ky}) \cos(kx)\, dk, \tag{E.14}$$

for $y > 0$, using the superposition principle for linear fields. We can safely assume that this integral converges because the filament and plasma currents form a closed 'go-and-return' current loop [16].

The magnetic field in air can be directly calculated using $\mathbf{B} = \nabla \times (A_z \hat{\mathbf{z}})$:

$$B_x = \frac{\partial A_z}{\partial y} = \int_0^\infty k(Q_k e^{ky} - R_k e^{-ky}) \cos(kx)\, dk, \tag{E.15}$$

$$B_y = -\frac{\partial A_z}{\partial x} = \int_0^\infty k(Q_k e^{ky} + R_k e^{-ky}) \sin(kx)\, dk. \tag{E.16}$$

Common sense tells us that the magnetic field in air, in figure E.1, is the sum of the filament current's magnetic field, which falls off with distance away from the filament, and the magnetic field due to the eddy currents, which decreases in both directions $+y$ and $-y$ away from the air/plasma interface.

Comparison with the filament magnetic field in air

To help find the unknowns Q_k and/or R_k, we now compare (E.15) and (E.16) with the well known Biot–Savart filament field $B = \frac{\mu_0 I}{2\pi}\frac{1}{r}$, where $r = (x^2 + (h - y)^2)^{1/2}$ is the radial distance from the filament. First, we concentrate on the air slab region between the filament and the surface, $0 < y < h$. By elementary trigonometry for the filament B-field Cartesian components, and then using standard converging Fourier integrals [18], we have

$$B_x = \left[\frac{\mu_0 I}{2\pi}\right] \frac{h - y}{x^2 + (h - y)^2} = \left[\frac{\mu_0 I}{2\pi}\right] \int_0^\infty e^{-k|h-y|} \cos(kx)\, dk, \tag{E.17}$$

$$B_y = \left[\frac{\mu_0 I}{2\pi}\right] \frac{x}{x^2 + (h - y)^2} = \left[\frac{\mu_0 I}{2\pi}\right] \int_0^\infty e^{-k|h-y|} \sin(kx)\, dk. \tag{E.18}$$

These integrals can be verified by direct integration [7]. By comparing our general solution (E.15) and (E.16), with the filament field (E.17) and (E.18) in the present region of interest $0 < y < h$ so that $|h - y| = h - y$, we can identify the Q_k term as follows [16]:

$$\begin{aligned} kQ_k e^{ky} &= \left[\frac{\mu_0 I}{2\pi}\right] e^{-k(h-y)}, \\ \text{hence} \quad Q_k &= \left[\frac{\mu_0 I}{2\pi}\right] \frac{e^{-kh}}{k}, \end{aligned} \tag{E.19}$$

for both the cos and sin integrands. By extending this argument similarly to $y > h$, the magnetic vector potential is found to be [16]

$$A_z(x, y) = \int_0^\infty \left(\frac{\mu_0 I}{2\pi} \frac{e^{-k|h-y|}}{k} + R_k e^{-ky} \right) \cos(kx)\, dk, \tag{E.20}$$

where the first term in the integrand represents the filament current, so that the second term must represent the contribution from the induced current in the plasma, which remains to be found. It was necessary to switch to B_x, B_y and back to $A(x, y)$ because the magnetic vector potential cannot, alone, be expressed as a convergent Fourier integral [16]. It is the return current via the plasma, forming a closed loop, which ensures a convergent integral, as stated after (E.14).

Separation of variables for plasma

Next we consider the general expression for the plasma currents. From (E.10):

$$\begin{aligned} \nabla^2 A_z = \frac{\partial^2 A_z}{\partial x^2} + \frac{\partial^2 A_z}{\partial y^2} = Y\frac{d^2 X}{dx^2} + X\frac{d^2 Y}{\partial y^2} = -k_c^2 A_z = -\, k_c^2 XY, \\ \frac{1}{X}\frac{d^2 X}{dx^2} + \frac{1}{Y}\frac{d^2 Y}{dy^2} = -\, k_c^2, \end{aligned} \tag{E.21}$$

$$\text{hence} \quad \frac{1}{X}\frac{d^2 X}{dx^2} = -k^2; \qquad \frac{1}{Y}\frac{d^2 Y}{\partial y^2} = k^2 - k_c^2 = q^2, \tag{E.22}$$

where the separation constant $-k^2$ for $X(x)$ is the same as in (E.12) for reasons of symmetry and boundary matching, and q^2 is another separation constant. Using $k_c = j/p_c$ from (D.41) or (D.42), we can write q in terms of the complex collisional skin depth p_c :

$$q^2 = k^2 - k_c^2 = k^2 + \frac{1}{p_c^2}. \tag{E.23}$$

Following the same procedure as for air, the general solution for A_z in plasma is

$$A_z(x, y) = \int_0^\infty P_k\, e^{qy} \cos(kx)\, dk, \tag{E.24}$$

applicable for $y < 0$ so that A_z tends to zero towards minus infinity in the source-free plasma.

Combining the expressions for air and plasma

The remaining unknown spectral functions R_k and P_k, which are both due to the magnetic field of the current induced in the plasma by the source, can be found by applying the boundary conditions for continuity of $A_z(x, y)$ and $\partial A_z(x, y)/\partial y$ at the air/plasma interface $y = 0$ [10]. Performing these two conditions for the integrands of (E.20) and (E.24), yields the simultaneous equations

$$\left[\frac{\mu_0 I}{2\pi}\right]\frac{e^{-kh}}{k} + R_k = P_k, \tag{E.25}$$

$$\left[\frac{\mu_0 I}{2\pi}\right]e^{-kh} - kR_k = qP_k, \tag{E.26}$$

whence

$$P_k = \left[\frac{\mu_0 I}{\pi}\right]\frac{e^{-kh}}{q + k}, \tag{E.27}$$

$$R_k = -\left[\frac{\mu_0 I}{2\pi}\right]\frac{e^{-kh}}{k}\left(\frac{q - k}{q + k}\right). \tag{E.28}$$

Finally, by substitution of R_m, (E.28), into (E.20), the magnetic vector potential in air is

$$A_z(x, y) = \left[\frac{\mu_0 I}{2\pi}\right]\int_0^{\infty}\left\{\frac{e^{-k|h-y|}}{k} - \left(\frac{q - k}{q + k}\right)\frac{e^{-k(h+y)}}{k}\right\}\cos(kx)\, dk, \tag{E.29}$$

where the first term of the integrand is due to the filament source current, and the second term is due to the plasma induced current. In the spirit of image theory, and in contrast to [16], we are interested in this field solution in air because of its image effect on the source, rather than in the induced current solution in the plasma, given by (E.24) and (E.27).

Interpreting the complex image current and location

The exact solution (insofar as it is limited by the magneto-quasistatic approximation) for the magnetic vector potential in air of figure E.1 is given by (E.29). By inspection, this solution could be interpreted as being due to the filament current source I at $y = h$ (the first term), and a current 'image' $-I \times \frac{q-k}{q+k}$ at $y = -h$ (the second term). Therefore, this 'image' corresponds to a complex current $-I \times \frac{q-k}{q+k}$ at the position of the classical image theory, $y = -h$. Note that, in the special case where the plasma is infinitely conducting, σ and k_c tend to infinity, hence $\frac{q-k}{q+k} \to 1$ and the image current becomes $-I$ at $y = -h$, identical to the ideal screen situation already presented in sections 4.7.1 and 5.4.3.

However, for finite plasma conductivity, this interpretation is not satisfying because we know that exactly the same current, $-I$, returns via the plasma. Therefore, we would prefer to find an image position for a current $-I$ which reproduces the same magnetic field in the medium of the source current, in order to simplify calculations of mutual partial inductance. Such an image is, at least, conceivable, because the boundary conditions imposed by the source usually constrain the induced currents to have similar characteristics.

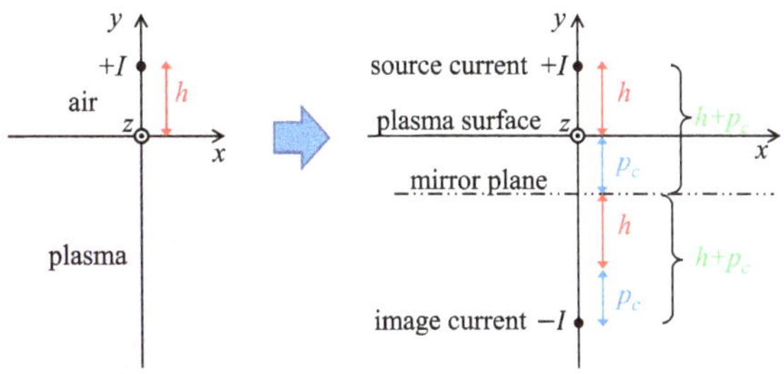

Figure E.2. Left: A repeat of figure E.1. Right: the image current $-I$ at the calculated position $y = -\{h + 2p_c\}$ in air, according to (E.30), which approximately reproduces an equivalent magnetic field in the source zone $y > 0$. The equivalent image interpretation is for a mirror plane at complex skin depth p_c below the surface of the resistive plasma; the current source $+I$ is a complex distance $(h + p_c)$ above the mirror plane, and the image current $-I$ is an equal (complex) distance $(h + p_c)$ below the mirror plane [10]. See also figure 6.12.

Empirically, we search to convert the 'inconvenient' fraction $\frac{q-k}{q+k}$ into unity, by converting it into an exponential whose argument could be combined into the $e^{-k(h+y)}$ term in (E.29); this extra term would then add to the depth term $-h$ and not modify the current itself. This approach was carried out in [2, 10] by expanding the term $\frac{q-k}{q+k}$ as a Taylor series about $k = 0$. Wait and Spies [2] obtained $\frac{q-k}{q+k} \approx e^{-2kp_c}$ to third order in kp_c, referring to (E.23). This approximation is good for $h^3 \gg |p_c|^3$, i.e. for current elevation h greater than the collisional skin depth δ_c in (D.42). Substituting this approximation into (E.29) yields

$$A_z(x, y) \approx \left[\frac{\mu_0 I}{2\pi}\right] \int_0^\infty \left\{\frac{e^{-k|h-y|}}{k} - \frac{e^{-k(\{h+2p_c\}+y)}}{k}\right\} \cos(kx)\, dk. \tag{E.30}$$

By inspection, the first term still corresponds to the filament current I at $y = h$, but the second term now corresponds to an image current $-I$ at a position $y = -\{h + 2p_c\}$, as shown schematically in figure E.2, where p_c is the complex skin depth in (D.42) for the limit of collisional plasma. This is the principal result necessary for section 6.4.2, which is that the effective mirror plane for the image current reflection is one complex skin depth p_c below the plasma surface [10].

The name 'complex image method' is somewhat of a misnomer because the image current $-I$ itself is not complex; it is its image *position* which is complex. In this section, the possibility of interchanging the magnitude of the current for the position of the current, goes to show that the complex image method is a mathematical construct for computational convenience [9]: a physical interpretation of a 'complex depth' is neither possible nor required.

Note that $k = 0$ corresponds to a one-dimensional sheet oscillating current, since $\cos 0$ is constant [10], generating a plane wave which is damped over the $1/e$ skin depth δ_c, as seen previously in section D.5. Also, for the special case of infinite-conductivity plasma representing an ideal screen, $p_c \rightarrow 0$, and the image current becomes $-I$ at $y = -h$, again, as expected from classical image theory.

References

[1] Wait J R 1961 On the impedance of long wire suspended over the ground *Proc. IRE* **49** 1576

[2] Wait J R and Spies K P 1969 On the image representation of the quasi-static fields of a line current source above the ground *Can. J. Phys.* **47** 2731

[3] Bannister P R 1970 Utilization of image theory techniques in determining the mutual coupling between elevated long horizontal line sources *Radio Sci.* **5** 1375

[4] Déri A, Tevan G, Semlyen A and Castanheira A 1981 The complex ground return plane: a simplified model for homogeneous and multi-layer earth return *IEEE Trans. Power Appar. Syst.* **PAS-100** 3686

[5] Déri A and Tevan G 1981 Mathematical verification of Dubanton's simplified calculation of overhead transmission line parameters and its physical interpretation *Arch. Elektrotech.* **63** 191

[6] Bannister P R 1986 Applications of complex image theory *Radio Sci.* **21** 605

[7] Boteler D H and Pirjola R J 1998 The complex-image method for calculating the magnetic and electric fields produced at the surface of the Earth by the auroral electrojet *Geophys. J. Int.* **132** 31

[8] Park D 1973 Magnetic field of a horizontal current above a conducting earth *J. Geophys. Res.* **78** 3040

[9] Thomson D J and Weaver J T 1975 The complex image approximation for induction in a multilayered earth *J. Geophys. Res.* **80** 123

[10] Weisshaar A, Lan H and Luoh A 2002 Accurate closed-form expressions for the frequency-dependent line parameters of on-chip interconnects on lossy silicon substrate *IEEE Trans. Adv. Packag.* **25** 288

[11] Jiang R, Fu W and Chen C C- 2005 EPEEC: comprehensive SPICE-compatible reluctance extraction for high-speed interconnects above lossy multilayer substrates *IEEE Trans. Comput.-Aided Des. Integr. Circuits Syst.* **24** 1562

[12] Kang K, Shi J, Yin W-Y, Li L-W, Zouhdi S, Rustagi S C and Mouthaan K 2007 Analysis of frequency- and temperature-dependent substrate eddy currents in on-chip spiral inductors using the complex image method *IEEE Trans. Magn.* **43** 3243

[13] Howling A A, Guittienne Ph, Jacquier R and Furno I 2015 Complex image method for RF antenna–plasma inductive coupling calculation in planar geometry. Part I: basic concepts *Plasma Sources Sci. Technol.* **24** 065014

[14] Jackson J D 1999 *Classical Electrodynamics* 3rd edn (New York: Wiley)

[15] Bleaney B I and Bleaney B 1976 *Electricity and Magnetism* (London: Oxford University Press) 3rd edn p 1

[16] Tegopoulos J A and Kriezis E E 1985 *Eddy Currents in Linear Conducting Media* (Amsterdam: Elsevier)

[17] Carson J R 1926 Wave propagation in overhead wires with ground return *Bell Syst. Tech. J.* **5** 539

[18] Gradshteyn I S and Ryzhik I M 2007 *Table of Integrals, Series and Products* (New York: Elsevier Academic) 7th edn

Appendix F

Solution of the MTL equations for the EMCP antenna source

The large ladder antenna was presented as an electromagnetically coupled plasma (EMCP) source in chapter 7. The equations to be solved relate to a multi-conductor transmission line (MTL) model. The uniform multi-conductor transmission line wave equation (7.9), analogous to the single transmission line [1–3], is

$$d^2\mathbf{V}(z)/dz^2 = -\omega^2(\hat{M}\ \hat{C})\mathbf{V}(z), \tag{F.1}$$

and identically for $\mathbf{I}(z)$, where $\hat{M}$ and $\hat{C}$ are the per-unit-length matrices for mutual partial inductance, and capacitance, respectively. Note that the presence or absence of plasma is determined by the choices of $\hat{M}$ and $\hat{C}$. The equations can be decoupled by a similarity transformation [4]

$$\mathbf{V}(z) = \bar{T}_V \mathbf{V}_{\mathrm{m}}(z), \tag{F.2}$$

where $\bar{T}_V$ is the matrix whose N columns are the eigenvectors of $(-\omega^2\hat{M}\hat{C})$ and $\mathbf{V}_{\mathrm{m}}(z)$ is the voltage vector for mode m. The N decoupled wave equations for the mode m voltage amplitudes are

$$d^2\mathbf{V}_{\mathrm{m}}(z)/dz^2 = \bar{\Gamma}^{\ 2}\,\mathbf{V}_{\mathrm{m}}(z), \tag{F.3}$$

where $\bar{\Gamma}^{\ 2} = \bar{T}_V^{-1}(-\omega^2\hat{M}\hat{C})\bar{T}_V$ is the $N \times N$ diagonal matrix of propagation constant eigenvalues. By analogy with single transmission lines [1–3], the N mode voltages and currents have the solution [4]

$$\begin{aligned}\mathbf{V}_{\mathrm{m}}(z) &= \overline{e^{-\Gamma z}}\ \ \mathbf{V}_{\mathrm{m}}^{+} + \overline{e^{\Gamma z}}\ \ \mathbf{V}_{\mathrm{m}}^{-},\\ \mathbf{I}_{\mathrm{m}}(z) &= \hat{Y}_{\mathrm{c}}\,(\overline{e^{-\Gamma z}}\ \ \mathbf{V}_{\mathrm{m}}^{+} - \overline{e^{\Gamma z}}\ \ \mathbf{V}_{\mathrm{m}}^{-}),\end{aligned} \tag{F.4}$$

doi:10.1088/978-0-7503-5296-3ch20

where the diagonal matrix exponentials are defined as

$$\overline{e^{\pm\Gamma z}} = \begin{pmatrix} e^{\pm\Gamma_1 z} & 0 & \cdots & 0 \\ 0 & e^{\pm\Gamma_2 z} & \ddots & \vdots \\ \vdots & \ddots & \ddots & 0 \\ 0 & \cdots & 0 & e^{\pm\Gamma_N z} \end{pmatrix}. \tag{F.5}$$

Here, $\mathbf{V}_{\mathrm{m}}^{+}$ and $\mathbf{V}_{\mathrm{m}}^{-}$ are vectors of complex phasor amplitudes of forward and backward waves, to be determined by boundary conditions (i.e. the termination impedances in F.1), and $\bar{Y}_{\mathrm{c}} = (j\omega\hat{M})^{-1}\bar{T}_V\,\bar{\Gamma}\,\bar{T}_V^{-1}$ is the characteristic admittance matrix of the line [4] which depends on plasma coupling.

F.1 MTL solution for the EMCP resonant network antenna

The boundary conditions, determined by the termination impedances appropriate to the resonant network shown in figures 7.4 and 7.6, are applied here. With regard to appendix B, instead of assuming constant current and Ohm's law in the antenna legs, the voltage and current of each leg are now related by the transmission line equations. The solution consists in finding the values of the phasor amplitude vectors $\mathbf{V}_{\mathrm{m}}^{+}$ and $\mathbf{V}_{\mathrm{m}}^{-}$ for the N legs for each mode m. The lumped-element model for purely inductive coupling in chapter 6 is recovered when capacitive coupling is neglected, i.e. in the limit $\hat{C} \to \bar{0}$.

Taking the origin for z at the middle of the leg length $2H$, the voltages $\mathbf{V}(z)$ and currents $\mathbf{I}(z)$ at both ends of the legs are $\mathbf{A} = \mathbf{V}(H)$, $\mathbf{B} = \mathbf{V}(-H)$, $\mathbf{I}_A = \mathbf{I}(H)$, $\mathbf{I}_B = \mathbf{I}(-H)$. These voltages and currents are determined by the termination impedances in the stringers. Using the terminology of appendix B, Ohm's law for the stringers gives

$$\begin{aligned} \bar{U}_T\mathbf{A} &= \bar{Z}_{\mathrm{line}}\mathbf{J} + \bar{Z}_{\mathrm{opp}}\mathbf{K}, \\ \bar{U}_T\mathbf{B} &= \bar{Z}_{\mathrm{line}}\mathbf{K} + \bar{Z}_{\mathrm{opp}}\mathbf{J}, \end{aligned} \tag{F.6}$$

where $\mathbf{J}$, $\mathbf{K}$ are the currents in each line of stringers, $\bar{Z}_{\mathrm{line}}$ is the impedance matrix of in-line stringers, including mutual partial inductance to all the other in-line stringers, and $\bar{Z}_{\mathrm{opp}}$ is the mutual partial inductance matrix with all opposing stringers. For the resonant network in this paper, $\bar{Z}_{\mathrm{line}}$ includes the stringer capacitors C_{str}, but other ICP linear array types can be modelled using termination impedances Z_{Ai} and Z_{Bi} appropriate to the stringers of each particular antenna. For example, referring to figure 7.6:

1. The Helyssen antenna has termination impedances $Z_{Ai} = Z_{Bi} \approx 1/(j\omega C_{\mathrm{str}})$, where C_{str} represents the stringer capacitor assembly;
2. The ladder (non-resonant) antenna [5–17] has $Z_{Ai} = Z_{Bi} \approx j\omega L_{\mathrm{str}}$, where $L_{\mathrm{str}} \sim 0$ is the self-inductance of short stringer conductors where appropriate;
3. The serpentine antenna [2, 18, 19] has alternate $Z_{Ai}, Z_{Bi} \sim 0$ and $Z_{Bi}, Z_{Ai} \to \infty$ to define the serpentine sequence of antenna legs;
4. The double-comb antenna [20–22] resembles the serpentine antenna, but with separate sources to the independent combs.

5. The U-type antenna [23–25] consists of parallel pairs of tubes connected together at one end ($Z_{Bi} \approx j\omega L_{\rm str}$, where $L_{\rm str} \sim 0$ is the self-inductance of a short stringer), with the impedance Z_{Ai} determined by the output impedance of the RF power circuit.

Current conservation at the nodes [26] requires

$$\begin{aligned} \bar{U}\mathbf{J} &= \mathbf{I}_A + \mathbf{S}_A, \\ \bar{U}\mathbf{K} &= -\mathbf{I}_B + \mathbf{S}_B, \end{aligned} \tag{F.7}$$

where $\mathbf{S}_{A,B}$ are the source current vectors. A relation between the node voltages and currents can be found by eliminating $\mathbf{J}$, $\mathbf{K}$ to obtain

$$\begin{aligned} \mathbf{A} &= \bar{m}_1(\mathbf{I}_A + \mathbf{S}_A) - \bar{m}_2(\mathbf{I}_B - \mathbf{S}_B), \\ \mathbf{B} &= \bar{m}_2(\mathbf{I}_A + \mathbf{S}_A) - \bar{m}_1(\mathbf{I}_B - \mathbf{S}_B), \end{aligned} \tag{F.8}$$

where $\bar{m}_1 = \bar{U}_T^{-1}\,\bar{Z}_{\rm line}\,\bar{U}^{-1}$ and $\bar{m}_2 = \bar{U}_T^{-1}\,\bar{Z}_{\rm opp}\,\bar{U}^{-1}$. The solution is facilitated by defining

$$\begin{aligned} \bar{C}_h &= \overline{e^{\Gamma H}} + \overline{e^{-\Gamma H}}, \\ \bar{S}_h &= \overline{e^{\Gamma H}} - \overline{e^{-\Gamma H}}, \end{aligned} \tag{F.9}$$

with a change of variables:

$$\begin{aligned} \Delta &= \mathbf{V}_{\rm m}^{+} - \mathbf{V}_{\rm m}^{-}, \\ \sigma &= \mathbf{V}_{\rm m}^{+} + \mathbf{V}_{\rm m}^{-}. \end{aligned} \tag{F.10}$$

With these definitions, according to (F.2) and (F.4),

$$\begin{aligned} \mathbf{A} &= \bar{T}_V(\bar{C}_h\sigma - \bar{S}_h\Delta)/2, \\ \mathbf{B} &= \bar{T}_V(\bar{C}_h\sigma + \bar{S}_h\Delta)/2, \\ \mathbf{I}_A &= \bar{Y}_{\rm c}\bar{T}_V(\bar{C}_h\Delta - \bar{S}_h\sigma)/2, \\ \mathbf{I}_B &= \bar{Y}_{\rm c}\bar{T}_V(\bar{C}_h\Delta + \bar{S}_h\sigma)/2. \end{aligned} \tag{F.11}$$

Introducing these expressions into (F.8) for ($\mathbf{B} - \mathbf{A}$) and ($\mathbf{B} + \mathbf{A}$), respectively, gives

$$\begin{aligned} [(\bar{m}_1 - \bar{m}_2)^{-1}\bar{T}_V\bar{S}_h + \bar{Y}_{\rm c}\bar{T}_V\bar{C}_h]\Delta &= \mathbf{S}_B - \mathbf{S}_A = \bar{O}_D\Delta, \\ [(\bar{m}_1 + \bar{m}_2)^{-1}\bar{T}_V\bar{C}_h + \bar{Y}_{\rm c}\bar{T}_V\bar{S}_h]\sigma &= \mathbf{S}_B + \mathbf{S}_A = \bar{O}_S\sigma, \end{aligned} \tag{F.12}$$

which define the matrices $\bar{O}_{D,S}$ as shorthand pre-factors of Δ, σ, respectively. The source current vectors $\mathbf{S}_A$ and $\mathbf{S}_B$ must account for the imposed input currents as well as for possible connections to ground. Considering an input current $I_{\rm in}$ injected into a single node $A_{\rm in}$ (node A13 in figure 7.4), we set

$$\begin{aligned} \mathbf{S}_A &= \mathbf{I}_{\rm in} - \bar{Y}_A\,\mathbf{A}, \\ \mathbf{S}_B &= -\bar{Y}_B\,\mathbf{B}, \end{aligned} \tag{F.13}$$

where the vector $\mathbf{I}_{\text{in}}$ has zeros for all components except for I_{in} at the RF connection node A13. For the case of the grounded antenna, $\bar{Y}_{A,B}$ are diagonal matrices with the ground stub admittance inserted at the relevant indices, nodes A10 and A16 in figures 7.4 and 7.6, with $\bar{Y}_B = \bar{0}$. On the other hand, for the floating antenna, $\bar{Y}_A = 0$ and $\bar{Y}_B = 0$ everywhere, corresponding to infinite impedance (open circuit) for every node.

The expressions for $\mathbf{S}_B \pm \mathbf{S}_A$ in (F.12), which were derived from the network currents and voltages, can be compared with expressions (F.13) for the source currents by using (F.11) to write

$$\begin{aligned} \mathbf{S}_B - \mathbf{S}_A &= (\bar{Y}_A - \bar{Y}_B)\bar{T}_V\bar{C}_h\sigma/2 - (\bar{Y}_A + \bar{Y}_B)\bar{T}_V\bar{S}_h\Delta/2 - \mathbf{I}_{\text{in}} \\ &= \bar{O}_1\sigma + \bar{O}_2\Delta - \mathbf{I}_{\text{in}}, \\ \mathbf{S}_B + \mathbf{S}_A &= -(\bar{Y}_A + \bar{Y}_B)\bar{T}_V\bar{C}_h\sigma/2 + (\bar{Y}_A - \bar{Y}_B)\bar{T}_V\bar{S}_h\Delta/2 + \mathbf{I}_{\text{in}} \\ &= \bar{O}_3\sigma + \bar{O}_4\Delta + \mathbf{I}_{\text{in}}, \end{aligned} \tag{F.14}$$

where the matrices $\bar{O}_{1-4}$ are shorthand pre-factors of Δ, σ. Using the pre-factor simplifications, equating (F.12) and (F.14) gives the following simultaneous equations for Δ and σ

$$\begin{aligned} (\bar{O}_D - \bar{O}_2)\Delta - \bar{O}_1\,\sigma &= -\mathbf{I}_{\text{in}}, \\ (\bar{O}_S - \bar{O}_3)\sigma - \bar{O}_4\,\Delta &= +\mathbf{I}_{\text{in}}, \end{aligned} \tag{F.15}$$

whose solution gives the required voltage solution using (F.10) and (F.2). See appendix K for a link to an example.

References

[1] Lamm A J 1997 Observations of standing waves on an inductive plasma coil modeled as a uniform transmission line *J. Vac Sci. Technol.* A **15** 2615

[2] Wu Y and Lieberman M A 2000 The influence of antenna configuration and standing wave effects on density profile in a large-area inductive plasma source *Plasma Sources Sci. Technol.* **9** 210

[3] Paul C R and Nasar S A 1987 *Introduction to Electromagnetic Fields* (New York: McGraw-Hill)

[4] Paul C R 2008 *Analysis of Multiconductor Transmission Lines* 2nd edn (Hoboken, NJ: Wiley)

[5] Takeuchi Y, Nawata Y, Ogawa K, Serizawa A, Yamauchi Y and Murata M 2001 Preparation of large uniform amorphous silicon films by VHF-PECVD using a ladder-shaped antenna *Thin Solid Films* **386** 133

[6] Takeuchi Y, Kawasaki I, Mashima H, Murata M and Kawai Y 2001 Characteristics of VHF excited hydrogen plasmas using a ladder-shaped electrode *Thin Solid Films* **390** 217

[7] Mashima H, Takeuchi Y, Noda M, Murata M, Naitou H, Kawasaki I and Kawai Y 2003 Uniformity of VHF plasma produced with ladder shaped electrode *Surf. Coat. Technol.* **171** 167

[8] Takatsuka H, Noda M, Yonekura Y, Takeuchi Y and Yamauchi Y 2004 Development of high efficiency large area silicon thin film modules using VHF-PECVD *Sol. Energy* **77** 951

[9] Takatsuka H, Yamauchi Y, Kawamura K, Mashima H and Takeuchi Y 2006 World's largest amorphous silicon photovoltaic module *Thin Solid Films* **506-7** 13

[10] Nishimiya T, Takeuchi Y, Yamauchi Y, Takatsuka H, Kai Y, Muta H and Kawai Y 2007 Large area SiH_4/H_2 VHF plasma produced with multi-rod electrode *Plasma Process. Polym.* **4** S991

[11] Nishimiya T, Takeuchi Y, Yamauchi Y, Takatsuka H, Shioya T, Muta H and Kawai Y 2008 Large area VHF plasma production by a balanced power feeding method *Thin Solid Films* **516** 4430

[12] Yamauchi Y, Takeuchi Y, Takatsuka H, Yamashita H, Muta H and Kawai Y 2008 Characteristics of VHF H_2 plasma produced at high pressure Contrib *Plasma Phys.* **48** 326

[13] Murata M, Takeuchi Y, Sasagawa E and Hamamoto K 1996 Inductively coupled radio frequency plasma chemical vapor deposition using a ladder-shaped antenna *Rev. Sci. Instrum.* **67** 1542

[14] Wendt A E and Mahoney L J 1996 Radio frequency inductive discharge source design for large area processing *Pure Appl. Chem.* **68** 1055

[15] Murata M, Mashima H, Yoshioka M, Nishida S, Morita S and Kawai Y 1997 Production of inductively coupled RF plasma using a ladder-shaped antenna *Jpn. J. Appl. Phys.* **36** 4563

[16] Kawai Y, Yoshioka M, Yamane T, Takeuchi Y and Murata M 1999 Radio-frequency plasma production using a ladder-shaped antenna *Surf. Coat. Technol.* **116–9** 662

[17] Mashima H, Murata M, Takeuchi Y, Yamakoshi H, Horioka T, Yamane T and Kawai Y 1999 Characteristics of very high frequency plasma produced using a ladder-shaped electrode *Jpn. J. Appl. Phys.* **38** 4305

[18] Park S E, Cho B U, Lee J K, Lee Y J and Yeom G Y 2003 The characteristics of large area processing plasmas *IEEE Trans. Plasma Sci.* **31** 628

[19] Kim K N, Lee Y J, Kyong S J and Yeom G Y 2004 Effects of multipolar magnetic fields on the characteristics of plasma and photoresist etching in an internal linear inductively coupled plasma system *Surf. Coat. Technol.* **177–8** 752

[20] Kim K N, Lee Y J, Jung S J and Yeom G Y 2004 Characteristics of parallel internal-type inductively coupled plasmas for large area flat panel display processing *Jpn. J. Appl. Phys.* **43** 4373

[21] Lim J H, Kim K N, Park J K, Lim J T and Yeom G Y 2008 Uniformity of internal linear-type inductively coupled plasma source for flat panel display processing *Appl. Phys. Lett.* **92** 051504

[22] Lim J H, Kim K N, Gweon G H, Park J B and Yeom G Y 2009 Study of internal linear inductively coupled plasma source for ultra large-scale flat panel display processing *Plasma Chem. Plasma Process.* **29** 251

[23] Lim J H, Kim K N and Yeom G Y 2007 Inductively coupled plasma source using internal multiple U-type antenna for ultra large-area plasma processing *Plasma Process. Polym.* **4** S999

[24] Takagi T, Ueda M, Ito N, Watabe Y and Kondo M 2006 Microcrystalline silicon solar cells fabricated using array-antenna-type very high frequency plasma-enhanced chemical vapor deposition system *Jpn. J. Appl. Phys.* **45** 4003

[25] Takagi T, Ueda M, Ito N, Watabe Y, Sato H and Sawaya K 2006 Large area VHF plasma sources *Thin Solid Films* **502** 50

[26] Guittienne Ph, Jacquier R, Howling A A and Furno I 2015 Complex image method for RF antenna–plasma inductive coupling calculation in planar geometry. Part II: measurements on a resonant network *Plasma Sources Sci. Technol.* **24** 065015

IOP Publishing

Resonant Network Antennas for Radio-Frequency Plasma Sources

Theory, technology and applications

Philippe Guittienne, Alan Howling and Ivo Furno

Appendix G

Maxwell's potential coefficient matrix and the partial image method

Section 7.6.2 uses Maxwell's potential coefficient matrix $\hat{P}$ to calculate the per-unit-length capacitance matrix $\hat{C}$ employed in the multi-conductor transmission line (MTL) calculations of chapter 7. In this appendix, we first describe the basics of Maxwell's potential coefficient matrix, and then apply it to a ladder antenna within a stripline. The same technique is then extended, using the partial image method, to the antenna on an inhomogeneous microstrip, consisting of the plasma behind a glass protection window in figures 7.11 and 7.12.

G.1 Maxwell's potential coefficient matrix

Consider a set of N, equally spaced ladder antenna legs, arranged parallel to a PEC ground reference plate. The simplest approach [1] is to apply a per-unit-length (p.u.l.) charge q_j on the jth leg, with zero charge on every other leg. An image charge $-q_j$ is induced on the reference plate, which remains at ground potential, 0 V, as shown in figure G.1. It is important to note from the start—concerning the calculation of electric potential in the domain of the source charge (i.e. above the plate)—that the system of source and image charge is equivalent to the system of source charge and the induced charge on the ground plate. This is one example of the equivalence principle [2]. The Maxwell coefficient of potential, p_{ij}, is defined as the resulting voltage V_i at leg i, per unit charge on leg j, i.e.

$$p_{ij} = \frac{V_i}{q_j}, \quad q_{(1,\ldots,N)\neq j} = 0. \tag{G.1}$$

Repeating this approach for a single lone charge separately on each leg in turn, by superposition of the voltages for a general charge distribution across the N legs, we have

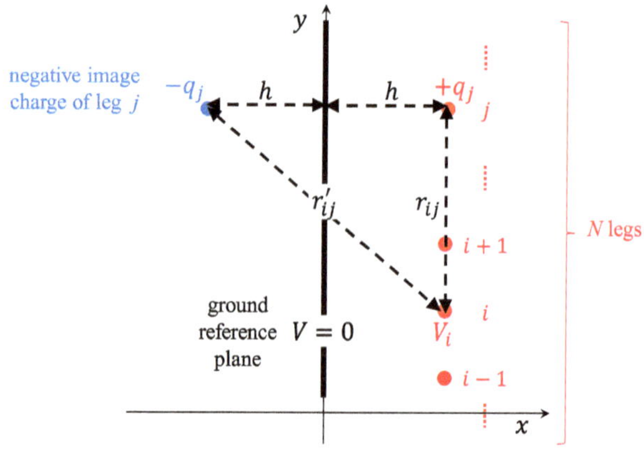

Figure G.1. An end-view schematic of an N-leg ladder antenna (red circles), distance h from a ground reference plate (black solid line), to indicate the calculation of Maxwell's potential coefficient $p_{ij} = V_i/q_j$ in (G.1) and (G.6). The potential V_i of the ith leg depends on the p.u.l. charge $+q_j$ on the jth leg and its image charge $-q_j$ (blue circle) in the plate, and their respective separation distances r_{ij} and r'_{ij}. All the other legs are uncharged.

$$\begin{pmatrix} V_1 \\ V_2 \\ \vdots \\ V_N \end{pmatrix} = \begin{pmatrix} p_{11} & p_{12} & \cdots & p_{1N} \\ p_{21} & p_{22} & \cdots & p_{2N} \\ \vdots & \vdots & \ddots & \vdots \\ p_{N1} & p_{N2} & \cdots & p_{NN} \end{pmatrix} \begin{pmatrix} q_1 \\ q_2 \\ \vdots \\ q_N \end{pmatrix}. \tag{G.2}$$

For (G.2) in matrix notation,

$$\begin{aligned} \mathbf{V} &= \hat{P}\,\mathbf{Q}, \\ \mathbf{Q} &= \hat{P}^{-1}\,\mathbf{V} = \hat{C}\,\mathbf{V}, \\ \hat{C} &= \hat{P}^{-1}, \end{aligned} \tag{G.3}$$

where the required capacitance matrix $\hat{C}$ is the inverse of Maxwell's potential coefficient matrix $\hat{P}$, although the interpretation in terms of individual leg-to-leg capacitances is not as trivial as might first be thought [1].

To calculate the potential coefficients, we start with Gauss's law to find the potential at radius r from a single infinite line charge, $V = -\frac{q}{2\pi\varepsilon_0}\ln r + V_0$, for a p.u.l. charge q, where V_0 is a constant which cannot be set to zero at infinity because the line charge itself extends to infinity [3]. From figure G.1 and the definition (G.1), the potential coefficient p_{ij}, at the position of leg i, at a distance r_{ij} from wire j having p.u.l. charge q_j, is

$$p_{ij} = -\left(\frac{1}{2\pi\varepsilon_0}\right)\ln r_{ij} - \left(\frac{-1}{2\pi\varepsilon_0}\right)\ln r'_{ij} \tag{G.4}$$

$$= \left(\frac{1}{2\pi\varepsilon_0}\right)\ln\left(\frac{r'_{ij}}{r_{ij}}\right) = \left(\frac{1}{4\pi\varepsilon_0}\right)\ln\left(\frac{(r'_{ij})^2}{(r_{ij})^2}\right) \tag{G.5}$$

$$= \left(\frac{1}{4\pi\varepsilon_0}\right)\ln\left(\frac{(y_j - y_i)^2 + 4h^2}{(y_j - y_i)^2}\right). \tag{G.6}$$

The last term in (G.4) is due to the image charge $-q_j$ at distance r'_{ij}, which is part and parcel of the definition, and the constant V_0 is eliminated by the potential subtraction. The potential at the position of the reference plane is zero, as required, because it is equidistant from every charge and its image. The switch from r to r^2 in (G.5) permits a direct use of Pythagoras's theorem in (G.6) which reduces the calculation to a simple geometrical expression. Every leg has been considered as a wire filament at its axis position h above the reference plate, which is justifiable for the leg spacing and leg height much greater than the leg radius r_{leg}. For the particular case of a leg's own self charge q_i, its voltage is $V_i = -\frac{q_i}{2\pi\varepsilon_0}\ln r_{\text{leg}} + V_0$ and its self potential coefficient $p_{ii} = \left(\frac{1}{4\pi\varepsilon_0}\right)\ln\left(4h^2/r_{\text{leg}}^2\right)$; the only case where the wire radius is taken into account. Note that p_{ij} is always greater than zero, owing to the closer proximity of the positive charge q_j to the wire j, on the same side of the ground reference plate, compared to the image charge $-q_j$.

For this regular array of legs, $\hat{P}$ is a symmetric Toeplitz matrix, which means that each descending diagonal from left to right is constant, with symmetry about both of its main diagonals (bisymmetric and centrosymmetric). Consequently, $\hat{P}$ can be computed from a single row (or column) of potential coefficients such as the first row, $\mathbf{p} = [p_{11}, p_{12}, \ldots, p_{1N}]$. In Matlab [4], $\hat{P} = \text{toeplitz}(\mathbf{p}, \mathbf{p})$. The matrix equation (G.7) shows an example of (G.2) for a five-leg antenna:

$$\begin{pmatrix} V_1 \\ V_2 \\ V_3 \\ V_4 \\ V_5 \end{pmatrix} = \begin{pmatrix} p_{11} & p_{12} & p_{13} & p_{14} & p_{15} \\ p_{12} & p_{11} & p_{12} & p_{13} & p_{14} \\ p_{13} & p_{12} & p_{11} & p_{12} & p_{13} \\ p_{14} & p_{13} & p_{12} & p_{11} & p_{12} \\ p_{15} & p_{14} & p_{13} & p_{12} & p_{11} \end{pmatrix}\begin{pmatrix} q_1 \\ q_2 \\ q_3 \\ q_4 \\ q_5 \end{pmatrix}. \tag{G.7}$$

The capacitance matrix is bisymmetric but not Toeplitz.

This capacitance matrix applies to the microstrip configuration (wire-above-plate) shown in figure G.1, supposing a homogeneous dielectric (vacuum in this case). It is applicable to the antenna above the baseplate in the open reactor of section 7.5, although the mutual partial inductance p.u.l. matrix $\hat{M}$ was used there instead to deduce $\hat{C}$, using $\hat{M}\hat{C} = \mu_0\varepsilon_0\bar{1}$ for a homogeneous dielectric.

G.2 Capacitance matrix for a stripline

For the closed reactor shown in figure 7.11, without plasma, the ladder antenna is now suspended between the top and bottom plates. If we neglect the thin glass plate in comparison with the antenna-to-plate spacings, and treat the highly porous foam as vacuum, the antenna electrical configuration presents as a homogeneous stripline, i.e. wires within parallel plates. The simple microstrip figure G.1 is consequently

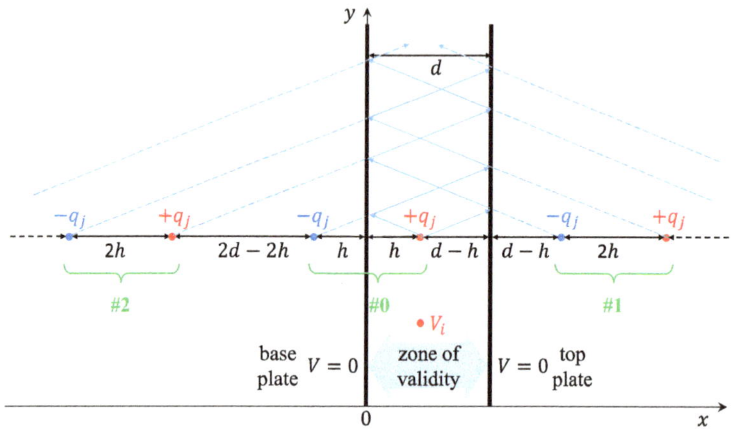

Figure G.2. A stripline schematic of legs i and j of a ladder antenna, distance h above the baseplate, and $(d - h)$ under the top-plate, to calculate p_{ij} in (G.8). The potential due to the source charge $+q_j$ between the plates, is equivalent to the source charge plus the infinite series of image charges without the plates. Bracket #0 shows the source charge and its image in the baseplate, as in figure G.1. Bracket #1 shows the first reflection of this pair in the top-plate. Bracket #2 shows the second reflection of this pair, this time in the baseplate. Charge signs inverse at each reflection. The blue lines are a geometric construction to help identify the image charges associated with the reflections inside the plates.

modified by an infinity of images reflected between the plates as shown in the stripline figure G.2.

Again, it is important to recognize—concerning the calculation of electric potential in the domain of the source charge [2]—that the system of source charge with its infinite reflections is equivalent to the system of source charge and the induced charges on the two plates. The domain of validity is between the plates, which is sufficient for calculating all of the p_{ij}, because all of the legs are also between the plates.

By inspection of figure G.2, it can be deduced that all of the positive charges are at distances $(\pm 2nd)$, $n = 0, 1, \ldots, \infty$, along the x-axis relative to the x position of leg i. Similarly, the negative charges are at relative x distances $(2h \pm 2nd)$. The positive and negative charges occur in pairs so the source-plus-image charges are globally neutral; this is consistent with the original system of a positive source charge and negative induced charges on the ground plates. Performing similar operations as in (G.4) to (G.6) for the infinite series, we have [5, 6]

$$
\begin{aligned}
p_{ij} &= -\left(\frac{1}{4\pi\varepsilon_0}\right)\sum_{n=0}^{\infty} \ln\left[(y_j - y_i)^2 + (\pm 2nd)^2\right] \\
&\quad -\left(\frac{-1}{4\pi\varepsilon_0}\right)\sum_{n=0}^{\infty} \ln\left[(y_j - y_i)^2 + (2h \pm 2nd)^2\right] \\
&= \left(\frac{1}{4\pi\varepsilon_0}\right)\sum_{n=-\infty}^{\infty} \ln\left(\frac{\left[(y_j - y_i)^2 + (2h - 2nd)^2\right]}{\left[(y_j - y_i)^2 + (2nd)^2\right]}\right),
\end{aligned}
\tag{G.8}
$$

for the potential coefficients of a ladder antenna in a vacuum stripline. This was used to calculate the capacitance matrix in vacuum in section 7.6.2. The homogeneous microstrip result (G.6) is obtained for $n = 0$, as expected.

G.3 The partial image method for electric charge

The previous sections G.1 and G.2 showed how to calculate the capacitance matrix for a homogeneous microstrip and homogeneous stripline, respectively, in a vacuum. In the presence of plasma, however, the antenna capacitive coupling becomes much stronger across the glass protection window towards the plasma, as shown in figures 7.12 and G.3. In this configuration, the inhomogeneous microstrip (wire-above-dielectric-on-plane) is the pertinent configuration, consisting of the ladder antenna embedded in highly porous foam (an effective vacuum), in the vicinity of the glass dielectric slab, which is backed by the conducting plasma. For simplification, the thin RF plasma sheath is taken to be assimilated into the dielectric layer. Capacitive coupling to the top-plate is considered to be negligible in comparison, and besides, there appears to be no commonly agreed analytical solution in the literature for inhomogeneous striplines, which would be necessary for this case.

G.3.1 Basis of the partial image method

We have already met two image methods for currents earlier in this book.: The classical image method for electrical current is used in chapters 4 and 7 on the understanding that the induced current is reflected in a perfect electric conductor (PEC) plane surface, with equal and opposite image current at an equal distance behind the PEC surface, obeying an optical mirror-image analogy. The complex image method in chapters 6, 7, and appendix E, uses the magneto-quasistatic approximation to estimate the complex depth of an equal and opposite line image current induced in a resistive conductor. The classical image method was employed for electric charges and PEC plates in sections G.1 and G.2 above. In contrast, the

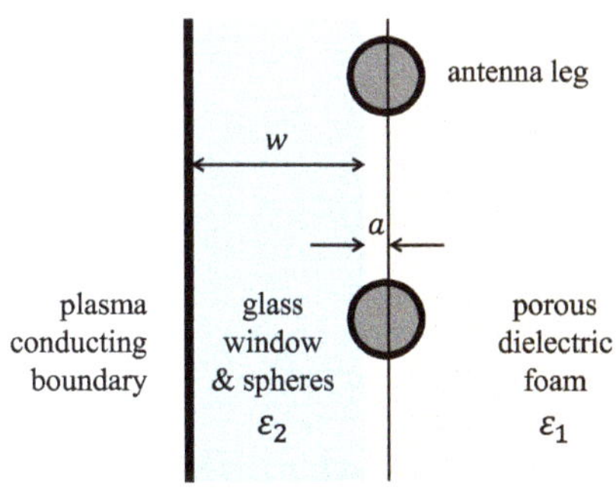

Figure G.3. A schematic cross-section of the ladder antenna legs mounted in a porous dielectric foam of relative permittivity ε_1. The legs are in contact with the glass protection window (relative permittivity ε_2). The interstices between the window, and channels cut in the foam to accommodate the legs, are filled with glass micro-spheres to preclude arcs igniting in any empty pockets of gas. The result is that the legs are partly embedded in the glass dielectric, effective thickness w, and the leg axes are a distance a above the glass.

partial image method pertains to the electrostatic reflection of electric charge in a dielectric, not a PEC. We will see that the image depth is equal to the source distance, as in the optical analogy, but this time, the induced charge is only a partial fraction of the source charge; hence the name partial image method.

Figure G.4(a) shows a line charge source q (Coulomb p.u.l.), adjacent to a dielectric interface at distance a. Figures G.4(b) and (c) show image configurations postulated to represent the resulting electric field in the region of dielectric ε_1, or in the region of dielectric ε_2, respectively. In each case (b) or (c), a line image charge q' or q'', respectively, is proposed at the same distance a, in the anticipation that their functional form, identical to the source field, will provide a self-consistent solution [2]. If the schemas (b) and (c) also satisfy the boundary conditions of the original field in (a), then the equivalence principle and the uniqueness theorem state that the partial image charges generate fields which are indeed equivalent to the original field, in their respective domains of validity [2]. It remains to find q' and q'' in terms of q.

The conventional boundary conditions across a dielectric interface are continuity of the tangential electric field strength $E_{\|}$, and continuity of the perpendicular electric flux density $D_{\perp}$ for an uncharged interface [1, 3]. Applying these conditions to the common interface in figures G.4(b) and (c) gives

$$\begin{aligned} E_{\|}^{\mathrm{RHS}} &= E_{\|}^{\mathrm{LHS}}, \\ \left(q + q'\right)\frac{\cos\alpha}{2\pi\varepsilon_1 r} &= q''\frac{\cos\alpha}{2\pi\varepsilon_2 r}, \\ \frac{q + q'}{\varepsilon_1} &= \frac{q''}{\varepsilon_2}; \end{aligned} \tag{G.9}$$

$$\begin{aligned} \text{and}\quad D_{\perp}^{\mathrm{RHS}} &= D_{\perp}^{\mathrm{LHS}}, \\ (q' - q)\frac{\sin\alpha}{2\pi r} &= -\,q''\frac{\sin\alpha}{2\pi r}, \\ q - q' &= q''; \end{aligned} \tag{G.10}$$

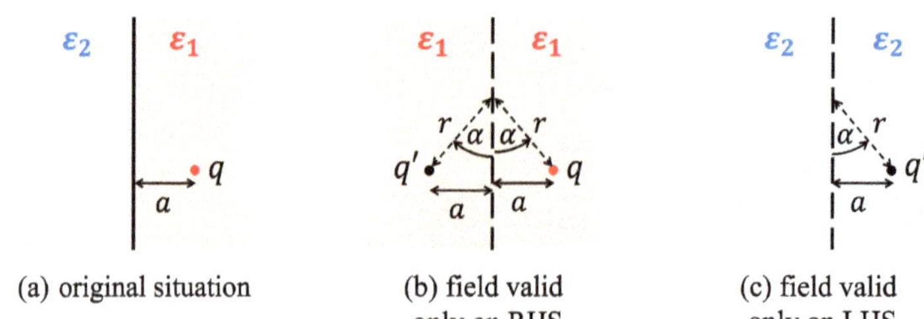

Figure G.4. (a) Line charge q parallel to a plane interface between two dielectrics of relative permittivity ε_1 and ε_2. (b) For field calculations valid only on the right-hand side (RHS) of the interface, the field due to induced charge in the dielectric ε_2 is replaced by an image charge q', in a uniform dielectric ε_1. (c) For field calculations valid only on the left-hand side (LHS) of the interface, the field due to induced charge in the dielectric ε_2 and the source q in ε_1, are both replaced by a single partial charge q'', in a uniform dielectric ε_2. A general position on the interface is denoted by distance r and angle α.

$$\text{hence} \quad q' = Kq \tag{G.11}$$

$$\text{and} \quad q'' = (1 - K)q, \tag{G.12}$$

$$\text{where} \quad K = \frac{\varepsilon_1 - \varepsilon_2}{\varepsilon_1 + \varepsilon_2}. \tag{G.13}$$

The boundary conditions are thereby satisfied over the whole interface[1]. Therefore, by the equivalence principle [2], the original source charge q and the partial charge Kq in ε_1 of figure G.4(b), represent the electric field and potential on the RHS of figure G.4(a). Likewise, charge $(1 - K)q$ in ε_2 of figure G.4(c) represents the electric field and potential on the LHS of figure G.4(a).

G.3.2 Partial image method for an inhomogeneous microstrip

For the final step to the case in figure G.3, we consider a line charge q a distance a above a dielectric slab of thickness w in figure G.5. Similarly to figure G.4(b), the potential above the slab is due to the source charge q and its primary image Kq from the reflection in the top dielectric interface. Figure G.5 is drawn for the relevant case

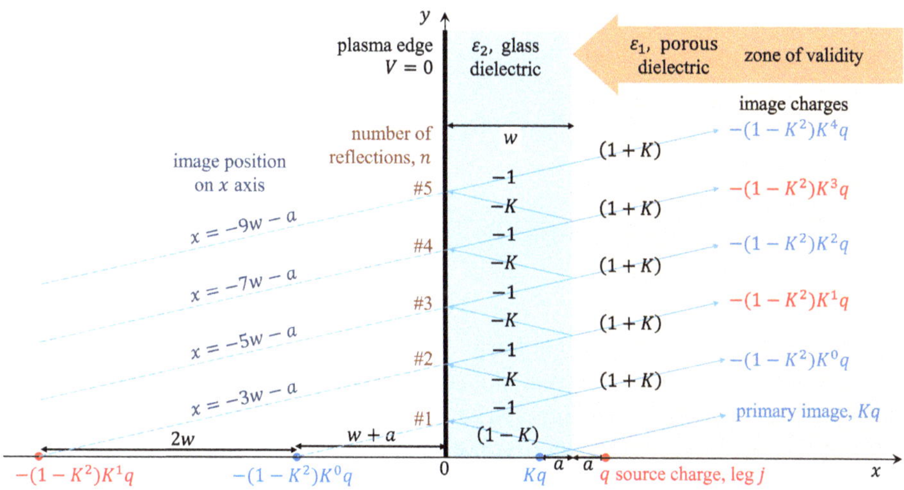

Figure G.5. Diagram of the image calculation for charge q in a porous dielectric, distance a above the glass thickness w (see figure G.3). The plasma behind the glass acts as a mirror, changing the sign of image charges. The blue lines are geometric constructions of successive paths starting from the source charge q. The factors $(1 - K)$, -1, $-K$, $(1 + K)$ are successively multiplied in the labelled transmissions and reflections, to arrive at the series of image charges in the porous dielectric. The dashed lines extrapolate to the position of each image charge on the x-axis. The source charge q is taken to be positive (red). The primary image charge Kq (blue) corresponds to $K < 0$, which is appropriate for vacuum, ε_1, and glass, $\varepsilon_2 > \varepsilon_1$, in (G.13).

[1] Note that k in [1, 2] equals $-K$ in [7].

of vacuum and glass, i.e. $\varepsilon_1 < \varepsilon_2$, i.e. $-1 < K < 0$. However, the calculation is more complicated [6] due to the infinite mirror reflections within the slab itself, which contribute an infinite series of image charges [1, 7].

The magnitude of the relative permittivity of unmagnetized RF plasma in (D.27) is typically so large, for plasma densities above 10^{15} m^{-3}, that $K = -1$ to all intents and purposes at the glass–plasma interface (corresponding to classical image theory), so the plasma will be assumed to behave as a PEC boundary behind the glass dielectric window in figures G.3 and G.5. From footnote 4 in chapter 10, the glass refractive index N is related to its relative permittivity ε_2, by $N^2 = \varepsilon_2$, measured at RF frequency.

The initial passage into, and the final passage out of, the glass top surface multiplies the apparent image charge by a factor $(1 - K)$ and $(1 + K)$, respectively, according to (G.12) and (G.13) when the roles of ε_1 and ε_2 are interchanged. Every internal reflection brings a factor $-K$ according to (G.11). Each successive pair of internal reflections returns the image to the source region, and distances the image further away from the source by $2w$. The result is that, after n reflections at the plasma interface, the nth image charge, perceived in the source region and with the same permittivity ε_1 as the zone of validity, is $(1 - K) \times K^{n-1} \times (1 + K) \times q$, at corresponding coordinate $x_n = -w(2n - 1) - a$, which is at a distance $w + a - x_n = (2nw + 2a)$ from the source charge q, for $n = 1, 2, \ldots, \infty$.

The total effective charge Q in the region of validity, using the partial image method, can be read directly off figure G.5; it is the sum total of the source, its primary image, and the infinite image series, namely

$$Q = q + Kq - (1 - K^2)q\sum_{n=1}^{\infty} K^{n-1} \tag{G.14}$$

$$\begin{aligned} &= q + Kq - (1 - K^2)q\frac{1}{(1 - K)} \\ &= 0, \end{aligned} \tag{G.15}$$

using the sum of an infinite geometric series, $\sum_{m=0}^{\infty} K^m = 1/(1 - K)$ for $|K| < 1$. Therefore, the partial image system is globally neutral, consistent with the original charge q and its induced opposite charge in the glass and PEC plasma interface.

Finally, the required Maxwell potential coefficient p_{ij} is obtained by summing the potential due to each charge of (G.14) in the zone of validity ε_1, accounting for the respective distance of leg i from the source leg j and its images, by extension of (G.6):

$$\begin{aligned} p_{ij} = &-\frac{\ln\left[(y_j - y_i)^2\right]}{4\pi\varepsilon_1} - \frac{K\ln\left[(2a)^2 + (y_j - y_i)^2\right]}{4\pi\varepsilon_1} \\ &+ \frac{(1 - K^2)}{4\pi\varepsilon_1}\sum_{n=1}^{\infty} K^{n-1}\ln\left[(2nw + 2a)^2 + (y_j - y_i)^2\right]. \end{aligned} \tag{G.16}$$

This expression was used to calculate Maxwell's potential coefficient matrix $\bar{P}$ in section 7.6.2. After mathematical re-grouping, (G.16) is identical to expressions derived by Silvester [7] and Weeks [6], who used different approaches to the analogous problem of inhomogeneous thick microstrip.

References

[1] Paul C R 2008 *Analysis of Multiconductor Transmission Lines* 2nd edn (Hoboken, NJ: Wiley)

[2] Harrington R F 1958 *Introduction to Electromagnetic Engineering* (New York: McGraw-Hill)

[3] Bleaney B I and Bleaney B 1976 *Electricity and Magnetism* (London: Oxford University Press) 3rd edn p 1

[4] The MathWorks Inc 2024 MATLAB v9.5.0 (R2018b), Natick, Massachusetts: https://www.mathworks.com

[5] Kammler D W 1968 Calculation of characteristic admittances and coupling coefficients for strip transmission lines *IEEE Trans. Microw. Theory Tech.* **MTT-16** 925

[6] Weeks W T 1970 Calculation of coefficients of capacitance of multiconductor transmission lines in the presence of a dielectric interface *IEEE Trans. Microw. Theory Tech.* **MTT-18** 35

[7] Silvester P 1968 TEM wave properties of microstrip transmission lines *Proc. IEE* **115** 430

Appendix H

Impedance of a hybrid antenna with parasitic capacitance

This appendix is concerned with the inductive probe diagnostic in chapter 7, section 7.7.2. With reference to figure 7.20 and [1], the antenna node voltages **A**, **B**, leg currents **I**, stringer currents **J**, **K**, and source currents $\mathbf{S}^K$, $\mathbf{S}^J$ are written as vectors.

Current conservation at the network nodes is

$$J_n = J_{n-1} + I_n + S_n^J - A_n\, j\omega C_\mathrm{p}, \tag{H.1}$$

$$K_n = K_{n-1} - I_n + S_n^K - B_n\, j\omega C_\mathrm{p}, \tag{H.2}$$

where the capacitive currents to ground are now included. Introducing the lower shift matrix $\bar{U}_L$, which has ones on the subdiagonal and zeroes elsewhere, and writing $\bar{U} = \bar{1} - \bar{U}_L$, current conservation expressed in matrix form becomes

$$\bar{U}\mathbf{J} = \mathbf{I} + \mathbf{S}^J - j\omega C_\mathrm{p}\mathbf{A}, \tag{H.3}$$

$$\bar{U}\mathbf{K} = -\mathbf{I} + \mathbf{S}^K - j\omega C_\mathrm{p}\mathbf{B}. \tag{H.4}$$

Ohm's law for the legs in matrix form is

$$\mathbf{B} - \mathbf{A} = \bar{Z}_\mathrm{leg}\mathbf{I}, \tag{H.5}$$

where $\bar{Z}_\mathrm{leg} = (R + \frac{1}{j\omega C_\mathrm{leg}})\bar{1} + j\omega\hat{M}$ is the leg impedance of the hybrid network, using $\hat{M}$ for the combined mutual partial inductance matrix for the legs, accounting for image currents in the screen and plasma.

Ohm's law for the stringers is unchanged with respect to [1], i.e.

$$\bar{U}^T(\mathbf{B} - \mathbf{A}) = \bar{Z}_\mathrm{str}(\mathbf{K} - \mathbf{J}), \tag{H.6}$$

doi:10.1088/978-0-7503-5296-3ch22

where we define a stringer impedance as

$$\mathbf{Z}_{\text{str}} = \left(r - \frac{j}{\omega C}\right)\mathbf{1} + j\omega(\mathbf{M}^{\text{line}} - \mathbf{M}^{\text{opp}}). \tag{H.7}$$

A solution for the leg current vector $\mathbf{I}$ is found first by eliminating the node voltage difference using (H.5) and (H.6) to give

$$\bar{U}^T \bar{Z}_{\text{leg}} \mathbf{I} = \bar{Z}_{\text{str}}(\mathbf{K} - \mathbf{J}), \tag{H.8}$$

and then using (H.4) minus (H.3) to eliminate the stringer current difference in (H.8), to finally obtain

$$\mathbf{I} = \left(\bar{U}\, \bar{Z}_{\text{str}}^{-1}\, \bar{U}^T\, \bar{Z}_{\text{leg}} + (2 + j\omega C_{\text{p}}\, \bar{Z}_{\text{leg}})\bar{1}\right)^{-1}(\mathbf{S}^K - \mathbf{S}^J). \tag{H.9}$$

This defines the leg currents from which all the other antenna parameters can be deduced [1].

At the required frequency of operation ω_{res}, the leg capacitor C_{leg} is chosen to cancel the leg self inductance according to $\omega_{\text{res}}^2 L_{\text{leg}} C_{\text{leg}} = 1$. The leading diagonal of the leg impedance matrix is now strongly reduced to R, the small resistance of the legs. From (H.9), this means that the influence of the parasitic capacitance term is minimized. Also, the relative importance of the off-diagonal, mutual impedance terms now dominate the solution for the leg currents, thereby maintaining the sensitivity of the probe to the plasma coupling.

Reference

[1] Guittienne Ph, Jacquier R, Howling A A and Furno I 2015 Complex image method for RF antenna-plasma inductive coupling calculation in planar geometry. Part II: measurements on a resonant network *Plasma Sources Sci. Technol.* **24** 065015

Appendix I

Cylindrical wave function constants

Starting from the boundary conditions in section 8.4, equations (8.46), here is a fully worked solution for the five constants A, B, C, D, E so the reader can check their derivation against this proposition. Having divided (I.4) and (I.5) throughout by k_0, there remains only one occurrence of wavenumbers k_d and k_0, namely $\frac{k_d}{k_0}$, in the pre-factors:

$$AJ_1(k_d a) - BJ_1(k_0 a) - CY_1(k_0 a) = 0, \tag{I.1}$$

$$BJ_1(k_0 b) + CY_1(k_0 b) - DJ_1(k_0 b) - EY_1(k_0 b) = 0, \tag{I.2}$$

$$DJ_1(k_0 c) + EY_1(k_0 c) = 0, \tag{I.3}$$

$$A\frac{k_d}{k_0}J_1'(k_d a) - BJ_1'(k_0 a) - CY_1'(k_0 a) = 0, \tag{I.4}$$

$$BJ_1'(k_0 b) + CY_1'(k_0 b) - DJ_1'(k_0 b) - EY_1'(k_0 b) = jZ_0 i_0, \tag{I.5}$$

where we have substituted the impedance of free space $Z_0 = \frac{\omega\mu_0}{k_0} = \sqrt{\frac{\mu_0}{\varepsilon_0}}$ in (I.5). There is now a straightforward but tedious process to solve these five simultaneous equations, involving the Abel identity for the Bessel Wronskian given in (8.28). It is obvious first to eliminate E using (I.3)

$$E = -D\,\frac{J_1(k_0 c)}{Y_1(k_0 c)}. \tag{I.6}$$

Substituting for E in (I.2) and (I.5)

$$BJ_1(k_0 b) + CY_1(k_0 b) - \frac{D}{Y_1(k_0 c)}\left[J_1(k_0 b)Y_1(k_0 c) - J_1(k_0 c)Y_1(k_0 b)\right] = 0, \tag{I.7}$$

doi:10.1088/978-0-7503-5296-3ch23

$$BJ_1'(k_0b) + CY_1'(k_0b) - \frac{D}{Y_1(k_0c)}\left[J_1'(k_0b)Y_1(k_0c) - J_1(k_0c)Y_1'(k_0b)\right] = jZ_0 i_0, \quad \text{(I.8)}$$

from which we eliminate D by subtraction. After some cancellation, the left-hand side has a common factor which can surprisingly be simplified using Abel's identity (8.28) for the Wronskian of Bessel functions:

$$J_1(k_0b)Y_1'(k_0b) - J_1'(k_0b)Y_1(k_0b) = \frac{2}{\pi k_0 b}, \quad \text{(I.9)}$$

so the subtraction results in

$$BJ_1(k_0c) + CY_1(k_0c) = \left[J_1(k_0b)Y_1(k_0c) - J_1(k_0c)Y_1(k_0b)\right] jZ_0 i_0 \pi k_0 b/2. \quad \text{(I.10)}$$

Performing a similar subtraction for (I.1) and (I.4) to eliminate B or C gives

$$A\left[J_1(k_da)Y_1'(k_0a) - \frac{k_d}{k_0}J_1'(k_da)Y_1(k_0a)\right] = B\frac{2}{\pi k_0 a}, \quad \text{(I.11)}$$

and

$$A\left[J_1(k_da)J_1'(k_0a) - \frac{k_d}{k_0}J_1'(k_da)J_1(k_0a)\right] = -C\frac{2}{\pi k_0 a}. \quad \text{(I.12)}$$

Substituting (I.11) and (I.12), respectively, for B and C in (I.10) finally gives an expression for A [1, 2]:

$$\begin{aligned} A &= \left(\frac{b}{a}\right)\frac{J_1(k_0b)Y_1(k_0c) - J_1(k_0c)Y_1(k_0b)}{(k_d/k_0)J_1'(k_da)F(k_0, a, c) - J_1(k_da)G(k_0, a, c)}(jZ_0 i_0), \\ \text{where} \quad & F(k_0, a, c) = J_1(k_0a)Y_1(k_0c) - J_1(k_0c)Y_1(k_0a), \\ \text{and} \quad & G(k_0, a, c) = J_1'(k_0a)Y_1(k_0c) - J_1(k_0c)Y_1'(k_0a). \end{aligned} \quad \text{(I.13)}$$

The constant A is more complicated than the two-domain vacuum solutions because now there are three domains and two permittivities [1, 2]. In the vacuum limit $k_d \to k_0$, this expression reduces to the vacuum constant A_1 in (8.27). The other constants B, C, D, E are obtained by back-substitution for A in these equations.

References

[1] Jin J-M 1998 *Electromagnetic Analysis and Design in Magnetic Resonance Imaging* (Boca Raton, FL: CRC Press)

[2] Jin J-M 2010 Practical electromagnetic modeling methods *eMagRes* **10**

IOP Publishing

Appendix J

Helicon mode derivations and methods

J.1 Tensor invariance in the cylindrical transformation

Consider, for example, the transformation of the wave's current vector in the Cartesian coordinates of chapter 10, $\mathbf{J}_{xyz} = J_x\hat{\mathbf{x}} + J_y\hat{\mathbf{y}} + J_z\hat{\mathbf{z}}$, into cylindrical coordinates in chapter 11, $\mathbf{J}_{r\phi z} = J_r\hat{\mathbf{r}} + J_\phi\hat{\boldsymbol{\phi}} + J_z\hat{\mathbf{z}}$, corresponding to figure 11.2. Although the conductivity and permittivity tensors are not used in chapter 11 (see figure J.1), their transformation into cylindrical geometry is of general interest. The question in section 11.1.1 is: do the tensors $\bar{\sigma}_p$ (10.14) and $\bar{\varepsilon}_p$ (10.18) remain the same in the transformation between Cartesian and cylindrical coordinates? The vector $\mathbf{J}_{xyz}$ and $\mathbf{J}_{r\phi z}$ is the same vector, just labelled with coordinates in the different systems. The transformation is performed by means of the matrix $\bar{R}$ [1]

$$\mathbf{J}_{r\phi z} = \bar{R} \cdot \mathbf{J}_{xyz},$$
$$\begin{pmatrix} J_r \\ J_\phi \\ J_z \end{pmatrix} = \begin{pmatrix} \cos\phi & \sin\phi & 0 \\ -\sin\phi & \cos\phi & 0 \\ 0 & 0 & 1 \end{pmatrix} \begin{pmatrix} J_x \\ J_y \\ J_z \end{pmatrix}, \tag{J.1}$$

which is effectively a two-dimensional coordinate rotational transform in the plasma cross-section because the $\hat{\mathbf{x}}$ and $\hat{\mathbf{y}}$ unit vectors are turned through angle ϕ to become the $\hat{\mathbf{r}}$ and $\hat{\boldsymbol{\phi}}$ unit vectors, respectively[1], in figure 11.2; the $\hat{\mathbf{z}}$ unit vector is invariant for the Cartesian and cylindrical systems. The plasma conductivity tensor $\bar{\sigma}_p$ (10.14) was defined by $\mathbf{J}_{xyz} = \bar{\sigma}_p \cdot \mathbf{E}_{xyz}$ in the Cartesian chapter 10. Using the transformation $\bar{R}$, we see that

[1] If the signs of $\sin\phi$ are exchanged in $\bar{R}$, then this rotates the vector $\mathbf{J}_{xyz}$ anti-clockwise through angle ϕ to $\mathbf{J}_{x'y'z'}$ in the Cartesian system.

doi:10.1088/978-0-7503-5296-3ch24

Figure J.1. 'Appendix relevance.' (Illustration by Alex Howling. Copyright 2023 Alex Howling.)

$$\begin{aligned} \mathbf{J}_{r\phi z} &= \bar{R} \cdot \mathbf{J}_{xyz}, \\ &= \bar{R} \cdot \bar{\sigma}_p \cdot \mathbf{E}_{xyz}. \end{aligned} \tag{J.2}$$

The same transformation applies to every vector, therefore we also have

$$\begin{aligned} \mathbf{E}_{r\phi z} &= \bar{R} \cdot \mathbf{E}_{xyz}, \\ \mathbf{E}_{xyz} &= \bar{R}^{-1} \cdot \mathbf{E}_{r\phi z}, \end{aligned} \tag{J.3}$$

where $\bar{R}^{-1}$ is the matrix inverse of $\bar{R}$. Substituting into (J.2)

$$\mathbf{J}_{r\phi z} = \bar{R} \cdot \bar{\sigma}_p \cdot \bar{R}^{-1} \cdot \mathbf{E}_{r\phi z}. \tag{J.4}$$

If the conductivity tensor is to be the same in both the Cartesian and cylindrical systems, then we also have $\mathbf{J}_{r\phi z} = \bar{\sigma}_p \cdot \mathbf{E}_{r\phi z}$ in (J.4), which requires that

$$\begin{aligned} \bar{\sigma}_p &= \bar{R} \cdot \bar{\sigma}_p \cdot \bar{R}^{-1}, \\ \text{i. e.} \quad \bar{\sigma}_p \cdot \bar{R} &= \bar{R} \cdot \bar{\sigma}_p \end{aligned} \tag{J.5}$$

where the last equation uses a post-multiplication by $\bar{R}$, and $\bar{R}^{-1} \cdot \bar{R} = \bar{1}$, the unity matrix. Invariance of $\bar{\sigma}_p$ therefore implies that its coordinate transformation must be commutative. Performing the matrix multiplication of $\bar{\sigma}_p$ (10.14) and $\bar{R}$ (J.1), and vice versa, shows that (J.5) is indeed commutative. In fact, the commutative condition is trivially satisfied because $\bar{\sigma}_p$ has the same symmetry as $\bar{R}$. Finally, therefore, the invariance between Cartesian and cylindrical systems of the tensors $\bar{\sigma}_p$ (10.14) and $\bar{\varepsilon}_p$ (10.18) is due to the inherent symmetry of the magnetic force $-q\mathbf{u} \times \mathbf{B}_0$ in the electron equation of motion (10.6).

J.2 Boundary conditions and wavenumber quantization

With regard to section 11.2.2, we explain here in detail the significance of the constraint $\mathrm{Cond}(k_z) = 0$ for the algebraically amenable case of a uniform plasma with a PEC boundary, supporting H and T–G waves. The following treatment applies to collisionless as well as collisional plasma. This is the simplest case apart from the helicon approximation, section 11.2.1, which has only an H wave, with $\mathrm{Cond}(k_z) = B_r(a) = 0$.

We start by considering all of the six electromagnetic field components and their role in the plasma/PEC boundary conditions at $r = a$. All of the boundary conditions depend on the fact that electromagnetic wavefields cannot penetrate into a perfect conductor.

1. $B_r(a) = 0$ due to continuity at the field-free PEC wall.
2. $B_\phi(a)$ continuity is assumed due to a J_z surface current in the PEC wall.
3. $B_z(a)$ continuity is assumed due to a J_ϕ surface current in the PEC wall.
4. $E_r(a)$ continuity is assumed due to a surface charge on the PEC wall.
5. $E_\phi(a) = 0$ due to continuity at the field-free PEC wall.
6. $E_z(a) = 0$ due to continuity at the field-free PEC wall.

It is important to note that these electromagnetic fields refer to the total, net resultant wavefield of the superposition of the {H,T–G} modes, *not* to the mode wavefields individually. For the three 'assumed' boundary conditions, the PEC surface current and charge are automatically satisfied and do not concern us here, although the values can be calculated after the complete solution of the wavefields, if required. The other three boundary conditions are related (see (11.43)), so one is redundant. Finally, therefore, there are only two boundary conditions to be satisfied, and here we choose $B_r(a) = 0$ and $E_z(a) = 0$.

J.2.1 The radial magnetic field

From (11.22), $C_2 = 0$ for H and T–G waves to prevent a singularity of the second kind Y_m Bessel functions at the axis. Therefore, we retain only the first kind Bessel functions with proportionality constant C_1, and the total B_r field is a superposition of the B_r for each of the H and T–G waves. Hence, at the PEC wall $r = a$, we have

$$\begin{aligned} B_r(a) = {} & \frac{jC_{1\mathrm{H}}}{2k_{\perp\mathrm{H}}}[(k_\mathrm{H} - k_z)J_{m+1}(k_{\perp\mathrm{H}}a) + (k_\mathrm{H} + k_z)J_{m-1}(k_{\perp\mathrm{H}}a)] \\ & + \frac{jC_{1\mathrm{T}}}{2k_{\perp\mathrm{T}}}[(k_\mathrm{T} - k_z)J_{m+1}(k_{\perp\mathrm{T}}a) + (k_\mathrm{T} + k_z)J_{m-1}(k_{\perp\mathrm{T}}a)] = 0. \end{aligned} \tag{J.6}$$

It is crucial at this point to note that k_H and k_T are the wavenumbers of the helicon and Trivelpiece–Gould modes known from (10.96), (10.107), or (11.9), but that k_z *is the single axial wavenumber common to both* in the resultant wavefield solution (see also section 8.1) of the superposed {H,T–G} modes. If you like, we can write $k_z = k_{z\mathrm{H}} = k_{z\mathrm{T}}$. The perpendicular wavenumbers are derived from these as usual, namely, $k_{\perp H}^2 = k_H^2 - k_z^2$, and $k_{\perp T}^2 = k_T^2 - k_z^2$. The reader will sympathize if we prefer to abbreviate (J.6) as

$$B_r(a) = C_H b_{rH} + C_T b_{rT} = 0, \tag{J.7}$$

where we also drop the redundant subscript '1' in C_H and C_T.

J.2.2 The axial electric field

Starting from the axial magnetic field (11.20) and again retaining only the first kind Bessel of functions for H and T–G waves, the total axial magnetic wavefield at the PEC wall is

$$B_z(a) = C_{1H} J_m(k_{\perp H} a) + C_{1T} J_m(k_{\perp T} a),$$

which we abbreviate in a similar manner to B_r to obtain

$$B_z(a) = C_H b_{zH} + C_T b_{zT}. \tag{J.8}$$

We can relate $B_z(a)$ to the total axial plasma current $J_z(a)$, in the quasi-magnetostatic approximation (neglecting the vacuum displacement current), by using $\nabla \times \mathbf{B} = \mu_0 \mathbf{J} = k\mathbf{B}$ for H and T–G waves, so that

$$\mu_0 J_z(a) = k_H C_H b_{zH} + k_T C_T b_{zT}. \tag{J.9}$$

Taking the z-component of the electron fluid force equation (11.2), the axial electric field consistent with both H and T–G waves is proportional to J_z:

$$E_z = -j \frac{\hat{\omega}}{\varepsilon_0 \omega_p^2} J_z \tag{J.10}$$

because the curl term $\mathbf{J} \times \hat{\mathbf{z}}$ is zero, which was the reason for opting for E_z instead of E_ϕ, although the final result is identical by taking either. The constants can be substituted in terms of the wavenumber roots by recalling (10.106), namely, $k_H k_T = \frac{\omega \omega_p^2}{c^2 \hat{\omega}}$, to obtain

$$E_z(a) = -j \frac{\omega}{k_H k_T} \mu_0 J_z(a) = -j\omega \left[\frac{C_H}{k_T} b_{zH} + \frac{C_T}{k_H} b_{zT} \right] = 0, \tag{J.11}$$

by using (J.9), and the boundary condition $E_z = 0$.

J.2.3 The condition on the axial wavenumber

We now solve the simultaneous equations (J.7) and (J.11) for the two boundary conditions, namely

$$C_H b_{rH} + C_T b_{rT} = 0, \tag{J.12}$$

$$\text{and} \quad \frac{C_H}{k_T} b_{zH} + \frac{C_T}{k_H} b_{zT} = 0. \tag{J.13}$$

It is worth reminding ourselves what are the known, and unknown, constants and variables at this point:

- C_{H} and C_{T} are the two unknown constants of mode amplitude in the boundary conditions $B_r(a) = 0$ and $E_z(a) = 0$.
- k_{H} and k_{T} are the wavenumbers of the H and T–G modes, functions of k_z, which are known from (10.96), (10.107), or (11.9).
- $b_{r\mathrm{H}}$, $b_{r\mathrm{T}}$, $b_{z\mathrm{H}}$, and $b_{z\mathrm{T}}$ are, in fact, constants, defined by the corresponding wavefields at $r = a$, but note from sections J.2.1 and J.2.2 that they depend on k_{H}, k_{T}, which in turn depend on k_z.
- Finally, the only variable is k_z; it is the single axial wavenumber which is common to both {H,T–G} modes in the resultant wavefield as described in section J.2.1—it is hidden from view in k_{H}, k_{T}, and also in the wavefield abbreviated expressions $b_{r\mathrm{H}}$, $b_{r\mathrm{T}}$, $b_{z\mathrm{H}}$, and $b_{z\mathrm{T}}$.

We now set about the resolution of the three unknowns C_{H}, C_{T}, and k_z. There is no excitation amplitude to satisfy in this mode study (in contrast to chapter 8), therefore we can arbitrarily normalize any wavefield amplitude, for example, $C_{\mathrm{H}} = 1$. Then $C_{\mathrm{T}} = -b_{r\mathrm{H}}/b_{r\mathrm{T}}$ directly from the first equation (J.12) for $B_z(a) = 0$. The amplitude constants are therefore completely defined. The remaining boundary condition (J.13) for $E_z(a) = 0$ serves as the constraint condition on k_z—defined in section 11.1.7 as $\mathrm{Cond}(k_z) = 0$—which leads to a quantization of the possible longitudinal wavenumbers k_z. In the relatively simple example of this appendix, substitution for C_{H} and C_{T} into (J.13) yields

$$\frac{b_{z\mathrm{H}}}{k_{\mathrm{T}}} - \frac{b_{r\mathrm{H}}b_{z\mathrm{T}}}{k_{\mathrm{H}}b_{r\mathrm{T}}} = 0, \tag{J.14}$$

$$\text{i. e.} \quad \mathrm{Cond}(k_z) = k_{\mathrm{H}}b_{z\mathrm{H}}b_{r\mathrm{T}} - k_{\mathrm{T}}b_{z\mathrm{T}}b_{r\mathrm{H}} = 0. \tag{J.15}$$

The roots of the plot of $\mathrm{Cond}(k_z)$ versus k_z in figure 11.9 correspond to the allowed values of k_z for the resultant wavefield of superposed {H,T–G} modes, appropriate to the specific case of section 11.2.2, namely a uniform, collisionless plasma with a PEC boundary, supporting {H,T–G} waves. The roots of the plot of $\mathrm{Cond}(k_z)$ versus k_z in the complex plane of figure 11.13 correspond to a uniform, *collisional* plasma with a PEC boundary, where k_z is complex in (J.15) due to the complex wavenumber roots k_{H} and k_{T} from (10.96), (10.107), or (11.9) with $\hat{\omega} = \omega + j\nu$ and $\nu \neq 0$.

The expressions for $\mathrm{Cond}(k_z)$ become much more cumbersome with improvements to the model, such as a dielectric boundary with the plasma, and/or plasma non-uniformity.

Reference

[1] Sadiku M N O 2018 *Elements of Electromagnetics* 7th edn (New York: Oxford University Press)

IOP Publishing

Resonant Network Antennas for Radio-Frequency Plasma Sources

Theory, technology and applications

Philippe Guittienne, Alan Howling and Ivo Furno

Appendix K

Link to programs

The following link, https://gitlab.epfl.ch/spc/ResonantAntennaBook is provided for readers to obtain Matlab [1] programs—according to availability—which were used to calculate many of the figures throughout the book. We hope that these can be useful for adapting the parameter values to new, individual projects. These files are by no means intended as examples of good programming. The library of routines will be updated from time to time after publication of the book itself.

Reference

[1] The MathWorks Inc. 2024 MATLAB v9.5.0 (R2018b), Natick, Massachusetts https://www.mathworks.com

www.ingramcontent.com/pod-product-compliance
Ingram Content Group UK Ltd.
Pitfield, Milton Keynes, MK11 3LW, UK
UKHW051909100926
8556IPUK00005B/823